MECHANICAL LOGIC IN THREE-DIMENSIONAL SPACE

Gennaro Auletta

MECHANICAL LOGIC IN THREE-DIMENSIONAL SPACE

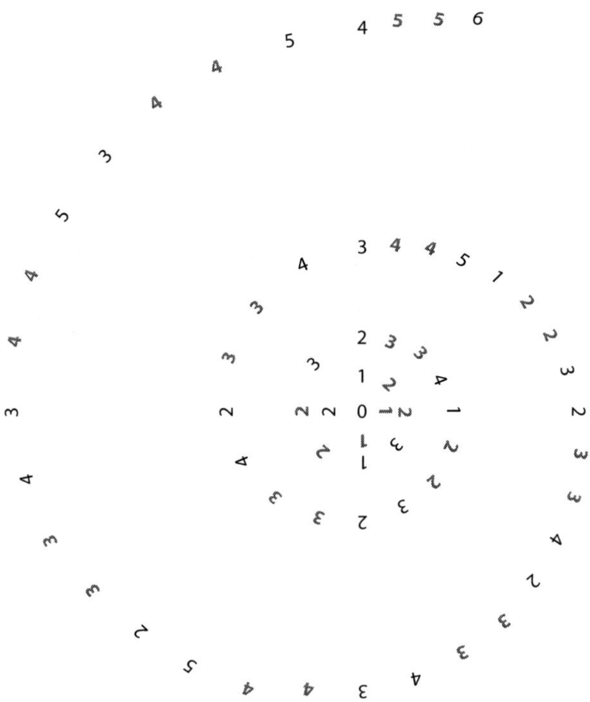

PAN STANFORD PUBLISHING

Published by

Pan Stanford Publishing Pte. Ltd.
Penthouse Level, Suntec Tower 3
8 Temasek Boulevard
Singapore 038988

Email: editorial@panstanford.com
Web: www.panstanford.com

British Library Cataloguing-in-Publication Data
A catalogue record for this book is available from the British Library.

Mechanical Logic in Three-Dimensional Space

ISBN 978-981-4411-50-9 (Hardcover)
ISBN 978-981-4411-51-6 (eBook)

Printed in the USA

Contents

Introduction

The dream of most of modern philosophers (but with roots already in the Middle Ages: see (*Llul*, 1305)) has been to generate all possible logical propositions by means of a mechanical calculation (see Fig. 1). Leibniz, following the research program presented in (*Hobbes*, 1655), introduced the idea that this can be fulfilled by means of a combinatorial calculus (*Leibniz*, 1666) (see Fig. 2). However, himself and some of his followers, like Frege, conceived this program as starting from a set of primitive axioms (*Frege*, 1884; Tarski, 1935b; Tarski, 1936). The further developments of mathematics and logic in the 20th century have shown that many different and alternative logical calculi are possible depending on the axioms that one assumes or rejects. Moreover, the fundamental results of Gödel seemed to exclude that even in primitive recursive arithmetic we can derive all true statement from a set of axioms (Gödel, 1931). In this way, also the possibility to establish a mechanical calculus of all possible propositions seemed to many banned for ever.

In the following I shall show that a full mechanization of logic is possible by renouncing the idea to start from a set of axioms. This can appear as a weakness. At the opposite, this is the greatest strength of the following proposal as far as the logical space that I shall build contains in itself several possible logical systems according to the path that is chosen inside such a space. Indeed, any possible path can recognize certain logical rules without being forced to assume other ones.

Moreover, such a logical space is comparable to the space of arithmetic, in which each number occupies a well defined position relative to its predecessors and followers and is therefore univocally identified and symbolized. It is precisely this property to represent the strength of mathematics since it allows the application of

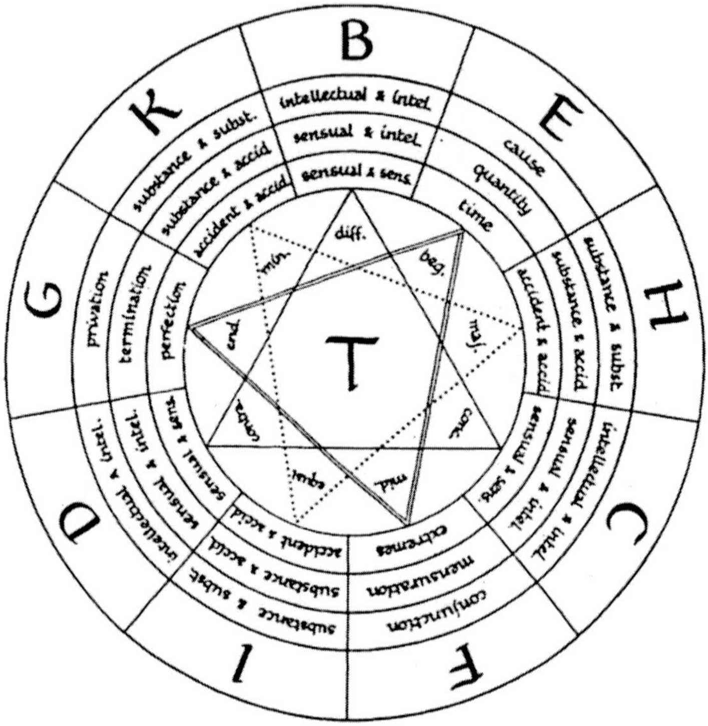

Figure 1 Llul's so-called Figure T from *ars magna*, showing the possible combinations determined by the triangles difference–concordance–contrariety (BCD), beginning–middle–end (EFG), and majority–equality–minority (HIK), where each vertex represents an elementary concept.

arithmetic operations to any number. In this way we can perform calculations not only in pure mathematics but also in physics, engineering, and so on. At the opposite, the current state of logic is like a mathematical algebra without an arithmetic. It is true that an abstract algebra still allows to compute some formulas (like the square of a binomial). However, without arithmetic we would not be able to perform calculations. Here, each proposition is univocally identified and denoted and therefore this logical space constitute to a certain extent an analogue with respect to arithmetics.

The assignation of a number to every proposition is not unprecedented. Gödel already succeeded in assigning to any proposition

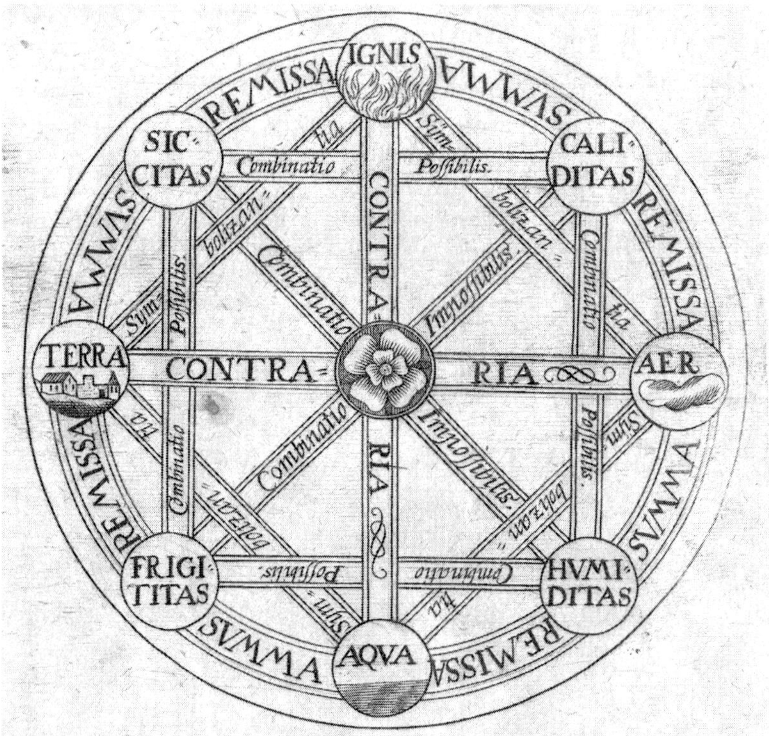

Figure 2 Leibniz's *ars combinatoria*, displaying the four traditional elements determined by the combination of four elementary properties (fire is dry and hot, air is hot and humid, water is humid and cold, earth is cold and dry).

a number (the so-called Gödel's number) (Gödel, 1931). However, these numbers are difficult to calculate and to use. Moreover and most importantly, no relations among logical propositions are univocally assigned in that context (since there is no definite space of propositions). At the opposite, the notation used in the following, assigning both identification numbers and mutual relations among propositions, makes possible (as it happens for arithmetic) that properties of these propositions may be explored and mechanical calculation be performed. In the following I shall devote indeed a large part of the exposition for showing such properties, but further research will certainly point out more aspects.

Part I

Logical Spaces

Chapter 1

Structural Description

1.1 One-Dimensional Space

Let me first introduce some of the logical connectives that will be used in the following:

Table 1.1 Logical connectives

Inputs		Inclusive disjunction	Implication	Conjunction
p	q	$p \vee q$	$p \to q$	$p \wedge q$
0	0	0	1	0
0	1	1	1	0
1	0	1	0	0
1	1	1	1	1

Here, 0 and 1 mean false and true, respectively, and p, q are arbitrary propositions. To these connectives we should add the negation $\neg$ that inverts the truth value of proposition (to deny a true proposition is false and vice versa). The equivalence $\leftrightarrow$ will be introduced below.

Let us now make some elementary consideration. Suppose that we know that a certain proposition p is true or we hypothetically assume that it is true. Logically speaking, we can also deny it

Mechanical Logic in Three-Dimensional Space
Gennaro Auletta

ISBN 978-981-4411-50-9 (Hardcover), 978-981-4411-51-6 (eBook)
www.panstanford.com

(or deny its truth), that is, produce a proposition $\neg p$. These two propositions obviously denote alternative state of affairs. We can express this by building a disjunction of the two. The resulting proposition $(p \vee \neg p)$ is obviously always true. We can reverse this order of consideration and start from the latter proposition as a composed one and allowing the possibility to deny each of its components. By denying $\neg p$ we obtain p whilst by denying p we obtain $\neg p$. In other words, we can obtain the two atomic propositions by denying one of the components of the proposition $p \vee \neg p$. However, it is in principle also possible to deny more than one component of a molecular proposition. However, in such a case (in which we only have p and $\neg p$) we run immediately (after the first step) into the contradiction $p \wedge \neg p$ (which is indeed the negation of $p \vee \neg p$). Reached this point, we can again reverse this order of consideration and start from the proposition $p \wedge \neg p$ and build other propositions by affirming one of its components. In this case, we again obtain the propositions p and $\neg p$.

We can depict this very elementary logical space as in Fig. 1.1. This is not a true logic since with this set of propositions we cannot build logical rules or do not have possibilities to perform any kind of derivation. Actually, apart from the tautology $p \vee \neg p$ and the contradiction $p \wedge \neg p$, we only have the assertion of two alternative (contingent) states of affairs denoted by p and $\neg p$. We need higher levels of complexity for dealing with logical rules and logical inferences. Nevertheless, we have set the general principles for dealing with any order of complexity:

Principle 1.1 *To assert any statement also implies the hypothetically possibility to deny the truth of some or one of its alternative components (of some of its disjoints). In this way, we build the conjunctive order.*

Principle 1.2 *To assert any statement always implies the possibility to hypothetically assert the truth of some or all of its components (conjoints) as alternative possibilities. In this way, we build the disjunctive order.*

The first rule corresponds to a sort of selection act since we single some more specific statement out of one that is more general

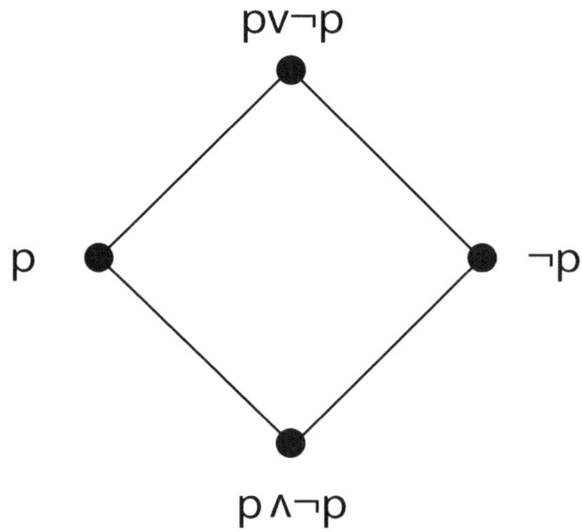

Figure 1.1 One-dimensional logical space.

and in this way we establish when statements are jointly true, the second one enables us to build hypothetical correlations between statements since by grouping them we are able also to establish when they are jointly true or jointly false (Peirce, 1885a, p. 169). As a consequence, there are only two kinds of basic molecular propositions, namely those constituted by either disjunctions or conjunctions (atomic propositions are the zeroth case for both). Obviously, also negation needs to be considered. Due to these very general rules, there is a full reversibility of such a logical space, that is, we can start with any arbitrary propositions and apply reiteratively these two rules and always go back to this initial proposition. In fact, although structural relations cannot be identified with inferences or logical derivations (as I shall show below), it remains true that we can pass from any point of the space to another through some kind of inference. Finally, I remark that here in the following it is not necessary to assume a principle of bivalence, according to which any proposition is either true or false.

The one-dimensional case is too trivial to spend too many words. The main combinatorial and logical properties are much better expressed when going to higher orders of logical space.

1.2 Two-Dimensional Space

When we pass the bidimensional logical space, we can depict it as in Fig. 1.2. I shall expand the traditional Lindenbaum–Tarski algebra into a two-dimensional (and tridimensional) logical space. In accordance with the two principles formulated in the

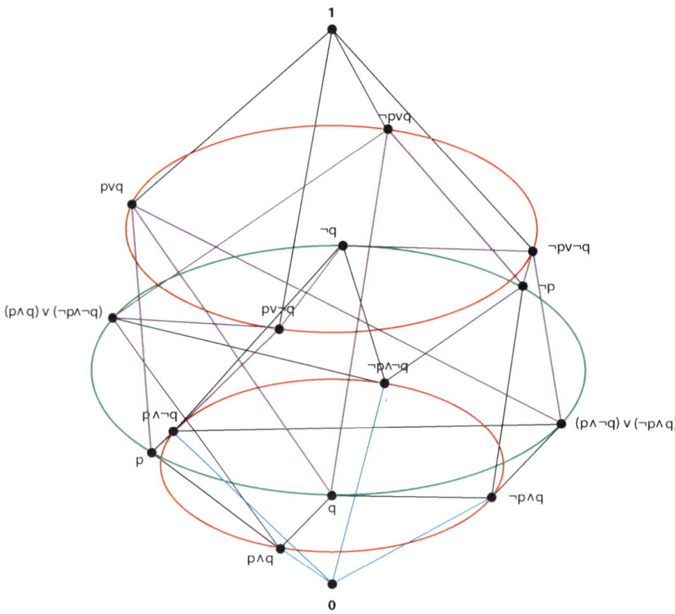

Figure 1.2 The so-called Lindenbaum–Tarksy algebra in a pictorial representation. As before, the lower nodes are constituted through conjunction of upper nodes while upper nodes through disjunction of lower nodes. Again, at the bottom we have the universal falsity **0**, at the top the universal truth **1**. Note that we have three levels between these two nodes, indicated by a bottom red circle, a middle green circle and a top red circle. All connections between **0** and the bottom red circle are in blue, all connections between the latter and the middle green circle are in black, all connections between the latter and top red circle are in purple, all connections between the latter and **1** are in black. The expression $\neg p \vee q$ is logically equivalent to $p \rightarrow q$ (in the Lindenbaum–Tarski algebra only disjunctions and conjunctions are used). Formula $(p \wedge q) \vee (\neg p \wedge \neg q)$ expresses the equivalence between p and q $(p \leftrightarrow q)$ whilst formula $(p \wedge \neg q) \vee (\neg p \wedge q)$ the exclusive disjunction (XOR).

previous chapter, we can represent the Lindenbaum–Tarski algebra as constituted through (*Boole*, 1854; Tarski, 1935a):

(1) An ascending (raising) proposition-building through disjunctions of truths. Here, we start from the bottom level (universal falsehood), and the levels 0, 1, 2, 3, 4 mean both the number of truths in the sequence of the truth values and the number of disjoints.
(2) A descending (lowering) proposition-building through conjunctions of falsities. Here, we start from the upper level (universal truth), and the levels 0, 1, 2, 3, 4 mean here both the number of falsehoods in the sequence of truth values and the number of elements that are conjoint.

Here and in the following, all levels are therefore denoted by a couple of numbers: the first number of the couple always means disjunctive order whilst the second one conjunctive order. For instance, Level 3-1 means Level 3 according to the disjunctive (raising) order and Level 1 according to the conjunctive (lowering) order. This couple of numbers could also be interpreted as connected with the number of relations that each statement has with statements above and statements below, respectively. Let us consider the number of combinations both of propositions and relations among propositions. The number of levels and of propositions contained at each level is computed according to the combinatorial calculus without permutation:

$$C\,(n, k) = \frac{n!}{k!(n-k)!}\,. \tag{1.1}$$

The number of levels and the repartition of propositions is determined by the numbers n and $0 \leq k \leq n$, where $n = 2^m$ and m is the number of the involved elementary (or atomic) propositions. In our case, since the number of atomic propositions is $m = 2$ (the logical space is two-dimensional), we have $n = 4$. We now constitute a series given by

$$C\,(n, k=0),\, C\,(n, k=1),\, C\,(n, k=2)\ldots,\, C\,(n, k=n). \tag{1.2}$$

If the series is read from the left to the right we have the raising order, if it is read from the right to the left we have the lowering one. In our case, $C(4, 0) = 1$, $C(4, 1) = 4$, $C(4, 2) = 6$, $C(4, 3) = 4$, $C(4, 4) = 1$. We have therefore 5 levels and a total number of $1 + 4 + 6 + 4 + 1 = 16$ propositions, as anticipated. In short, the whole number of the involved propositions across all levels is given by 2^n and the number l of levels is given by $n + 1$.

The number of relations among propositions of the different levels is given by the series

$$\frac{n!}{(n-k)!(k-1)!}, \tag{1.3}$$

with k going from 1 to n. Note that k gives the raising order whilst $n - k$ the lowering one with increasing values of k. In our case, we have: $4!/3!0! = 4$ relations among propositions of Level 0-4 and propositions of Level 1-3; $4!/2!1! = 12$ relations among propositions of Level 1-3 and propositions of Level 2-2; $4!/1!2! = 12$ relations among propositions of Level 2-2 and propositions of Level 1-3; $4!/0!3! = 4$ relations among propositions of Level 1-3 and propositions of Level 4-0. In the two-dimensional case, this makes $4 + 12 + 12 + 4 = 32$ relations as a whole.

Such a logical system is only built according to the two Principles 1–2 and, as mentioned, does not have axioms or rules (as we shall show they can on the contrary be derived). We write the possible truth values of the all expressions occurring in the Lindenbaum–Tarski algebra as in the following table:

Table 1.2 Inputs and outputs of two-dimensional logical space

Inputs	X	0	0	1	1
	Y	0	1	0	1
Outputs		a	b	c	d

This allows us to build the logical space as in Table 1.3, where each statement is represented with a sequence of 4 numbers (according to the series abcd). This numbers can be 0 (for falsehood) and 1 (for truth).

Here and in the following I shall use X, Y, Z instead of p, q, r since the symbols X, Y, Z, can mean both sets of objects (which

Table 1.3 Two-dimensional logical space

Level	Truth value	Disjunctions	Conjunctions	Simp. form
4-0	1111	$X \vee \neg X \vee Y \vee \neg Y$	$\neg(X \wedge \neg X \wedge Y \wedge \neg Y)$	
3-1	1110	$\neg X \vee \neg Y \vee [(X \wedge \neg Y) \vee (\neg X \wedge Y)]$	$\neg(X \wedge Y)$	$\neg X \vee \neg Y$
	1101	$\neg X \vee Y \vee [(X \wedge Y) \vee (\neg X \wedge \neg Y)]$	$\neg(X \wedge \neg Y)$	$\neg X \vee Y$
	1011	$X \vee \neg Y \vee [(X \wedge Y) \vee (\neg X \wedge \neg Y)]$	$\neg(\neg X \wedge Y)$	$X \vee \neg Y$
	0111	$X \vee Y \vee [(X \wedge \neg Y) \vee (\neg X \wedge Y)]$	$\neg(\neg X \wedge \neg Y)$	$X \vee Y$
2-2	1100	$(\neg X \wedge Y) \vee (\neg X \wedge \neg Y)$	$(\neg X \vee \neg Y) \wedge (\neg X \vee Y)$	$\neg X$
	1010	$(X \wedge \neg Y) \vee (\neg X \wedge \neg Y)$	$(X \vee \neg Y) \wedge (\neg X \vee \neg Y)$	$\neg Y$
	1001	$X \wedge Y) \vee (\neg X \wedge \neg Y)$	$(\neg X \vee Y) \wedge (X \vee \neg Y)$	$X \leftrightarrow Y$
	0110	$(X \wedge \neg Y) \vee (\neg X \wedge Y)$	$(\neg X \vee \neg Y) \wedge (X \vee Y)$	$\neg(X \leftrightarrow Y)$
	0101	$(X \wedge Y) \vee (\neg X \wedge Y)$	$(X \vee Y) \wedge (\neg X \vee Y)$	Y
	0011	$(X \wedge Y) \vee (X \wedge \neg Y)$	$(X \vee \neg Y) \wedge (X \vee Y)$	X
1-3	1000	$(\neg X \wedge \neg Y)$	$\neg X \wedge \neg Y \wedge [(\neg X \vee Y) \wedge (X \vee \neg Y)]$	$\neg(X \vee Y)$
	0100	$(\neg X \wedge Y)$	$X \wedge [(\neg X \vee \neg Y) \wedge (X \vee Y)] \wedge Y$	$\neg(X \vee \neg Y)$
	0010	$(X \wedge \neg Y)$	$\neg Y \wedge [(\neg X \vee \neg Y) \wedge (X \vee Y)] \wedge X$	$\neg(\neg X \vee Y)$
	0001	$(X \wedge Y)$	$[(\neg X \vee Y) \wedge (X \vee \neg Y)] \wedge Y \wedge X$	$\neg(\neg X \vee \neg Y)$
0-4	0000	$\neg(X \vee \neg X \vee Y \vee \neg Y)$	$X \wedge \neg X \wedge Y \wedge \neg Y$	

is more congruent with the Venn diagrams) and the relative propositions. Following the first interpretation, according to the previous truth-table, the whole of the Lindenbaum–Tarski algebra can be represented as in Fig. 1.3 or (more synthetically) can be built by all combination of the four regions shown in Fig. 1.4 that corresponds to the above outputs a, b, c, d. This convention, which will be fully justified below, will allow a simplified connection between sets and propositions. The simplified forms (especially those in the last column on the right) in Table 1.3 can be derived by using the distribution rule and dropping all tautologies in conjunctions and all contradictions in disjunctions (Peirce, 1883a, p. 457; Peirce, 1885a, p. 184)

I stress that $\neg X \vee Y$ could be understood as the traditional (Middle-Age) Aristotelian statement **A** (universal affirmation) whilst $\neg X \vee \neg Y$ could be understood as the analogous of the statement **E** (universal denial) (*Aristotle An. Pr., An. Post.*; *Petrus Hispanus SL*; *William of Shyreswood IL*; Auletta, 2013). A statement of hypothetical form expresses a hypothesis such that if the consequent (here Y) is true, the hypothetical statement in its logical sense ought to be regarded as true as well, while if the antecedent (here X) is false, it is a matter of indifference if we understand the hypothesis as true or not although it is simpler to class it among the true propositions (Peirce, 1885a, p. 170). On the other hand, the set d could be understood as the analogous of statement **I** (particular affirmation) and the set c as the analogue of statement **O** (particular denial). All the universal and particular propositions can be easily translated in the calculus of predicates (any particular at the bottom level contradicts the universal statement in the upper row in the same column), as shown in Table 1.4.

Table 1.4 Quantified logical expressions

$(\forall t)(Xt \to Yt)$	$(\forall t)(Xt \to \neg Yt)$	$(\forall t)(\neg Xt \to Yt)$	$(\forall t)(\neg Xt \to \neg Yt)$
$(\exists t)(Xt \wedge \neg Yt)$	$(\exists t)(Xt \wedge Yt)$	$(\exists t)(\neg Xt \wedge \neg Yt)$	$(\exists t)(\neg Xt \wedge Yt)$

However, the use of explicit quantification is unnecessary in the context of this language (if we confine our examination to the predicate calculus of the first order, which is indeed the case for what

0000	$\mathbf{0}$	(contradiction)
0001	$X \wedge Y$	(conjunction)
0010	$X \wedge \neg Y$	(non-implication, Boole's except)
0100	$\neg X \wedge Y$	(converse non-implication)
1000	$\neg X \wedge \neg Y$	(non-disjunction, Peirce functor)
0011	X	(left projection)
0101	Y	(right projection)
0110	$(X \wedge \neg Y) \vee (\neg X \wedge Y)$	(exclusive disjunction, XOR)
1001	$(X \wedge Y) \vee (\neg X \wedge \neg Y)$	(equivalence)
1010	$\neg Y$	(right complementation)
1100	$\neg X$	(left complementation)
0111	$X \vee Y$	(inclusive disjunction, OR)
1011	$X \vee \neg Y$	(converse implication)
1101	$\neg X \vee Y$	(implication)
1110	$\neg X \vee \neg Y$	**(non-conjunction, Sheffer functor)**
1111	$\mathbf{1}$	(tautology)

Figure 1.3 All logical propositions (nodes) of the Lindenbaum–Tarski algebra (two-dimensional space) shown in terms of Venn diagrams (sets). In red the value true, in white the value false. The series of four numbers in the column on the right of the Venn diagrams is the output according to the following assignations for X, Y : 00, 01, 10, 11 (note that we have indeed four set areas to cover). The outputs (propositions) are shown in the column on the right of these numbers. These propositions can be also understood as the definitions of logical connectives: their names are shown in the right column.

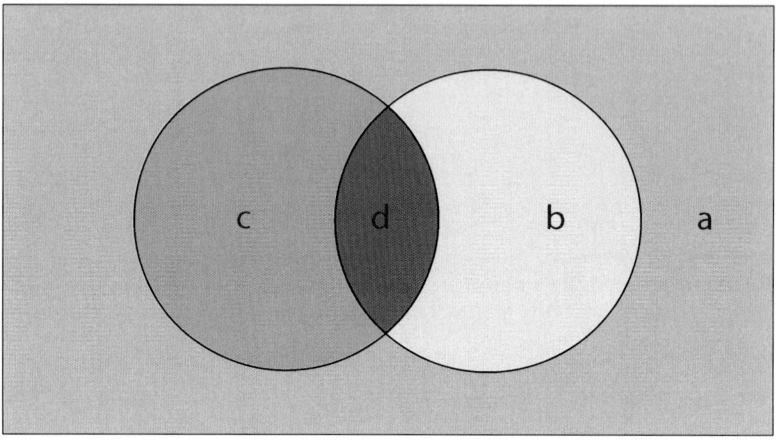

Figure 1.4 Venn diagram for the two-dimensional space. The four regions a, b, c, d represent possible outputs when X (represented here by the union of areas c and d) and Y (represented here by the union of areas b and d) are somehow combined.

follows), so that we can use $X \to Y$ as a shorthand of $(\forall t)(Xt \to Yt)$ (meaning that if all objects t pertain to the class X, it follows that they pertain to the class Y) and $X \wedge Y$ as a shorthand of $(\exists t)(Xt \wedge Yt)$ (meaning that some object t pertain to both the class X and the class Y). In other words, when the symbols X, Y, Z mean propositions, they could be taken to be shorthand for statements like: "The object t is member of the class X" (or also "t is X"), "The object t is member of the class Y", or "The object t is member of the class Z", where t is any arbitrary object. We may say with Peirce that this is a grammatical difference and not a logical one (Peirce, 1885a, p. 168). I also remark that no expressions about individuals are necessary at a logical level (no statement about individuals will be presented in this book). Obviously, I have made use of the ordinary rules for quantifiers exchange (*Schröder*, 1890–95, iii v., p. 37), i.e.:

$$\forall t = \neg\exists\neg t \text{ and } \exists t = \neg\forall\neg t\,. \tag{1.4}$$

When equivocation problems may arise, like in expressions of the type

$$(\exists t)(Xt \wedge Yt) \wedge (\exists t)(Xt \wedge Yt)\,, \qquad (1.5)$$

we can solve this problem by making use of indices, like $X_1 \wedge Y_1 \wedge X_2 \wedge Y_2$, where it is understood that X_1 and X_2 represent non-empty subsets of X as well as Y_1 and Y_2 non-empty subsets of Y, respectively. I wish to stress that the distinction between universal and particular statements is crucial: if the theory of inferences does not deal with the relation among classes of less or greater generality, it becomes an empty game. The general and abstract structures of logic can still be very useful for dealing with very general classes of problems (and I shall make use of this possibility in the following), but are less helpful when dealing with inferences that can be applied also in empirical or mathematical sciences as well as in practical matters.

For reasons that will be clarified later on, the statements $X \vee Y$ and $X \vee \neg Y$ of Table 1.3 are associated with symbols **L** and **M**, respectively, whilst b and a with symbols **S** and **T**, respectively. We may observe that each proposition of Level 3-1 represents the negation of a proposition of Level 1-3 and vice versa. Obviously, the tautology negates the contradiction and vice versa. As a general rule, it can be noted that to negate a proposition inverts the 1s and 0s that are present in its identifying sequence. In particular, we have the contradictory pairs (where a′ denotes the complement set of a and similarly for b, c and d) shown in Table 1.5.

Table 1.5 Contradictory pairs

$\neg X \vee \neg Y$	d′	$X \wedge Y$	d
$\neg X \vee \neg Y$	c′	$X \wedge \neg Y$	c
$X \vee \neg Y$	b′	$\neg X \wedge Y$	b
$X \vee Y$	a′	$\neg X \wedge \neg Y$	a

Note that each of paired terms (in the same row) both do not share a common area and sum to tautology. This allows us to better understand the proper distinction between contradictory pairs and contrary or subcontrary pairs: in Aristotelian parlance, all pairs at Level 3-1 displayed in Table 1.3 are contrary among them (they share two areas couple-wise and sum to tautology) whilst all pairs

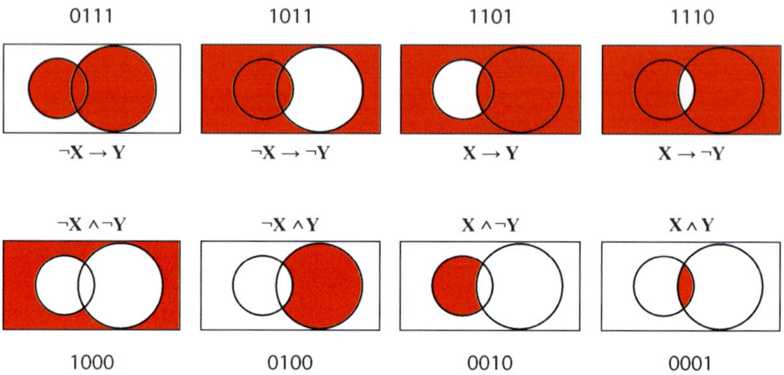

Figure 1.5 The top four cases represent universal statements whilst the bottom one particular statements. Each column presents contradictory pairs. Truth is represented in gray, falsehood in white. The correct relations of contrariety and subcontrariety can be established only among these complete sets.

at Level 1-3 are subcontrary among them (pairwise they both share no area and do not sum to tautology). However, when dealing with contrariety and subcontrariety, we need to consider the whole set of four propositions in both cases (and not only two, as it is customary in the Aristotelian tradition). This forbids the relation of subalternity and therefore the conversion from **A** to **I** or from **E** to **O**. This is shown in Fig. 1.5.

Another remarkable property is that all statements of (the intermediate) Level 2-2 can be divided in two subsets such that any statement of one subset is a negation of a statement pertaining to the other subset, as shown in Table 1.6.

Table 1.6 Partition in subsets

X	$\neg X$
Y	$\neg Y$
$(X \wedge Y) \vee (\neg X \wedge \neg Y)$	$(X \wedge \neg Y) \vee (\neg X \wedge Y)$

Note that in the first row (or, alternatively, in the second row) there are the two statements of the one-dimensional logical space (there, they have been expressed as p and $\neg p$) apart from contradiction and tautology. Last row presents the equivalence $X \leftrightarrow Y$ on the left and its negation (countervalence or XOR) on the right. Summarizing, the whole of all statements of any n-level logical space can be divided in two subset each of which is the specular negative of the other: all statements of the middle level are divided in two subset while any statement of any j-k level finds its negation in a statement of the k-j level. This means that for the two-dimensional case we have 8 statements and their negations.

The previous examination allows us to make use of a simple representation of the logical space: having identified statements with a sequence of 4 numbers (according to the series abcd), we can represent disjunctions by simply adding four-numbers strings number by number and conjunctions by simply multiplying four-numbers strings number by number—see also (*Hobbes*, 1655) (*Boole*, 1854) (*Schröder*, 1890–95, v. iii, p. 17). In other words, in each column:

Table 1.7 Sum and product

$0+0=0$	$0+1=1$	$1+0=1$	$1+1=1$
$0\times 0=0$	$0\times 1=0$	$1\times 0=0$	$1\times 1=1$

This turns out to be just the ordinary arithmetic product and also the ordinary sum apart from the sum $1+1=1$: this, however, does not depend on the definition of the operation but is due to the fact that the set of allowed numbers is represented by just 0 and 1, so that, being 1 the highest number, we cannot add anything to it. I also recall that Peirce had defined the addition in this way (Peirce, 1883a, p. 455). Then, we may redraw Table 1.3 as in Table 1.8 (where, e.g. a′b′ means the product of the complement sets of a and b).

In Table 1.8 I have indicated explicitly the products of conjunctive couples. On the contrary, I have omitted to indicate that triples of disjoints are themselves the sum of three couples of disjoints. For instance, a+b+c = (a+b)+(a+c)+(b+c). With such a system at this (second) level of complexity any logical rule can be derived.

Table 1.8 Arithmetic form of the two-dimensional logical space

ID	Disjunctions		Conjunctions	
1111	1111	a+b+c+d	1110 + 1101 + 1011 + 0111	a′+ b′+ c′+ d′
1110	0010 + 0100 + 1000	a+b+c	1110	d′
1101	0001 + 0100 + 1000	a+b+d	1101	c′
1011	0001 + 0010 + 1000	a+c+d	1011	b′
0111	0001 + 0010 + 0100	b+c+d	0111	a′
1100	0100 + 1000	a+b	1101 × 1110	c′ d′
1010	0010 + 1000	a+c	1011 × 1110	b′ d′
1001	0001 + 1000	a+d	1011 × 1101	b′ c′
0110	0010 + 0100	b+c	0111 × 1110	a′ d′
0101	0001 + 0100	b+d	0111 × 1101	a′ c′
0011	0001 + 0010	c+d	0111 × 1011	a′ b′
1000	1000	a	1110 × 1101 × 1011	b′ c′ d′=(c′ d′)(b′ d′)(b′ c′)
0100	0100	b	1110 × 1101 × 0111	a′ c′ d′=(c′ d′)(a′ d′)(a′c)
0010	0010	c	1110 × 1011 × 0111	a′ b′ d′=(b′ d′)(a′ d′)(a′ b′)
0001	0001	d	1101 × 1011 × 0111	a′ b′ c′=(b′ c′)(a′ c′)(a′ b′)
0000	1000 × 0100 × 0010 × 0001	abcd	0000	a′ b′ c′ d′

All equivalence rules are simply a game of product or sum. For instance,

Table 1.9 Distribution

Logical input	Calculation	Logical output
$X \vee (\neg X \wedge Y)$	0011 +	
	0100 =	
	0111	$X \vee Y$

The result is obviously equivalent to $(X \vee Y) \wedge (X \vee \neg X)$, since, as explained a conjunction with a tautology reduces to the other conjoint. Obviously, it would be better to have three propositions but I shall develop this system below. A system of three propositions will allow us not only to derive any logical rule through sums and products but also to perform any logical inference. It is known that there are some logics (like the family of the so-called quantum logic) that do not acknowledge the rule of distribution. However, this is not a problem for this logical system since we are not obliged to assume any logical rule in advance apart from Principles 1–2. This means that a variety of sub-logics can be generated according to the allowed "paths" across the logical space, as anticipated in the Introduction.

Let us derive now other logical rules (which can also be obtain by direct comparison of some formulation of the raising and the correspondent ones of the lowering order):

Table 1.10 Absorption and De Morgan

Logical input	Calculation	Logical output
$X \vee (X \wedge Y)$	0011 +	
	0001 =	
	0011	X
$X \wedge Y$	0001 = 0011 × 0101	
$\neg X$	1100 +	
$\neg Y$	1010 =	
$\neg X \vee \neg Y$	1110	
negation	0001	$\neg(\neg X \vee \neg Y)$

In the De Morgan, I have shown that we can arrive to the same result (whose ID is 0001) through both ways: (i) the conjunction of X and Y or (ii) the negation of the disjunction between $\neg X$ and $\neg Y$. Obviously, the complementary De Morgan (or similar forms) can be trivially derived. About all rules concerning the implication (like transposition), they can be easily derived by these means once that we define the implication $X \rightarrow Y$ as $\neg X \vee Y$.

More complex are the rules of derivation. All the forms of the so-called disjunctive syllogism (including all forms of *Modus ponens* or *tollens* if the implication is used instead of the disjunction) can be reduced to this general rule: The conclusion is the product of two statements: the first one is one of the premises (the disjunction itself) and the other is shared with the second premise. For instance, let us consider the case in which we wish to derive $\neg X$ (the Conclusion) from both $\neg X \vee Y$ (Premise 1) and $\neg Y$ (Premise 2). In the ordinary case, we do not know the conclusion but only the premises. However, the Conclusion can mechanical be computed in this way:

$$\begin{array}{lcll} \text{Premise} & 1 & 1101 & \times \\ \text{Premise} & 2 & 1010 & = \\ \text{Conclusion} & & 1000 & \end{array}$$

Now the conclusion corresponds to the statement $\neg X \wedge \neg Y$. However, this statement is redundant since we already had $\neg Y$ as a premise, so that the conclusion is necessarily $\neg X$ (in the context of a three-dimensional logical space we shall explicitly deal with the rule of Simplification). Note that we take Premise 1 from the Level 3-1 whilst Premise 2 and the Conclusion pertain to Level 2-2. Therefore, the latter two statements consist in the product of two statements pertaining to Level 3-1 (we could also do the reverse and consider statements of Level 3-1 as sums of statements of Level 2-2). Then, the desired Conclusion can be analyzed in this way:

$$\begin{array}{lcl} 1101 & \times & \\ 1110 & = & \\ 1100 & & \neg X \end{array}$$

Now, 1101 is the identification number of the first premise ($\neg X \vee Y$). On the other hand, Premise 2 can be analyzed as follows:

$$\begin{array}{lll} 1011 & \times & \\ 1110 & = & \\ 1010 & & \neg Y \end{array}$$

Therefore, the Conclusion and Premise 2 share the statement 1110 (which corresponds to the disjunction $\neg X \lor \neg Y$). It may also be noted that the three involved propositions (Premises 1 and 2 and Conclusion) share a truth (the first number of the sequence). Therefore, the only possible conclusion is the one that is the product of two statements, the one shared with the first premise (actually it coincides with the latter since this pertains to an upper Level), and the other shared with the second premise. It is easy to see that we also have:

Table 1.11 Disjunctive syllogisms

P1	1101×	0111×	0111×	1011×	1011×	1110×	1110×
P2	0011=	1100=	1010=	1100=	0101=	0011=	0101=
P2 ×C	0001	0100	0010	1000	0001	0010	0100

Obviously, the rules of conjunction and addition are already embedded in the lowering and raising order, respectively.

In conclusion, I remark that the system (and this is true for ever level of complexity) has another and much more amazing property: any combination (either through sum or product) of a couple of statements will produce a result that is always logically sound. It is precisely this basic property (due to the fact that the position of the statements and their relations are univocally defined) that allows a full mechanization of logic. Indeed, as in arithmetic any operation of any set of numbers will produce again other numbers (and therefore can be computed mechanically) the same is true here for statements.

1.3 Three-Dimensional Space

I shall present now an expansion of the Lindenbaum–Tarski algebra into a three-dimensional logical space. The Venn diagrams are shown in Fig. 1.6 while the relative truth-value assignation in Table 1.12. Let us compute the combinatory of this logical space,

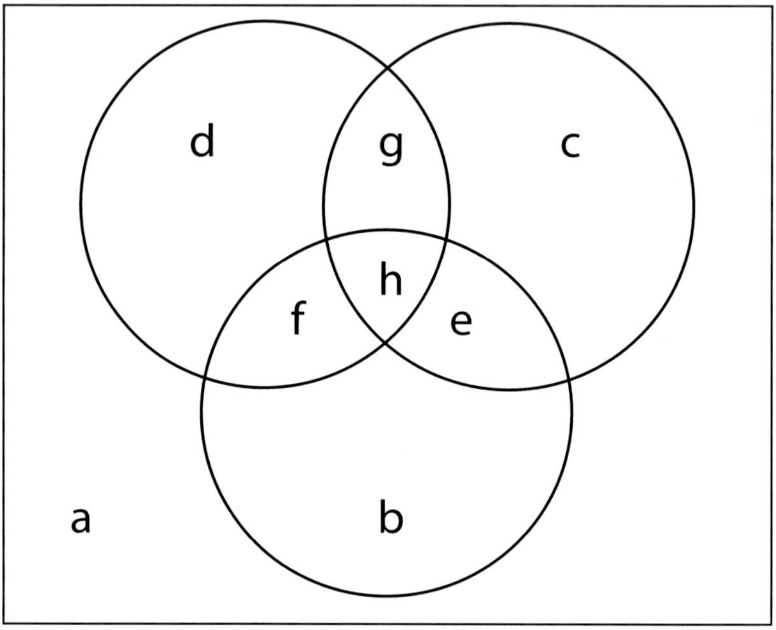

Figure 1.6 The Venn diagrams when the relations among three statements are considered. Note that we have: X = d+f+g+h, Y = c+e+g+h, Z = b+e+f+h.

Table 1.12 Inputs and outputs of three-dimensional logic

X	0	0	0	1	0	1	1	1
Y	0	0	1	0	1	0	1	1
Z	0	1	0	0	1	1	0	1
output	a	b	c	d	e	f	g	h

considering that we have 8 truth-value assignation and therefore statements defined through a string of 8 numbers.

A combinatorial calculus according to what established in the previous section gives the result:

$$C(8,0)=1,\quad C(8,1)=8,\quad C(8,2)=28,\quad C(8,3)=56,\quad C(8,4)=70$$
$$C(8,5)=56,\quad C(8,6)=28,\quad C(8,7)=8,\quad C(8,8)=1.$$

This allows us to establish the following:

- Number of levels: 9.
- Number of statements: $1+8+28+56+70+56+28+8+1 = 256 = 2^8$.
- Number of relations: $8+56+168+280+280+168+56+8 = 1024 = 2^{10}$.

Let us now represent all the propositions at the different levels through the following tables (I recall that here and in the following, e.g. the expression f′h′ means the product of the complement sets of f and h):

Table 1.13 $C(8, 8) = C(8, 0)$

11111111	a+b+c+d+e+f+g+h

Table 1.14 $C(8, 7) = 8$ disjunctive eptaplets, $C(8, 1) = 8$ conjunctive singlets

1.	11111110	a+b+c+d+e+f+g	h′	2.	11111101	a+b+c+d+e+f+h	g′
3.	11111011	a+b+c+d+e+g+h	f′	4.	11110111	a+b+c+d+f+g+h	e′
5.	11101111	a+b+c+e+f+g+h	d′	6.	11011111	a+b+d+e+f+g+h	c′
7.	10111111	a+c+d+e+f+g+h	b′	8.	01111111	b+c+d+e+f+g+h	a′

Table 1.15 $C(8, 6) = 28$ disjunctive esaplets, $C(8, 2) = 28$ conjunctive duplets

1.	11111100	a+b+c+d+e+f	g′h′	2.	11111010	a+b+c+d+e+g	f′h′
3.	11111001	a+b+c+d+e+h	f′g′	4.	11110110	a+b+c+d+f+g	e′h′
5.	11110101	a+b+c+d+f+h	e′g′	6.	11110011	a+b+c+d+g+h	e′f′
7.	11101110	a+b+c+e+f+g	d′h′	8.	11101101	a+b+c+e+f+h	d′g′
9.	11101011	a+b+c+e+g+h	d′f′	10.	11100111	a+b+c+f+g+h	d′e′
11.	11011110	a+b+d+e+f+g	c′h′	12.	11011101	a+b+d+e+f+h	c′g′
13.	11011011	a+b+d+e+g+h	c′f′	14.	11010111	a+b+d+f+g+h	c′e′
15.	11001111	a+b+e+f+g+h	c′d′	16.	10111110	a+c+d+e+f+g	b′h′
17.	10111101	a+c+d+e+f+h	b′g′	18.	10111011	a+c+d+e+g+h	b′f′
19.	10110111	a+c+d+f+g+h	b′e′	20.	10101111	a+c+e+f+g+h	b′d′
21.	10011111	a+d+e+f+g+h	b′c′	22.	01111110	b+c+d+e+f+g	a′h′
23.	01111101	b+c+d+e+f+h	a′g′	24.	01111011	b+c+d+e+g+h	a′f′
25.	01110111	b+c+d+f+g+h	a′e′	26.	01101111	b+c+e+f+g+h	ad′
27.	01011111	b+d+e+f+g+h	a′c′	28.	00111111	c+d+e+f+g+h	a′b′

Table 1.16 $C(8, 5) = 56$ disjunctive pentaplets, $C(8, 3) = 56$ conjunctive triplets

1.	11111000	a+b+c+d+e	f'g'h'	2.	11110100	a+b+c+d+f	e'g'h'
3.	11110010	a+b+c+d+g	e'f'h'	4.	11110001	a+b+c+d+h	e'f'g'
5.	11101100	a+b+c+e+f	d'g'h'	6.	11101010	a+b+c+e+g	d'f'h'
7.	11101001	a+b+c+e+h	d'f'g'	8.	11100110	a+b+c+f+g	d'e'h'
9.	11100101	a+b+c+f+h	d'e'g'	10.	11100011	a+b+c+g+h	d'e'f'
11.	11011100	a+b+d+e+f	c'g'h'	12.	11011010	a+b+d+e+g	c'f'h'
13.	11011001	a+b+d+e+h	c'f'g'	14.	11010110	a+b+d+f+g	c'e'h'
15.	11010101	a+b+d+f+h	c'e'g'	16.	11010011	a+b+d+g+h	c'e'f'
17.	11001110	a+b+e+f+g	c'd'h'	18.	11001101	a+b+e+f+h	c'd'g'
19.	11001011	a+b+e+g+h	c'd'f'	20.	11000111	a+b+f+g+h	c'd'e'
21.	10111100	a+c+d+e+f	b'g'h'	22.	10111010	a+c+d+e+g	b'f'h'
23.	10111001	a+c+d+e+h	b'f'g'	24.	10110110	a+c+d+f+g	b'e'h'
25.	10110101	a+c+d+f+h	b'e'g'	26.	10110011	a+c+d+g+h	b'e'f'
27.	10101110	a+c+e+f+g	b'd'h'	28.	10101101	a+c+e+f+h	b'd'g'
29.	10101011	a+c+e+g+h	b'd'f'	30.	10100111	a+c+f+g+h	b'd'e'
31.	10011110	a+d+e+f+g	b'c'h'	32.	10011101	a+d+e+f+h	b'c'g'
33.	10011011	a+d+e+g+h	b'c'f'	34.	10010111	a+d+f+g+h	b'c'e'
35.	10001111	a+e+f+g+h	b'c'd'	36.	01111100	b+c+d+e+f	a'g'h'
37.	01111010	b+c+d+e+g	a'f'h'	38.	01111001	b+c+d+e+h	a'f'g'
39.	01110110	b+c+d+f+g	a'e'h'	40.	01110101	b+c+d+f+h	a'e'g'
41.	01110011	b+c+d+g+h	a'e'f'	42.	01101110	b+c+e+f+g	a'd'h'
43.	01101101	b+c+e+f+h	a'd'g'	44.	01101011	b+c+e+g+h	a'd'f'
45.	01100111	b+c+f+g+h	a'd'e'	46.	01011110	b+d+e+f+g	a'c'h'
47.	01011101	b+d+e+f+h	a'c'g'	48.	01011011	b+d+e+g+h	a'c'f'
49.	01010111	b+d+f+g+h	a'c'e'	50.	01001111	b+e+f+g+h	a'c'd'
51.	00111110	c+d+e+f+g	a'b'h'	52.	00111101	c+d+e+f+h	a'b'g'
53.	00111011	c+d+e+g+h	a'b'f'	54.	00110111	c+d+f+g+h	a'b'e'
55.	00101111	c+e+f+g+h	a'b'd'	56.	00011111	d+e+f+g+h	a'b'c'

Table 1.17 $C(8,\ 4) = 70$ disjunctive quadruplets, $C(8,\ 4) = 70$ conjunctive quadruplets

1.	11110000	a+b+c+d	e′f′g′h′	2.	11101000	a+b+c+e	d′f′g′h′
3.	11100100	a+b+c+f	d′e′g′h′	4.	11100010	a+b+c+g	d′e′f′h′
5.	11100001	a+b+c+h	d′e′f′g′	6.	11011000	a+b+d+e	c′f′g′h′
7.	11010100	a+b+d+f	c′e′g′h′	8.	11010010	a+b+d+g	c′e′f′h′
9.	11010001	a+b+d+h	c′e′f′g′	10.	11001100	a+b+e+f	c′d′g′h′
11.	11001010	a+b+e+g	c′d′f′h′	12.	11001001	a+b+e+h	c′d′f′g′
13.	11000110	a+b+f+g	c′d′e′h′	14.	11000101	a+b+f+h	c′d′e′g′
15.	11000011	a+b+g+h	c′d′e′f′	16.	10111000	a+c+d+e	b′f′g′h′
17.	10110100	a+c+d+f	b′e′g′h′	18.	10110010	a+c+d+g	b′e′f′h′
19.	10110001	a+c+d+h	b′e′f′g′	20.	10101100	a+c+e+f	b′d′g′h′
21.	10101010	a+c+e+g	b′d′f′h′	22.	10101001	a+c+e+h	b′d′f′g′
23.	10100110	a+c+f+g	b′d′e′h′	24.	10100101	a+c+f+h	b′d′e′g′
25.	10100011	a+c+g+h	b′d′e′f′	26.	10011100	a+d+e+f	b′c′g′h′
27.	10011010	a+d+e+g	b′c′f′h′	28.	10011001	a+d+e+h	b′c′f′g′
29.	10010110	a+d+f+g	b′c′e′h′	30.	10010101	a+d+f+h	b′c′e′g′
31.	10010011	a+d+g+h	b′c′e′f′	32.	10001110	a+e+f+g	b′c′d′h′
33.	10001101	a+e+f+h	b′c′d′g′	34.	10001011	a+e+g+h	b′c′d′f′
35.	10000111	a+f+g+h	b′c′d′e′	36.	01111000	b+c+d+e	a′f′g′h′
37.	01110100	b+c+d+f	a′e′g′h′	38.	01110010	b+c+d+g	a′e′f′h′
39.	01110001	b+c+d+h	a′e′f′g′	40.	01101100	b+c+e+f	a′d′g′h′
41.	01101010	b+c+e+g	a′d′f′h′	42.	01101001	b+c+e+h	a′d′f′g′
43.	01100110	b+c+f+g	a′d′e′h′	44.	01100101	b+c+f+h	a′d′e′g′
45.	01100011	b+c+g+h	a′d′e′f′	46.	01011100	b+d+e+f	a′c′g′h′
47.	01011010	b+d+e+g	a′c′f′h′	48.	01011001	b+d+e+h	a′c′f′g′
49.	01010110	b+d+f+g	a′c′e′h′	50.	01010101	b+d+f+h	a′c′e′g′
51.	01010011	b+d+g+h	a′c′e′f′	52.	01001110	b+e+f+g	a′c′d′h′
53.	01001101	b+e+f+h	a′c′d′g′	54.	01001011	b+e+g+h	a′c′d′f′
55.	01000111	b+f+g+h	a′c′d′e′	56.	00111100	c+d+e+f	a′b′g′h′
57.	00111010	c+d+e+g	a′b′f′h′	58.	00111001	c+d+e+h	a′b′f′g′
59.	00110110	c+d+f+g	a′b′e′h′	60.	00110101	c+d+f+h	a′b′e′g′
61.	00110011	c+d+g+h	a′b′e′f′	62.	00101110	c+e+f+g	a′b′d′h′
63.	00101101	c+e+f+h	a′b′d′g′	64.	00101011	c+e+g+h	a′b′d′f′
65.	00100111	c+f+g+h	a′b′d′e′	66.	00011110	d+e+f+g	a′b′c′h′
67.	00011101	d+e+f+h	a′b′c′g′	68.	00011011	d+e+g+h	a′b′c′f′
69.	00010111	d+f+g+h	a′b′c′e′	70.	00001111	e+f+g+h	a′b′c′d′

Table 1.18 $C\,(8,3) = 56$ disjunctive triplets, $C\,(8,5) = 56$ conjunctive pentaplets

1.	11100000	a+b+c	d′e′f′g′h′	2.	11010000	a+b+d	c′e′f′g′h′
3.	11001000	a+b+e	c′d′f′g′h′	4.	11000100	a+b+f	c′d′e′g′h′
5.	11000010	a+b+g	c′d′e′f′h′	6.	11000001	a+b+h	c′d′e′f′g′
7.	10110000	a+c+d	b′e′f′g′h′	8.	10101000	a+c+e	b′d′f′g′h′
9.	10100100	a+c+f	b′d′e′g′h′	10.	10100010	a+c+g	b′d′e′f′h′
11.	10100001	a+c+h	b′d′e′f′g′	12.	10011000	a+d+e	b′c′f′g′h′
13.	10010100	a+d+f	b′c′e′g′h′	14.	10010010	a+d+g	b′c′e′f′h′
15.	10010001	a+d+h	b′c′e′f′g′	16.	10001100	a+e+f	b′c′d′g′h′
17.	10001010	a+e+g	b′c′d′f′h′	18.	10001001	a+e+h	b′c′d′f′g′
19.	10000110	a+f+g	b′c′d′e′h′	20.	10000101	a+f+h	b′c′d′e′g′
21.	10000011	a+g+h	b′c′d′e′f′	22.	01110000	b+c+d	a′e′f′g′h′
23.	01101000	b+c+e	a′d′f′g′h′	24.	01100100	b+c+f	a′d′e′g′h′
25.	01100010	b+c+g	a′d′e′f′h′	26.	01100001	b+c+h	a′d′e′f′g′
27.	01011000	b+d+e	a′c′f′g′h′	28.	01010100	b+d+f	a′c′e′g′h′
29.	01010010	b+d+g	a′c′e′f′h′	30.	01010001	b+d+h	a′c′e′f′g′
31.	01001100	b+e+f	a′c′d′g′h′	32.	01001010	b+e+g	a′c′d′f′h′
33.	01001001	b+e+h	a′c′d′f′g′	34.	01000110	b+f+g	a′c′d′e′h′
35.	01000101	b+f+h	a′c′d′e′g′	36.	01000011	b+g+h	a′c′d′e′f′
37.	00111000	c+d+e	a′b′f′g′h′	38.	00110100	c+d+f	a′b′e′g′h′
39.	00110010	c+d+g	a′b′e′f′h′	40.	00110001	c+d+h	a′b′e′f′g′
41.	00101100	c+e+f	a′b′d′g′h′	42.	00101010	c+e+g	a′b′d′f′h′
43.	00101001	c+e+h	a′b′d′f′g′	44.	00100110	c+f+g	a′b′d′e′h′
45.	00100101	c+f+h	a′b′d′e′g′	46.	00100011	c+g+h	a′b′d′e′f′
47.	00011100	d+e+f	a′b′c′g′h′	48.	00011010	d+e+g	a′b′c′f′h′
49.	00011001	d+e+h	a′b′c′f′g′	50.	00010110	d+f+g	a′b′c′e′h′
51.	00010101	d+f+h	a′b′c′e′g′	52.	00010011	d+g+h	a′b′c′e′f′
53.	00001110	e+f+g	a′b′c′d′h′	54.	00001101	e+f+h	a′b′c′d′g′
55.	00001011	e+g+h	a′b′c′d′f′	56.	00000111	f+g+h	a′b′c′d′e′

Table 1.19 $C(8, 2) = 28$ disjunctive duplets, $C(8, 6) = 28$ conjunctive esaplets

1.	11000000	a+b	c′d′e′f′g′h′	2.	10100000	a+c	b′d′e′f′g′h′
3.	10010000	a+d	b′c′e′f′g′h′	4.	10001000	a+e	b′c′d′f′g′h′
5.	10000100	a+f	b′c′d′e′g′h′	6.	10000010	a+g	b′c′d′e′f′h′
7.	10000001	a+h	b′c′d′e′f′g′	8.	01100000	b+c	a′d′e′f′g′h′
9.	01010000	b+d	a′c′e′f′g′h′	10.	01001000	b+e	a′c′d′f′g′h′
11.	01000100	b+f	a′c′d′e′g′h′	12.	01000010	b+g	a′c′d′e′f′h′
13.	01000001	b+h	a′c′d′e′f′g′	14.	00110000	c+d	a′b′e′f′g′h′
15.	00101000	c+e	a′b′d′f′g′h′	16.	00100100	c+f	a′b′d′e′g′h′
17.	00100010	c+g	a′b′d′e′f′h′	18.	00100001	c+h	a′b′d′e′f′g′
19.	00011000	d+e	a′b′c′f′g′h′	20.	00010100	d+f	a′b′c′e′g′h′
21.	00010010	d+g	a′b′c′e′f′h′	22.	00010001	d+h	a′b′c′e′f′g′
23.	00001100	e+f	a′b′c′d′g′h′	24.	00001010	e+g	a′b′c′d′f′h′
25.	00001001	e+h	a′b′c′d′f′g′	26.	00000110	f+g	a′b′c′d′e′h′
27.	00000101	f+h	a′b′c′d′e′g′	28.	00000011	g+h	a′b′c′d′e′f′

Table 1.20 $C(8, 1) = 8$ disjunctive singlets, $C(8, 7) = 8$ conjunctive eptaplets

1.	10000000	a	b′c′d′e′f′g′h′	2.	01000000	b	a′c′d′e′f′g′h′
3.	00100000	c	a′b′d′e′f′g′h′	4.	00010000	d	a′b′c′e′f′g′h′
5.	00001000	e	a′b′c′d′f′g′h′	6.	00000100	f	a′b′c′d′e′g′h′
7.	00000010	g	a′b′c′d′e′f′h′	8.	00000001	h	a′b′c′d′e′f′g′

Table 1.21 $C(8, 0) = C(8, 8)$

00000000	a′ b′c′d′e′f′g′h′

The relations between all the levels considered here can be summarized as in Figs. 1.7–1.9.

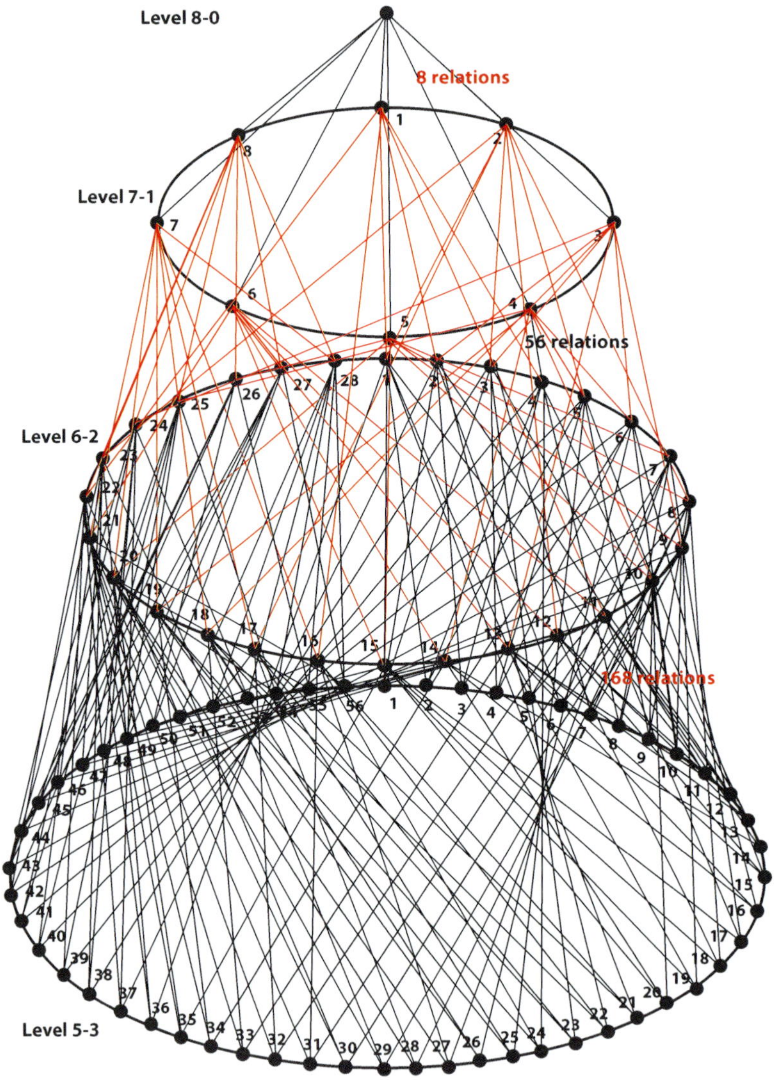

Figure 1.7 The relations among the first 4 levels of the three-dimensional logical space. Note that there are 8 relations among Level 8-0 and Level 7-1, 56 relations (in red for avoiding confusion) between the latter and Level 6-2, and 168 between the latter and Level 5-3.

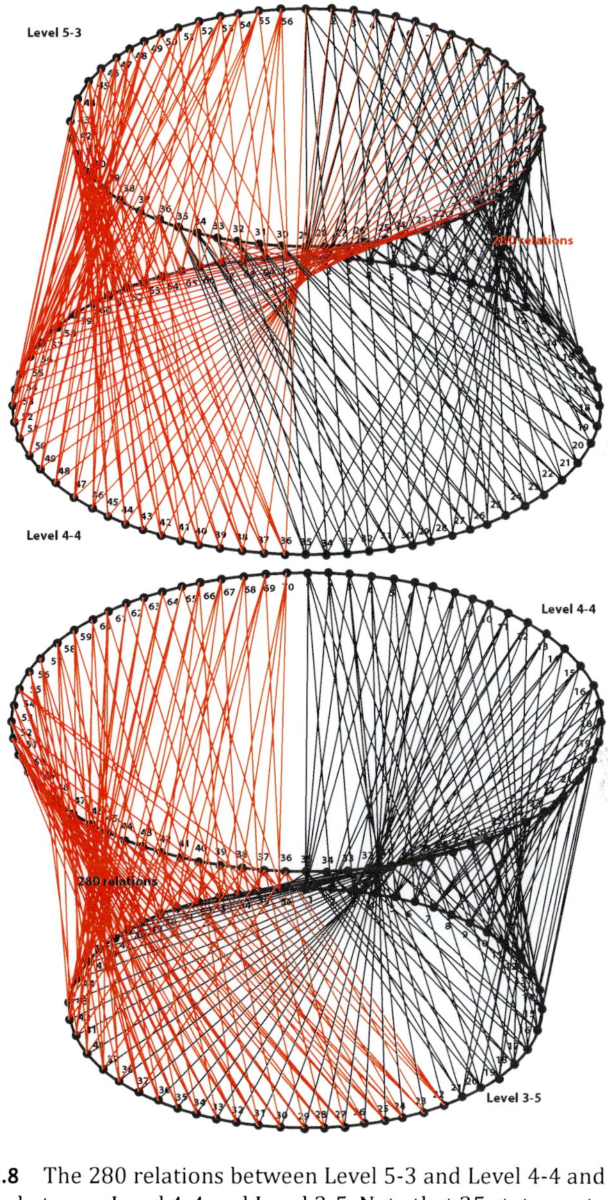

Figure 1.8 The 280 relations between Level 5-3 and Level 4-4 and the 280 relations between Level 4-4 and Level 3-5. Note that 35 statements of Level 4-4 are the negation of the other 35. For this reason, I have drawn half of the relations in red and half in black, according to these subsets.

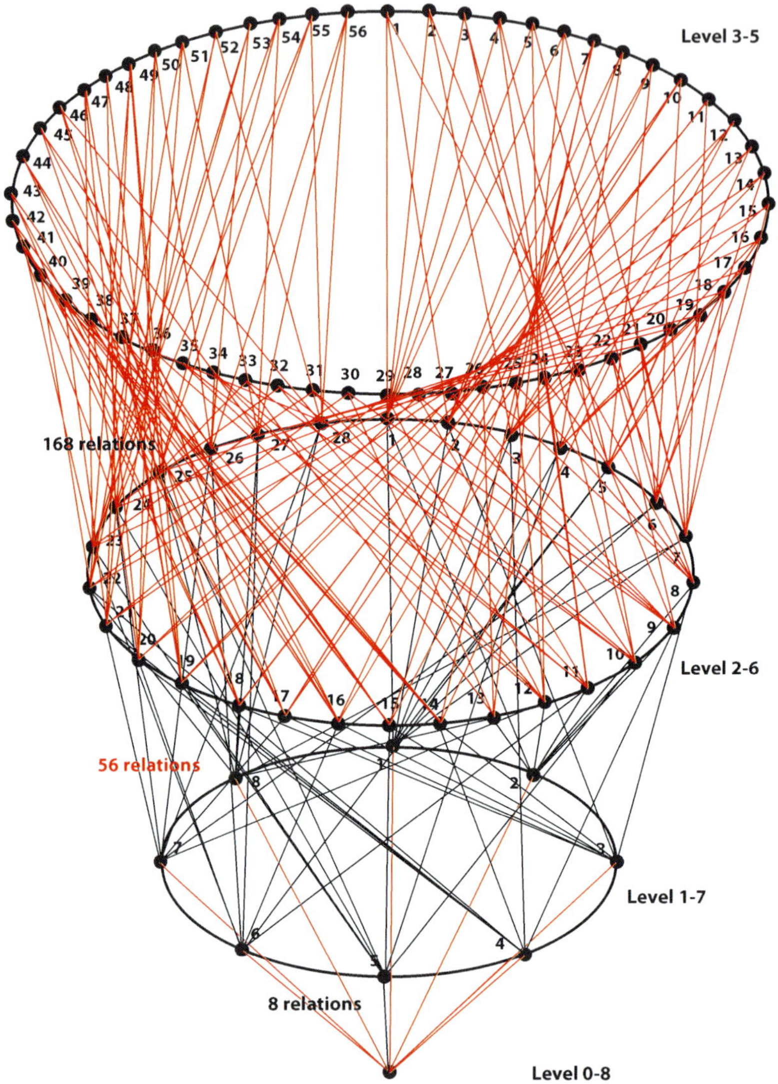

Figure 1.9 The 168 relations between Level 3-5 and Level 2-6, the 56 relations (shown in black to avoid confusion) between Level 2-6 and Level 1-7, and the 8 relations between the latter and Level 0-8.

In the following exposition, for the sake of simplicity I shall consider conjunctive forms (products) from Level 7-1 to Level 4-4 and disjunctive forms (sums) from Level 4-4 to Level 1-7. This means that the former step is descendent whilst the latter is ascendant. Moreover, I have chosen each time the representation in current logical language that turns out to be the simplest one.

The meaning of the 8 disjunctive eptaplets or conjunctive singlets is quite simple: they deny the truth of a specific area, as shown in Table 1.22.

Table 1.22 The 8 statements of Level 7-1

1	11111110	$\neg(X \wedge Y \wedge Z)$	$\neg X \vee \neg Y \vee \neg Z$
2	11111101	$\neg(X \wedge Y \wedge \neg Z)$	$\neg X \vee \neg Y \vee Z$
3	11111011	$\neg(X \wedge \neg Y \wedge Z)$	$\neg X \vee Y \vee \neg Z$
4	11110111	$\neg(\neg X \wedge Y \wedge Z)$	$X \vee \neg Y \vee \neg Z$
5	11101111	$\neg(X \wedge \neg Y \wedge \neg Z)$	$\neg X \vee Y \vee Z$
6	11011111	$\neg(\neg X \wedge Y \wedge \neg Z)$	$X \vee \neg Y \vee Z$
7	10111111	$\neg(\neg X \wedge \neg Y \wedge Z)$	$X \vee Y \vee \neg Z$
8	01111111	$\neg(\neg X \wedge \neg Y \wedge \neg Z)$	$X \vee Y \vee Z$

It is therefore trivial to write the derivations of these 8 statements from above, but it is quite interesting to show the derivation from statements of the lower level (although I did not perform the derivation explicitly), as displayed in Table 1.23.

Table 1.23 Derivation of statements of Level 7-1

Level 7-1	**1**	**2**	**3**	**4**
Level 6-2	1,2,4,7,11,16,22	1,3,5,8,12,17,23	2,3,6,9,13,18,24	4,5,6,10,14,19,25
Level 7-1	**5**	**6**	**7**	**8**
Level 6-2	7,8,9,10,15,20,26	11,12,13,14,15,21,27	16,17,18,19,20,21,28	22,23,24,25,26,27,28

As I shall say in the following the generating set of 7 statements of Level 6-2 for each statement of Level 7-1 can be thought of to be constituted by couples each of which sufficient for deriving the latter. Let us now consider the meaning of the 28 disjunctive esaplets or conjunctive duplets, as displayed in Table 1.24.

Table 1.24 The 28 statements of Level 6-2

1	11111100	11111110 × 11111101	$\neg X \vee \neg Y$	$X \to \neg Y$
2	11111010	11111110 × 11111011	$\neg Z \vee \neg X$	$Z \to \neg X$
3	11111001	11111101 × 11111011	$(\neg X \vee Y \vee \neg Z) \wedge (\neg X \vee \neg Y \vee Z)$	$X \to (Y \leftrightarrow Z)$
4	11110110	11111110 × 11110111	$\neg Z \vee \neg Y$	$Z \to \neg Y$
5	11110101	11111101 × 11110111	$(X \vee \neg Y \vee \neg Z) \wedge (\neg X \vee \neg Y \vee Z)$	$Y \to (X \leftrightarrow Z)$
6	11110011	11111011 × 11110111	$(X \vee \neg Y \vee \neg Z) \wedge (\neg X \vee Y \vee \neg Z)$	$Z \to (X \leftrightarrow Y)$
7	11101110	11111110 × 11101111	$(\neg X \vee Y \vee Z) \wedge (\neg X \vee \neg Y \vee \neg Z)$	$X \to \neg(Y \leftrightarrow Z)$
8	11101101	11111101 × 11101111	$Z \vee \neg X$	$\neg Z \to \neg X$
9	11101011	11111011 × 11101111	$\neg X \vee Y$	$X \to Y$
10	11100111	11110111 × 11101111	$(\neg X \vee Y \vee Z) \wedge (X \vee \neg Y \vee \neg Z)$	$\neg(\neg X \leftrightarrow Y \leftrightarrow Z)$
11	11011110	11111111 × 11011111	$(X \vee \neg Y \vee Z) \wedge (\neg X \vee \neg Y \vee \neg Z)$	$Y \to \neg(X \leftrightarrow Z)$
12	11011101	11111101 × 11011111	$\neg Y \vee Z$	$\neg Z \to \neg Y$
13	11011011	11111011 × 11011111	$(X \vee \neg Y \vee Z) \wedge (\neg X \vee Y \vee \neg Z)$	$\neg(X \leftrightarrow \neg Y \leftrightarrow Z)$
14	11010111	11110111 × 11011111	$\neg Y \vee X$	$\neg X \to \neg Y$
15	11001111	11101111 × 11011111	$(X \vee \neg Y \vee Z) \wedge (\neg X \vee Y \vee Z)$	$\neg Z \to (X \leftrightarrow Y)$
16	10111110	11111110 × 10111111	$(X \vee Y \vee \neg Z) \wedge (\neg X \vee \neg Y \vee \neg Z)$	$Z \to \neg(X \leftrightarrow Y)$
17	10111101	11111101 × 10111111	$(X \vee Y \vee \neg Z) \wedge (\neg X \vee \neg Y \vee Z)$	$\neg(X \leftrightarrow Y \leftrightarrow \neg Z)$
18	10111011	11111011 × 10111111	$\neg Z \vee Y$	$Z \to Y$
19	10110111	11110111 × 10111111	$\neg Z \vee X$	$Z \to X$
20	10101111	11101111 × 10111111	$(X \vee Y \vee \neg Z) \wedge (\neg X \vee Y \vee Z)$	$\neg Y \to (X \leftrightarrow Z)$
21	10011111	11011111 × 10111111	$(X \vee Y \vee \neg Z) \wedge (X \vee \neg Y \vee Z)$	$\neg X \to (Y \leftrightarrow Z)$
22	01111110	11111110 × 01111111	$(X \vee Y \vee Z) \wedge (\neg X \vee \neg Y \vee \neg Z)$	$\neg(X \leftrightarrow Y \leftrightarrow Z)$
23	01111101	11111101 × 01111111	$(X \vee Y \vee Z) \wedge (\neg X \vee \neg Y \vee Z)$	$\neg Z \to \neg(X \leftrightarrow Y)$
24	01111011	11111011 × 01111111	$(X \vee Y \vee Z) \wedge (\neg X \vee Y \vee \neg Z)$	$\neg Y \to \neg(X \leftrightarrow Z)$
25	01110111	11110111 × 01111111	$(X \vee Y \vee Z) \wedge (X \vee \neg Y \vee \neg Z)$	$\neg X \to \neg(Y \leftrightarrow Z)$
26	01101111	11101111 × 01111111	$Y \vee Z$	$\neg Z \to Y$
27	01011111	11011111 × 01111111	$X \vee Z$	$\neg Z \to X$
28	00111111	10111111 × 01111111	$X \vee Y$	$\neg X \to Y$

We may summarize all the previous forms as in Table 1.25 (I focus on the set X but similar forms are true for Y and Z). This means that we have:

- 12 universal statements: 1, 2. 4, 8, 9, 12, 14, 18, 19, 26, 27, 28,
- 12 implications of equivalences or cross statements: 3, 5, 6, 7, 11, 15, 16, 20, 21, 23, 24, 25, and
- 4 negation of equivalences: 10,13, 17, 22.

In other words, any of the forms at this level of the logical space is an universal statement (or at most a conjunction of universal statements) and should be considered as a premise of an inference.

Table 1.25 General forms of Level 6-2

$X \to Y$	$X \to \neg Y$	$\neg X \to Y$	$\neg X \to \neg Y$
$X \to (Y \leftrightarrow Z)$	$X \to \neg(Y \leftrightarrow Z)$	$\neg X \to (Y \leftrightarrow Z)$	$\neg X \to \neg(Y \leftrightarrow Z)$
$(X \leftrightarrow Y \leftrightarrow Z)$	$\neg(X \leftrightarrow Y \leftrightarrow Z)$	$\neg(\neg X \leftrightarrow Y \leftrightarrow Z)$	

Table 1.26 Derivations of statements of Level 6-2

Level 7-1	1,2	1,3	2,3	1,4
Level 6-2	**1**	**2**	**3**	**4**
Level 5-3	1,2,5,11,21,36	1,3,6,12,22,37	1,4,7,13,23,38	2,3,8,14,24,39
Level 7-1	2,4	3,4	1,5	2,5
Level 6-2	**5**	**6**	**7**	**8**
Level 5-3	2,4,9,15,25,40	3,4,10,16,26,41	5,6,8,17,27,42	5,7,9,18,28,43
Level 7-1	3,5	4,5	1,6	2,6
Level 6-2	**9**	**10**	**11**	**12**
Level 5-3	6,7,10,19,29,44	8,9,10,20,30,45	11,12,14,17,31,46	11,13,15,18,32,47
Level 7-1	3,6	4,6	5,6	1,7
Level 6-2	**13**	**14**	**15**	**16**
Level 5-3	12,13,16,19,33,48	14,15,16,20,34,49	17,18,19,20,35,50	21,22,24,27,31,51
Level 7-1	2,7	3,7	4,7	5,7
Level 6-2	**17**	**18**	**19**	**20**
Level 5-3	21,23,25,28,32,52	22,23,26,29,33,53	24,25,26,30,34,54	27,28,29,30,35,55
Level 7-1	6,7	1,8	2,8	3,8
Level 6-2	**21**	**22**	**23**	**24**
Level 5-3	31,32,33,34,35,56	36,37,39,42,46,51	36,38,40,43,47,52	37,38,41,44,48,53
Level 7-1	4,8	5,8	6,8	7,8
Level 6-2	**25**	**26**	**27**	**28**
Level 5-3	39,40,41,45,49,54	42,43,44,45,50,55	46,47,48,49,50,56	51,52,53,54,55,56

It is worth noting that each statement of Level 7-1 contributes to 7 statements of Level 6-2 (as it is also clear from Table 1.23), which means $7 \times 8 = 56$ relations between these two levels as already mentioned. Since each statement of Level 6-2 is generated through two statements of Level 7-1, we have $56/2 = 28$ statements (this expresses a general rule that is valid for any level and any logical space of the kind treated here). Table 1.26 summarizes how the 28 statements of Level 6-2 derive from above and below (again I recall both that I do not perform the derivation from below and that any couple out of the generating statements will suffice).

According to Table 1.26, the generating set of each statement at any level is composed always of 8 statements (i.e. 2^3) and can be divided in two subsets: the Lowering Generating Set (LGS) from above and the Raising Generating Set (RGS) from below. For instance, let us consider the case of Statement 15 of Level 6-2. Its Generating Set (GS) is:

$$\begin{aligned}\text{GS(Level 6-2, \#15)} &= \{\text{LGS, RGS}\}_{15,6-2}\\ &= \{\{5, 6\}, \{17, 18, 19, 20, 35, 50\}\}_{15,6-2}\,.\end{aligned} \tag{1.6}$$

Note that while each statement from above in Table 1.26 contributes to 7 distinct statements of Level 6-2, any statement from below contributes to 3 different statements of Level 6-2.

Let us now consider the 56 disjunctive esaplets or conjunctive triplets as in Table 1.27. All these statements, which could be considered as results of inferences, follow out of a combination of three premises, one of which can be shown to be redundant.

A general consideration is appropriate here. As we have seen, each sequence of numbers is associated to specific values of the areas a, b, c, d, e, f, g, h. Let us consider Statement 1 of Level 5-3 in particular, which I write here in the form f′g′h′, that is, as $\neg X \vee (\neg Y \wedge \neg Z)$. This can be represented as in the left part of Fig. 1.10. In other words, we can represent it by making use of products among columns of 3-value assignments and therefore as:

$$\begin{array}{lll} \neg X & 11101000 & + \\ \neg Y \wedge \neg Z & 10010000 & = \\ \neg X \vee (\neg Y \wedge \neg Z) & 11111000 & \end{array}$$

Alternatively, we can represent it as as a sum a+b+c+d+e among columns of 3.-value assignments, that is, $\neg[X \wedge (Y \vee Z)]$, as on the right of the same figure. In such a case, we have:

Table 1.27 The 56 statements of Level 5-3

1	11111000	11111100 × 11111010 × 11111001	$X \to (\neg Y \wedge \neg Z)$
2	11110100	11111100 × 11110110 × 11110101	$Y \to (\neg X \wedge \neg Z)$
3	11110010	11110110 × 11110110 × 11110011	$Z \to (\neg X \wedge \neg Y)$
4	11110001	11111001 × 11110101 × 11110011	$(X \leftrightarrow Y \leftrightarrow Z) \vee (\neg X \wedge \neg Y \wedge \neg Z)$
5	11101100	11111100 × 11101110 × 11101101	$X \to (\neg Y \wedge Z)$
6	11101010	11111010 × 11101110 × 11101011	$X \to (Y \wedge \neg Z)$
7	11101001	11111001 × 11101110 × 11101011	$X \to (Y \wedge Z)$
8	11100110	11110110 × 11101110 × 11100111	$(\neg Y \vee \neg Z) \wedge (\neg X \vee Y \vee Z)$
9	11100101	11110101 × 11101101 × 11100111	$(\neg X \vee Z) \wedge (X \vee \neg Y \vee \neg Z)$
10	11100011	11110011 × 11101011 × 11100111	$(\neg X \vee Y) \wedge (X \vee \neg Y \vee \neg Z)$
11	11011100	11111100 × 11011110 × 11011101	$Y \to (\neg X \wedge Z)$
12	11011010	11111010 × 11011110 × 11011011	$(\neg X \vee \neg Z) \wedge (X \vee \neg Y \vee Z)$
13	11011001	11111001 × 11011101 × 11010111	$(\neg Y \vee Z) \wedge (\neg X \vee Y \vee \neg Z)$
14	11010110	11110110 × 11011110 × 11010111	$Y \to (X \wedge \neg Z)$
15	11010101	11110101 × 11011101 × 11010111	$Y \to (X \wedge Z)$
16	11010011	11110101 × 11011101 × 11010111	$(X \vee \neg Y) \wedge (\neg X \vee Y \vee \neg Z)$
17	11001110	11101110 × 11011110 × 11001111	$\neg(X \leftrightarrow Y \leftrightarrow Z) \vee (\neg X \wedge \neg Y \wedge Z)$
18	11001101	11101101 × 11011101 × 11001111	$\neg Z \to (\neg X \wedge \neg Y)$
19	11001011	11101011 × 11011011 × 11001111	$(\neg X \vee Y) \wedge (X \vee \neg Y \vee Z)$
20	11000111	11100111 × 11010111 × 11001111	$(X \vee \neg Y) \wedge (\neg X \vee Y \vee Z)$
21	10111100	10111110 × 11111100 × 10111101	$(\neg X \vee \neg Y) \wedge (X \vee Y \vee \neg Z)$
22	10111010	11111010 × 10111110 × 10111011	$Z \to (\neg X \wedge Y)$
23	10111001	11111001 × 10111101 × 10111011	$(Y \vee \neg Z) \wedge (\neg X \vee \neg Y \vee Z)$
24	10110110	11110110 × 10111110 × 10110111	$Z \to (X \wedge \neg Y)$
25	10110101	11110101 × 10111101 × 10110111	$(X \vee \neg Z) \wedge (\neg X \vee \neg Y \vee Z)$
26	10110011	11110011 × 10111011 × 10110111	$Z \to (X \wedge Y)$
27	10101110	11101110 × 10111011 × 10101111	$\neg(X \leftrightarrow Y \leftrightarrow Z) \vee (\neg X \wedge Y \wedge \neg Z)$
28	10101101	11101101 × 10111101 × 10101111	$(\neg X \vee Z) \wedge (X \vee Y \vee \neg Z)$
29	10101011	11101011 × 10111011 × 10101111	$\neg Y \to (\neg X \wedge \neg Z)$
30	10100111	11100111 × 10110111 × 10101111	$(X \vee \neg Z) \wedge (\neg X \vee Y \vee Z)$
31	10011110	11011110 × 10111110 × 10011111	$\neg(X \leftrightarrow Y \leftrightarrow Z) \vee (X \wedge \neg Y \wedge \neg Z)$
32	10011101	11011101 × 10111101 × 10011111	$(\neg Y \vee Z) \wedge (X \vee Y \vee \neg Z)$
33	1011011	11011011 × 10111011 × 10011111	$(Y \vee \neg Z) \wedge (X \vee \neg Y \vee Z)$
34	10010111	11010111 × 10110111 × 10011111	$\neg X \to (\neg Y \wedge \neg Z)$
35	10001111	11001111 × 10101111 × 10011111	$\neg(X \leftrightarrow Y \leftrightarrow Z) \vee (X \wedge Y \wedge Z)$
36	01111100	11111100 × 01111110 × 01111101	$(\neg X \vee \neg Y) \wedge (X \vee Y \vee Z)$
37	01111010	11111010 × 01111110 × 01111011	$(\neg X \vee \neg Z) \wedge (X \vee Y \vee Z)$
38	01111001	11111001 × 01111101 × 01111011	$(X \leftrightarrow Y \leftrightarrow Z) \vee (\neg X \wedge Y \wedge Z)$
39	01110110	11110110 × 01111110 × 01110111	$(\neg Y \vee \neg Z) \wedge (X \vee Y \vee Z)$
40	01110101	11110101 × 01111101 × 01110111	$(X \leftrightarrow Y \leftrightarrow Z) \vee (X \wedge \neg Y \wedge Z)$
41	01110011	11110011 × 01111011 × 01110111	$(X \leftrightarrow Y \leftrightarrow Z) \vee (X \wedge Y \wedge \neg Z)$

(*Continued*)

Table 1.27 (*Continued*)

42	01101110	11101110 × 01101111 × 01101111	$(Y \vee Z) \wedge (\neg X \vee \neg Y \vee \neg Z)$
43	01101101	11101101 × 01111101 × 01101111	$\neg Z \rightarrow (\neg X \wedge Y)$
44	01101011	11101011 × 01111011 × 01101111	$\neg Y \rightarrow (\neg X \wedge Z)$
45	01100111	11100111 × 01110111 × 01101111	$(Y \vee Z) \wedge (X \vee \neg Y \vee \neg Z)$
46	01011110	11011110 × 01111110 × 01011111	$(X \vee Z) \wedge (\neg X \vee \neg Y \vee \neg Z)$
47	01011101	11011101 × 01111101 × 01011111	$\neg Z \rightarrow (X \wedge \neg Y)$
48	01011011	11011011 × 01111011 × 01011111	$(X \vee Z) \wedge (\neg X \vee Y \vee \neg Z)$
49	01010111	11010111 × 01110111 × 01011111	$\neg X \rightarrow (\neg Y \wedge Z)$
50	01001111	11001111 × 01101111 × 01011111	$\neg Z \rightarrow (X \wedge Y)$
51	00111110	10111110 × 01111110 × 00111111	$(X \vee Y) \wedge (\neg X \vee \neg Y \vee \neg Z)$
52	00111101	10111101 × 01111101 × 00111111	$(X \vee Y) \wedge (\neg X \vee \neg Y \vee Z)$
53	00111011	10111011 × 01111011 × 00111111	$\neg Y \rightarrow (X \wedge \neg Z)$
54	00110111	10110111 × 01110111 × 00111111	$\neg X \rightarrow (Y \wedge \neg Z)$
55	00101111	10101111 × 01101111 × 00111111	$\neg Y \rightarrow (X \wedge Z)$
56	00011111	10011111 × 01011111 × 00111111	$\neg X \rightarrow (Y \wedge Z)$

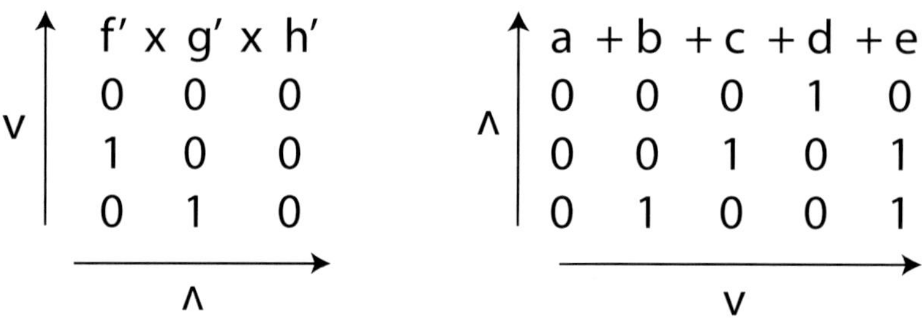

Figure 1.10 The same statement $\neg X \vee (\neg Y \wedge \neg Z)$ represented on the left as a conjunctions and on the right as a disjunction.

X	00010111	×
$Y \vee Z$	01101111	=
$X \wedge (Y \vee Z)$	00000111	
$\neg[X \wedge (Y \vee Z)]$	11111000	

Some general forms of the above 56 statements may be distinguished as in Table 1.28, where again, I focus on the X statement.

Many variations of the formulae on the latter two rows are possible, according to where the negations are placed. I give here

Table 1.28 General forms of Level 5-3

$X \to (Y \wedge Z)$	$\neg X \to (Y \wedge Z)$
$(X \vee Y) \wedge (\neg X \vee \neg Y \vee \neg Z)$	$(X \vee Y) \wedge (\neg X \vee \neg Y \vee Z)$
$(X \leftrightarrow Y \leftrightarrow Z) \vee (\neg X \wedge \neg Y \wedge \neg Z)$	$\neg(X \leftrightarrow Y \leftrightarrow Z) \vee (\neg X \wedge Y \wedge \neg Z)$

a complete list of all forms (consider that the formulae of the first and second row actually pertain to the same set and have been distinguished only for practical purposes):

- Variations of the first and second forms: 1, 2, 3, 4, 5, 6, 7, 11, 14, 15, 18, 22, 24, 26, 29, 34, 43, 44, 47, 49, 50, 53, 54, 55, 56.
- Variations of the third and fourth forms: 8, 9, 10, 12, 13. 16. 19, 20, 21, 23. 25, 28, 30, 32, 33, 36, 37, 39, 42, 45, 46, 48, 51, 52.
- Variations of the fifth and sixth forms: 4, 17, 27, 31, 35, 38, 40, 41.

This makes $24 + 24 + 8 = 56$ (the variations of the last two lines are only 8 because each line already encompasses three forms). It may be noted that all forms in the second row are a conjunction between a statement of Level 6-2 (e.g. Statement 28: $X \vee Y$) and two statements of Level 7-1 (Statements 1 and 2: $\neg X \vee \neg Y \vee \neg Z$ and $\neg X \vee \neg Y \vee Z$, respectively). It may be further noted that the latter two generate Statement 1 of Level 6-2, i.e. $\neg X \vee \neg Y$. To take another example (Statements 13 and 32 of Level 5-3) we see that Statement 12 of Level 6-2, i.e. $\neg Y \vee Z$, is conjoint with Statements 3 ($\neg X \vee Y \vee \neg Z$) and 7 ($X \vee Y \vee \neg Z$) of Level 7-1, respectively. The latter two statements generate Statement 18 of Level 6-2, i.e. $Y \vee \neg Z$. This structure is a general rule that the reader may check. Another interesting question is the kind of premises from which these general forms follow. Let us take one representative for both the first and second kind of Table 1.28 (the reader can himself/herself try to derive some of the propositions the listed above from the corresponding premises):

P1:	$X \to (Y \leftrightarrow Z)$	(Statement 3 of Level 6-2)
P2:	$X \to Z$	(Statement 8 of Level 6-2)
P3:	$X \to Y$	(Statement 9 of Level 6-22)
C:	$X \to (Y \wedge Z)$	(Statement 7 of Level 5-3)

In other words, $P1 \times P2 \times P3 = C$. Statements of the form displayed in the third kind may have different kinds of premises, for instance:

P1:	$Y \to (\neg X \wedge Z)$	(Statement 11 of Levell 6-2)
P2:	$\neg X \to (Y \leftrightarrow Z)$	(Statement 22 of Level 6-2)
P3:	$\neg Z \to X$	(Statement 27 of Level 6-2)
C:	$(X \vee Z) \wedge (\neg X \vee \neg Y \vee \neg Z)$	(Statement 46 of Level 5-3)

Similarly for the fourth (or fifth) kind. For instance:

P1:	$X \to \neg(Y \leftrightarrow Z)$	(Statement 7 of Level 6-2)
P2:	$Y \to (\neg X \wedge Z)$	(Statement 11 of Level 6-2)
P3:	$\neg Z \to (X \leftrightarrow Y)$	(Statement 15 of Level 6-2)
C:	$[X \to \neg(Y \leftrightarrow Z)]$	
	$\wedge(X \vee \neg Y \vee Z)$	(Statement 17 of l Level 5-3)

Table 1.29 summarizes which of the 28 statements of Level 6-2 generate one of the 56 statements of Level 5-3, constituting their LGS, as well as the statements of Level 4-4 generate statements of Level 5-3, constituting their RGS from below (note that each statement of Level 6-2 contributes to 6 distinct statements whilst each statement of Level 4-4 contributes to 4 statements of Level 5-3).

Since each statement of Level 6-2 contributes to 6 statements of Level 5-3, this means $28 \times 6 = 168$ relations between these two levels. However, since each statement of Level 5-3 has a generating set of three statements of Level 6-2, this means 168/3=56 different propositions for Level 5-3. I also recall that any statement of Level 5-3 can be generated by any pair out of the LGS comprehending the three corresponding statements of Level 6-2 listed above. For instance, let us consider Statement 25: it can be generated by the following couples (of statements pertaining to Level 6-2): 5 and 17, 5 and 19, or 17 and 19, as shown in Table 1.30.

Let us now consider the 70 disjunctive or conjunctive quadruplets (Level 4-4) as shown in Table 1.31.

As already remarked in the bidimensional case, the whole set of statements at the middle level (here Level 4-4) is constituted of two subsets such that any statement of one subset is the negation of a statement of the other subset. In particular, we have a progressive

Table 1.29 Derivations of statements of Level 5-3

Level 6-2	1,2,3	1,4,5	2,4,6	3,5,6
Level 5-3	**1**	**2**	**3**	**4**
Level 4-4	1,2,6,16,36	1,3,7,17,37	1,4,8,18,38	1,5,9,19,39
Level 6-2	1,7,8	2,7,9	3,8,9	4,7,10
Level 5-3	**5**	**6**	**7**	**8**
Level 4-4	2,3,10,20,40	2,4,11,21,41	2,5,12,22,42	3,4,13,23,43
Level 6-2	5,8,10	6,9,10	1,11,12	2,11,13
Level 5-3	**9**	**10**	**11**	**12**
Level 4-4	3,5,14,24,44	4,5,15,25,45	6,7,10,26,46	6,8,11,27,47
Level 6-2	3,12,13	4,11,14	5,12,14	6,13,14
Level 5-3	**13**	**14**	**15**	**16**
Level 4-4	6,9,12,28,48	7,8,13,29,49	7,9,14,30,50	8,9,15,31,51
Level 6-2	7,11,15	8,12,15	9,13,15	10,14,15
Level 5-3	**17**	**18**	**19**	**20**
Level 4-4	10,11,13,32,52	10,12,14,33,53	11,12,15,34,54	13,14,15,35,55
Level 6-2	1,16,17	2,16,18	3,17,18	4,16,19
Level 5-3	**21**	**22**	**23**	**24**
Level 4-4	16,17,20,26,56	16,18,21,27,57	16,19,22,28,58	17,18,23,29,59
Level 6-2	5,17,19	6,18,19	7,16,20	8,17,20
Level 5-3	**25**	**26**	**27**	**28**
Level 4-4	17,19,24,30,60	18,19,25,31,61	20,21,23,32,62	20,22,24,33,63
Level 6-2	9,18,20	10,19,20	11,16,21	12,17,21
Level 5-3	**29**	**30**	**31**	**32**
Level 4-4	21,22,25,34,64	23,24,25,35,65	26,27,29,32,66	26,28,30,33,67
Level 6-2	13,18,21	14,19,21	15,20,21	1,22,23
Level 5-3	**33**	**34**	**35**	**36**
Level 4-4	27,28,31,34,68	29,30,31,35,69	32,33,34,35,70	36,37,40,46,56
Level 6-2	2,22,24	3,23,24	4,22,25	5,23,25
Level 5-3	**37**	**38**	**39**	**40**
Level 4-4	36,38,41,47,57	36,39,42,48,58	37,38,43,49,59	37,39,44,50,60
Level 6-2	6,24,25	7,22,27	8,23,26	6,24,25
Level 5-3	**41**	**42**	**43**	**44**
Level 4-4	38,39,45,51,61	40,41,43,52,62	40,42,44,53,63	41,42,45,54,64
Level 6-2	10,25,26	11,22,27	12,23,27	13,24,27
Level 5-3	**45**	**46**	**47**	**48**
Level 4-4	43,44,45,55,65	46,47,49,52,66	46,48,50,53,67	47,48,51,54,68
Level 6-2	14,25,27	15,26,27	16,22,28	17,23,28
Level 5-3	**49**	**50**	**51**	**52**
Level 4-4	49,50,51,55,69	52,53,54,55,70	56,57,59,62,66	56,58,60,63,67
Level 6-2	18,24,28	19,25,28	20,26,28	21,27,28
Level 5-3	**53**	**54**	**55**	**56**
Level 4-4	57,58,61,64,68	59,60,61,65,69	62,63,64,65,70	66,67,68,69,70

Table 1.30 Generation of Statement 25

11110101×	11110101×	10111101×
10111101 =	10110111 =	10110111 =
10110101	10110101	10110101

order in the first subset from Statement 1 until Statement 35, and then a reverse order in the second subset, such that Statement 36 is the negation of Statement 35, until Statement 70, which is the negation of Statement 1.

From here on I shall consider the ascendant movement, that is, I shall start from the bottom and consider disjunctive forms until the Level 4-4. However, in the exposition I shall revert the logical order and start from Level 4-4 instead of starting from Level 1-7. The reader interested in following the latter order can begin from below where the 8 elementary disjunctive forms are presented. The reason of my choice is the following: we can immediately verify whether the derivation of the 70 statements of Level 4-4 as derived from Level 3-5 are in accordance with those derived from Level 5-3. This comparison is highly instructive. This is the reason why I prefer a certain redundancy here. Indeed, since all of the 70 statements of Level 4-4 are divided in two subsets both of 35 statements, for economical reasons we could consider the first 35 ones as generated from above and the second 35 as generated from below. Let us therefore consider how the 70 statements of Level 4-4 can be derived from below (Level 3-5) instead of from above (Level 5-3), as displayed in Table 1.32.

It may be interesting to note that Statements 32 and its negation, namely Statement 39, are the result of the product of Statements 4, 17, 27, 31, 35, 38, 40, 41 of Level 5-3, which all have the same form (and exhaust their number), as well as the result of the sum of Statements 16, 17, 19, 22, 26, 30, 40, 53 of Level 3-5, which again have the same form and exhaust their number (they are actually the contradictory statements relative to the mentioned ones of Level 5-3). Let us now consider the relations among LGSs and RGSs as in Table 1.33.

It is not very helpful here to indicate general forms of the statements (we shall deal with this problem later on). The reader

Table 1.31 The 70 statements of Level 4-4 (derived from above)

1	11110000	11111000 × 11110100 × 11110010 × 11110001	$\neg\{(X \vee Y) \wedge [Z \vee \neg(X \leftrightarrow Y)]\}$
2	11101000	11111000 × 11101100 × 11101010 × 11101001	$\neg X$
3	11100100	11110100 × 11101100 × 11100110 × 11100101	$(Z \rightarrow \neg Y) \wedge (X \rightarrow Z) \wedge (X \rightarrow \neg Y)$
4	11100010	11110010 × 11101010 × 11100110 × 11100011	$(Y \rightarrow \neg Z) \wedge (X \rightarrow Y) \wedge (X \rightarrow \neg Z)$
5	11100001	11110001 × 11101001 × 11100101 × 11100011	$(X \leftrightarrow Y \leftrightarrow Z) \vee [\neg X \wedge \neg(Y \leftrightarrow Z)]$
6	11011000	11111000 × 11011100 × 11011010 × 11011001	$(Z \rightarrow \neg X) \wedge (Y \rightarrow Z) \wedge (Y \rightarrow \neg X)$
7	11010100	11110100 × 11011100 × 11010110 × 11010101	$\neg Y$
8	11010010	11110010 × 11011010 × 11010110 × 11010011	$(\neg X \rightarrow \neg Y) \wedge (Z \rightarrow \neg X) \wedge (Z \rightarrow \neg Y)$
9	11010001	11110001 × 11011001 × 11010101 × 11010011	$(X \leftrightarrow Y \leftrightarrow Z) \vee [\neg Y \wedge \neg(X \leftrightarrow Z)]$
10	11001100	11101100 × 11011100 × 11001110 × 11001101	$\neg\{(X \vee Y) \wedge [\neg Z \vee (X \leftrightarrow Y)]\}$
11	11001010	11101010 × 11011010 × 11001110 × 11001011	$\neg\{(X \vee Y) \wedge [\neg Y \vee (X \leftrightarrow Z)]\}$
12	11001001	11101001 × 11011001 × 11001101 × 11001011	$(Y \rightarrow Z) \wedge (X \rightarrow Y) \wedge (X \rightarrow Z)$
13	11000110	11100110 × 11010110 × 11001110 × 11000111	$\neg\{(X \vee Y) \wedge [\neg X \vee (Y \leftrightarrow Z)]\}$
14	11000101	11100101 × 11010101 × 11001101 × 11000111	$(X \rightarrow Z) \wedge (Y \rightarrow X) \wedge (Y \rightarrow Z)$
15	11000011	11100011 × 11010011 × 11001011 × 11000111	$X \leftrightarrow Y$
16	10111000	11111000 × 10111100 × 10111010 × 10111001	$(Y \rightarrow \neg X) \wedge (Z \rightarrow Y) \wedge (Z \rightarrow \neg X)$
17	10110100	11110100 × 10111100 × 10110110 × 10110101	$(X \rightarrow \neg Y) \wedge (Z \rightarrow X) \wedge (Z \rightarrow \neg Y)$
18	10110010	11110010 × 10111010 × 10110110 × 10110011	$\neg Z$
19	10110001	11110001 × 10111001 × 10110101 × 10110011	$(X \leftrightarrow Y \leftrightarrow Z) \vee [\neg Z \wedge \neg(X \leftrightarrow Y)]$
20	10101100	11101100 × 10111100 × 10101110 × 10101101	$\neg\{(X \vee Z) \wedge [\neg Z \vee (X \leftrightarrow Y)]\}$
21	10101010	11101010 × 10111010 × 10101110 × 10101011	$\neg\{(X \vee Z) \wedge [\neg Y \vee (X \leftrightarrow Z)]\}$
22	10101001	11101001 × 10111001 × 10101101 × 10101011	$(Z \rightarrow Y) \wedge (X \rightarrow Z) \wedge (X \rightarrow Y)$
23	10100110	11100110 × 10110110 × 10101110 × 10100111	$\neg\{(X \vee Z) \wedge [\neg X \vee (Y \leftrightarrow Z)]\}$

(Continued)

Table 1.31 (*Continued*)

24	10100101	11100101 × 10110101 × 10101101 × 10100111	$X \leftrightarrow Z$
25	10100011	11100011 × 10110011 × 10101011 × 10100111	$(X \to Y) \wedge (Z \to X) \wedge (Z \to Y)$
26	10011100	11011100 × 10111100 × 10011110 × 10011101	$\neg\{(Y \vee Z) \wedge [\neg Z \vee (X \leftrightarrow Y)]\}$
27	10011010	11011010 × 10111010 × 10011110 × 10011011	$\neg\{(Y \vee Z) \wedge [\neg Y \vee (X \leftrightarrow Z)]\}$
28	10011001	11011001 × 10111001 × 10011101 × 10011011	$Y \leftrightarrow Z$
29	10010110	11010110 × 10110110 × 10011110 × 10010111	$\neg\{(Y \vee Z) \wedge [\neg X \vee (Y \leftrightarrow Z)]\}$
30	10010101	11010101 × 10110101 × 10011101 × 10010111	$(Z \to X) \wedge (Y \to Z) \wedge (Y \to X)$
31	10010011	11010011 × 10110011 × 10011011 × 10010111	$(Y \to X) \wedge (Z \to Y) \wedge (Z \to X)$
32	10001110	11001110 × 10101110 × 10011110 × 10001111	$\neg[X \leftrightarrow (Y \leftrightarrow Z)] \wedge \neg[Y \leftrightarrow (X \leftrightarrow Z)]$ $\wedge\neg[Z \leftrightarrow (X \leftrightarrow Y)]$
33	10001101	11001101 × 10101101 × 10011101 × 10001111	$(X \leftrightarrow Y \leftrightarrow Z) \vee [Z \wedge \neg(X \leftrightarrow Y)]$
34	10001011	11001011 × 10101011 × 10011011 × 10001111	$(X \leftrightarrow Y \leftrightarrow Z) \vee [Y \wedge \neg(X \leftrightarrow Z)]$
35	10000111	11000111 × 10100111 × 10010111 × 10001111	$(X \leftrightarrow Y \leftrightarrow Z) \vee [X \wedge \neg(Y \leftrightarrow Z)]$
36	01111000	11111000 × 01111100 × 01111010 × 01111001	$\neg(X \leftrightarrow Y \leftrightarrow Z) \wedge [\neg X \vee (Y \leftrightarrow Z)]$
37	01110100	11110100 × 01111100 × 01110110 × 01110101	$\neg(X \leftrightarrow Y \leftrightarrow Z) \wedge [\neg Y \vee (X \leftrightarrow Z)]$
38	01110010	11110010 × 01111010 × 01110110 × 01110011	$\neg(X \leftrightarrow Y \leftrightarrow Z) \wedge [\neg Z \vee (X \leftrightarrow Y)]$
39	01110001	11110001 × 01111001 × 01110101 × 01110011	$[X \leftrightarrow (Y \leftrightarrow Z)] \vee [Y \leftrightarrow (X \leftrightarrow Z)]$ $\vee[Z \leftrightarrow (X \leftrightarrow Y)]$
40	01101100	11101100 × 01111100 × 01101110 × 01101101	$(\neg Y \to Z) \wedge (X \to \neg Y) \wedge (X \to Z)$
41	01101010	11101010 × 01111010 × 01101110 × 01101011	$(\neg Y \to \neg X) \wedge (\neg Y \to Z) \wedge (Z \to \neg X)$
42	01101001	11101001 × 01111001 × 01101101 × 01101011	$(Y \vee Z) \wedge [\neg X \vee (Y \leftrightarrow Z)]$
43	01100110	11100110 × 01110110 × 01101110 × 01100111	$\neg(Y \leftrightarrow Z)$
44	01100101	11100101 × 01110101 × 01101101 × 01100111	$(Y \vee Z) \wedge [\neg Y \vee (X \leftrightarrow Z)]$
45	01100011	11100011 × 01110011 × 01101011 × 01100111	$(Y \vee Z) \wedge [\neg Z \vee (X \leftrightarrow Y)]$

46	01011100	11011100 × 01111100 × 01011110 × 01011101	$(X \to \neg Y) \wedge (\neg Z \to X) \wedge (\neg Z \to \neg Y)$
47	01011010	11011010 × 01111010 × 01011110 × 01011011	$\neg(X \leftrightarrow Z)$
48	01011001	11011001 × 01111001 × 01011101 × 01011011	$(X \vee Z) \wedge [\neg X \vee (Y \leftrightarrow Z)]$
49	01010110	11010110 × 01110110 × 01011110 × 01010111	$(Z \to \neg Y) \wedge (\neg X \to Z) \wedge (\neg X \to \neg Y)$
50	01010101	11010101 × 01110101 × 01011101 × 01010111	$(X \vee Z) \wedge [\neg Y \vee (X \leftrightarrow Z)]$
51	01010011	11010011 × 01110011 × 01011011 × 01010111	$(X \vee Z) \wedge [\neg Z \vee (X \leftrightarrow Y)]$
52	01001110	11001110 × 01101110 × 01011110 × 01001111	$\neg(X \leftrightarrow Y \leftrightarrow Z) \wedge [\neg Z \vee (X \leftrightarrow Y)]$
53	01001101	11001101 × 01101101 × 01011101 × 01001111	Z
54	01001011	11001011 × 01101011 × 01011011 × 01001111	$(X \to Y) \wedge (\neg Z \to X) \wedge (\neg Z \to Y)$
55	01000111	11000111 × 01100111 × 01010111 × 01001111	$(Y \to X) \wedge (\neg Z \to Y) \wedge (\neg Z \to X)$
56	00111100	10111100 × 01111100 × 00111110 × 00111101	$\neg(X \leftrightarrow Y)$
57	00111010	10111010 × 01111010 × 00111110 × 00111011	$(X \to \neg Z) \wedge (\neg Y \to X) \wedge (\neg Y \to \neg Z)$
58	00111001	10111001 × 01111001 × 00111101 × 00111011	$(X \vee Y) \wedge [\neg X \vee (Y \leftrightarrow Z)]$
59	00110110	10110110 × 01110110 × 00111110 × 00110111	$(Y \to \neg Z) \wedge (\neg X \to Y) \wedge (\neg X \to \neg Z)$
60	00110101	10110101 × 01110101 × 00111101 × 00110111	$(X \vee Y) \wedge [\neg Y \vee (X \leftrightarrow Z)]$
61	00110011	10110011 × 01110011 × 00111011 × 00110111	$(X \vee Y) \wedge [\neg Z \vee (X \leftrightarrow Y)]$
62	00101110	10101110 × 01101110 × 00111110 × 00101111	$\neg(X \leftrightarrow Y \leftrightarrow Z) \wedge [Y \vee (X \leftrightarrow Z)]$
63	00101101	10101101 × 01101101 × 00111101 × 00101111	$(X \to Z) \wedge (\neg Y \to X) \wedge (\neg Y \to Z)$
64	00101011	10101011 × 01101011 × 00111011 × 00101111	Y
65	00100111	10100111 × 01100111 × 00110111 × 00101111	$(Z \to X) \wedge (\neg Y \to Z) \wedge (\neg Y \to X)$
66	00011110	01111110 × 01011110 × 00111110 × 00011111	$\neg(X \leftrightarrow Y \leftrightarrow Z) \wedge [X \vee (Y \leftrightarrow Z)]$
67	00011101	10011101 × 01011101 × 00111101 × 00011111	$(Y \to Z) \wedge (\neg X \to Y) \wedge (\neg X \to Z)$
68	00011011	10011011 × 01011011 × 00111011 × 00011111	$(\neg X \to Y) \wedge (\neg X \to Z) \wedge (Z \to Y)$
69	00010111	10010111 × 01010111 × 00110111 × 00011111	X
70	00001111	10001111 × 01001111 × 00101111 × 00011111	$(X \vee Y) \wedge [Z \vee \neg(X \leftrightarrow Y)]$

Table 1.32 The 70 statements of Level 4-4 (derived from below)

1	11110000	11100000 + 11010000 + 10110000 + 01110000	$\neg\{(X \vee Y) \wedge [Z \vee \neg(X \leftrightarrow Y)]\}$
2	11101000	11100000 + 11001000 + 10101000 + 01101000	$\neg X$
3	11100100	11100000 + 11000100 + 10100100 + 01100100	$(\neg X \wedge \neg Z) \vee (\neg Y \wedge Z)$
4	11100010	11100000 + 11000010 + 10100010 + 01100010	$(\neg X \wedge \neg Y) \vee (Y \wedge \neg Z)$
5	11100001	11100000 + 11000001 + 10100001 + 01100001	$(X \leftrightarrow Y \leftrightarrow Z) \vee [\neg X \wedge \neg(Y \leftrightarrow Z)]$
6	11011000	11010000 + 11001000 + 10011000 + 01011000	$(\neg X \wedge Z) \vee (\neg Y \wedge \neg Z)$
7	11010100	11010000 + 11000100 + 10010100 + 01010100	$\neg Y$
8	11010010	11010000 + 11000010 + 10010010 + 01010010	$(X \wedge \neg Z) \vee (\neg X \wedge \neg Y)$
9	11010001	11010000 + 11000001 + 10010001 + 01010001	$(X \leftrightarrow Y \leftrightarrow Z) \vee [\neg Y \wedge \neg(X \leftrightarrow Z)]$
10	11001100	11001000 + 11000100 + 10001100 + 01001100	$(\neg X \wedge \neg Y) \vee [Z \wedge \neg(X \leftrightarrow Y)]$
11	11001010	11001000 + 11000010 + 10001010 + 01001010	$(\neg X \wedge \neg Y) \vee [Y \wedge \neg(X \leftrightarrow Z)]$
12	11001001	11001000 + 11000001 + 10001001 + 01001001	$(\neg X \wedge \neg Y) \vee (Y \wedge Z)$
13	11000110	11000100 + 11000010 + 10000110 + 01000110	$(\neg X \wedge \neg Y) \vee [X \wedge \neg(Y \leftrightarrow Z)]$
14	11000101	11000100 + 11000001 + 10000101 + 01000101	$(\neg X \wedge \neg Y) \vee (X \wedge Z)$
15	11000011	11000010 + 11000001 + 10000011 + 01000011	$X \leftrightarrow Y$
16	10111000	10110000 + 10101000 + 10011000 + 00111000	$(\neg X \wedge Y) \vee (\neg Y \wedge \neg Z)$
17	10110100	10110000 + 10100100 + 10010100 + 00110100	$(X \wedge \neg Y) \vee (\neg X \wedge \neg Z)$
18	10110010	10110000 + 10100010 + 10010010 + 00110010	$\neg Z$
19	10110001	10110000 + 10100001 + 10010001 + 00110001	$(X \leftrightarrow Y \leftrightarrow Z) \vee [\neg Z \wedge \neg(X \leftrightarrow Y)]$
20	10101100	10101000 + 10100100 + 10001100 + 00101100	$(\neg X \wedge \neg Z) \vee [Z \wedge \neg(X \leftrightarrow Y)]$
21	10101010	10101000 + 10100010 + 10001010 + 00101010	$(\neg X \wedge \neg Z) \vee [Y \wedge \neg(X \leftrightarrow Z)]$
22	10101001	10101000 + 10100001 + 10001001 + 00101001	$(\neg X \wedge \neg Z) \vee (Y \wedge Z)$
23	10100110	10100100 + 10100010 + 10000110 + 00100110	$(\neg X \wedge \neg Z) \vee [X \wedge \neg(Y \leftrightarrow Z)]$

24	10100101	10100100 + 10100001 + 10000100 + 00100101	$X \leftrightarrow Z$
25	10100011	10100010 + 10100001 + 10000011 + 00100011	$(X \wedge Y) \vee (\neg X \wedge \neg Z)$
26	10011100	10011000 + 10010100 + 10001100 + 00011100	$(\neg Y \wedge \neg Z) \vee [Z \wedge \neg(X \leftrightarrow Y)]$
27	10011010	10011000 + 10010010 + 10001010 + 00011010	$(\neg Y \wedge \neg Z) \vee [Y \wedge \neg(X \leftrightarrow Z)]$
28	10011001	10011000 + 10010001 + 10001001 + 00011001	$Y \leftrightarrow Z$
29	10010110	10010100 + 10010010 + 10000110 + 00010110	$(\neg Y \wedge \neg Z) \vee [X \wedge \neg(Y \leftrightarrow Z)]$
30	10010101	10010100 + 10010001 + 10000101 + 00010101	$(X \wedge Z) \vee (\neg Y \wedge \neg Z)$
31	10010011	10010010 + 10010001 + 10000011 + 00010011	$(X \wedge Y) \vee (\neg Y \wedge \neg Z)$
32	10001110	10001100 + 10001010 + 10000110 + 00001110	$\neg[X \leftrightarrow (Y \leftrightarrow Z)] \wedge \neg[Y \leftrightarrow (X \leftrightarrow Z)]$ $\wedge \neg[Z \leftrightarrow (X \leftrightarrow Y)]$
33	10001101	10001100 + 10001001 + 10000101 + 00001101	$(X \leftrightarrow Y \leftrightarrow Z) \vee [Z \wedge \neg(X \leftrightarrow Y)]$
34	10001011	10001010 + 10001001 + 10000011 + 00001011	$(X \leftrightarrow Y \leftrightarrow Z) \vee [Y \wedge \neg(X \leftrightarrow Z)]$
35	10000111	10000110 + 10000101 + 10000011 + 00000111	$(X \leftrightarrow Y \leftrightarrow Z) \vee [X \wedge \neg(Y \leftrightarrow Z)]$
36	01111000	01110000 + 01101000 + 01011000 + 00111000	$\neg(X \leftrightarrow Y \leftrightarrow Z) \wedge [\neg X \vee (Y \leftrightarrow Z)]$
37	01110100	01110000 + 01100100 + 01010100 + 00110100	$\neg(X \leftrightarrow Y \leftrightarrow Z) \wedge [\neg Y \vee (X \leftrightarrow Z)]$
38	01110010	01110000 + 01100010 + 01010010 + 00110010	$\neg(X \leftrightarrow Y \leftrightarrow Z) \wedge [\neg Z \vee (X \leftrightarrow Y)]$
39	01110001	01110000 + 01100001 + 01010001 + 00110001	$[X \leftrightarrow (Y \leftrightarrow Z)] \vee [Y \leftrightarrow (X \leftrightarrow Z)]$ $\vee[Z \leftrightarrow (X \leftrightarrow Y)]$
40	01101100	01101000 + 01100100 + 01001100 + 00101100	$(\neg X \wedge Y) \vee (\neg Y \wedge Z)$
41	01101010	01101000 + 01100010 + 01001010 + 00101010	$(\neg X \wedge Z) \vee (Y \wedge \neg Z)$
42	01101001	01101000 + 01100001 + 01001001 + 00101001	$\neg\{(\neg Y \wedge \neg Z) \vee [X \wedge \neg(Y \leftrightarrow Z)]\}$
43	01100110	01100100 + 01100010 + 01000110 + 00100110	$\neg(Y \leftrightarrow Z)$
44	01100101	01100100 + 01100001 + 01000101 + 00100101	$\neg\{(\neg Y \wedge \neg Z) \vee [Y \wedge \neg(X \leftrightarrow Z)]\}$
45	01100011	01100010 + 01100001 + 01000011 + 00100011	$\neg\{(\neg Y \wedge \neg Z) \vee [Z \wedge \neg(X \leftrightarrow Y)]\}$

(Continued)

Table 1.32 (*Continued*)

46	01011100	01011000 + 01010100 + 01001100 + 00011100	$(X \wedge \neg Y) \vee (\neg X \wedge Z)$
47	01011010	01011000 + 01010010 + 01001010 + 00011010	$\neg(X \leftrightarrow Z)$
48	01011001	01011000 + 01010001 + 01001001 + 00011001	$\neg\{(\neg X \wedge \neg Z) \vee [X \wedge \neg(Y \leftrightarrow Z)]\}$
49	01010110	01010100 + 01010010 + 01000110 + 00010110	$(X \wedge \neg Z) \vee (\neg Y \wedge Z)$
50	01010101	01010100 + 01010001 + 01000101 + 00010101	$\neg\{(\neg X \wedge \neg Z) \vee [Y \wedge \neg(X \leftrightarrow Z)]\}$
51	01010011	01010010 + 01010001 + 01000011 + 00010011	$\neg\{(\neg X \wedge \neg Z) \vee [Z \wedge \neg(X \leftrightarrow Y)]\}$
52	01001110	01001100 + 01001010 + 01000110 + 00001110	$\neg(X \leftrightarrow Y \leftrightarrow Z) \wedge [\neg Z \vee (X \leftrightarrow Y)]$
53	01001101	01001100 + 01001001 + 01000101 + 00001101	Z
54	01001011	01001010 + 01001001 + 01000011 + 00001011	$(X \wedge Y) \vee (\neg X \wedge Z)$
55	01000111	01000110 + 01000101 + 01000011 + 00000111	$(X \wedge Y) \vee (\neg Y \wedge Z)$
56	00111100	00111000 + 00110100 + 00101100 + 00011100	$\neg(X \leftrightarrow Y)$
57	00111010	00111000 + 00110010 + 00101010 + 00011010	$(\neg X \wedge Y) \vee (X \wedge \neg Z)$
58	00111001	00111000 + 00110001 + 00101001 + 00011001	$\neg\{(\neg X \wedge \neg Y) \vee [X \wedge \neg(Y \leftrightarrow Z)]\}$
59	00110110	00110100 + 00110010 + 00100110 + 00010110	$(X \wedge \neg Y) \vee (Y \wedge \neg Z)$
60	00110101	00110100 + 00110001 + 00100101 + 00010101	$\neg\{(\neg X \wedge \neg Y) \vee [Y \wedge \neg(X \leftrightarrow Z)]\}$
61	00110011	00110010 + 00110001 + 00100011 + 00010011	$\neg\{(\neg X \wedge \neg Y) \vee [Z \wedge \neg(X \leftrightarrow Y)]\}$
62	00101110	00101100 + 00101010 + 00100110 + 00001110	$\neg(X \leftrightarrow Y \leftrightarrow Z) \wedge [Y \vee (X \leftrightarrow Z)]$
63	00101101	00101100 + 00101001 + 00100101 + 00001101	$(\neg X \wedge Y) \vee (X \wedge Z)$
64	00101011	00101010 + 00101001 + 00100011 + 00001011	Y
65	00100111	00100110 + 00100101 + 00100011 + 00000111	$(X \wedge Z) \vee (Y \wedge \neg Z)$
66	00011110	00011100 + 00011010 + 00010110 + 00001110	$\neg(X \leftrightarrow Y \leftrightarrow Z) \wedge [X \vee (Y \leftrightarrow Z)]$
67	00011101	00011100 + 00011001 + 00010101 + 00001101	$(X \wedge \neg Y) \vee (Y \wedge Z)$
68	00011011	00011010 + 00011001 + 00010011 + 00001011	$(X \wedge \neg Z) \vee (Y \wedge Z)$
69	00010111	00010110 + 00010101 + 00010011 + 00000111	X
70	00001111	00001110 + 00001101 + 00001011 + 00000111	$(X \vee Y) \wedge [Z \vee \neg(X \leftrightarrow Y)]$

Table 1.33 Derivation of the 70 statements of Level 4-4

Level 5-3	1,2,3,4	1,5,6,7	2,5,8,9	3,6,8,10	4,7,8,10	1,11,12,13	2,11,14,15
Level 4-4	**1**	**2**	**3**	**4**	**5**	**6**	**7**
Level 3-5	1,2,7,22	1,3,8,23	1,4,9,24	1,5,10,25	1,6,11,26	2,3,12,27	2,4,13,28
Level 5-3	3,12,14,16	4,13,15,16	5,11,17,18	6,12,17,19	7,13,18,19	8,14,17,20	9,15,18,20
Level 4-4	**8**	**9**	**10**	**11**	**12**	**13**	**14**
Level 3-5	2,5,14,29	2,6,15,30	3,4,16,31	3,5,17,32	3,6,18,33	4,5,19,34	4,6,20,35
Level 5-3	10,16,19,20	1,21,22,23	2,21,24,25	3,22,24,26	4,23,25,26	5,21,27,28	6,22,27,29
Level 4-4	**15**	**16**	**17**	**18**	**19**	**20**	**21**
Level 3-5	5,6,21,36	7,8,12,37	7,9,13,38	7,10,14,39	7,11,15,40	8,9,16,41	8,10,17,42
Level 5-3	7,23,28,29	8,24,27,30	9,25,28,30	10,26,29,30	11,21,31,32	12,22,31,33	13,23,32,33
Level 4-4	**22**	**23**	**24**	**25**	**26**	**27**	**28**
Level 3-5	8,11,18,43	9,10,19,44	9,11,20,45	10,11,21,46	12,13,16,47	12,14,17,48	12,15,18,49
Level 5-3	14,24,31,34	15,25,32,34	16,26,33,34	17,27,31,35	18,28,32,35	19,29,33,35	20,30,34,35
Level 4-4	**29**	**30**	**31**	**32**	**33**	**34**	**35**
Level 3-5	13,14,19,50	13,15,20,51	14,15,21,52	16,17,19,53	16,18,20,54	17,18,21,55	19,20,21,56
Level 5-3	1,36,37,38	2,36,39,40	3,37,39,41	4,38,40,41	5,36,42,43	6,37,42,44	7,38,43,44
Level 4-4	**36**	**37**	**38**	**39**	**40**	**41**	**42**
Level 3-5	22,23,27,37	22,24,28,38	22,25,29,39	22,26,30,40	23,24,31,41	23,25,32,42	23,26,33,43
Level 5-3	8,39,42,45	9,40,43,45	10,41,44,45	11,36,46,47	12,37,46,48	13,38,47,48	14,39,46,49
Level 4-4	**43**	**44**	**45**	**46**	**47**	**48**	**49**
Level 3-5	24,25,34,44	24,26,35,45	25,26,36,46	27,28,31,47	27,29,32,48	27,30,33,49	28,29,34,50

(Continued)

Table 1.33 (*Continued*)

Level 5-3	15,40,47,49	16,41,48,49	17,42,46,50	18,43,47,50	19,44,48,50	20,45,49,50	21,36,51,52
Level 4-4	**50**	**51**	**52**	**53**	**54**	**55**	**56**
Level 3-5	28,30,35,51	29,30,36,52	31,32,34,53	31,33,35,54	32,33,36,55	34,35,36,56	37,38,41,47
Level 5-3	22,37,51,53	23,38,52,53	24,39,51,54	25,40,52,54	26,41,53,54	27,42,51,55	28,43,52,55
Level 4-4	**57**	**58**	**59**	**60**	**61**	**62**	**63**
Level 3-5	37,39,42,48	37,40,43,49	38,39,44,50	38,40,45,51	39,40,46,52	41,42,44,53	41,43,45,54
Level 5-3	29,44,53,55	30,45,54,55	31,46,51,56	32,47,52,56	33,48,53,56	34,49,54,56	35,50,55,56
Level 4-4	**64**	**65**	**66**	**67**	**68**	**69**	**70**
Level 3-5	42,43,46,55	44,45,46,56	47,48,50,53	47,49,51,54	48,49,52,55	50,51,52,56	53,54,55,56

may have a look at alternative formulations and compare in particular when there are three and when are two: he may see some interesting structural recurrences. I limit myself to indicate very general forms of recurrence:

- Atomic statements like X, Y or Z and their negations.
- Conjunctions (in the conjunctive formulation) among three implications.
- Equivalences and their negations.
- Disjunctions or conjunctions that are relatively complex.

Also in the case of Level 4-4, all the 70 statements can be generated by any couple of statements either of Level 5-3 constituting the LGS (of 4 statements) of the relative statement or of Level 3-5 constituting the RGS (again of 4 statements) of the relative statement.

The previous examination shows a considerable advantage of my notation on the current logical one: there are several cases in which different formulations in the current logical language (when all logical symbols are used) are possible (a case that can produce some equivocation or ambiguity in the procedures to follow). At the opposite, the formulation in the proposed language is always univocal. Obviously, if only negation, disjunction and conjunction are used (or also the so-called normal disjunctive form or normal conjunctive form) in the standard language the problem disappears. However, this can make calculations more lengthy. A very easy way to find standard logical formulae when starting with the arrays of numbers shown in Fig. 1.10 is the following: we try to find which is the maximal sequence either horizontally or vertically of 0s or 1s or which are the systematic opposite values. Then, we try to build blocks constituted out of rows or columns, as shown in Fig. 1.11.

The chief rule here is that any subdivision of this matricial space (either sum or product) needs to cover the whole of rows and columns. Obviously, we are authorized to discard any row and column when its sum or product is a tautology (exhausts all alternatives) relative to the remnant rows or columns. Another easy example is represented in Fig. 1.12. It is not difficult to see that the sum-form of Statement 24 (last panel on the right) represents values of X and Z that are always coincident: either negative (first

	Statement 69, Level 4-4		Statement 2, Level 4-4	
	a′x b′x c′x e′	d + f + g + h	a + b + c + e	d′x f′x g′x h′
X	1 1 1 1	1 1 1 1	0 0 0 0	0 0 0 0
	∨	∧	∧	∨
Y	1 1 0 0	0 0 1 1	0 0 1 1	1 1 0 0
Z	1 ∧ 0 ∧ 1 ∧ 0	0 ∨ 1 ∨ 0 ∨ 1	0 ∨ 1 ∨ 0 ∨ 1	1 ∧ 0 ∧ 1 ∧ 0

Figure 1.11 Two statements taken from Level 4-4 that are the negation of each other. I have presented both the conjunctive and disjunctive form (see also Fig. 1.10). As it can be seen, in the first table on the left X (first row) is always true. Since it can be disjoint with four conjunctive terms (the four columns in the last two rows) that display any possible value-combination (again conjunctions) of Y and Z, this amounts to assert that X is true. The second panel shows the same proposition but expressed as a conjunction between X and a disjunction presenting any possible values of Y and Z (again the last two rows). Again, this shows that X is true independently from the values of Y and Z. The two panels on the right show the negation of Statement 69 of Level 4-4 (i.e. Statement 2). Indeed, in both formulations X (first row in both cases) is false independently of the values of Y and Z (last two rows in both cases).

	Statement 47, Level 4-4		Statement 24, Level 4-4	
	b + d + e + g	a′x c′ x f′x h′	b′x d′x e′x g′	a + c + f + h
X	0 1 0 1	1 1 0 0	1 0 1 0	0 0 1 1
Y	0 0 1 1	1 0 1 0	1 1 0 0	0 1 0 1
Z	1 0 1 0	1 1 0 0	0 1 0 1	0 0 1 1

Figure 1.12 Two other statements from Level 4-4. The code is the same as in Fig. 1.11. Statement 24 is $X \leftrightarrow Z$ whilst Statement 47 is its negation.

and second column), whatever the value of Y (second row) is or affirmative (third and fourth column), again whatever the value of Y is. Since we have here a disjunction (horizontal dimension) of conjunctions (vertical dimension) of negative and affirmative values of both X and Z, that is,

$$(0 \wedge 0) \vee (1 \wedge 1)\,, \tag{1.7}$$

the meaning of the whole expression is an equivalence. The same is expressed by the product form but by considering columns 1-3 and

2-4. In this case we have:

$$(1 \vee 0) \wedge (0 \vee 1)\,, \tag{1.8}$$

which again expresses an equivalence. The sum-form of Statement 47 (represented on the left of Fig. 1.12) tells that the values of X and Z are either 0 and 1, respectively (first and third column) or 1 and 0, respectively (second and fourth column). In other words, we have:

$$(0 \wedge 1) \vee (1 \wedge 0)\,, \tag{1.9}$$

which expresses the counter-valence (or the negation of the equivalence) between X and Z. The same logical statement is expressed by the product-form (but by considering columns 1-2 and 3-4):

$$(1 \vee 1) \wedge (0 \vee 0)\,. \tag{1.10}$$

This examination shows that an opportune combination of the methods can promptly lead to the easiest or shortest standard logical form of the statement. The advantage of this methodology is that one avoids any reference to whatever premises are considered and focuses only on the logical meaning.

Let us now consider the 56 disjunctive triplets or conjunctive pentaplets of Level 3-5 as displayed in Table 1.34.

Table 1.34 The 56 statements of Level 3-5

1	11100000	11000000 + 10100000 + 01100000	$\neg X \wedge (\neg Y \vee \neg Z)$
2	11010000	11000000 + 10010000 + 01010000	$\neg Y \wedge (\neg X \vee \neg Z)$
3	11001000	11000000 + 10001000 + 01001000	$\neg X \wedge (\neg Y \vee Z)$
4	11000100	11000000 + 10000100 + 01000100	$\neg Y \wedge (\neg X \vee Z)$
5	11000010	11000000 + 10000010 + 01000010	$(\neg X \wedge \neg Y) \vee (X \wedge Y \wedge \neg Z)$
6	11000001	11000000 + 10000001 + 01000001	$(\neg X \wedge \neg Y) \vee (X \wedge Y \wedge Z)$
7	10110000	10100000 + 10010000 + 00110000	$\neg Z \wedge (\neg X \vee \neg Y)$
8	10101000	10100000 + 10001000 + 00101000	$\neg X \wedge (Y \vee \neg Z)$
9	10100100	10100000 + 10000100 + 00100100	$(\neg X \wedge \neg Z) \vee (X \wedge \neg Y \wedge Z)$
10	10100010	10100000 + 10000010 + 00100010	$\neg Z \wedge (\neg X \vee Y)$
11	10100001	10100000 + 10000001 + 00100001	$(\neg X \wedge \neg Z) \vee (X \wedge Y \wedge Z)$
12	10011000	10010000 + 10001000 + 00011000	$(\neg Y \wedge \neg Z) \vee (\neg X \wedge Y \wedge Z)$
13	10010100	10010000 + 10000100 + 00010100	$\neg Y \wedge (X \vee \neg Z)$
14	10010010	10010000 + 10000010 + 00010010	$\neg Z \wedge (X \vee \neg Y)$
15	10010001	10010000 + 10000001 + 00010001	$(\neg Y \wedge \neg Z) \vee (X \wedge Y \wedge Z)$

(*Continued*)

Table 1.34 (*Continued*)

16	10001100	10001000 + 10000100 + 00001100	$\neg(X \leftrightarrow Y \leftrightarrow Z) \wedge (\neg X \vee \neg Y \vee Z)$
17	10001010	10001000 + 10000010 + 00001010	$\neg(X \leftrightarrow Y \leftrightarrow Z) \wedge (\neg X \vee Y \vee \neg Z)$
18	10001001	10001000 + 10000001 + 00001001	$(Y \wedge Z) \vee (\neg X \wedge \neg Y \wedge \neg Z)$
19	10000110	10000100 + 10000010 + 00000110	$\neg(X \leftrightarrow Y \leftrightarrow Z) \wedge (X \vee \neg Y \vee \neg Z)$
20	10000101	10000100 + 10000001 + 00000101	$(X \wedge Z) \vee (\neg X \wedge \neg Y \wedge \neg Z)$
21	10000011	10000010 + 10000001 + 00000011	$(X \wedge Y) \vee (\neg X \wedge \neg Y \wedge \neg Z)$
22	01110000	01100000 + 01010000 + 00110000	$(X \leftrightarrow Y \leftrightarrow Z) \wedge (\neg X \vee \neg Y \vee \neg Z)$
23	01101000	01100000 + 01001000 + 00101000	$\neg X \wedge (Y \vee Z)$
24	01100100	01100000 + 01000100 + 00100100	$(\neg Y \wedge Z) \vee (\neg X \wedge Y \wedge \neg Z)$
25	01100010	01100000 + 01000010 + 00100010	$(Y \wedge \neg Z) \vee (\neg X \wedge \neg Y \wedge Z)$
26	01100001	01100000 + 01000001 + 00100001	$(X \leftrightarrow Y \leftrightarrow Z) \wedge (\neg X \vee Y \vee Z)$
27	01011000	01010000 + 01001000 + 00011000	$(\neg X \wedge Z) \vee (X \wedge \neg Y \wedge \neg Z)$
28	01010100	01010000 + 01000100 + 00010100	$\neg Y \wedge (X \vee Z)$
29	01010010	01010000 + 01000010 + 00010010	$(X \wedge \neg Z) \vee (\neg X \wedge \neg Y \wedge Z)$
30	01010001	01010000 + 01000001 + 00010001	$(X \leftrightarrow Y \leftrightarrow Z) \wedge (X \vee \neg Y \vee Z)$
31	01001100	01001000 + 01000100 + 00001100	$Z \wedge (\neg X \vee \neg Y)$
32	01001010	01001000 + 01000010 + 00001010	$(\neg X \wedge Z) \vee (X \wedge Y \wedge \neg Z)$
33	01001001	01001000 + 01000001 + 00001001	$Z \wedge (\neg X \vee Y)$
34	01000110	01000100 + 01000010 + 00000110	$(\neg Y \wedge Z) \vee (X \wedge Y \wedge \neg Z)$
35	01000101	01000100 + 01000001 + 00000101	$Z \wedge (X \vee \neg Y)$
36	01000011	01000100 + 01000001 + 00000011	$(X \wedge Y) \vee (\neg X \wedge \neg Y \wedge Z)$
37	00111000	00110000 + 00101000 + 00011000	$(\neg X \wedge Y) \vee (X \wedge \neg Y \wedge \neg Z)$
38	00110100	00110000 + 00100100 + 00010100	$(X \wedge \neg Y) \vee (\neg X \wedge Y \wedge \neg Z)$
39	00110010	00110000 + 00100010 + 00010010	$\neg Z \wedge (X \vee Y)$
40	00110001	00110000 + 00100001 + 00010001	$(X \leftrightarrow Y \leftrightarrow Z) \wedge (X \vee Y \vee \neg Z)$
41	00101100	00101000 + 00100100 + 00001100	$(\neg X \wedge Y) \vee (X \wedge \neg Y \wedge Z)$
42	00101010	00101000 + 00100010 + 00001010	$Y \wedge (\neg X \vee \neg Z)$
43	00101001	00101000 + 00100001 + 00001001	$Y \wedge (\neg X \vee Z)$
44	00100110	00100100 + 00100010 + 00000110	$(Y \wedge \neg Z) \vee (X \wedge \neg Y \wedge Z)$
45	00100101	00100100 + 00100001 + 00000101	$(X \wedge Z) \vee (\neg X \wedge Y \wedge \neg Z)$
46	00100011	00100010 + 00100001 + 00000011	$Y \wedge (X \vee \neg Z)$
47	00011100	00011000 + 00010100 + 00001100	$(X \wedge \neg Y) \vee (\neg X \wedge Y \wedge Z)$
48	00011010	00011000 + 00010010 + 00001010	$(X \wedge \neg Z) \vee (\neg X \wedge Y \wedge Z)$
49	00011001	00011000 + 00010001 + 00001001	$(Y \wedge Z) \vee (X \wedge \neg Y \wedge \neg Z)$
50	00010110	00010100 + 00010010 + 00000110	$X \wedge (\neg Y \vee \neg Z)$
51	00010101	00010100 + 00010001 + 00000101	$X \wedge (\neg Y \vee Z)$
52	00010011	00010010 + 00010001 + 00000011	$X \wedge (Y \vee \neg Z)$
53	00001110	00001100 + 00001010 + 00000110	$\neg(X \leftrightarrow Y \leftrightarrow Z) \wedge (X \vee Y \vee Z)$
54	00001101	00001100 + 00001001 + 00000101	$Z \wedge (X \vee Y)$
55	00001011	00001010 + 00001001 + 00000011	$Y \wedge (X \vee Z)$
56	00000111	00000110 + 00000101 + 00000011	$X \wedge (Y \vee Z)$

It should be obvious by now that any of the above 56 statements of Level 3-5 is the negation of a statement of Level 5-3 but in inverse order: Statement 1 of Level 3-5 is the negation of Statement 56 of Level 5-3, Statement 2 of Level 3-5 is the negation of Statement 55 of Level 5-3, and so on. The general forms of Level 3-5 are therefore negations of the correspondent ones of Level 5.3. They are displayed in Table 1.35.

Table 1.35 The general forms of Level 3-5

$X \wedge (Y \vee Z)$	$\neg X \wedge (Y \vee Z)$
$(X \wedge Y) \vee (\neg X \wedge \neg Y \wedge Z)$	$(X \leftrightarrow Y \leftrightarrow Z) \wedge (X \vee Y \vee \neg Z)$

Summarizing,

- Specimens of the first row are Statements 1, 2, 3, 4, 7, 8, 10, 13, 14, 23, 28, 31, 33, 35, 39, 42, 43, 46, 50, 51, 52, 54, 55, 56..
- Specimens of the third form are Statements 5, 6, 9, 11, 12, 15, 18, 20, 21, 24, 25, 27, 29, 32, 34, 36, 37, 38, 41, 44, 45, 47, 48, 49.
- Specimens of the last form are Statements 16, 17, 19, 22, 26, 30, 40, 53.

Note that in analogy with Level 5-3, Statements 21 and 36 of Level 3-5 can be thought of to be a disjunction between Statement 28 of Level 2-6 (i.e. $X \wedge Y$) with Statement 1 $(\neg X \wedge \neg Y \wedge \neg Z)$ and Statement 2 $(\neg X \wedge \neg Y \wedge Z)$ of Level 1-7, respectively; the latter two statements generate Statement 1 of Level 2-6, i.e. $\neg X \wedge \neg Y$. Similarly for the other cases. As remarked for similar cases, any of the previous 56 statements can be derived from any combination of two out of either the RGS or the LGS, as displayed in Table 1.36.

The 28 disjunctive duplets or conjunctive esaplets of Level 2-6 are shown in Table 1.37.

Table 1.36 Derivation of the statements of Level 3-5

Level 4-4	1,2,3,4,5	1,6,7,8,9,	2,6,10,11,12	3,7,10,13,14
Level 3-5	**1**	**2**	**3**	**4**
Level 2-6	1,2,8	1,3,9	1,4,10	1,5,11
Level 4-4	4,8,11,13,15	5,9,12,14,15	1,16,17,18,19	2,16,20,21,22
Level 3-5	**5**	**6**	**7**	**8**
Level 2-6	1,6,12	1,7,13	2,3,14	2,4,15
Level 4-4	3,17,20,23,24	4,18,21,23,25	5,19,22,24,25	6,16,26,27,28
Level 3-5	**9**	**10**	**11**	**12**
Level 2-6	2,5,16	2,6,17	2,7,18	3,4,19
Level 4-4	7,17,26,29,30	8,18,27,29,31	9,19,28,30,31	10,20,26,32,33
Level 3-5	**13**	**14**	**15**	**16**
Level 2-6	3,5,20	3,6,21	3,7,22	4,5,23
Level 4-4	11,21,27,32,34	12,22,28,33,34	13,23,29,32,35	14,24,30,33,35
Level 3-5	**17**	**18**	**19**	**20**
Level 2-6	4,6,24	4,7,25	5,6,26	5,7,27
Level 4-4	15,25,31,34,35	1,36,37,38,39	2,36,40,41,42	3,37,40,43,44
Level 3-5	**21**	**22**	**23**	**24**
Level 2-6	6,7,28	8,9,14	8,10,15	8,11,16
Level 4-4	4,38,41,43,45	5,39,42,44,45	6,36,46,47,48	7,37,46,49,50
Level 3-5	**25**	**26**	**27**	**28**
Level 2-6	8,12,17	8,13,18	9,10,19	9,11,20
Level 4-4	8,38,47,49,51	9,39,48,50,51	10,40,46,52,53	11,40,46,52,53
Level 3-5	**29**	**30**	**31**	**32**
Level 2-6	9,12,21	9,13,22	10,11,23	10,12,24
Level 4-4	12,42,48,53,54	13,43,49,52,55	14,44,50,53,55	15,45,51,54,55
Level 3-5	**33**	**34**	**35**	**36**
Level 2-6	10,13,25	11,12,26	11,13,27	12,13,28
Level 4-4	16,36,56,57,58	17,37,56,59,60	18,38,57,59,61	19,39,58,60,61
Level 3-5	**37**	**38**	**39**	**40**
Level 2-6	14,15,19	14,16,20	14,17,21	14,18,22
Level 4-4	20,40,56,62,63	21,41,57,62,64	22,42,58,63,64	23,43,59,62,65
Level 3-5	**41**	**42**	**43**	**44**
Level 2-6	15,16,23	15,17,24	15,18,25	16,17,26
Level 4-4	24,44,60,63,65	25,45,61,64,65	26,46,56,66,67	27,47,57,66,68
Level 3-5	**45**	**46**	**47**	**48**
Level 2-6	16,18,27	17,18,28	19,20,23	19,21,24
Level 4-4	28,48,58,67,68	29,49,59,66,69	30,50,60,67,69	31,51,61,68,69
Level 3-5	**49**	**50**	**51**	**52**
Level 2-6	19,22,25	20,21,26	20,22,27	21,22,28
Level 4-4	32,52,62,66,70	33,53,63,67,70	34,54,64,68,70	35,55,65,69,70
Level 3-5	**53**	**54**	**55**	**56**
Level 2-6	23,24,26	23,25,27	24,25,28	26,27,28

Table 1.37 The 28 statements of Level 2-6

1	11000000	10000000 + 01000000	$\neg X \wedge \neg Y$
2	10100000	10000000 + 00100000	$\neg Z \wedge \neg X$
3	10010000	10000000 + 00010000	$\neg Z \wedge \neg Y$
4	10001000	10000000 + 00001000	$\neg X \wedge (Y \leftrightarrow Z)$
5	10000100	10000000 + 00000100	$\neg Y \wedge (X \leftrightarrow Z)$
6	10000010	10000000 + 00000010	$\neg Z \wedge (X \leftrightarrow Y)$
7	10000001	10000000 + 00000001	$X \leftrightarrow Y \leftrightarrow Z$
8	01100000	01000000 + 00100000	$\neg X \wedge \neg(Y \leftrightarrow Z)$
9	01010000	01000000 + 00010000	$\neg Y \wedge \neg(X \leftrightarrow Z)$
10	01001000	01000000 + 00001000	$Z \wedge \neg X$
11	01000100	01000000 + 00000100	$Z \wedge \neg Y$
12	01000010	01000000 + 00000010	$X \leftrightarrow Y \leftrightarrow \neg Z$
13	01000001	01000000 + 00000001	$Z \wedge (X \leftrightarrow Y)$
14	00110000	00100000 + 00010000	$\neg Z \wedge \neg(X \leftrightarrow Y)$
15	00101000	00100000 + 00001000	$\neg X \wedge Y$
16	00100100	00100000 + 00000100	$X \leftrightarrow \neg Y \leftrightarrow Z$
17	00100010	00100000 + 00000010	$\neg Z \wedge Y$
18	00100001	00100000 + 00000001	$Y \wedge (X \leftrightarrow Z)$
19	00011000	00010000 + 00001000	$\neg X \leftrightarrow Y \leftrightarrow Z$
20	00010100	00010000 + 00000100	$X \wedge \neg Y$
21	00010010	00010000 + 00000010	$\neg Z \wedge X$
22	00010001	00010000 + 00000001	$X \wedge (Y \leftrightarrow Z)$
23	00001100	00001000 + 00000100	$Z \wedge \neg(X \leftrightarrow Y)$
24	00001010	00001000 + 00000010	$Y \wedge \neg(X \leftrightarrow Z)$
25	00001001	00001000 + 00000001	$Z \wedge Y$
26	00000110	00000100 + 00000010	$X \wedge \neg(Y \leftrightarrow Z)$
27	00000101	00000100 + 00000001	$Z \wedge X$
28	00000011	00000010 + 00000001	$X \wedge Y$

In Table 1.38 some general forms are shown.

Table 1.38 General forms of Level 2-6

$X \wedge Y$	$X \wedge \neg Y$	$\neg X \wedge Y$	$\neg X \wedge \neg Y$
$X \wedge (Y \leftrightarrow Z)$	$X \wedge \neg(Y \leftrightarrow Z)$	$\neg X \wedge (Y \leftrightarrow Z)$	$\neg X \wedge \neg(Y \leftrightarrow Z)$
$X \leftrightarrow Y \leftrightarrow Z$	$X \leftrightarrow Y \leftrightarrow \neg Z$	$X \leftrightarrow \neg Y \leftrightarrow Z$	$\neg X \leftrightarrow Y \leftrightarrow Z$

Then, there are:

- 12 statements of the kind of the first row: 1, 2, 3, 10, 11, 15, 17, 20, 21, 25, 27, 28,

- 12 statements of the kind of the second row: 4, 5, 6, 8, 9, 13, 14, 18, 22, 23, 24, 26, and
- The four statement presented in the last row: 7, 12, 16, 19.

It is also easy to verify that each statement of Level 2-6 represents the negation of a statement of Level 6-2, but with inverse order (Statement 1 of Level 2-6 corresponds to the negation of Statement 28 of Level 6-2, Statement 2 of Level 2-6 corresponds to the negation of Statement 27 of Level 6-2, and so on). Let us consider now the connections of Level 2-6 with Levels 3-5 and 1-7:

Table 1.39 Derivation of the statements of Level 2-6

Level 3-5	1,2,3,4,5,6	1,7,8,9,10,11	2,7,12,13,14,15	3,8,12,16,17,18
Level 2-6	**1**	**2**	**3**	**4**
Level 1-7	1,2	1,3	1,4	1,5
Level 3-5	4,9,13,16,19,20	5,10,14,17,19,21	6,11,15,18,20,21	1,22,23,24,25,26
Level 2-6	**5**	**6**	**7**	**8**
Level 1-7	1,6	1,7	1,8	2,3
Level 3-5	2,22,27,28,29,30	3,23,27,31,32,33	4,24,28,31,34,35	5,25,29,32,34,36
Level 2-6	**9**	**10**	**11**	**12**
Level 1-7	2,4	2,5	2,6	2,7
Level 3-5	6,26,30,33,35,36	7,22,37,38,39,40	8,23,37,41,42,43	9,24,38,41,44,45
Level 2-6	**13**	**14**	**15**	**16**
Level 1-7	2,8	3,4	3,5	3,6
Level 3-5	10,25,39,42,44,46	11,26,40,43,45,46	12,27,37,47,48,49	13,28,38,47,50,51
Level 2-6	**17**	**18**	**19**	**20**
Level 1-7	3,7	3,8	4,5	4,6
Level 3-5	14,29,39,48,50,52	15,30,40,49,51,52	16,31,41,47,53,54	17,32,42,48,53,55
Level 2-6	**21**	**22**	**23**	**24**
Level 1-7	4,7	4,8	5,6	5,7
Level 3-5	18,33,43,49,54,55	19,34,44,50,53,56	20,35,45,51,54,56	21,36,46,52,55,56
Level 2-6	**25**	**26**	**27**	**28**
Level 1-7	5,8	6,7	6,8	7,8

Level 1-7 is shown in Table 1.40. The derivations of these statements from the raising complete sets is shown in Table 1.41.

I have shown how to orderly generate all statements of the three-dimensional space. Nevertheless, it is allowed to take short cuts when dealing with some specific statements. For instance, Statement

Table 1.40 The 8 statements of Level 1-7

1	10000000	$\neg(X \vee Y \vee Z)$	$\neg X \wedge \neg Y \wedge \neg Z$
2	01000000	$\neg(X \vee Y \vee \neg Z)$	$\neg X \wedge \neg Y \wedge Z$
3	00100000	$\neg(X \vee \neg Y \vee Z)$	$\neg X \wedge Y \wedge \neg Z$
4	00010000	$\neg(\neg X \vee Y \vee Z)$	$X \wedge \neg Y \wedge \neg Z$
5	00001000	$\neg(X \vee \neg Y \vee \neg Z)$	$\neg X \wedge Y \wedge Z$
6	00000100	$\neg(\neg X \vee Y \vee \neg Z)$	$X \wedge \neg Y \wedge Z$
7	00000010	$\neg(\neg X \vee \neg Y \vee Z)$	$X \wedge Y \wedge \neg Z$
8	00000001	$\neg(\neg X \vee \neg Y \vee \neg Z)$	$X \wedge Y \wedge Z$

Table 1.41 Derivation of the statements of Level 1-7

Level 2-6	1,2,3,4,5,6,7	1,8,9,10,11,12,13	2,8,14,15,16,17,18	3,9,14,19,20,21,22
Level 1-7	**1**	**2**	**3**	**4**
Level 2-6	4,10,15,19,23,24,25	5,11,16,20,23,26,27	6,12,17,21,24,26,28	7,13,18,22,25,27,28
Level 1-7	**5**	**6**	**7**	**8**

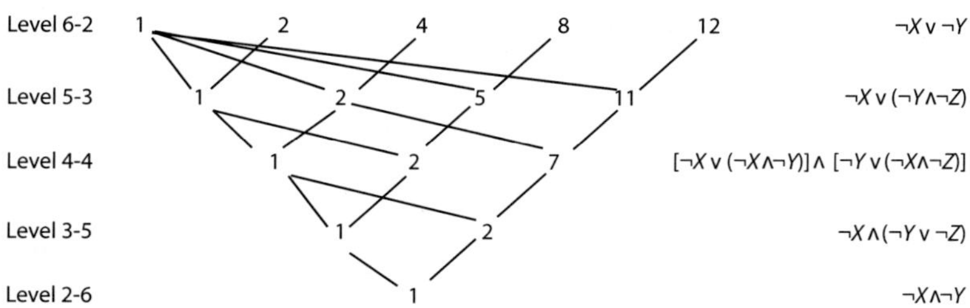

Figure 1.13 Short cuts across the space: how Statement 1 of Level 2-6 is generated through an array of statements of the higher levels. The general logical forms of these statements are shown on the right.

1 of Level 2-6, i.e. $\neg X \wedge \neg Y$, can be directly obtained by making the product of Statements 2 ($\neg X$) and 7 ($\neg Y$) of Level 4-4. Indeed, $11101000 \times 11010100 = 11000000$. In this way we can jump any level we like, provided that the computation is always well executed. We can even take short cuts across the whole space. An instance is shown in Fig. 1.13. It is understood that the product of Statements 1 and 2 of Level 6-2 generates Statement 1 of Level 5-3, the product of Statements 1 and 4 of Level 6-2 generates Statement 2 of Level 5-3, and so on, up to the product of Statements 1 and 2 of Level 3-5 generates Statement 1 of Level 2-6. Now, we can reformulate this by

Table 1.42 Summary of all statements divided into two subsets

8-0	11111111				0-8	00000000			
7-1	11111110	11111101	11111011	11110111	1-7	00000001	00000010	00000100	00001000
	11101111	11011111	10111111	01111111		00010000	00100000	01000000	10000000
6-2	11111100	11111010	11111001	11110110	2-6	00000011	00000101	00000110	00001001
	11110101	11110011	11101110	11101101		00001010	00001100	00010001	00010010
	11101011	11100111	11011110	11011101		00010100	00011000	00100001	00100010
	11011011	11010111	11001111	10111110		00100100	00101000	00110000	01000001
	10111101	10111011	10110111	10101111		01000010	01000100	01001000	01010000
	10011111	01111110	01111101	01111011		01100000	10000001	10000010	10000100
	01110111	01101111	01011111	00111111		10001000	10010000	10100000	11000000
5-3	11111000	11110100	11110010	11110001	3-5	00000111	00001011	00001101	00001101
	11101100	11101010	11101001	11100110		00010011	00010101	00010110	00011001
	11100101	11100011	11011100	11011010		00011010	00011100	00100011	00100101
	11011001	11010110	11010101	11010011		00100110	00101001	00101010	00101100
	11001110	11001101	11001011	11000111		00110001	00110010	00110100	00111000
	10111100	10111010	10111001	10110110		01000011	01000101	01000110	01001001
	10110101	10110011	10101110	10101101		01001010	01001100	01010001	01010010
	10101011	10100111	10011110	10011101		01010100	01011000	01100001	01100010
	10011011	10010111	10001111	01111100		01100100	01101000	01110000	10000011
	01111010	01111001	01110110	01110101		10000101	10000110	10001001	10001010
	01110011	01101110	01101101	01101011		10001100	10010001	10010010	10010100
	01100111	01011110	01011101	01011011		10011000	10100001	10100010	10100100
	01010111	01001111	00111110	00111101		10101000	10110000	11000001	11000010
	00111011	00110111	00101111	00011111		11000100	11001000	11010000	11100000

4-4	11110000	11101000	11100100	11100010	4-4	00001111	00010111	00011011	00011101
	11100001	11011000	11010100	11010010		00011110	00100111	00101011	00101101
	11010001	11001100	11001010	11001001		00101110	00110011	00110101	00110110
	11000110	11000101	11000011	10111000		00111001	00111010	00111100	01000111
	10110100	10110010	10110001	10101100		01001011	01001101	01001110	01010011
	10101010	10101001	10100110	10100101		01010101	01010110	01011001	01011010
	10100011	10011100	10011010	10011001		01011100	01100011	01100101	01100110
	10010110	10010101	10010011	10001110		01101001	01101010	01101100	01110001
	10001101	10001011	10000111			01110010	01110100	01111000	

affirming that the product of Statements (i) 1, 2, 4, 8 and 12 of Level 6-2 or (ii) 1, 2, 5, 11 of Level 5-3 or (iii) 1, 2 and 7 of Level 4-4 or (iv) 1 and 2 of Level 3-5, generate Statement 1 of Level 2-6. Obviously, also some combinations of these options are possible.

It is possible to summarize all of the statements of the tridimensional logical space as in Table 1.42; on the left we have all disjunctive formulations (four for each row) starting from the top (Level 8-0) in a progressive order whilst on the right we have their negations expressed in conjunctive form (always four for each row) starting from the bottom (Level 0-8) in the inverse order (note that the whole of the statements of Level 4-4 are divided in two subsets).

We could also have formulated all statements on the left following the conjunctive ordering and all statements on the right following the disjunctive ordering, what I have also done when derived the above formulae, since it is easier to proceed in a conjunctive way from the top level as well as in a disjunctive way from the bottom. However, what is good for derivation is not necessarily good for a synoptic table. Here, it is better to consider the whole logical space as starting from two opposite extremes (11111111 and 00000000) and building any statements by progressively reducing the number of 1s and 0s, respectively, and by displacing the zeros or the ones from the right to the left. This analysis also shows that the whole logical space is only determined by the number of 0s and 1s independently of the number of conjoint or disjoint statements.

Part II

"Closed" Inferences

Chapter 2

Product Inferences

2.1 Introduction

This system easily and mechanically covers the whole space of traditional syllogisms, which (by following C. Peirce) I consider as a basis sufficient to cover any possible kind of triadic and "closed" inference, that is, inferences constituted by three variables as a whole and by statements always involving 2 variables (this will be proved in Chapter 8). Here, from two premises obeying specific classical rules a conclusion may be derived. However, I shall also introduce other forms of inference, which although deducible from the former through some substitution nevertheless enlarge the space of traditional logic. Then, I shall also consider forms of inferences that seem to do not have been considered in the past. Indeed, I shall derive all the fundamental inferences in four different forms: the product form, the sum form, the subtraction form, and the division form. I shall identify any statement in an univocal way (which is a necessary requirement for mechanical calculation). I shall begin with the former. First of all I shall preserve the traditional Aristotelian notation (Auletta, 2013):

$$\mathbf{A}: X \to Y, \quad \mathbf{E}: X \to \neg Y, \quad \mathbf{I}: X \wedge Y, \quad \mathbf{O}: X \wedge \neg Y,$$

Mechanical Logic in Three-Dimensional Space
Gennaro Auletta

ISBN 978-981-4411-50-9 (Hardcover), 978-981-4411-51-6 (eBook)
www.panstanford.com

and consider these as basic (I recall that they can be formulated in explicit quantified form as in Table 1.4 but also that this is unnecessary in such a context). Obviously, other choices are possible but this changes nothing in the system (due to its symmetry). It is also worth mentioning that these statements already occur in the two-dimensional space (see Table 1.3). Then, I shall try to derive these statements from more elementary ones. Moreover, I recall that all statements have both a LGS and RGS and that the sum of these two gives precisely to 8 generating elements for any statement considered; for instance, {**A**} = {{A1, A2}, {A3, A4, A5, A6, A7, A8}}, where the first subset is the LGS whilst the second one is the RGS. Some of these elements will be introduced in this section, some later. However, to avoid any confusion I present here three summary tables that gives the connection between the number of the statement and its associated symbol.

Table 2.1 Summary of all symbols of all statements: Levels 7-1–5-3

Level	#	Symbol	#	Symbol	#	Symbol	#	Symbol
7-1	1	E1=D1=F1	2	E2=P1=R1	3	A1=C1=F2	4	M1=B1=D2
	5	A2=N1=R2	6	M2=P2=Q1	7	L1=B2=C2	8	L2=N2=Q2
6-2	1	**E**	2	**F**	3	E2xA1	4	**D**
	5	E2xM1	6	A1xM1	7	E1xA2	8	**R**
	9	**A**	10	M1xA2	11	E1xM2	12	**P**
	13	A1xM2	14	**M**	15	A2xM2	16	E1xL1
	17	E2xL1	18	**C**	19	**B**	20	A2xL1
	21	M2xL1	22	E1xL2	23	E2xL2	24	A1xL2
	25	M1xL2	26	**N**	27	**Q**	28	**L**
5-3	1	E3=F3	2	E4=D3	3	D4=F4	4	—
	5	E5=R3	6	A3=F5	7	A4=R4	8	D5
	9	R5	10	A5	11	E6=P3	12	F6
	13	P4	14	M3=D6	15	M4=P5	16	M5
	17	—	18	P6=R6	19	A6	20	M6
	21	E7	22	C3=F7	23	C4	24	B3=D7
	25	B4	26	B5=C5	27	—	28	R7
	29	A7=C6	30	B6	31	—	32	P7
	33	C7	34	M7=B7	35	—	36	E8
	37	F8	38	—	39	D8	40	—
	41	—	42	N3	43	N4=R8	44	A8=N5
	45	N6	46	Q3	47	P8=Q4	48	Q5
	49	M8=Q6	50	N7=Q7	51	L3	52	L4
	53	L5=C8	54	L6=B8	55	L7=N8	56	L8=Q8

Table 2.2 Summary of all symbols of all statements: Level 4-4

#	Symbol	#	Symbol	#	Symbol	#	Symbol
1	E3xE4=T1+T2	2	E3xE5=A3xA4= F3xF5=R3xR4= T1+T3=S1+S2= K1+K2=Ω1+Ω3	3	E4xE5=T1+T4	4	A3xA5=T1+T5
5	A4xA5=T1+T6	6	E3xE6=T2+T3	7	E4xE6=M3xM4= D3xD6=P3xP5= T2+T4=O1+O2= H1+H3=V1+V4	8	M3xM5=T2+T5
9	M4xM5=T2+T6	10	E5xE6=T3+T4	11	A3xA6=T3+T5	12	A4xA6=T3+T6
13	M3xM6=T4+T5	14	M4xM6=T4+T6	15	A5xA6=M5xM6= T5+T6=I1+I2	16	E3xE7=S1+S3
17	E4xE7=O1+O3	18	B3xB5=C3xC5= D4xD7=F4xF7= U1+U3=V2+V5= W1+W3=Ω2+Ω5	19	B4xB5=C4xC5= V2+V6= Ω2+Ω6	20	E5xE7=S1+S4
21	A3xA7=S1+S5	22	A4xA7=S1+S6	23	B3xB6=D5xD7= U1+U5=Ω4+Ω5	24	B4xB6=R5xR7= J1+J3=Ω4+Ω6
25	A5xA7=I1+I3	26	E6xE7=O1+O4	27	C3xC7=F6xF7= V3+V5=W1+W4	28	C4xC7=P4xP7= G1+G4=V3+V6
29	M3xM7=O1+O5	30	M4xM7=O1+O6	31	M5xM7=I1+I4	32	—
33	R6xR7=P6xP7= G1+G5+J1+J5	34	A6xA7=I1+I5	35	M6xM7=I1+I6	36	E3xE8=S2+S3
37	E4xE8=O2+O3	38	D4xD8=F4xF8= U2+U3=W2+W3	39	—	40	E5xE8=S2+S4
41	A3xA8=S2+S5	42	A4xA8=S2+S6	43	D5xD8=N3xN6= H2+H5=U2+U5	44	N4xN6=R5xR8= H2+H6=J2+J3
45	A5xA8=I2+I3	46	E6xE8=O2+O4	47	F6xF8=Q3xQ5= K2+K3=V2+V4	48	P4xP8=Q4xQ5= G2+G4=K3+K6
49	M3xM8=O2+O5	50	M4xM8=O2+O6	51	M5xM8=I2+I4	52	N3xN7=Q3xQ7= H4+H5=K2+K4
53	P6xP8=N4xN7= Q4xQ7=R6xR8= G2+G5=H4+H6= J2+J5=K4+K6	54	A6xA8=I2+I5	55	M6xM8=I2+I6	56	E7xE8=L3xL4= S3+S4=O3+O4
57	L3xL5=S3+S5	58	L4xL5=S3+S6	59	L3xL6=O3+O5	60	L4xL6=O3+O6

(Continued)

Table 2.2 (*Continued*)

61	L5xL6=I3+I4	62	L3xL7=S4+S5	63	L4xL7=S4+S6	64	A7xA8=L5xL7= C6xC8=N5xN8= S5+S6=I3+I5= G3+G6=U4+U6
65	L6xL7=I3+I6	66	L3xL8=O4+O5	67	L4xL8=O4+O6	68	L5xL8=I4+I5
69	M7xM8=L6xL8= B7xB8=Q6xQ8= O5+O6=I4+I6= J4+J6=W5+W6	70	L7xL8=I5+I6				

Table 2.3 Summary of all symbols of all statements: Levels 3-5–1-7

Level	#	Symbol	#	Symbol	#	Symbol	#	Symbol
3-5	1	T1=Ω1	2	T2=V1	3	T3=K1	4	T4=H1
	5	T5	6	T6	7	V2=Ω2	8	S1=Ω3
	9	Ω4	10	U1=Ω5	11	Ω6	12	V3
	13	O1=V4	14	V5=W1	15	V6	16	—
	17	—	18	G1	19	—	20	J1
	21	I1	22	—	23	S2=K2	24	H2
	25	U2	26	—	27	K3	28	O2=H3
	29	W2	30	—	31	H4=K4	32	K5
	33	G2=K6	34	H5	35	H6=J2	36	I2
	37	S3	38	O3	39	U3=W3	40	—
	41	S4	42	S5=U4	43	S6=G3	44	U5
	45	J3	46	I3=U6	47	O4	48	W4
	49	G4	50	O5=W5	51	O6=J4	52	I4=W6
	53	—	54	G5=J5	55	I5=G6	56	I6=J6
2-6	1	**T**	2	**Ω**	3	**V**	4	T7+S8
	5	T7+O8	6	T7+I7	7	T7+I8	8	T8+S7
	9	T8+O7	10	**K**	11	**H**	12	T8+I7
	13	T8+I8	14	S7+O7	15	**S**	16	S7+O8
	17	**U**	18	S7+I8	19	O7+S8	20	**O**
	21	**W**	22	O7+I8	23	S8+O8	24	S8+I7
	25	**G**	26	O8+I7	27	**J**	28	**I**
1-7	1	T7=V7=Ω7	2	T8=H7=K7	3	S7=U7=Ω8	4	O7=V8=W7
	5	S8=G7=K8	6	O8=H8=J7	7	I7=U8=W8	8	I8=G8=J8

As mentioned in Section 1.2, I take Statements **L, M, S, T** also as basic (see again Table 1.3). Again, other choices are possible but this is immaterial to the system due to its symmetry. Also the meaning of other symbols will be clear in due course (they are due to the expansion from a two-dimensional to a three-dimensional space). Moreover, I have considered combinations of levels higher than Level 4-4 as products whilst combinations of levels lower than Level 4-4 as additions, since this is the derivation path that I have followed in the previous chapter. Since the statements of Level 4-4 will rarely occur in the next derivations, allow me to spend a couple of words. At this level, any statement pertaining to a raising or lowering generating set is combined with any other statement of the same set (for instance, E3 is combined with E4, E5, E6, E7, E8, and the same is true for any other statement pertaining to the LGS of E). Moreover, no statement is present in connection with its contradictory (for instance, we never have Ex $\times$ Ey $=$ Iw $+$ Iz, where x, y, can vary from 3 to 8 and w, z from 1 to 6).

First, I shall deal with the two universal statements **A** and **E** (pertaining to Level 6-2), which are easier to handle (for all what follows in this section consider Section 1.3), as shown in Table 2.4.

Table 2.4 A and E

Level	#	Symbol	ID		Traditional logical form
7-1	3	A1	11111011	$\times$	$\neg X \vee Y \vee \neg Z$
7-1	5	A2	11101111	=	$\neg X \vee Y \vee Z$
6-2	9	**A**	11101011		$X \rightarrow Y$
7-1	1	E1	11111110	$\times$	$\neg X \vee \neg Y \vee \neg Z$
7-1	2	E2	11111101	=	$\neg X \vee \neg Y \vee Z$
6-2	1	**E**	11111100		$X \rightarrow \neg Y$

As mentioned, these two universal statements can be considered as derived from more elementary ones that turn out to be triadic (i.e. they all involve the three variables X, Y, Z). In other words, Table 2.4 shows all couples of statements involving all the three variables that can generate through product the traditional statements **A** and **E**. We shall consider later the consequences of this. In analogy with the above formulation, Table 2.5 deals with the other universal statements involved in traditional syllogistic treatment (however, traditionally they were not denoted by particular letters as here).

Table 2.5 B, C, D, and F

Level	#	Symbol	ID		Traditional logical form
7-1	4	B1=M1	11110111	×	$X \vee \neg Y \vee \neg Z$
7-1	7	B2=L1	10111111	=	$X \vee Y \vee \neg Z$
6-2	19	**B**	10110111		$Z \rightarrow X$
7-1	3	C1=A1	11111011	×	$\neg X \vee Y \vee \neg Z$
7-1	7	C2=L1	10111111	=	$X \vee Y \vee \neg Z$
6-2	18	**C**	10111011		$Z \rightarrow Y$
7-1	1	D1=E1	11111110	×	$\neg X \vee \neg Y \vee \neg Z$
7-1	4	D2=M1	11110111	=	$X \vee \neg Y \vee \neg Z$
6-2	4	**D**	11110110		$Z \rightarrow \neg Y$
7-1	1	F1=E1	11111110	×	$\neg X \vee \neg Y \vee \neg Z$
7-1	3	F2=A1	11111011	=	$\neg X \vee Y \vee \neg Z$
6-2	2	**F**	11111010		$Z \rightarrow \neg X$

Also here the Table shows all couples of statements involving X, Y, Z capable to generate through product the statements **B, C, D** and **F**. It is possible to obtain directly the result of an inference involving only universal statements like those considered so far through a product of their composed form. For instance, the product of **A** and **B** will give us Statement 25 of Level 4-4 that is a conjunction of **A, B** and **C**, and therefore it yields the correct result that is **C**. However, this will not work when particular statements like **I** or **O** are involved. Indeed, the product of **A** and **I** gives rise to **I**, which is not the correct result in this language. In other words, we cannot find the correct result in a general way through mechanical product of the statements themselves alone. Moreover, there is an additional reason that will be shown later.

Therefore, in order to perform a correct calculation we need now to express also the particular statements in terms of products which make our task more difficult since those statements pertain to Level 2-6, which means that they can be generated by the product of any couple out of the relative LGS of 6 propositions of Level 3-5 (whilst the RGS of particular statements only encompasses 2 statements), what makes 15 alternative ways to derive them. Let us start with the traditional **I** and **O** as displayed in Tables 2.6–2.7 (only the first derivation is shown explicitly in each table). Note that the LGS of **I** is represented by Statements 21, 36, 46, 52, 55, 56 whilst the LGS of **O** is represented by statements 13, 28, 38, 47, 50, 51.

Table 2.6 Derivations of **I**

Level	#	Symbol	ID		Traditional logical form
3-5	56	I6	00000111	×	$X \wedge (Y \vee Z)$
3-5	55	I5	00001011	=	$Y \wedge (X \vee Z)$
2-6	28	**I**	00000011		$X \wedge Y$
3-5	56	I6	00000111		$X \wedge (Y \vee Z)$
3-5	52	I4	00010011		$X \wedge (Y \vee \neg Z)$
3-5	56	I6	00000111		$X \wedge (Y \vee Z)$
3-5	46	I3	00100011		$Y \wedge (X \vee \neg Z)$
3-5	56	I6	00000111		$X \wedge (Y \vee Z)$
3-5	36	I2	01000011		$(X \wedge Y) \vee (\neg X \wedge \neg Y \wedge Z)$
3-5	56	I6	00000111		$X \wedge (Y \vee Z)$
3-5	21	I1	10000011		$(X \wedge Y) \vee (\neg X \wedge \neg Y \wedge \neg Z)$
3-5	55	I5	00001011		$Y \wedge (X \vee Z)$
3-5	52	I4	00010011		$X \wedge (Y \vee \neg Z)$
3-5	55	I5	00001011		$Y \wedge (X \vee Z)$
3-5	46	I3	00100011		$Y \wedge (X \vee \neg Z)$
3-5	55	I5	00001011		$Y \wedge (X \vee Z)$
3-5	36	I2	01000011		$(X \wedge Y) \vee (\neg X \wedge \neg Y \wedge Z)$
3-5	55	I5	00001011		$Y \wedge (X \vee Z)$
3-5	21	I1	10000011		$(X \wedge Y) \vee (\neg X \wedge \neg Y \wedge \neg Z)$
3-5	52	I4	00010011		$X \wedge (Y \vee \neg Z)$
3-5	46	I3	00100011		$Y \wedge (X \vee \neg Z)$
3-5	52	I4	00010011		$X \wedge (Y \vee \neg Z)$
3-5	36	I2	01000011		$(X \wedge Y) \vee (\neg X \wedge \neg Y \wedge Z)$
3-5	52	I4	00010011		$X \wedge (Y \vee \neg Z)$
3-5	21	I1	10000011		$(X \wedge Y) \vee (\neg X \wedge \neg Y \wedge \neg Z)$
3-5	46	I3	00100011		$Y \wedge (X \vee \neg Z)$
3-5	36	I2	01000011		$(X \wedge Y) \vee (\neg X \wedge \neg Y \wedge Z)$
3-5	46	I3	00100011		$Y \wedge (X \vee \neg Z)$
3-5	21	I1	10000011		$(X \wedge Y) \vee (\neg X \wedge \neg Y \wedge \neg Z)$
3-5	36	I2	01000011		$(X \wedge Y) \vee (\neg X \wedge \neg Y \wedge Z)$
3-5	21	I1	10000011		$(X \wedge Y) \vee (\neg X \wedge \neg Y \wedge \neg Z)$

Table 2.7 Derivations of **O**

Level	#	Symbol	ID		Traditional logical form
3-5	51	06	00010101	×	$X \wedge (\neg Y \vee Z)$
3-5	50	05	00010110	=	$X \wedge (\neg Y \vee \neg Z)$
2-6	20	**O**	00010100		$X \wedge \neg Y$
3-5	51	06	00010101		$X \wedge (\neg Y \vee Z)$
3-5	47	04	00011100		$(X \wedge \neg Y) \vee (\neg X \wedge Y \wedge Z)$
3-5	51	06	00010101		$X \wedge (\neg Y \vee Z)$
3-5	38	03	00110100		$(X \wedge \neg Y) \vee (\neg X \wedge Y \wedge \neg Z)$
3-5	51	06	00010101		$X \wedge (\neg Y \vee Z)$
3-5	28	02	01010100		$\neg Y \wedge (X \vee Z)$
3-5	51	06	00010101		$X \wedge (\neg Y \vee Z)$
3-5	13	01	10010100		$\neg Y \wedge (X \vee \neg Z)$
3-5	50	05	00010110		$X \wedge (\neg Y \vee \neg Z)$
3-5	47	04	00011100		$(X \wedge \neg Y) \vee (\neg X \wedge Y \wedge Z)$
3-5	50	05	00010110		$X \wedge (\neg Y \vee \neg Z)$
3-5	38	03	00110100		$(X \wedge \neg Y) \vee (\neg X \wedge Y \wedge \neg Z)$
3-5	50	05	00010110		$X \wedge (\neg Y \vee \neg Z)$
3-5	28	02	01010100		$\neg Y \wedge (X \vee Z)$
3-5	50	05	00010110		$X \wedge (\neg Y \vee \neg Z)$
3-5	13	01	10010100		$\neg Y \wedge (X \vee \neg Z)$
3-5	47	04	00011100		$(X \wedge \neg Y) \vee (\neg X \wedge Y \wedge Z)$
3-5	38	03	00110100		$(X \wedge \neg Y) \vee (\neg X \wedge Y \wedge \neg Z)$
3-5	47	04	00011100		$(X \wedge \neg Y) \vee (\neg X \wedge Y \wedge Z)$
3-5	28	02	01010100		$\neg Y \wedge (X \vee Z)$
3-5	47	04	00011100		$(X \wedge \neg Y) \vee (\neg X \wedge Y \wedge Z)$
3-5	13	01	10010100		$\neg Y \wedge (X \vee \neg Z)$
3-5	38	03	00110100		$(X \wedge \neg Y) \vee (\neg X \wedge Y \wedge \neg Z)$
3-5	28	02	01010100		$\neg Y \wedge (X \vee Z)$
3-5	38	03	00110100		$(X \wedge \neg Y) \vee (\neg X \wedge Y \wedge \neg Z)$
3-5	13	01	10010100		$\neg Y \wedge (X \vee \neg Z)$
3-5	28	02	01010100		$\neg Y \wedge (X \vee Z)$
3-5	13	01	10010100		$\neg Y \wedge (X \vee \neg Z)$

We proceed now in a similar way for the particular statements **G, H, J, K**, as shown in Tables 2.8–2.11.

We may remark that all statements that are not shared among several particular statements (like K5 or K3) have the form of disjunctions of conjunctions (at the opposite of e.g. K1 or K2). I shall now proceed to the derivation of the classical Aristotelian syllogisms (Auletta, 2013) by means of this logical system.

Table 2.8 Derivations of **G**

Level	#	Symbol	ID		Traditional logical form
3-5	55	G6=I5	00001011	×	$Y \wedge (X \vee Z)$
3-5	54	G5=J5	00001101	=	$Z \wedge (X \vee Y)$
2-6	25	**G**	00001001		$Z \wedge Y$
3-5	55	G6=I5	00001011		$Y \wedge (X \vee Z)$
3-5	49	G4	00011001		$(Y \wedge Z) \vee (X \wedge \neg Y \wedge \neg Z)$
3-5	55	G6=I5	00001011		$Y \wedge (X \vee Z)$
3-5	43	G3=S6	00101001		$Y \wedge (\neg X \vee Z)$
3-5	55	G6=I5	00001011		$Y \wedge (X \vee Z)$
3-5	33	G2=K6	01001001		$Z \wedge (\neg X \vee Y)$
3-5	55	G6=I5	00001011		$Y \wedge (X \vee Z)$
3-5	18	G1	10001001		$(Y \wedge Z) \vee (\neg X \wedge \neg Y \wedge \neg Z)$
3-5	54	G5=J5	00001101		$Z \wedge (X \vee Y)$
3-5	49	G4	00011001		$(Y \wedge Z) \vee (X \wedge \neg Y \wedge \neg Z)$
3-5	54	G5=J5	00001101		$Z \wedge (X \vee Y)$
3-5	43	G3=S6	00101001		$Y \wedge (\neg X \vee Z)$
3-5	54	G5=J5	00001101		$Z \wedge (X \vee Y)$
3-5	33	G2=K6	01001001		$Z \wedge (\neg X \vee Y)$
3-5	54	G5=J5	00001101		$Z \wedge (X \vee Y)$
3-5	18	G1	10001001		$(Y \wedge Z) \vee (\neg X \wedge \neg Y \wedge \neg Z)$
3-5	49	G4	00011001		$(Y \wedge Z) \vee (X \wedge \neg Y \wedge \neg Z)$
3-5	43	G3=S6	00101001		$Y \wedge (\neg X \vee Z)$
3-5	49	G4	00011001		$(Y \wedge Z) \vee (X \wedge \neg Y \wedge \neg Z)$
3-5	33	G2=K6	01001001		$Z \wedge (\neg X \vee Y)$
3-5	49	G4	00011001		$(Y \wedge Z) \vee (X \wedge \neg Y \wedge \neg Z)$
3-5	18	G1	10001001		$(Y \wedge Z) \vee (\neg X \wedge \neg Y \wedge \neg Z)$
3-5	43	G3=S6	00101001		$Y \wedge (\neg X \vee Z)$
3-5	33	G2=K6	01001001		$Z \wedge (\neg X \vee Y)$
3-5	43	G3=S6	00101001		$Y \wedge (\neg X \vee Z)$
3-5	18	G1	10001001		$(Y \wedge Z) \vee (\neg X \wedge \neg Y \wedge \neg Z)$
3-5	33	G2=K6	01001001		$Z \wedge (\neg X \vee Y)$
3-5	18	G1	10001001		$(Y \wedge Z) \vee (\neg X \wedge \neg Y \wedge \neg Z)$

Table 2.9 Derivations of **H**

Level	#	Symbol	ID		Traditional logical form
3-5	35	H6=J2	01000101	×	$Z \wedge (X \vee \neg Y)$
3-5	34	H5	01000110	=	$(\neg Y \wedge Z) \vee (X \wedge Y \wedge \neg Z)$
2-6	11	**H**	01000100		$Z \wedge \neg Y$
3-5	35	H6=J2	01000101		$Z \wedge (X \vee \neg Y)$
3-5	31	H4=K4	01001100		$Z \wedge (\neg X \vee \neg Y)$
3-5	35	H6=J2	01000101		$Z \wedge (X \vee \neg Y)$
3-5	28	H3=O2	01010100		$\neg Y \wedge (X \vee Z)$
3-5	35	H6=J2	01000101		$Z \wedge (X \vee \neg Y)$
3-5	24	H2	01100100		$(\neg Y \wedge Z) \vee (\neg X \wedge Y \wedge \neg Z)$
3-5	35	H6=J2	01000101		$Z \wedge (X \vee \neg Y)$
3-5	4	H1=T4	11000100		$\neg Y \wedge (\neg X \vee Z)$
3-5	34	H5	01000110		$(\neg Y \wedge Z) \vee (X \wedge Y \wedge \neg Z)$
3-5	31	H4=K4	01001100		$Z \wedge (\neg X \vee \neg Y)$
3-5	34	H5	01000110		$(\neg Y \wedge Z) \vee (X \wedge Y \wedge \neg Z)$
3-5	28	H3=O2	01010100		$\neg Y \wedge (X \vee Z)$
3-5	34	H5	01000110		$(\neg Y \wedge Z) \vee (X \wedge Y \wedge \neg Z)$
3-5	24	H2	01100100		$(\neg Y \wedge Z) \vee (\neg X \wedge Y \wedge \neg Z)$
3-5	34	H5	01000110		$(\neg Y \wedge Z) \vee (X \wedge Y \wedge \neg Z)$
3-5	4	H1=T4	11000100		$\neg Y \wedge (\neg X \vee Z)$
3-5	31	H4=K4	01001100		$Z \wedge (\neg X \vee \neg Y)$
3-5	28	H3=O2	01010100		$\neg Y \wedge (X \vee Z)$
3-5	31	H4=K4	01001100		$Z \wedge (\neg X \vee \neg Y)$
3-5	24	H2	01100100		$(\neg Y \wedge Z) \vee (\neg X \wedge Y \wedge \neg Z)$
3-5	31	H4=K4	01001100		$Z \wedge (\neg X \vee \neg Y)$
3-5	4	H1=T4	11000100		$\neg Y \wedge (\neg X \vee Z)$
3-5	28	H3=O2	01010100		$\neg Y \wedge (X \vee Z)$
3-5	24	H2	01100100		$(\neg Y \wedge Z) \vee (\neg X \wedge Y \wedge \neg Z)$
3-5	28	H3=O2	01010100		$\neg Y \wedge (X \vee Z)$
3-5	4	H1=T4	11000100		$\neg Y \wedge (\neg X \vee Z)$
3-5	24	H2	01100100		$(\neg Y \wedge Z) \vee (\neg X \wedge Y \wedge \neg Z)$
3-5	4	H1=T4	11000100		$\neg Y \wedge (\neg X \vee Z)$

Table 2.10 Derivations of **J**

Level	#	Symbol	ID		Traditional logical form
3-5	56	J6=I6	00000111	×	$X \wedge (Y \vee Z)$
3-5	54	J5=G5	00001101	=	$Z \wedge (X \vee Y)$
2-6	27	**J**	00000101		$Z \wedge X$
3-5	56	J6=I6	00000111		$X \wedge (Y \vee Z)$
3-5	51	J4=O6	00010101		$X \wedge (\neg Y \vee Z)$
3-5	56	J6=I6	00000111		$X \wedge (Y \vee Z)$
3-5	45	J3	00100101		$(X \wedge Z) \vee (\neg X \wedge Y \wedge \neg Z)$
3-5	56	J6=I6	00000111		$X \wedge (Y \vee Z)$
3-5	35	J2=H6	01000101		$Z \wedge (X \vee \neg Y)$
3-5	56	J6=I6	00000111		$X \wedge (Y \vee Z)$
3-5	20	J1	10000101		$(X \wedge Z) \vee (\neg X \wedge \neg Y \wedge \neg Z)$
3-5	54	J5=G5	00001101		$Z \wedge (X \vee Y)$
3-5	51	J4=O6	00010101		$X \wedge (\neg Y \vee Z)$
3-5	54	J5=G5	00001101		$Z \wedge (X \vee Y)$
3-5	45	J3	00100101		$(X \wedge Z) \vee (\neg X \wedge Y \wedge \neg Z)$
3-5	54	J5=G5	00001101		$Z \wedge (X \vee Y)$
3-5	35	J2=H6	01000101		$Z \wedge (X \vee \neg Y)$
3-5	54	J5=G5	00001101		$Z \wedge (X \vee Y)$
3-5	20	J1	10000101		$(X \wedge Z) \vee (\neg X \wedge \neg Y \wedge \neg Z)$
3-5	51	J4=O6	00010101		$X \wedge (\neg Y \vee Z)$
3-5	45	J3	00100101		$(X \wedge Z) \vee (\neg X \wedge Y \wedge \neg Z)$
3-5	51	J4=O6	00010101		$X \wedge (\neg Y \vee Z)$
3-5	35	J2=H6	01000101		$Z \wedge (X \vee \neg Y)$
3-5	51	J4=O6	00010101		$X \wedge (\neg Y \vee Z)$
3-5	20	J1	10000101		$(X \wedge Z) \vee (\neg X \wedge \neg Y \wedge \neg Z)$
3-5	45	J3	00100101		$(X \wedge Z) \vee (\neg X \wedge Y \wedge \neg Z)$
3-5	35	J2=H6	01000101		$Z \wedge (X \vee \neg Y)$
3-5	45	J3	00100101		$(X \wedge Z) \vee (\neg X \wedge Y \wedge \neg Z)$
3-5	20	J1	10000101		$(X \wedge Z) \vee (\neg X \wedge \neg Y \wedge \neg Z)$
3-5	35	J2=H6	01000101		$Z \wedge (X \vee \neg Y)$
3-5	20	J1	10000101		$(X \wedge Z) \vee (\neg X \wedge \neg Y \wedge \neg Z)$

Table 2.11 Derivations of **K**

Level	#	Symbol	ID		Traditional logical form
3-5	33	K6=G2	01001001	×	$Z \wedge (\neg X \vee Y)$
3-5	32	K5	01001010	=	$(\neg X \wedge Z) \vee (X \wedge Y \wedge \neg Z)$
2-6	10	**K**	01001000		$Z \wedge \neg X$
3-5	33	K6=G2	01001001		$Z \wedge (\neg X \vee Y)$
3-5	31	K4=H4	01001100		$Z \wedge (\neg X \vee \neg Y)$
3-5	33	K6=G2	01001001		$Z \wedge (\neg X \vee Y)$
3-5	27	K3	01011000		$(\neg X \wedge Z) \vee (X \wedge \neg Y \wedge \neg Z)$
3-5	33	K6=G2	01001001		$Z \wedge (\neg X \vee Y)$
3-5	23	K2=S2	01101000		$\neg X \wedge (Y \vee Z)$
3-5	33	K6=G2	01001001		$Z \wedge (\neg X \vee Y)$
3-5	3	K1=T3	11001000		$\neg X \wedge (\neg Y \vee Z)$
3-5	32	K5	01001010		$(\neg X \wedge Z) \vee (X \wedge Y \wedge \neg Z)$
3-5	31	K4=H4	01001100		$Z \wedge (\neg X \vee \neg Y)$
3-5	32	K5	01001010		$(\neg X \wedge Z) \vee (X \wedge Y \wedge \neg Z)$
3-5	27	K3	01011000		$(\neg X \wedge Z) \vee (X \wedge \neg Y \wedge \neg Z)$
3-5	32	K5	01001010		$(\neg X \wedge Z) \vee (X \wedge Y \wedge \neg Z)$
3-5	23	K2=S2	01101000		$\neg X \wedge (Y \vee Z)$
3-5	32	K5	01001010		$(\neg X \wedge Z) \vee (X \wedge Y \wedge \neg Z)$
3-5	3	K1=T3	11001000		$\neg X \wedge (\neg Y \vee Z)$
3-5	31	K4=H4	01001100		$Z \wedge (\neg X \vee \neg Y)$
3-5	27	K3	01011000		$(\neg X \wedge Z) \vee (X \wedge \neg Y \wedge \neg Z)$
3-5	31	K4=H4	01001100		$Z \wedge (\neg X \vee \neg Y)$
3-5	23	K2=S2	01101000		$\neg X \wedge (Y \vee Z)$
3-5	31	K4=H4	01001100		$Z \wedge (\neg X \vee \neg Y)$
3-5	3	K1=T3	11001000		$\neg X \wedge (\neg Y \vee Z)$
3-5	27	K3	01011000		$(\neg X \wedge Z) \vee (X \wedge \neg Y \wedge \neg Z)$
3-5	23	K2=S2	01101000		$\neg X \wedge (Y \vee Z)$
3-5	27	K3	01011000		$(\neg X \wedge Z) \vee (X \wedge \neg Y \wedge \neg Z)$
3-5	3	K1=T3	11001000		$\neg X \wedge (\neg Y \vee Z)$
3-5	23	K2=S2	01101000		$\neg X \wedge (Y \vee Z)$
3-5	3	K1=T3	11001000		$\neg X \wedge (\neg Y \vee Z)$

2.2 Derivation of Classical Inferences through Products

Suppose now that we wish to simulate mechanically the traditional inference called *Barbara*, that is, to derive the statement **C** from **A** and **B**. Our problem is how to single out mechanically the right combination of elementary statements that allows such an inference. As a matter of fact, abstractly speaking we have all the ($2\times2=4$) possible combinations, as shown in Table 2.12.

Table 2.12 *Barbara*

Derivations	Level	#	Symbol	ID		Traditional logical form
	7-1	3	A1=C1	11111011	×	$\neg X \vee Y \vee \neg Z$
1	7-1	4	B1=M1	11110111	=	$X \vee \neg Y \vee \neg Z$
	6-2	6	A1xM1	11111011		$Z \rightarrow (X \leftrightarrow Y)$
	7-1	5	A2	11101111	×	$\neg X \vee Y \vee Z$
2	7-1	4	B1=M1	11110111	=	$X \vee \neg Y \vee \neg Z$
	6-2	10	M1xA2	11100111		$\neg(\neg X \leftrightarrow Y \leftrightarrow Z)$
	7-1	3	A1=C1	11111011	×	$\neg X \vee Y \vee \neg Z$
3	7-1	7	B2=C2=L1	10111111	=	$X \vee Y \vee \neg Z$
	6-2	18	**C**	10111011		$Z \rightarrow Y$
	7-1	5	A2	11101111	×	$\neg X \vee Y \vee Z$
4	7-1	7	B2=L1	10111111	=	$X \vee Y \vee \neg Z$
	6-2	20	A2xL1	10101111		$\neg Y \rightarrow (X \leftrightarrow Z)$

As mentioned, a feature of this approach to logic is indeed that any possible sum or product (but also subtraction or division) among statements gives a result that is itself logically sound. However, I can reformulate our problem by saying that we wish to perform classical interference without thinking, that is, without knowing the logical rules that assist us in such cases. Therefore, although all the above derivations are logically sound, we are interested in picking up the third one in a way that also a machine could perform. This can be easily accomplished by satisfying two general requirements that hold for any derivation in the framework of the classical theory of derivation:

- The conclusion must pertain to the highest level allowed by the premises.
- The conclusion must be about two variables only and precisely those that are present in the premises only one time.

It is clear that any computer is able to follow these rules. Now, let us apply these rules to our case. About Requirement 1: since here the premises **A** and **B** both pertain to Level 6-2, also the conclusion must be of Level 6-2. This is ineffective here, since all conclusions pertain to Level 6-2. About Requirement 2: the conclusions of the first, second and fourth derivation must be discarded. It can be stressed that, concerning the last requirement especially, **C** is the easiest conclusion that is possible given A1, A2, B1, B2 as premises. Indeed, it is an implication of a single term by a single term or a disjunction between two single terms. Since its truth value only depends on two variables (Z and Y) out of three, it can be considered as equivalent to an information selection. Moreover, **C** is the most similar statement (from a structural point of view) to the premises **A** and **B**. This means that classical logic has precisely dealt with those statements that can be considered "natural" given certain premises. Another consideration is appropriate here. Any product as well as any sum of statements pertaining to the logical space gives a result that is not only logically sound but represents an equivalence. For instance, when we say that A1$\times$B2=**C**, the equality means a logical equivalence. Indeed, we have that

$$(\mathrm{A1} = \mathrm{C1}) \times (\mathrm{B2} = \mathrm{C2}) = (\mathrm{C1} \times \mathrm{C2}) = \mathbf{C}\,. \tag{2.1}$$

Now, logical equivalence can be interpreted in terms of both sufficient and necessary conditions. However, when going back to the main statements (**A, B, C**), the above equality (or logical equivalence) implies the result, i.e.

$$[(\mathrm{C1} \times \mathrm{A2}) \wedge (\mathrm{B1} \times \mathrm{C2})] \rightarrow (\mathrm{C1} \times \mathrm{C2})\,, \tag{2.2}$$

which is true (note that such an implication cannot be reversed). Then, the only admissible interpretation is in terms of sufficient conditions (in accordance with the classical theory of inferences). In other words, we are forced to the inference:

$$(\mathbf{A} \wedge \mathbf{B}) \rightarrow \mathbf{C}\,, \tag{2.3}$$

where if the conclusion is a logical consequence of the premises, this implication must be a tautology (by looking at the explicit forms of **A, B, C** the reader may easily verify that this is indeed the case). This is a general rule when dealing with inferences involving products: to be clear, any general result of the inference must be interpreted in terms of sufficient conditions. However, the true power of this

Table 2.13 *Celarent*

Derivations	Level	#	Symbol	ID		Traditional logical form
	7-1	1	E1= D1	11111110	×	$\neg X \vee \neg Y \vee \neg Z$
1	7-1	4	B1=D2=M1	11110111	=	$X \vee \neg Y \vee \neg Z$
	6-2	4	**D**	11110110		$Z \rightarrow \neg Y$
	7-1	2	E2	11111101	×	$\neg X \vee \neg Y \vee Z$
2	7-1	4	B1=M1	11110111	=	$X \vee \neg Y \vee \neg Z$
	6-2	5	E2xM1	11110101		$Y \rightarrow (X \leftrightarrow Z)$
	7-1	1	E1	11111110	×	$\neg X \vee \neg Y \vee \neg Z$
3	7-1	7	B2=L1	10111111	=	$X \vee Y \vee \neg Z$
	6-2	16	E1xL1	10111110		$Z \rightarrow \neg(X \leftrightarrow Y)$
	7-1	2	E2	11111101	×	$\neg X \vee \neg Y \vee Z$
4	7-1	7	B2=L1	10111111	=	$X \vee Y \vee \neg Z$
	6-2	17	E2xL1	10111101		$\neg(X \leftrightarrow Y \leftrightarrow \neg Z)$

language is that it also allows inferences when using the sum (what classically was not explored). In this case, as we shall see, we need to interpret the inference as expressing a necessary and not sufficient condition.

Similar considerations are true for the inference traditionally called *Celarent*, that is, the derivation of statement **D** from premises **E** and **B**. Again, Table 2.13 shows all possible derivations from a common set of premises (in this case E1, E2, B1, B2).

Again, I remark that the conclusion we are looking for (the first one) is the easiest given this set of 4 premises and also the most similar one to these premises. Moreover, D1=E1 and D2=B1. I shall take for granted that the previous proof of the soundness of the derivation of *Barbara* is also true in this case. As mentioned, a little bit more cumbersome is the derivation of **G** from premises **A** and **J**, which is traditionally called *Darii* (in this case we have $2 \times 6 = 12$ different combinations), as displayed in Table 2.14.

Derivations 2, 4, 8, 10, 12 need to be discarded because in conflict with the first requirement (the required level is indeed 2-6) whilst Statements 1, 5, 7, 9, 11 need to be discarded because in conflict with the second requirement. Only the third result remains which is what we looked for. I also remark that A1, J5=G5 and G share the

Table 2.14 *Darii*

Derivations	Level	#	Symbol	ID		Traditional logical form
	7-1	3	A1	11111011	×	$\neg X \vee Y \vee \neg Z$
1	3-5	56	J6=I6	00000111	=	$X \wedge (Y \vee Z)$
	2-6	28	**I**	00000011		$X \wedge Y$
	7-1	5	A2	11101111	×	$\neg X \vee Y \vee Z$
2	3-5	56	J6=I6	00000111	=	$X \wedge (Y \vee Z)$
	3-5	56	J6=I6	00000111		$X \wedge (Y \vee Z)$
	7-1	3	A1	11111011	×	$\neg X \vee Y \vee \neg Z$
3	3-5	54	J5=G5	00001101	=	$Z \wedge (X \vee Y)$
	2-6	25	**G**	00001001		$Z \wedge Y$
	7-1	5	A2	11101111	×	$\neg X \vee Y \vee Z$
4	3-5	54	J5=G5	00001101	=	$Z \wedge (X \vee Y)$
	3-5	54	J5=G5	00001101		$Z \wedge (X \vee Y)$
	7-1	3	A1	11111011	×	$\neg X \vee Y \vee \neg Z$
5	3-5	51	J4=O6	00010101	=	$X \wedge (\neg Y \vee Z)$
	2-6	22	O7+I8	00010001		$X \wedge (Y \leftrightarrow Z)$
	7-1	5	A2	11101111	×	$\neg X \vee Y \vee Z$
6	3-5	51	J4=O6	00010101	=	$X \wedge (\neg Y \vee Z)$
	2-6	27	**J**	00000101		$Z \wedge X$
	7-1	3	A1	11111011	×	$\neg X \vee Y \vee \neg Z$
7	3-5	45	J3	00100101	=	$(X \wedge Z) \vee (\neg X \wedge Y \wedge \neg Z)$
	2-6	18	S7+I8	00100001		$Y \wedge (X \leftrightarrow Z)$
	7-1	5	A2	11101111	×	$\neg X \vee Y \vee Z$
8	3-5	45	J3	00100101	=	$(X \wedge Z) \vee (\neg X \wedge Y \wedge \neg Z)$
	3-5	45	J3	00100101		$(X \wedge Z) \vee (\neg X \wedge Y \wedge \neg Z)$
	7-1	3	A1	11111011	×	$\neg X \vee Y \vee \neg Z$
9	3-5	35	J2=H6	01000101	=	$Z \wedge (X \vee \neg Y)$
	2-6	13	T8+I8	01000001		$Z \wedge (X \leftrightarrow Y)$
	7-1	5	A2	11101111	×	$\neg X \vee Y \vee Z$
10	3-5	35	J2=H6	01000101	=	$Z \wedge (X \vee \neg Y)$
	3-5	35	J2=H6	01000101		$Z \wedge (X \vee \neg Y)$
	7-1	3	A1	11111011	×	$\neg X \vee Y \vee \neg Z$
11	3-5	20	J1	10000101	=	$(X \wedge Z) \vee (\neg X \wedge \neg Y \wedge \neg Z)$
	2-6	7	T7+I8	10000001		$X \leftrightarrow Y \leftrightarrow Z$
	7-1	5	A2	11101111	×	$\neg X \vee Y \vee Z$
12	3-5	20	J1	10000101	=	$(X \wedge Z) \vee (\neg X \wedge \neg Y \wedge \neg Z)$
	3-5	20	J1	10000101		$(X \wedge Z) \vee (\neg X \wedge \neg Y \wedge \neg Z)$

values true for e and h. The reader may check that this is the case anytime we try to derive a particular statement from a general and a particular one.

It may be noted that all above statements pertaining to Level 2-6 that show some sum (like S7+I8) have a recurrent statement. In this case, it is I8. The reader may remark that this is common to all subsequent derivations having a particular premise. It is possible that the logical validity of this particular inference is questioned on the ground that the conclusion shares only one common statements with the premises (here, with J5) instead of two, as previously. Before dealing with this problem, let us consider another derivation involving particular premises, in particular let us consider the inference that is traditionally called *Ferio*, that is, the derivation of **H** from premises **E** and **J**:

Table 2.15 *Ferio*

Derivations	Level	#	Symbol	ID		Traditional logical form
	7-1	1	E1	11111110	×	$\neg X \vee \neg Y \vee \neg Z$
1	3-5	56	J6=I6	00000111	=	$X \wedge (Y \vee Z)$
	2-6	26	O8+I7	00000110		$X \wedge \neg(Y \leftrightarrow Z)$
	7-1	2	E2	11111101	×	$\neg X \vee \neg Y \vee Z$
2	3-5	56	J6=I6	00000111	=	$X \wedge (Y \vee Z)$
	2-6	27	**J**	00000101		$Z \wedge X$
	7-1	1	E1	11111110	×	$\neg X \vee \neg Y \vee \neg Z$
3	3-5	54	J5=G5	00001101	=	$Z \wedge (X \vee Y)$
	2-6	23	S8+O8	00001100		$Z \wedge \neg(X \leftrightarrow Y)$
	7-1	2	E2	11111101	×	$\neg X \vee \neg Y \vee Z$
4	3-5	54	J5=G5	00001101	=	$Z \wedge (X \vee Y)$
	3-5	54	J5=G5	00001101		$Z \wedge (X \vee Y)$
	7-1	1	E1	11111110	×	$\neg X \vee \neg Y \vee \neg Z$
5	3-5	51	J4=O6	00010101	=	$X \wedge (\neg Y \vee Z)$
	2-6	20	**O**	00010100		$X \wedge \neg Y$
	7-1	2	E2	11111101	×	$\neg X \vee \neg Y \vee Z$
6	3-5	51	J4=O6	00010101	=	$X \wedge (\neg Y \vee Z)$
	3-5	51	J4=O6	00010101		$X \wedge (\neg Y \vee Z)$
	7-1	1	E1	11111110	×	$\neg X \vee \neg Y \vee \neg Z$
7	3-5	45	J3	00100101	=	$(X \wedge Z) \vee (\neg X \wedge Y \wedge \neg Z)$
	2-6	16	S7+O8	00100100		$X \leftrightarrow \neg Y \leftrightarrow Z$

(*Continued*)

Table 2.15 (*Continued*)

Derivations	Level	#	Symbol	ID		Traditional logical form
	7-1	2	E2	11111101	×	$\neg X \vee \neg Y \vee Z$
8	3-5	45	J3	00100101	=	$(X \wedge Z) \vee (\neg X \wedge Y \wedge \neg Z)$
	3-5	45	J3	00100101		$(X \wedge Z) \vee (\neg X \wedge Y \wedge \neg Z)$
	7-1	1	E1	11111110	×	$\neg X \vee \neg Y \vee \neg Z$
9	3-5	35	J2=H6	01000101	=	$Z \wedge (X \vee \neg Y)$
	2-6	11	**H**	01000100		$Z \wedge \neg Y$
	7-1	2	E2	11111101	×	$\neg X \vee \neg Y \vee Z$
10	3-5	35	J2=H6	01000101	=	$Z \wedge (X \vee \neg Y)$
	3-5	35	J2=H6	01000101		$Z \wedge (X \vee \neg Y)$
	7-1	1	E1	11111110	×	$\neg X \vee \neg Y \vee \neg Z$
11	3-5	20	J1	10000101	=	$(X \wedge Z) \vee (\neg X \wedge \neg Y \wedge \neg Z)$
	2-6	5	T7+O8	10000100		$\neg Y \wedge (X \leftrightarrow Z)$
	7-1	2	E2	11111101	×	$\neg X \vee \neg Y \vee Z$
12	3-5	20	J1	10000101	=	$(X \wedge Z) \vee (\neg X \wedge \neg Y \wedge \neg Z)$
	3-5	20	J1	10000101		$(X \wedge Z) \vee (\neg X \wedge \neg Y \wedge \neg Z)$

We can remark that we need to discard the Derivations 4, 8, 10 and 12 because in conflict with Requirement 1 (the requested level is 2-6) and Derivations 1, 2, 3, 5, 6, 11 and 12 because in conflict with Requirement 2, so that what remains is Derivation 9, what we were looking for. There is, however, the mentioned difficulty that needs to be dealt with now. In the case of universal derivations like *Barbara*, the result is certainly logically sound since it is finally an identity (or a logical equivalence) as I have shown. However, what ensures us that also for inferences involving particular statements we are authorized e.g. (in the case of *Ferio*) to derive **H** from **E** and **J** given the identity displayed in Derivation 9, i.e. (E1 × J2=H6) = **H**? The problem is represented here by the fact that we can indeed establish an identity among statements (i.e. J2 and H6) generating (together with another statement each) the two particular statements involved in this derivation (i.e. **J** and **H**), but we cannot establish this kind of connection between any of the other statements of the LGS of **H** (i.e. H1, H2, H3, H4, H5) on the one hand, and the premise E2 (which is the other statement of the LGS of **E**) that is involved in Derivation 9, on the other: the reader may see that a similar situation also occurs for *Darii*. In order to prove that the above derivation is logically sound, we need to rewrite the previous identity (i.e. Derivation 9) as follows (I recall that any couple of statements of the LGS of a given bold statement (like **E**, **H**, **J**) can give rise to the latter):

$$\mathrm{E1} \times (\mathrm{J2} = \mathrm{H6}) = (\mathrm{H3} = \mathrm{O2}) \times (\mathrm{H6} = \mathrm{J2})\,. \tag{2.4}$$

Now, we may remark that the fifth derivation above gives an expected (and apparently not pertinent) result, namely the statement **O**. Far from being irrelevant, this is instead crucial for our proof. Indeed, Derivation 5 can be rewritten as:

$$\mathrm{E1} \times (\mathrm{J4} = \mathrm{O6}) = (\mathrm{O2} = \mathrm{H3}) \times (\mathrm{O6} = \mathrm{J4})\,. \tag{2.5}$$

These two identities (or logical equivalences) allow us to write (in ordinary logical language) the following two propositions, which are logical consequences of the two premises of *Ferio* and we can therefore consider as antecedents in our proof:

$$(\mathrm{E1} \wedge \mathrm{J2}) \rightarrow (\mathrm{H3} \wedge \mathrm{J2}) \text{ and } (\mathrm{E1} \wedge \mathrm{J4}) \rightarrow (\mathrm{H3} \wedge \mathrm{J4})\,. \tag{2.6}$$

Now, Derivation 2 above:

$$(\mathrm{E2} \times \mathrm{J6}) = \mathbf{J} = (\mathrm{J4} \times \mathrm{J6})\,, \tag{2.7}$$

allows the proof context-substitution of E2 by J4 and therefore also the following implication:

$$(\mathrm{E1} \wedge \mathrm{J2} \wedge \mathrm{J4} = \mathrm{E2}) \rightarrow (\mathrm{H6} = \mathrm{J2} \wedge \mathrm{H3})\,, \tag{2.8}$$

which is a reformulation of the desired conclusion, i.e. that **H** logically follows from both **E** and **J**. Now, it is easy to verify that the latter expression logically follows (i.e. is tautologically implied by) the two premises above, i.e.

$$\begin{aligned}&\{[(\mathrm{E1} \wedge \mathrm{J2}) \rightarrow (\mathrm{H3} \wedge \mathrm{J2})] \wedge [(\mathrm{E1} \wedge \mathrm{J4}) \rightarrow (\mathrm{H3} \wedge \mathrm{J4})]\}\\ &\qquad \rightarrow [(\mathrm{E1} \wedge \mathrm{J2} \wedge \mathrm{J4}) \rightarrow (\mathrm{H6} \wedge \mathrm{H3})]\,.\end{aligned} \tag{2.9}$$

Therefore, in the following I take for granted that all derivations are sound and focus only on the two derivation rules as valid also for particular inferences. At the end, I shall come back to this issue when summarizing the whole matter.

In this way we have concluded the derivations of the first set of 4 classical inferences. They correspond to the so-called Aristotelian first figure. However, in order to distinguish the actual procedure, which is more general, I prefer to use the word “first group”. Now, let us pass to the second group and consider the inference that is classically known as *Baroco*, that is, the inference of **K** from **A** and **H**:

Table 2.16 *Baroco*

Derivations	Level	#	Symbol	ID		Traditional logical form
	7-1	3	A1	11111011	×	$\neg X \vee Y \vee \neg Z$
1	3-5	35	H6=J2	01000101	=	$Z \wedge (X \vee \neg Y)$
	2-6	13	T8+I8	01000001		$Z \wedge (X \leftrightarrow Y)$
	7-1	5	A2	11101111	×	$\neg X \vee Y \vee Z$
2	3-5	35	H6=J2	01000101	=	$Z \wedge (X \vee \neg Y)$
	3-5	35	H6=J2	01000101		$Z \wedge (X \vee \neg Y)$
	7-1	3	A1	11111011	×	$\neg X \vee Y \vee \neg Z$
3	3-5	34	H5	01000110	=	$(\neg Y \wedge Z) \vee (X \wedge Y \wedge \neg Z)$
	2-6	12	T8+I7	01000010		$X \leftrightarrow Y \leftrightarrow \neg Z$
	7-1	5	A2	11101111	×	$\neg X \vee Y \vee Z$
4	3-5	34	H5	01000110	=	$(\neg Y \wedge Z) \vee (X \wedge Y \wedge \neg Z)$
	3-5	34	H5	01000110		$(\neg Y \wedge Z) \vee (X \wedge Y \wedge \neg Z)$
	7-1	3	A1	11111011	×	$\neg X \vee Y \vee \neg Z$
5	3-5	31	H4=K4	01001100	=	$Z \wedge (\neg X \vee \neg Y)$
	2-6	10	**K**	01001000		$Z \wedge \neg X$
	7-1	5	A2	11101111	×	$\neg X \vee Y \vee Z$
6	3-5	31	H4=K4	01001100	=	$Z \wedge (\neg X \vee \neg Y)$
	3-5	31	H4=K4	01001100		$Z \wedge (\neg X \vee \neg Y)$
	7-1	3	A1	11111011	×	$\neg X \vee Y \vee \neg Z$
7	3-5	28	H3=O2	01010100	=	$\neg Y \wedge (X \vee Z)$
	2-6	9	T8+O7	01010000		$\neg Y \wedge \neg(X \leftrightarrow Z)$
	7-1	5	A2	11101111	×	$\neg X \vee Y \vee Z$
8	3-5	28	H3=O2	01010100		$\neg Y \wedge (X \vee Z)$
	2-6	11	**H**	01000100		$Z \wedge \neg Y$
	7-1	3	A1	11111011	×	$\neg X \vee Y \vee \neg Z$
9	3-5	24	H2	01100100	=	$(\neg Y \wedge Z) \vee (\neg X \wedge Y \wedge \neg Z)$
	2-6	8	T8+S7	01100000		$\neg X \wedge \neg(Y \leftrightarrow Z)$
	7-1	5	A2	11101111	×	$\neg X \vee Y \vee Z$
10	3-5	24	H2	01100100	=	$(\neg Y \wedge Z) \vee (\neg X \wedge Y \wedge \neg Z)$
	3-5	24	H2	01100100		$(\neg Y \wedge Z) \vee (\neg X \wedge Y \wedge \neg Z)$
	7-1	3	A1	11111011	×	$\neg X \vee Y \vee \neg Z$
11	3-5	4	H1=T4	11000100	=	$\neg Y \wedge (\neg X \vee Z)$
	2-6	1	**T**	11000000		$\neg X \wedge \neg Y$
	7-1	5	A2	11101111	×	$\neg X \vee Y \vee Z$
12	3-5	4	H1=T4	11000100	=	$\neg Y \wedge (\neg X \vee Z)$
	3-5	4	H1=T4	11000100		$\neg Y \wedge (\neg X \vee Z)$

We have to discard Derivations 2, 4, 6, 10, 12 because in conflict with Requirement 1 (the requested level is 2-6) and Derivations 1, 3, 7, 8 9, 11 because in conflict with Requirement 2, so that only Derivation 5 remains. Now, I shall derive what is traditionally called *Festino*, i.e. statement **K** from premises **E** and **G**:

Table 2.17 *Festino*

Derivations	Level	#	Symbol	ID		Traditional logical form
	7-1	1	E1	11111110	×	$\neg X \vee \neg Y \vee \neg Z$
1	3-5	55	G6=I5	00001011	=	$Y \wedge (X \vee Z)$
	2-6	24	S8+I7	00001010		$Y \wedge \neg(X \leftrightarrow Z)$
	7-1	2	E2	11111101	×	$\neg X \vee \neg Y \vee Z$
2	3-5	55	G6=I5	00001011		$Y \wedge (X \vee Z)$
	2-6	25	**G**	00001001		$Z \wedge Y$
	7-1	1	E1	11111110	×	$\neg X \vee \neg Y \vee \neg Z$
3	3-5	54	G5=J5	00001101	=	$Z \wedge (X \vee Y)$
	2-6	23	S8+O8	00001100		$Z \wedge \neg(X \leftrightarrow Y)$
	7-1	2	E2	11111101	×	$\neg X \vee \neg Y \vee Z$
4	3-5	54	G5=J5	00001101	=	$Z \wedge (X \vee Y)$
	3-5	54	G5=J5	00001101	=	$Z \wedge (X \vee Y)$
	7-1	1	E1	11111110	×	$\neg X \vee \neg Y \vee \neg Z$
5	3-5	49	G4	00011001	=	$(Y \wedge Z) \vee (X \wedge \neg Y \wedge \neg Z)$
	2-6	19	O7+S8	00011000		$\neg X \leftrightarrow Y \leftrightarrow Z$
	7-1	2	E2	11111101	×	$\neg X \vee \neg Y \vee Z$
6	3-5	49	G4	00011001	=	$(Y \wedge Z) \vee (X \wedge \neg Y \wedge \neg Z)$
	3-5	49	G4	00011001		$(Y \wedge Z) \vee (X \wedge \neg Y \wedge \neg Z)$
	7-1	1	E1	11111110	×	$\neg X \vee \neg Y \vee \neg Z$
7	3-5	43	G3=S6	00101001	=	$Y \wedge (\neg X \vee Z)$
	2-6	15	**S**	00101000		$\neg X \wedge Y$
	7-1	2	E2	11111101	×	$\neg X \vee \neg Y \vee Z$
8	3-5	43	G3=S6	00101001	=	$Y \wedge (\neg X \vee Z)$
	3-5	43	G3=S6	00101001		$Y \wedge (\neg X \vee Z)$
	7-1	1	E1	11111110	×	$\neg X \vee \neg Y \vee \neg Z$
9	3-5	33	G2=K6	01001001	=	$Z \wedge (\neg X \vee Y)$
	2-6	10	**K**	01001000		$Z \wedge \neg X$
	7-1	2	E2	11111101	×	$\neg X \vee \neg Y \vee Z$
10	3-5	33	G2=K6	01001001	=	$Z \wedge (\neg X \vee Y)$
	3-5	33	G2=K6	01001001		$Z \wedge (\neg X \vee Y)$

(*Continued*)

Table 2.17 (*Continued*)

Derivations	Level	#	Symbol	ID		Traditional logical form
	7-1	1	E1	11111110	×	$\neg X \vee \neg Y \vee \neg Z$
11	3-5	18	G1	10001001	=	$(Y \wedge Z) \vee (\neg X \wedge \neg Y \wedge \neg Z)$
	2-6	4	T7+S8	10001000		$\neg X \wedge (Y \leftrightarrow Z)$
	7-1	2	E2	11111101	×	$\neg X \vee \neg Y \vee Z$
12	3-5	18	G1	10001001	=	$(Y \wedge Z) \vee (\neg X \wedge \neg Y \wedge \neg Z)$
	3-5	18	G1	10001001		$(Y \wedge Z) \vee (\neg X \wedge \neg Y \wedge \neg Z)$

Derivations 4, 6, 8, 10 and 12 conflict with Requirement 1 whilst Derivations 1, 2, 3, 5, 7, 11 conflict with Requirement 2 so that only Derivation 9 remains, as expected. The derivation of *Camestres*, that is, of **F** from **A** and **D**, is much easier since only universal statements are involved:

Table 2.18 *Camestres*

Derivations	Level	#	Symbol	ID		Traditional logical form
	7-1	3	A1=F2	11111011	×	$\neg X \vee Y \vee \neg Z$
1	7-1	1	D1=F1=E1	11111110	=	$\neg X \vee \neg Y \vee \neg Z$
	6-2	2	**F**	11111010		$Z \rightarrow \neg X$
	7-1	5	A2	11101111	×	$\neg X \vee Y \vee Z$
2	7-1	1	D1=E1	11111110	=	$\neg X \vee \neg Y \vee \neg Z$
	6-2	7	E1xA2	11101110		$X \rightarrow \neg(Y \leftrightarrow Z)$
	7-1	3	A1	11111011	×	$\neg X \vee Y \vee \neg Z$
3	7-1	4	D2=M1	11110111	=	$X \vee \neg Y \vee \neg Z$
	6-2	6	A1xM1	11110011		$Z \rightarrow (X \leftrightarrow Y)$
	7-1	5	A2	11101111	×	$\neg X \vee Y \vee Z$
4	7-1	4	D2=M1	11110111	=	$X \vee \neg Y \vee \neg Z$
	6-2	10	M1xA2	11100111		$\neg(\neg X \leftrightarrow Y \leftrightarrow Z)$

All the derivation but the first one need to be discarded because in conflict with Requirement 2. Also the derivation of *Cesare*, that is, of **F** from **E** and **C**, is relative easy:

Table 2.19 *Cesare*

Derivations	Level	#	Symbol	ID		Traditional logical form
	7-1	1	E1=F1	11111110	×	$\neg X \vee \neg Y \vee \neg Z$
1	7-1	3	C1=F2=A1	11111011	=	$\neg X \vee Y \vee \neg Z$
	6-2	2	**F**	11111010		$Z \to \neg X$
	7-1	2	E2	11111101	×	$\neg X \vee \neg Y \vee Z$
2	7-1	3	C1=A1	11111011	=	$\neg X \vee Y \vee \neg Z$
	6-2	3	E2xA1	11111001		$X \to (Y \leftrightarrow Z)$
	7-1	1	E1	11111110	×	$\neg X \vee \neg Y \vee \neg Z$
3	7-1	7	C2=L1	10111111	=	$X \vee Y \vee \neg Z$
	6-2	16	E1xL1	10111110		$Z \to \neg(X \leftrightarrow Y)$
	7-1	2	E2	11111101	×	$\neg X \vee \neg Y \vee Z$
4	7-1	7	C2=L1	10111111	=	$X \vee Y \vee \neg Z$
	6-2	17	E2xL1	10111101		$\neg(X \leftrightarrow Y \leftrightarrow \neg Z)$

It is clear that Derivations 2, 3, and 4 conflict with Requirement 2. This completes the derivation of the second group of classical inferences. Now, I shall deal with the third group and start with the derivation of *Bocardo*, i.e. of Statement **O** from **H** and **B**:

Table 2.20 *Bocardo*

Derivations	Level	#	Symbol	ID		Traditional logical form
	3-5	35	H6=J2	01000101	×	$Z \wedge (X \vee \neg Y)$
1	7-1	4	B1=M1	11110111	=	$X \vee \neg Y \vee \neg Z$
	3-5	35	H6=J2	01000101		$Z \wedge (X \vee \neg Y)$
	3-5	35	H6=J2	01000101	×	$Z \wedge (X \vee \neg Y)$
2	7-1	7	B2=L1	10111111	=	$X \vee Y \vee \neg Z$
	2-6	27	**J**	00000101		$Z \wedge X$
	3-5	34	H5	01000110	×	$(\neg Y \wedge Z) \vee (X \wedge Y \wedge \neg Z)$
3	7-1	4	B1=M1	11110111	=	$X \vee \neg Y \vee \neg Z$
	3-5	34	H5	01000110		$(\neg Y \wedge Z) \vee (X \wedge Y \wedge \neg Z)$
	3-5	34	H5	01000110	×	$(\neg Y \wedge Z) \vee (X \wedge Y \wedge \neg Z)$
4	7-1	7	B2=L1	10111111	=	$X \vee Y \vee \neg Z$
	2-6	26	O8+I7	00000110		$X \wedge \neg(Y \leftrightarrow Z)$
	3-5	31	H4=K4	01001100	×	$Z \wedge (\neg X \vee \neg Y)$
5	7-1	4	B1=M1	11110111	=	$X \vee \neg Y \vee \neg Z$
	2-6	11	**H**	01000100		$Z \wedge \neg Y$

(*Continued*)

Table 2.20 (*Continued*)

Derivations	Level	#	Symbol	ID		Traditional logical form
	3-5	31	H4=K4	01001100	×	$Z \wedge (\neg X \vee \neg Y)$
6	7-1	7	B2=L1	10111111	=	$X \vee Y \vee \neg Z$
	2-6	23	S8+O8	00001100		$Z \wedge \neg(X \leftrightarrow Y)$
	3-5	28	H3=O2	01010100	×	$\neg Y \wedge (X \vee Z)$
7	7-1	4	B1=M1	11110111	=	$X \vee \neg Y \vee \neg Z$
	3-5	28	H3=O2	01010100		$\neg Y \wedge (X \vee Z)$
	3-5	28	H3=O2	01010100	×	$\neg Y \wedge (X \vee Z)$
8	7-1	7	B2=L1	10111111	=	$X \vee Y \vee \neg Z$
	2-6	20	**O**	00010100		$X \wedge \neg Y$
	3-5	24	H2	01100100	×	$(\neg Y \wedge Z) \vee (\neg X \wedge Y \wedge \neg Z)$
9	7-1	4	B1=M1	11110111	=	$X \vee \neg Y \vee \neg Z$
	3-5	24	H2	01100100		$(\neg Y \wedge Z) \vee (\neg X \wedge Y \wedge \neg Z)$
	3-5	24	H2	01100100	×	$(\neg Y \wedge Z) \vee (\neg X \wedge Y \wedge \neg Z)$
10	7-1	7	B2=L1	10111111	=	$X \vee Y \vee \neg Z$
	2-6	16	S7+O8	00100100		$X \leftrightarrow \neg Y \leftrightarrow Z$
	3-5	4	H1=T4	11000100	×	$\neg Y \wedge (\neg X \vee Z)$
11	7-1	4	B1=M1	11110111	=	$X \vee \neg Y \vee \neg Z$
	3-5	4	H1=T4	11000100		$\neg Y \wedge (\neg X \vee Z)$
	3-5	4	H1=T4	11000100	×	$\neg Y \wedge (\neg X \vee Z)$
12	7-1	7	B2=L1	10111111	=	$X \vee Y \vee \neg Z$
	2-6	5	T7+O8	10000100		$\neg Y \wedge (X \leftrightarrow Z)$

Derivations 1, 3, 7, 9 and 11 conflict with Requirement 1 whilst Derivations 2, 4, 5, 6, 10 and 12 conflict with Requirement 2, so that Derivation 8 remains. The inference called *Disamis* is the derivation of **I** from **G** and **B**:

Table 2.21 *Disamis*

Derivations	Level	#	Symbol	ID		Traditional logical form
	3-5	55	G6=I5	00001011	×	$Y \wedge (X \vee Z)$
1	7-1	4	B1=M1	11110111	=	$X \vee \neg Y \vee \neg Z$
	2-6	28	**I**	00000011		$X \wedge Y$
	3-5	55	G6=I5	00001011	×	$Y \wedge (X \vee Z)$
2	7-1	7	B2=L1	10111111	=	$X \vee Y \vee \neg Z$
	3-5	55	G6=I5	00001011		$Y \wedge (X \vee Z)$

Derivations	Level	#	Symbol	ID		Traditional logical form
	3-5	54	G5=J5	00001101	×	$Z \wedge (X \vee Y)$
3	7-1	4	B1=M1	11110111	=	$X \vee \neg Y \vee \neg Z$
	2-6	27	**J**	00000101		$Z \wedge X$
	3-5	54	G5=J5	00001101	×	$Z \wedge (X \vee Y)$
4	7-1	7	B2=L1	10111111	=	$X \vee Y \vee \neg Z$
	3-5	54	G5=J5	00001101		$Z \wedge (X \vee Y)$
	3-5	49	G4	00011001	×	$(Y \wedge Z) \vee (X \wedge \neg Y \wedge \neg Z)$
5	7-1	4	B1=M1	11110111	=	$X \vee \neg Y \vee \neg Z$
	2-6	22	O7+I8	00010001		$X \wedge (Y \leftrightarrow Z)$
	3-5	49	G4	00011001	×	$(Y \wedge Z) \vee (X \wedge \neg Y \wedge \neg Z)$
6	7-1	7	B2=L1	10111111	=	$X \vee Y \vee \neg Z$
	3-5	49	G4	00011001		$(Y \wedge Z) \vee (X \wedge \neg Y \wedge \neg Z)$
	3-5	43	G3=S6	00101001	×	$Y \wedge (\neg X \vee Z)$
7	7-1	4	B1=M1	11110111	=	$X \vee \neg Y \vee \neg Z$
	2-6	18	S7+I8	00100001		$Y \wedge (X \leftrightarrow Z)$
	3-5	43	G3=S6	00101001	×	$Y \wedge (\neg X \vee Z)$
8	7-1	7	B2=L1	10111111	=	$X \vee Y \vee \neg Z$
	3-5	43	G3=S6	00101001		$Y \wedge (\neg X \vee Z)$
	3-5	33	G2=K6	01001001	×	$Z \wedge (\neg X \vee Y)$
9	7-1	4	B1=M1	11110111	=	$X \vee \neg Y \vee \neg Z$
	2-6	13	T8+I8	01000001		$Z \wedge (X \leftrightarrow Y)$
	3-5	33	G2=K6	01001001	×	$Z \wedge (\neg X \vee Y)$
10	7-1	7	B2=L1	10111111	=	$X \vee Y \vee \neg Z$
	2-6	25	**G**	00001001		$Z \wedge Y$
	3-5	18	G1	10001001	×	$(Y \wedge Z) \vee (\neg X \wedge \neg Y \wedge \neg Z)$
11	7-1	4	B1=M1	11110111	=	$X \vee \neg Y \vee \neg Z$
	2-6	7	T7+I8	10000001		$X \leftrightarrow Y \leftrightarrow Z$
	3-5	18	G1	10001001	×	$(Y \wedge Z) \vee (\neg X \wedge \neg Y \wedge \neg Z)$
12	7-1	7	B2=L1	10111111	=	$X \vee Y \vee \neg Z$
	3-5	18	G1	10001001		$(Y \wedge Z) \vee (\neg X \wedge \neg Y \wedge \neg Z)$

Derivations 2, 4, 6, 8, and 12 conflict with Requirement 1 whilst Derivations 3, 5, 7, 9, 10, and 11 conflict with Requirement 2, so that only Derivation 1 satisfies both. The classical inference known as *Ferison* is the derivation of **O** from **D** and **J**:

Table 2.22 *Ferison*

Derivations	Level	#	Symbol	ID		Traditional logical form
	7-1	1	D1=E1	11111110	×	$\neg X \vee \neg Y \vee \neg Z$
1	3-5	56	J6=I6	00000111	=	$X \wedge (Y \vee Z)$
	2-6	26	O8+I7	00000110		$X \wedge \neg(Y \leftrightarrow Z)$
	7-1	4	D2=M1	11110111	×	$X \vee \neg Y \vee \neg Z$
2	3-5	56	J6=I6	00000111	=	$X \wedge (Y \vee Z)$
	3-5	56	J6=I6	00000111		$X \wedge (Y \vee Z)$
	7-1	1	D1=E1	11111110	×	$\neg X \vee \neg Y \vee \neg Z$
3	3-5	54	J5=G5	00001101	=	$Z \wedge (X \vee Y)$
	2-6	23	S8+O8	00001100		$Z \wedge \neg(X \leftrightarrow Y)$
	7-1	4	D2=M1	11110111	×	$X \vee \neg Y \vee \neg Z$
4	3-5	54	J5=G5	00001101	=	$Z \wedge (X \vee Y)$
	2-6	27	**J**	00000101		$Z \wedge X$
	7-1	1	D1=E1	11111110	×	$\neg X \vee \neg Y \vee \neg Z$
5	3-5	51	J4=O6	00010101	=	$X \wedge (\neg Y \vee Z)$
	2-6	20	**O**	00010100		$X \wedge \neg Y$
	7-1	4	D2=M1	11110111	×	$X \vee \neg Y \vee \neg Z$
6	3-5	51	J4=O6	00010101	=	$X \wedge (\neg Y \vee Z)$
	3-5	51	J4=O6	00010101		$X \wedge (\neg Y \vee Z)$
	7-1	1	D1=E1	11111110	×	$\neg X \vee \neg Y \vee \neg Z$
7	3-5	45	J3	00100101	=	$(X \wedge Z) \vee (\neg X \wedge Y \wedge \neg Z)$
	2-6	16	S7+O8	00100100		$X \leftrightarrow \neg Y \leftrightarrow Z$
	7-1	4	D2=M1	11110111	×	$X \vee \neg Y \vee \neg Z$
8	3-5	45	J3	00100101	=	$(X \wedge Z) \vee (\neg X \wedge Y \wedge \neg Z)$
	3-5	45	J3	00100101		$(X \wedge Z) \vee (\neg X \wedge Y \wedge \neg Z)$
	7-1	1	D1=E1	11111110	×	$\neg X \vee \neg Y \vee \neg Z$
9	3-5	35	J2=H6	01000101	=	$Z \wedge (X \vee \neg Y)$
	2-6	11	**H**	01000100		$Z \wedge \neg Y$
	7-1	4	D2=M1	11110111	×	$X \vee \neg Y \vee \neg Z$
10	3-5	35	J2=H6	01000101	=	$Z \wedge (X \vee \neg Y)$
	3-5	35	J2=H6	01000101		$Z \wedge (X \vee \neg Y)$
	7-1	1	D1=E1	11111110	×	$\neg X \vee \neg Y \vee \neg Z$
11	3-5	20	J1	10000101	=	$(X \wedge Z) \vee (\neg X \wedge \neg Y \wedge \neg Z)$
	2-6	5	T7+O8	10000100		$\neg Y \wedge (X \leftrightarrow Z)$
	7-1	4	D2=M1	11110111	×	$X \vee \neg Y \vee \neg Z$
12	3-5	20	J1	10000101	=	$(X \wedge Z) \vee (\neg X \wedge \neg Y \wedge \neg Z)$
	3-5	20	J1	10000101		$(X \wedge Z) \vee (\neg X \wedge \neg Y \wedge \neg Z)$

Derivations 2, 6, 8, 10, and 12 conflict with Requirement 1 whilst Derivations 1, 3, 4, 7, 9 and 11 conflict with Requirement 2, so that only Derivation 5 satisfies both. Finally, *Datisi* consists in the derivation of **I** from **C** and **J**:

Table 2.23 *Datisi*

Derivations	Level	#	Symbol	ID		Traditional logical form
	7-1	3	C1=A1	11111011	×	$\neg X \vee Y \vee \neg Z$
1	3-5	56	J6=I6	00000111	=	$X \wedge (Y \vee Z)$
	2-6	28	**I**	00000011		$X \wedge Y$
	7-1	7	C2=L1	10111111	×	$X \vee Y \vee \neg Z$
2	3-5	56	J6=I6	00000111	=	$X \wedge (Y \vee Z)$
	3-5	56	J6=I6	00000111		$X \wedge (Y \vee Z)$
	7-1	3	C1=A1	11111011	×	$\neg X \vee Y \vee \neg Z$
3	3-5	54	J5=G5	00001101	=	$Z \wedge (X \vee Y)$
	2-6	25	**G**	00001001		$Z \wedge Y$
	7-1	7	C2=L1	10111111	×	$X \vee Y \vee \neg Z$
4	3-5	54	J5=G5	00001101	=	$Z \wedge (X \vee Y)$
	3-5	54	J5=G5	00001101		$Z \wedge (X \vee Y)$
	7-1	3	C1=A1	11111011	×	$\neg X \vee Y \vee \neg Z$
5	3-5	51	J4=O6	00010101	=	$X \wedge (\neg Y \vee Z)$
	2-6	22	O7+I8	00010001		$X \wedge (Y \leftrightarrow Z)$
	7-1	7	C2=L1	10111111	×	$X \vee Y \vee \neg Z$
6	3-5	51	J4=O6	00010101	=	$X \wedge (\neg Y \vee Z)$
	3-5	51	J4=O6	00010101		$X \wedge (\neg Y \vee Z)$
	7-1	3	C1=A1	11111011	×	$\neg X \vee Y \vee \neg Z$
7	3-5	45	J3	00100101	=	$(X \wedge Z) \vee (\neg X \wedge Y \wedge \neg Z)$
	2-6	18	S7+I8	00100001		$Y \wedge (X \leftrightarrow Z)$
	7-1	7	C2=L1	10111111	×	$X \vee Y \vee \neg Z$
8	3-5	45	J3	00100101	=	$(X \wedge Z) \vee (\neg X \wedge Y \wedge \neg Z)$
	3-5	45	J3	00100101		$(X \wedge Z) \vee (\neg X \wedge Y \wedge \neg Z)$
	7-1	3	C1=A1	11111011	×	$\neg X \vee Y \vee \neg Z$
9	3-5	35	J2=H6	01000101	=	$Z \wedge (X \vee \neg Y)$
	2-6	13	T8+I8	01000001		$Z \wedge (X \leftrightarrow Y)$
	7-1	7	C2=L1	10111111	×	$X \vee Y \vee \neg Z$
10	3-5	35	J2=H6	01000101	=	$Z \wedge (X \vee \neg Y)$
	2-6	27	**J**	00000101		$Z \wedge X$
	7-1	3	C1=A1	11111011	×	$\neg X \vee Y \vee \neg Z$
11	3-5	20	J1	10000101	=	$(X \wedge Z) \vee (\neg X \wedge \neg Y \wedge \neg Z)$
	2-6	7	T7+I8	10000001		$X \leftrightarrow Y \leftrightarrow Z$
	7-1	7	C2=L1	10111111	×	$X \vee Y \vee \neg Z$
12	3-5	20	J1	10000101	=	$(X \wedge Z) \vee (\neg X \wedge \neg Y \wedge \neg Z)$
	3-5	20	J1	10000101		$(X \wedge Z) \vee (\neg X \wedge \neg Y \wedge \neg Z)$

Derivations 2, 4, 6, 8, and 12 conflict with Requirement 1 whilst Derivations 3, 5, 7, 9, 10, and 11 conflict with Requirement 2, so that only the first derivation satisfies both requirements. We have completed the derivation also of the third group and therefore of all classical inferences.

2.3 Extension of Classical Inferences through Products

Traditionally, inferences in which the subject of statement is considered as no-existent or in a negative form have been discarded (see Section 1.2). Actually, the logical form of these latter statements is quite similar to their classical counterparts and one only needs a trivial substitution to obtain the other forms, as it was examined in (*De Morgan*, 1847, Chapter IV)—see also (Peirce, 1880, pp. 170 ff.). For instance, by replacing X with $\neg X$ we obtain $\neg X \rightarrow Y$ instead of $X \rightarrow Y$. I have also remarked (see Fig. 1.5) that a complete set of statements is necessary if we need to deal in a proper way with their relations. This is still more important if we like to have a full mechanization of logical derivations. Then, in the following I shall introduce the following universal statements that are the counterparts of **A, E, B, C, D, F** (see Tables 2.1–2.5).

Table 2.24 L, M, N, P, Q, R

Level	#	Symbol	ID		Traditional logical form
7-1	7	L1	10111111	×	$X \vee Y \vee \neg Z$
7-1	8	L2	01111111	=	$X \vee Y \vee Z$
6-2	28	**L**	00111111		$\neg X \rightarrow Y$
7-1	4	M1	11110111	×	$X \vee \neg Y \vee \neg Z$
7-1	6	M2	11011111	=	$X \vee \neg Y \vee Z$
6-2	14	**M**	11010111		$\neg X \rightarrow \neg Y$
7-1	5	N1=A2	11101111	×	$\neg X \vee Y \vee Z$
7-1	8	N2=L2	01111111	=	$X \vee Y \vee Z$
6-2	26	**N**	01101111		$\neg Z \rightarrow Y$
7-1	2	P1=E2	11111101	×	$\neg X \vee \neg Y \vee Z$
7-1	6	P2=M2	11011111	=	$X \vee \neg Y \vee Z$
6-2	12	**P**	11011101		$\neg Z \rightarrow \neg Y$

Level	#	Symbol	ID		Traditional logical form
7-1	6	Q1=M2	11011111	×	$X \vee \neg Y \vee Z$
7-1	8	Q2=L2	01111111	=	$X \vee Y \vee Z$
6-2	27	**Q**	01011111		$\neg Z \rightarrow X$
7-1	2	R1=E2	11111101	×	$\neg X \vee \neg Y \vee Z$
7-1	5	R2=A2	11101111	=	$\neg X \vee Y \vee Z$
6-2	8	**R**	11101101		$\neg Z \rightarrow \neg X$

In other words, we take **L** and **M** to be the counterparts of **A** and **E**, respectively. This is justified by the fact that all these statements already occur in the two-dimensional space (see Table 1.3). This justifies the introduction of these symbols already in the previous derivations. Table 2.24 shows all couples of statements involving three variables able to generate through product statements **L, M, N, P, Q, R**. Let us now consider particular statements as displayed in Table 2.25.

Table 2.25 S, T, U, V, W, Ω

Level	#	Symbol	ID	Traditional logical form
2-6	15	**S**	00101000	$\neg X \wedge Y$
	8	S1	10101000	$\neg X \wedge (Y \vee \neg Z)$
	23	S2=K2	01101000	$\neg X \wedge (Y \vee Z)$
3-5	37	S3	00111000	$(\neg X \wedge Y) \vee (X \wedge \neg Y \wedge \neg Z)$
	41	S4	00101100	$(\neg X \wedge Y) \vee (X \wedge \neg Y \wedge Z)$
	42	S5	00101010	$Y \wedge (\neg X \vee \neg Z)$
	43	S6=G3	00101001	$Y \wedge (\neg X \vee Z)$
2-6	1	**T**	11000000	$\neg X \wedge \neg Y$
	1	T1	11100000	$\neg X \wedge (\neg Y \vee \neg Z)$
	2	T2	11010000	$\neg Y \wedge (\neg X \vee \neg Z)$
3-5	3	T3	11001000	$\neg X \wedge (\neg Y \vee Z)$
	4	T4=H1	11000100	$\neg Y \wedge (\neg X \vee Z)$
	5	T5	11000010	$(\neg X \wedge \neg Y) \vee (X \wedge Y \wedge \neg Z)$
	6	T6	11000001	$(\neg X \wedge \neg Y) \vee (X \wedge Y \wedge Z)$

(*Continued*)

Table 2.25 (*Continued*)

Level	#	Symbol	ID	Traditional logical form
2-6	17	**U**	00100010	$\neg Z \wedge Y$
3-5	10	U1= Ω5	10100010	$\neg Z \wedge (\neg X \vee Y)$
	25	U2	01100010	$(\neg Z \wedge Y) \vee (\neg X \wedge \neg Y \wedge Z)$
	39	U3=W3	00110010	$\neg Z \wedge (X \vee Y)$
	42	U4=S5	00101010	$Y \wedge (\neg X \vee \neg Z)$
	44	U5	00100110	$(\neg Z \wedge Y) \vee (X \wedge \neg Y \wedge Z)$
	46	U6=I3	00100011	$Y \wedge (X \vee \neg Z)$
2-6	3	**V**	10010000	$\neg Z \wedge \neg Y$
3-5	2	V1=T2	11010000	$\neg Y \wedge (\neg X \vee \neg Z)$
	7	V2=Ω2	10110000	$\neg Z \wedge (\neg X \vee \neg Y)$
	12	V3	10011000	$(\neg Z \wedge \neg Y) \vee (\neg X \wedge Y \wedge Z)$
	13	V4=O1	10010100	$\neg Y \wedge (X \vee \neg Z)$
	14	V5=W1	10010010	$\neg Z \wedge (X \vee \neg Y)$
	15	V6	10010001	$(\neg Z \wedge \neg Y) \vee (X \wedge Y \wedge Z)$
2-6	21	**W**	00010010	$\neg Z \wedge X$
3-5	14	W1=V5	10010010	$\neg Z \wedge (X \vee \neg Y)$
	29	W2	01010010	$(\neg Z \wedge X) \vee (\neg X \wedge \neg Y \wedge Z)$
	39	W3=U3	00110010	$\neg Z \wedge (X \vee Y)$
	48	W4	00011010	$(\neg Z \wedge X) \vee (\neg X \wedge Y \wedge Z)$
	50	W5=O5	00010110	$X \wedge (\neg Y \vee \neg Z)$
	52	W6=I4	00010011	$X \wedge (Y \vee \neg Z)$
2-6	2	**Ω**	10100000	$\neg Z \wedge \neg X$
3-5	1	Ω1=T1	11100000	$\neg X \wedge (\neg Y \vee \neg Z)$
	7	Ω2=V2	10110000	$\neg Z \wedge (\neg X \vee \neg Y)$
	8	Ω3=S1	10101000	$\neg X \wedge (Y \vee \neg Z)$
	9	Ω4	10100100	$(\neg Z \wedge \neg X) \vee (X \wedge \neg Y \wedge Z)$
	10	Ω5=U1	10100010	$\neg Z \wedge (\neg X \vee Y)$
	11	Ω6	10100001	$(\neg Z \wedge \neg X) \vee (X \wedge Y \wedge Z)$

In other words, I take **S** and **T** to be the counterparts of **I** and **O**, respectively. This is justified by the fact that all of these statements already occur in the two-dimensional space (see Table 1.3). The reader can consider the insightful structural similarity among the statements of the generating set of each main statement. It is also insightful for all these derivations to have a look at Tables 2.6–2.11. Table 2.25 shows all couples of statements involving three variables able to generate through product statements **S, T, U, V, W** and **Ω**. The following table gives a summary of all kinds of statements:

Table 2.26 Kinds of Statements

Statement	Quality (predicate)		Subject		Quantity	
	Affirmative	Negative	Positive	Negative	Universal	Particular
A	•		•		•	
B	•		•		•	
C	•		•		•	
D		•	•		•	
E		•	•		•	
F		•	•		•	
G	•		•			•
H		•	•			•
I	•		•			•
J	•		•			•
K		•	•			•
L	•			•	•	
M		•		•	•	
N	•			•	•	
O		•	•			•
P		•		•	•	
Q	•			•	•	
R		•		•	•	
S	•			•		•
T		•		•		•
U	•			•		•
V		•		•		•
W	•			•		•
Ω		•		•		•

This allows us to build an octagon of relations (instead of the classical Aristotelian square) as shown in Fig. 2.1. Let us call these 24 statements *grounding* statements (or also *bold* statements, in short). What is remarkable is that in all the inferences as we have considered so far and also in all that will follow the three involved statements must lie each on a different octagon (since they must exhaust pairwise the three involved variable X, Y, Z). Moreover, at least two of them must be aligned, either on next octagons, like **A, B, C** for the classical *Barbara* or like **C, J, I** for the classical *Datisi*, or two of them must be on the opposite sides connected by a countervalent

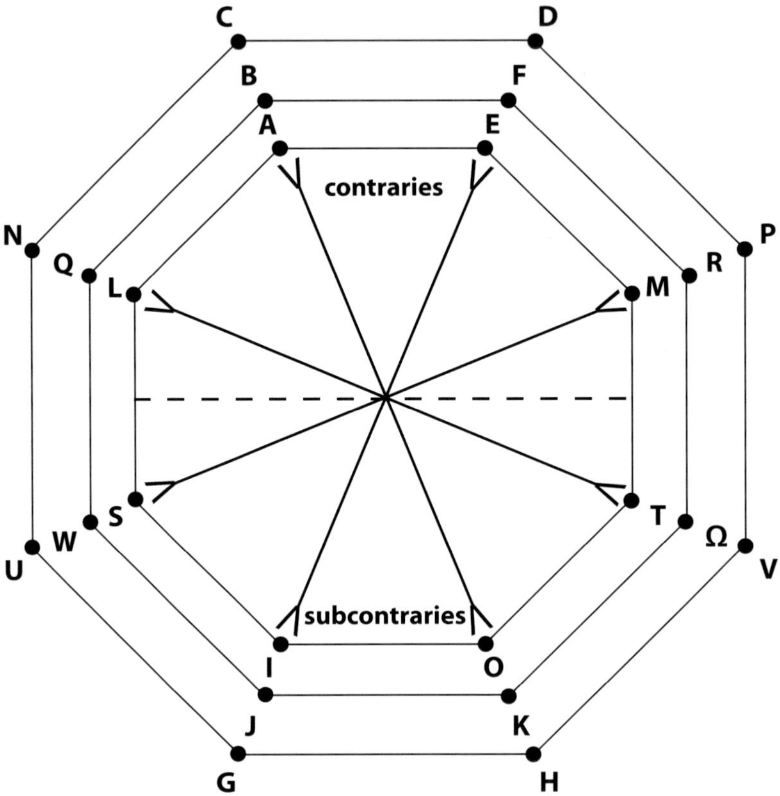

Figure 2.1 Above the dashed line any couple of statements represent a contrariety (both can be false) whilst below the dashed line any couple of statements can be contrary (both can be true). The lines $> - <$ connect contradictory statements. Obviously, we can also build three alternative octagons with statements **A-E-L-M-I-O-S-T**, **B-F-Q-R-J-K-W-Ω**, and **C-D-N-P-G-H-U-V**, respectively, depending on whether we are considering the relations between X and Y, between Z and X or between Z and Y.

line, as it happens for statements **N** and **Ω** in the derivation of **S** from the latter two statements.

Let us first consider the universal derivations—see also (*De Morgan*, 1847, Chapter V). The analogue of *Barbara* is:

Table 2.27 Derivation of **N** from **L** and **R**

Derivations	Level	#	Symbol	ID		Traditional logical form
	7-1	7	L1	10111111	×	$X \vee Y \vee \neg Z$
1	7-1	2	R1=E2	11111101	=	$\neg X \vee \neg Y \vee Z$
	6-2	17	E2xL1	10111101		$\neg(X \leftrightarrow Y \leftrightarrow \neg Z)$
	7-1	8	L2=N2	01111111	×	$X \vee Y \vee Z$
2	7-1	2	R1=E2	11111101	=	$\neg X \vee \neg Y \vee Z$
	6-2	23	E2xL2	01111101		$\neg Z \rightarrow \neg(X \leftrightarrow Y)$
	7-1	7	L1	10111111	×	$X \vee Y \vee \neg Z$
3	7-1	5	R2=N1=A2	11101111	=	$\neg X \vee Y \vee Z$
	6-2	20	A2xL1	10101111		$\neg Y \rightarrow (X \leftrightarrow Z)$
	7-1	8	L2=N2	01111111	×	$X \vee Y \vee Z$
4	7-1	5	R2=N1=A2	11101111	=	$\neg X \vee Y \vee Z$
	6-2	26	**N**	01101111		$\neg Z \rightarrow Y$

I do not need to recall here the reasons for selecting the fourth derivation. Allow me to give here a very elementary example of this kind of derivation that can turn out to be helpful for the understanding of what follows: If All objects that are not material are ideal and No non-physical entity is material, then All non-physical entities are ideal. The analogue of the mode *Celarent* is the following:

Table 2.28 Derivation of **P** from **M** and **R**

Derivations	Level	#	Symbol	ID		Traditional logical form
	7-1	4	M1	11110111	×	$X \vee \neg Y \vee \neg Z$
1	7-1	2	R1=P1=E2	11111101	=	$\neg X \vee \neg Y \vee Z$
	6-2	5	E2xM1	11110101		$Y \rightarrow (X \leftrightarrow Z)$
	7-1	6	M2	11011111	×	$X \vee \neg Y \vee Z$
2	7-1	2	R1=P1=E2	11111101	=	$\neg X \vee \neg Y \vee Z$
	6-2	12	**P**	11011101		$\neg Z \rightarrow \neg Y$
	7-1	4	M1	11110111	×	$X \vee \neg Y \vee \neg Z$
3	7-1	5	R2=A2	11101111	=	$\neg X \vee Y \vee Z$
	6-2	10	M1xA2	11100111		$\neg(\neg X \leftrightarrow Y \leftrightarrow Z)$
	7-1	6	M2	11011111	×	$X \vee \neg Y \vee Z$
4	7-1	5	R2=A2	11101111	=	$\neg X \vee Y \vee Z$
	6-2	15	A2xM2	11001111		$\neg Z \rightarrow (X \leftrightarrow Y)$

Also here it is quite clear why Derivation 2 can and must be selected. Derivations involving particular statements are always more cumbersome. The analogue of *Darii* is the following:

Table 2.29 Derivation of **U** from **L** and **Ω**

Derivations	Level	#	Symbol	ID		Traditional logical form
	7-1	7	L1	10111111	×	$X \vee Y \vee \neg Z$
1	3-5	1	Ω1=T1	11100000	=	$\neg X \wedge (\neg Y \vee \neg Z)$
	2-6	2	**Ω**	10100000		$\neg Z \wedge \neg X$
	7-1	8	L2	01111111	×	$X \vee Y \vee Z$
2	3-5	1	Ω1=T1	11100000	=	$\neg X \wedge (\neg Y \vee \neg Z)$
	2-6	8	T8+S7	01100000		$\neg X \wedge \neg(Y \leftrightarrow Z)$
	7-1	7	L1	10111111	×	$X \vee Y \vee \neg Z$
3	3-5	7	Ω2=V2	10110000	=	$\neg Z \wedge (\neg X \vee \neg Y)$
	3-5	7	Ω2=V2	10110000		$\neg Z \wedge (\neg X \vee \neg Y)$
	7-1	8	L2	01111111	×	$X \vee Y \vee Z$
4	3-5	7	Ω2=V2	10110000	=	$\neg Z \wedge (\neg X \vee \neg Y)$
	2-6	14	S7+07	00110000		$\neg Z \wedge \neg(X \leftrightarrow Y)$
	7-1	7	L1	10111111	×	$X \vee Y \vee \neg Z$
5	3-5	8	Ω3=S1	10101000	=	$\neg X \wedge (Y \vee \neg Z)$
	3-5	8	Ω3=S1	10101000		$\neg X \wedge (Y \vee \neg Z)$
	7-1	8	L2	01111111	×	$X \vee Y \vee Z$
6	3-5	8	Ω3=S1	10101000	=	$\neg X \wedge (Y \vee \neg Z)$
	2-6	15	**S**	00101000		$\neg X \wedge Y$
	7-1	7	L1	10111111	×	$X \vee Y \vee \neg Z$
7	3-5	9	Ω4	10100100	=	$(\neg Z \wedge \neg X) \vee (X \wedge \neg Y \wedge Z)$
	3-5	9	Ω4	10100100		$(\neg Z \wedge \neg X) \vee (X \wedge \neg Y \wedge Z)$
	7-1	8	L2	01111111	×	$X \vee Y \vee Z$
8	3-5	9	Ω4	10100100	=	$(\neg Z \wedge \neg X) \vee (X \wedge \neg Y \wedge Z)$
	2-6	16	S7+O8	00100100		$X \leftrightarrow \neg Y \leftrightarrow Z$
	7-1	7	L1	10111111	×	$X \vee Y \vee \neg Z$
9	3-5	10	Ω5=U1	10100010	=	$\neg Z \wedge (\neg X \vee Y)$
	3-5	10	Ω5=U1	10100010		$\neg Z \wedge (\neg X \vee Y)$
	7-1	8	L2	01111111	×	$X \vee Y \vee Z$
10	3-5	10	Ω5=U1	10100010	=	$\neg Z \wedge (\neg X \vee Y)$
	2-6	17	**U**	00100010		$\neg Z \wedge Y$
	7-1	7	L1	10111111	×	$X \vee Y \vee \neg Z$
11	3-5	11	Ω6	10100001	=	$(\neg Z \wedge \neg X) \vee (X \wedge Y \wedge Z)$
	3-5	11	Ω6	10100001		$(\neg Z \wedge \neg X) \vee (X \wedge Y \wedge Z)$
	7-1	8	L2	01111111	×	$X \vee Y \vee Z$
12	3-5	11	Ω6	10100001	=	$(\neg Z \wedge \neg X) \vee (X \wedge Y \wedge Z)$
	2-6	18	S7+I8	00100001		$Y \wedge (X \leftrightarrow Z)$

The analogue of *Ferio* is:

Table 2.30 Derivation of **V** from **M** and **Ω**

Derivations	Level	#	Symbol	ID		Traditional logical form
	7-1	4	M1	11110111	×	$X \vee \neg Y \vee \neg Z$
1	3-5	1	Ω1=T1	11100000	=	$\neg X \wedge (\neg Y \vee \neg Z)$
	3-5	1	Ω1=T1	11100000		$\neg X \wedge (\neg Y \vee \neg Z)$
	7-1	6	M2	11011111	×	$X \vee \neg Y \vee Z$
2	3-5	1	Ω1=T1	11100000	=	$\neg X \wedge (\neg Y \vee \neg Z)$
	2-6	1	**T**	11000000		$\neg X \wedge \neg Y$
	7-1	4	M1	11110111	×	$X \vee \neg Y \vee \neg Z$
3	3-5	7	Ω2=V2	10110000	=	$\neg Z \wedge (\neg X \vee \neg Y)$
	3-5	7	Ω2=V2	10110000		$\neg Z \wedge (\neg X \vee \neg Y)$
	7-1	6	M2	11011111	×	$X \vee \neg Y \vee Z$
4	3-5	7	Ω2=V2	10110000	=	$\neg Z \wedge (\neg X \vee \neg Y)$
	2-6	3	**V**	10010000		$\neg Z \wedge \neg Y$
	7-1	4	M1	11110111	×	$X \vee \neg Y \vee \neg Z$
5	3-5	8	Ω3=S1	10101000	=	$\neg X \wedge (Y \vee \neg Z)$
	3-5	2	**Ω**	10100000		$\neg Z \wedge \neg X$
	7-1	6	M2	11011111	×	$X \vee \neg Y \vee Z$
6	3-5	8	Ω3=S1	10101000	=	$\neg X \wedge (Y \vee \neg Z)$
	2-6	4	T7+S8	10001000		$\neg X \wedge (Y \leftrightarrow Z)$
	7-1	4	M1	11110111	×	$X \vee \neg Y \vee \neg Z$
7	3-5	9	Ω4	10100100	=	$(\neg Z \wedge \neg X) \vee (X \wedge \neg Y \wedge Z)$
	3-5	9	Ω4	10100100		$(\neg Z \wedge \neg X) \vee (X \wedge \neg Y \wedge Z)$
	7-1	6	M2	11011111	×	$X \vee \neg Y \vee Z$
8	3-5	9	Ω4	10100100	=	$(\neg Z \wedge \neg X) \vee (X \wedge \neg Y \wedge Z)$
	2-6	5	T7+O8	10000100		$\neg Y \wedge (X \leftrightarrow Z)$
	7-1	4	M1	11110111	×	$X \vee \neg Y \vee \neg Z$
9	3-5	10	Ω5=U1	10100010	=	$\neg Z \wedge (\neg X \vee Y)$
	3-5	10	Ω5=U1	10100010		$\neg Z \wedge (\neg X \vee Y)$
	7-1	6	M2	11011111	×	$X \vee \neg Y \vee Z$
10	3-5	10	Ω5=U1	10100010	=	$\neg Z \wedge (\neg X \vee Y)$
	2-6	6	T7+I7	10000010		$\neg Z \wedge (X \leftrightarrow Y)$
	7-1	4	M1	11110111	×	$X \vee \neg Y \vee \neg Z$
11	3-5	11	Ω6	10100001	=	$(\neg Z \wedge \neg X) \vee (X \wedge Y \wedge Z)$
	3-5	11	Ω6	10100001		$(\neg Z \wedge \neg X) \vee (X \wedge Y \wedge Z)$
	7-1	6	M2	11011111	×	$X \vee \neg Y \vee Z$
12	3-5	11	Ω6	10100001	=	$(\neg Z \wedge \neg X) \vee (X \wedge Y \wedge Z)$
	2-6	12	T8+I7	01000010		$X \leftrightarrow Y \leftrightarrow \neg Z$

We have completed the first group. The analogue of *Baroco* is:

Table 2.31 Derivation of **W** from **L** and **V**

Derivations	Level	#	Symbol	ID		Traditional logical form
	7-1	7	L1	10111111	×	$X \vee Y \vee \neg Z$
1	3-5	2	V1=T2	11010000	=	$\neg Y \wedge (\neg X \vee \neg Z)$
	2-6	3	**V**	10010000		$\neg Z \wedge \neg Y$
	7-1	8	L2	01111111	×	$X \vee Y \vee Z$
2	3-5	2	V1=T2	11010000	=	$\neg Y \wedge (\neg X \vee \neg Z)$
	2-6	9	T8+O7	01010000		$\neg Y \wedge \neg(X \leftrightarrow Z)$
	7-1	7	L1	10111111	×	$X \vee Y \vee \neg Z$
3	3-5	7	V2=Ω2	10110000	=	$\neg Z \wedge (\neg X \vee \neg Y)$
	3-5	7	V2=Ω2	10110000		$\neg Z \wedge (\neg X \vee \neg Y)$
	7-1	8	L2	01111111	×	$X \vee Y \vee Z$
4	3-5	7	V2=Ω2	10110000	=	$\neg Z \wedge (\neg X \vee \neg Y)$
	2-6	14	S7+O7	00110000		$\neg Z \wedge \neg(X \leftrightarrow Y)$
	7-1	7	L1	10111111	×	$X \vee Y \vee \neg Z$
5	3-5	12	V3	10011000	=	$(\neg Z \wedge \neg Y) \vee (\neg X \wedge Y \wedge Z)$
	3-5	12	V3	10011000		$(\neg Z \wedge \neg Y) \vee (\neg X \wedge Y \wedge Z)$
	7-1	8	L2	01111111	×	$X \vee Y \vee Z$
6	3-5	12	V3	10011000	=	$(\neg Z \wedge \neg Y) \vee (\neg X \wedge Y \wedge Z)$
	2-6	19	O7+S8	00011000		$\neg X \leftrightarrow Y \leftrightarrow Z$
	7-1	7	L1	10111111	×	$X \vee Y \vee \neg Z$
7	3-5	13	V4=O1	10010100	=	$\neg Y \wedge (X \vee \neg Z)$
	3-5	13	V4=O1	10010100		$\neg Y \wedge (X \vee \neg Z)$
	7-1	8	L2	01111111	×	$X \vee Y \vee Z$
8	3-5	13	V4=O1	10010100	=	$\neg Y \wedge (X \vee \neg Z)$
	2-6	20	**O**	00010100		$X \wedge \neg Y$
	7-1	7	L1	10111111	×	$X \vee Y \vee \neg Z$
9	3-5	14	V5=W1	10010010	=	$\neg Z \wedge (X \vee \neg Y)$
	3-5	14	V5=W1	10010010		$\neg Z \wedge (X \vee \neg Y)$
	7-1	8	L2	01111111	×	$X \vee Y \vee Z$
10	3-5	14	V5=W1	10010010	=	$\neg Z \wedge (X \vee \neg Y)$
	2-6	21	**W**	00010010		$\neg Z \wedge X$
	7-1	7	L1	10111111	×	$X \vee Y \vee \neg Z$
11	3-5	15	V6	10010001	=	$(\neg Z \wedge \neg Y) \vee (X \wedge Y \wedge Z)$
	3-5	15	V6	10010001		$(\neg Z \wedge \neg Y) \vee (X \wedge Y \wedge Z)$
	7-1	8	L2	01111111	×	$X \vee Y \vee Z$
12	3-5	15	V6	10010001	=	$(\neg Z \wedge \neg Y) \vee (X \wedge Y \wedge Z)$
	2-6	22	O7+I8	00010001		$X \wedge (Y \leftrightarrow Z)$

The analogue of *Festino* is:

Table 2.32 Derivation of **W** from **M** and **U**

Derivations	Level	#	Symbol	ID		Traditional logical form
	7-1	4	M1	11110111	×	$X \vee \neg Y \vee \neg Z$
1	3-5	10	U1= Ω5	10100010	=	$\neg Z \wedge (\neg X \vee Y)$
	3-5	10	U1= Ω5	10100010		$\neg Z \wedge (\neg X \vee Y)$
	7-1	6	M2	11011111	×	$X \vee \neg Y \vee Z$
2	3-5	10	U1= Ω5	10100010	=	$\neg Z \wedge (\neg X \vee Y)$
	2-6	6	T7+I7	10000010		$\neg Z \wedge (X \leftrightarrow Y)$
	7-1	4	M1	11110111	×	$X \vee \neg Y \vee \neg Z$
3	3-5	25	U2	01100010	=	$(\neg Z \wedge Y) \vee (\neg X \wedge \neg Y \wedge Z)$
	3-5	25	U2	01100010		$(\neg Z \wedge Y) \vee (\neg X \wedge \neg Y \wedge Z)$
	7-1	6	M2	11011111	×	$X \vee \neg Y \vee Z$
4	3-5	25	U2	01100010	=	$(\neg Z \wedge Y) \vee (\neg X \wedge \neg Y \wedge Z)$
	2-6	13	T8+I8	01000001		$Z \wedge (X \leftrightarrow Y)$
	7-1	4	M1	11110111	×	$X \vee \neg Y \vee \neg Z$
5	3-5	39	U3=W3	00110010	=	$\neg Z \wedge (X \vee Y)$
	3-5	39	U3=W3	00110010		$\neg Z \wedge (X \vee Y)$
	7-1	6	M2	11011111	×	$X \vee \neg Y \vee Z$
6	3-5	39	U3=W3	00110010	=	$\neg Z \wedge (X \vee Y)$
	2-6	21	**W**	00010010		$\neg Z \wedge X$
	7-1	4	M1	11110111	×	$X \vee \neg Y \vee \neg Z$
7	3-5	42	U4=S5	00101010	=	$Y \wedge (\neg X \vee \neg Z)$
	2-6	17	**U**	00100010		$\neg Z \wedge Y$
	7-1	6	M2	11011111	×	$X \vee \neg Y \vee Z$
8	3-5	42	U4=S5	00101010	=	$Y \wedge (\neg X \vee \neg Z)$
	2-6	24	S8+I7	00001010		$Y \wedge \neg(X \leftrightarrow Z)$
	7-1	4	M1	11110111	×	$X \vee \neg Y \vee \neg Z$
9	3-5	44	U5	00100110	=	$(\neg Z \wedge Y) \vee (X \wedge \neg Y \wedge Z)$
	3-5	44	U5	00100110		$(\neg Z \wedge Y) \vee (X \wedge \neg Y \wedge Z)$
	7-1	6	M2	11011111	×	$X \vee \neg Y \vee Z$
10	3-5	44	U5	00100110	=	$(\neg Z \wedge Y) \vee (X \wedge \neg Y \wedge Z)$
	2-6	26	O8+I7	00000110		$X \wedge \neg(Y \leftrightarrow Z)$
	7-1	4	M1	11110111	×	$X \vee \neg Y \vee \neg Z$
11	3-5	46	U6=I3	00100011	=	$Y \wedge (X \vee \neg Z)$
	3-5	46	U6=I3	00100011		$Y \wedge (X \vee \neg Z)$
	7-1	6	M2	11011111	×	$X \vee \neg Y \vee Z$
12	3-5	46	U6=I3	00100011	=	$Y \wedge (X \vee \neg Z)$
	2-6	28	**I**	00000011		$X \wedge Y$

The analogue of *Camestres* is:

Table 2.33 Derivation of **Q** from **L** and **P**

Derivations	Level	#	Symbol	ID		Traditional logical form
	7-1	7	L1	10111111	×	$X \vee Y \vee \neg Z$
1	7-1	2	P1=E2	11111101	=	$\neg X \vee \neg Y \vee Z$
	6-2	17	E2xL1	10111101		$\neg(X \leftrightarrow Y \leftrightarrow \neg Z)$
	7-1	8	L2=Q2	01111111	×	$X \vee Y \vee Z$
2	7-1	2	P1=E2	11111101	=	$\neg X \vee \neg Y \vee Z$
	6-2	23	E2xL2	01111101		$\neg Z \rightarrow \neg(X \leftrightarrow Y)$
	7-1	7	L1	10111111	×	$X \vee Y \vee \neg Z$
3	7-1	6	P2=Q1=M2	11011111	=	$X \vee \neg Y \vee Z$
	6-2	21	M2xL1	10011111		$\neg X \rightarrow (Y \leftrightarrow Z)$
	7-1	8	L2=Q2	01111111	×	$X \vee Y \vee Z$
4	7-1	6	P2=Q1=M2	11011111	=	$X \vee \neg Y \vee Z$
	6-2	27	**Q**	01011111		$\neg Z \rightarrow X$

The analogue of *Cesare* is:

Table 2.34 Derivation of **Q** from **M** and **N**

Derivations	Level	#	Symbol	ID		Traditional logical form
	7-1	4	M1	11110111	×	$X \vee \neg Y \vee \neg Z$
1	7-1	5	N1=A2	11101111	=	$\neg X \vee Y \vee Z$
	6-2	10	M1xA2	11100111		$\neg(\neg X \leftrightarrow Y \leftrightarrow Z)$
	7-1	6	M2	11011111	×	$X \vee \neg Y \vee Z$
2	7-1	5	N1=A2	11101111	=	$\neg X \vee Y \vee Z$
	6-2	15	A2xM2	11001111		$\neg Z \rightarrow (X \leftrightarrow Y)$
	7-1	4	M1	11110111	×	$X \vee \neg Y \vee \neg Z$
3	7-1	8	N2=Q2=L2	01111111	=	$X \vee Y \vee Z$
	6-2	25	M1xL2	01110111		$\neg X \rightarrow \neg(Y \leftrightarrow Z)$
	7-1	6	M2	11011111	×	$X \vee \neg Y \vee Z$
4	7-1	8	N2=Q2=L2	01111111	=	$X \vee Y \vee Z$
	6-2	27	**Q**	01011111		$\neg Z \rightarrow X$

We have completed the second group. The analogue of *Bocardo* is:

Table 2.35 Derivation of **T** from **V** and **R**

Derivations	Level	#	Symbol	ID		Traditional logical form
	3-5	2	V1=T2	11010000	×	$\neg Y \wedge (\neg X \vee \neg Z)$
1	7-1	2	R1=E2	11111101	=	$\neg X \vee \neg Y \vee Z$
	3-5	2	V1=T2	11010000		$\neg Y \wedge (\neg X \vee \neg Z)$
	3-5	2	V1=T2	11010000	×	$\neg Y \wedge (\neg X \vee \neg Z)$
2	7-1	5	R2=A2	11101111	=	$\neg X \vee Y \vee Z$
	2-6	1	**T**	11000000		$\neg X \wedge \neg Y$
	3-5	7	V2=Ω2	10110000	×	$\neg Z \wedge (\neg X \vee \neg Y)$
3	7-1	2	R1=E2	11111101	=	$\neg X \vee \neg Y \vee Z$
	3-5	7	V2=Ω2	10110000		$\neg Z \wedge (\neg X \vee \neg Y)$
	3-5	7	V2=Ω2	10110000	×	$\neg Z \wedge (\neg X \vee \neg Y)$
4	7-1	5	R2=A2	11101111	=	$\neg X \vee Y \vee Z$
	2-6	2	**Ω**	10100000		$\neg Z \wedge \neg X$
	3-5	12	V3	10011000	×	$(\neg Z \wedge \neg Y) \vee (\neg X \wedge Y \wedge Z)$
5	7-1	2	R1=E2	11111101	=	$\neg X \vee \neg Y \vee Z$
	3-5	12	V3	10011000		$(\neg Z \wedge \neg Y) \vee (\neg X \wedge Y \wedge Z)$
	3-5	12	V3	10011000	×	$(\neg Z \wedge \neg Y) \vee (\neg X \wedge Y \wedge Z)$
6	7-1	5	R2=A2	11101111	=	$\neg X \vee Y \vee Z$
	2-6	4	T7+S8	10001000		$\neg X \wedge (Y \leftrightarrow Z)$
	3-5	13	V4=O1	10010100	×	$\neg Y \wedge (X \vee \neg Z)$
7	7-1	2	R1=E2	11111101	=	$\neg X \vee \neg Y \vee Z$
	3-5	13	V4=O1	10010100		$\neg Y \wedge (X \vee \neg Z)$
	3-5	13	V4=O1	10010100	×	$\neg Y \wedge (X \vee \neg Z)$
8	7-1	5	R2=A2	11101111	=	$\neg X \vee Y \vee Z$
	2-6	5	T7+O8	10000100		$\neg Y \wedge (X \leftrightarrow Z)$
	3-5	14	V5=W1	10010010	×	$\neg Z \wedge (X \vee \neg Y)$
9	7-1	2	R1=E2	11111101	=	$\neg X \vee \neg Y \vee Z$
	2-6	3	**V**	10010000		$\neg Z \wedge \neg Y$
	3-5	14	V5=W1	10010010	×	$\neg Z \wedge (X \vee \neg Y)$
10	7-1	5	R2=A2	11101111	=	$\neg X \vee Y \vee Z$
	2-6	6	T7+I7	10000010		$\neg Z \wedge (X \leftrightarrow Y)$
	3-5	15	V6	10010001	×	$(\neg Z \wedge \neg Y) \vee (X \wedge Y \wedge Z)$
11	7-1	2	R1=E2	11111101	=	$\neg X \vee \neg Y \vee Z$
	3-5	15	V6	10010001		$(\neg Z \wedge \neg Y) \vee (X \wedge Y \wedge Z)$
	3-5	15	V6	10010001	×	$(\neg Z \wedge \neg Y) \vee (X \wedge Y \wedge Z)$
12	7-1	5	R2=A2	11101111	=	$\neg X \vee Y \vee Z$
	2-6	7	T7+I8	10000001		$X \leftrightarrow Y \leftrightarrow Z$

The analogue of *Disamis* is:

Table 2.36 Derivation of **S** from **U** and **R**

Derivations	Level	#	Symbol	ID		Traditional logical form
	3-5	10	U1= Ω5	10100010	×	$\neg Z \wedge (\neg X \vee Y)$
1	7-1	2	R1=E2	11111101	=	$\neg X \vee \neg Y \vee Z$
	2-6	2	**Ω**	10100000		$\neg Z \wedge \neg X$
	3-5	10	U1= Ω5	10100010	×	$\neg Z \wedge (\neg X \vee Y)$
2	7-1	5	R2=A2	11101111	=	$\neg X \vee Y \vee Z$
	3-5	10	U1= Ω5	10100010		$\neg Z \wedge (\neg X \vee Y)$
	3-5	25	U2	01100010	×	$(\neg Z \wedge Y) \vee (\neg X \wedge \neg Y \wedge Z)$
3	7-1	2	R1=E2	11111101	=	$\neg X \vee \neg Y \vee Z$
	2-6	8	T8+S7	01100000		$\neg X \wedge \neg(Y \leftrightarrow Z)$
	3-5	25	U2	01100010	×	$(\neg Z \wedge Y) \vee (\neg X \wedge \neg Y \wedge Z)$
4	7-1	5	R2=A2	11101111	=	$\neg X \vee Y \vee Z$
	3-5	25	U2	01100010		$(\neg Z \wedge Y) \vee (\neg X \wedge \neg Y \wedge Z)$
	3-5	39	U3=W3	00110010	×	$\neg Z \wedge (X \vee Y)$
5	7-1	2	R1=E2	11111101	=	$\neg X \vee \neg Y \vee Z$
	2-6	14	S7+O7	00110000		$\neg Z \wedge \neg(X \leftrightarrow Y)$
	3-5	39	U3=W3	00110010	×	$\neg Z \wedge (X \vee Y)$
6	7-1	5	R2=A2	11101111	=	$\neg X \vee Y \vee Z$
	2-6	17	**U**	00100010		$\neg Z \wedge Y$
	3-5	42	U4=S5	00101010	×	$Y \wedge (\neg X \vee \neg Z)$
7	7-1	2	R1=E2	11111101	=	$\neg X \vee \neg Y \vee Z$
	2-6	15	**S**	00101000		$\neg X \wedge Y$
	3-5	42	U4=S5	00101010	×	$Y \wedge (\neg X \vee \neg Z)$
8	7-1	5	R2=A2	11101111	=	$\neg X \vee Y \vee Z$
	3-5	42	U4=S5	00101010		$Y \wedge (\neg X \vee \neg Z)$
	3-5	44	U5	00100110	×	$(\neg Z \wedge Y) \vee (X \wedge \neg Y \wedge Z)$
9	7-1	2	R1=E2	11111101	=	$\neg X \vee \neg Y \vee Z$
	2-6	16	S7+O8	00100100		$X \leftrightarrow \neg Y \leftrightarrow Z$
	3-5	44	U5	00100110	×	$(\neg Z \wedge Y) \vee (X \wedge \neg Y \wedge Z)$
10	7-1	5	R2=A2	11101111	=	$\neg X \vee Y \vee Z$
	3-5	44	U5	00100110		$(\neg Z \wedge Y) \vee (X \wedge \neg Y \wedge Z)$
	3-5	46	U6=I3	00100011	×	$Y \wedge (X \vee \neg Z)$
11	7-1	2	R1=E2	11111101	=	$\neg X \vee \neg Y \vee Z$
	2-6	18	S7+I8	00100001		$Y \wedge (X \leftrightarrow Z)$
	3-5	46	U6=I3	00100011	×	$Y \wedge (X \vee \neg Z)$
12	7-1	5	R2=A2	11101111	=	$\neg X \vee Y \vee Z$
	3-5	46	U6=I3	00100011		$Y \wedge (X \vee \neg Z)$

Let us now consider the analogue of *Ferison*:

Table 2.37 Derivation of **T** from **P** and **Ω**

Derivations	Level	#	Symbol	ID		Traditional logical form
	7-1	2	P1=E2	11111101	×	$\neg X \vee \neg Y \vee Z$
1	3-5	1	Ω1=T1	11100000	=	$\neg X \wedge (\neg Y \vee \neg Z)$
	3-5	1	Ω1=T1	11100000		$\neg X \wedge (\neg Y \vee \neg Z)$
	7-1	6	P2=M2	11011111	×	$X \vee \neg Y \vee Z$
2	3-5	1	Ω1=T1	11100000	=	$\neg X \wedge (\neg Y \vee \neg Z)$
	2-6	1	**T**	11000000		$\neg X \wedge \neg Y$
	7-1	2	P1=E2	11111101	×	$\neg X \vee \neg Y \vee Z$
3	3-5	7	Ω2=V2	10110000	=	$\neg Z \wedge (\neg X \vee \neg Y)$
	3-5	7	Ω2=V2	10110000		$\neg Z \wedge (\neg X \vee \neg Y)$
	7-1	6	P2=M2	11011111	×	$X \vee \neg Y \vee Z$
4	3-5	7	Ω2=V2	10110000	=	$\neg Z \wedge (\neg X \vee \neg Y)$
	2-6	3	**V**	10010000		$\neg Z \wedge \neg Y$
	7-1	2	P1=E2	11111101	×	$\neg X \vee \neg Y \vee Z$
5	3-5	8	Ω3=S1	10101000	=	$\neg X \wedge (Y \vee \neg Z)$
	3-5	8	Ω3=S1	10101000		$\neg X \wedge (Y \vee \neg Z)$
	7-1	6	P2=M2	11011111	×	$X \vee \neg Y \vee Z$
6	3-5	8	Ω3=S1	10101000	=	$\neg X \wedge (Y \vee \neg Z)$
	2-6	4	T7+S8	10001000		$\neg X \wedge (Y \leftrightarrow Z)$
	7-1	2	P1=E2	11111101	×	$\neg X \vee \neg Y \vee Z$
7	3-5	9	Ω4	10100100	=	$(\neg Z \wedge \neg X) \vee (X \wedge \neg Y \wedge Z)$
	3-5	9	Ω4	10100100		$(\neg Z \wedge \neg X) \vee (X \wedge \neg Y \wedge Z)$
	7-1	6	P2=M2	11011111	×	$X \vee \neg Y \vee Z$
8	3-5	9	Ω4	10100100	=	$(\neg Z \wedge \neg X) \vee (X \wedge \neg Y \wedge Z)$
	2-6	5	T7+O8	10000100		$\neg Y \wedge (X \leftrightarrow Z)$
	7-1	2	P1=E2	11111101	×	$\neg X \vee \neg Y \vee Z$
9	3-5	10	Ω5=U1	10100010	=	$\neg Z \wedge (\neg X \vee Y)$
	3-5	10	Ω5=U1	10100010		$\neg Z \wedge (\neg X \vee Y)$
	7-1	6	P2=M2	11011111	×	$X \vee \neg Y \vee Z$
10	3-5	10	Ω5=U1	10100010	=	$\neg Z \wedge (\neg X \vee Y)$
	2-6	6	T7+I7	10000010		$\neg Z \wedge (X \leftrightarrow Y)$
	7-1	2	P1=E2	11111101	×	$\neg X \vee \neg Y \vee Z$
11	3-5	11	Ω6	10100001	=	$(\neg Z \wedge \neg X) \vee (X \wedge Y \wedge Z)$
	3-5	11	Ω6	10100001		$(\neg Z \wedge \neg X) \vee (X \wedge Y \wedge Z)$
	7-1	6	P2=M2	11011111	×	$X \vee \neg Y \vee Z$
12	3-5	11	Ω6	10100001	=	$(\neg Z \wedge \neg X) \vee (X \wedge Y \wedge Z)$
	2-6	7	T7+I8	10000001		$X \leftrightarrow Y \leftrightarrow Z$

The analogue of *Datisi* is:

Table 2.38 Derivation of **S** from **N** and **Ω**

Derivations	Level	#	Symbol	ID		Traditional logical form
	7-1	5	N1=A2	11101111	×	$\neg X \vee Y \vee Z$
1	3-5	1	Ω1=T1	11100000	=	$\neg X \wedge (\neg Y \vee \neg Z)$
	3-5	1	Ω1=T1	11100000		$\neg X \wedge (\neg Y \vee \neg Z)$
	7-1	8	N2=L2	01111111	×	$X \vee Y \vee Z$
2	3-5	1	Ω1=T1	11100000	=	$\neg X \wedge (\neg Y \vee \neg Z)$
	2-6	8	T8+S7	01100000		$\neg X \wedge \neg(Y \leftrightarrow Z)$
	7-1	5	N1=A2	11101111	×	$\neg X \vee Y \vee Z$
3	3-5	7	Ω2=V2	10110000	=	$\neg Z \wedge (\neg X \vee \neg Y)$
	2-6	2	**Ω**	10100000		$\neg Z \wedge \neg X$
	7-1	8	N2=L2	01111111	×	$X \vee Y \vee Z$
4	3-5	7	Ω2=V2	10110000	=	$\neg Z \wedge (\neg X \vee \neg Y)$
	2-6	14	S7+O7	00110000		$\neg Z \wedge \neg(X \leftrightarrow Y)$
	7-1	5	N1=A2	11101111	×	$\neg X \vee Y \vee Z$
5	3-5	8	Ω3=S1	10101000	=	$\neg X \wedge (Y \vee \neg Z)$
	3-5	8	Ω3=S1	10101000		$\neg X \wedge (Y \vee \neg Z)$
	7-1	8	N2=L2	01111111	×	$X \vee Y \vee Z$
6	3-5	8	Ω3=S1	10101000	=	$\neg X \wedge (Y \vee \neg Z)$
	2-6	15	**S**	00101000		$\neg X \wedge Y$
	7-1	5	N1=A2	11101111	×	$\neg X \vee Y \vee Z$
7	3-5	9	Ω4	10100100	=	$(\neg Z \wedge \neg X) \vee (X \wedge \neg Y \wedge Z)$
	3-5	9	Ω4	10100100		$(\neg Z \wedge \neg X) \vee (X \wedge \neg Y \wedge Z)$
	7-1	8	N2=L2	01111111	×	$X \vee Y \vee Z$
8	3-5	9	Ω4	10100100	=	$(\neg Z \wedge \neg X) \vee (X \wedge \neg Y \wedge Z)$
	2-6	16	S7+O8	00100100		$X \leftrightarrow \neg Y \leftrightarrow Z$
	7-1	5	N1=A2	11101111	×	$\neg X \vee Y \vee Z$
9	3-5	10	Ω5=U1	10100010	=	$\neg Z \wedge (\neg X \vee Y)$
	3-5	10	Ω5=U1	10100010		$\neg Z \wedge (\neg X \vee Y)$
	7-1	8	N2=L2	01111111	×	$X \vee Y \vee Z$
10	3-5	10	Ω5=U1	10100010	=	$\neg Z \wedge (\neg X \vee Y)$
	2-6	17	**U**	00100010		$\neg Z \wedge Y$
	7-1	5	N1=A2	11101111	×	$\neg X \vee Y \vee Z$
11	3-5	11	Ω6	10100001	=	$(\neg Z \wedge \neg X) \vee (X \wedge Y \wedge Z)$
	3-5	11	Ω6	10100001		$(\neg Z \wedge \neg X) \vee (X \wedge Y \wedge Z)$
	7-1	8	N2=L2	01111111	×	$X \vee Y \vee Z$
12	3-5	11	Ω6	10100001	=	$(\neg Z \wedge \neg X) \vee (X \wedge Y \wedge Z)$
	2-6	18	S7+I8	00100001		$Y \wedge (X \leftrightarrow Z)$

2.4 Derivation of the Inferences of the First Mixed Mode through Products

Two main mixed modes may distinguished according to whether we take two statements from classical inferences and one from its extension provided in the previous chapter or we do the opposite. I shall first briefly summarize the first mixed mode. The counterpart of *Barbara* in this context is the following:

Table 2.39 Derivation of **C** from **L** and **F**

Derivations	Level	#	Symbol	ID		Traditional logical form
	7-1	7	L1=C2	10111111	×	$X \vee Y \vee \neg Z$
1	7-1	1	F1=E1	11111110	=	$\neg X \vee \neg Y \vee \neg Z$
	6-2	16	E1xL1	10111110		$Z \to \neg(X \leftrightarrow Y)$
	7-1	8	L2	01111111	×	$X \vee Y \vee Z$
2	7-1	1	F1=E1	11111110	=	$\neg X \vee \neg Y \vee \neg Z$
	6-2	22	E1xL2	01111110		$\neg(X \leftrightarrow Y \leftrightarrow Z)$
	7-1	7	L1=C2	10111111	×	$X \vee Y \vee \neg Z$
3	7-1	3	F2=A1	11111011	=	$\neg X \vee Y \vee \neg Z$
	6-2	18	**C**	10111011		$Z \to Y$
	7-1	8	L2	01111111	×	$X \vee Y \vee Z$
4	7-1	3	F2=A1	11111011	=	$\neg X \vee Y \vee \neg Z$
	6-2	24	A1xL2	01111011		$\neg Y \to \neg(X \leftrightarrow Z)$

The counterpart of *Celarent* is:

Table 2.40 Derivation of **D** from **M** and **F**

Derivations	Level	#	Symbol	ID		Traditional logical form
	7-1	4	M1=D2	11110111	×	$X \vee \neg Y \vee \neg Z$
1	7-1	1	F1=D1=E1	11111110	=	$\neg X \vee \neg Y \vee \neg Z$
	6-2	4	**D**	11110110		$Z \to \neg Y$
	7-1	6	M2	11011111	×	$X \vee \neg Y \vee Z$
2	7-1	1	F1=D1=E1	11111110	=	$\neg X \vee \neg Y \vee \neg Z$
	6-2	11	E1xM2	11011110		$Y \to \neg(X \leftrightarrow Z)$
	7-1	4	M1=D2	11110111	×	$X \vee \neg Y \vee \neg Z$
3	7-1	3	F2=A1	11111011	=	$\neg X \vee Y \vee \neg Z$
	6-2	6	A1xM1	11110011		$Z \to (X \leftrightarrow Y)$
	7-1	6	M2	11011111	×	$X \vee \neg Y \vee Z$
4	7-1	3	F2=A1	11111011	=	$\neg X \vee Y \vee \neg Z$
	6-2	13	A1xM2	11011011		$\neg(X \leftrightarrow \neg Y \leftrightarrow Z)$

The counterpart of *Darii* is:

Table 2.41 Derivation of **G** from **L** and **K**

Derivations	Level	#	Symbol	ID		Traditional logical form
	7-1	7	L1	10111111	×	$X \vee Y \vee \neg Z$
1	3-5	33	K6=G2	01001001	=	$Z \wedge (\neg X \vee Y)$
	2-6	25	**G**	00001001		$Z \wedge Y$
	7-1	8	L2	01111111	×	$X \vee Y \vee Z$
2	3-5	33	K6=G2	01001001	=	$Z \wedge (\neg X \vee Y)$
	3-5	33	K6=G2	01001001		$Z \wedge (\neg X \vee Y)$
	7-1	7	L1	10111111	×	$X \vee Y \vee \neg Z$
3	3-5	32	K5	01001010	=	$(\neg X \wedge Z) \vee (X \wedge Y \wedge \neg Z)$
	2-6	24	S8+I7	00001010		$Y \wedge \neg(X \leftrightarrow Z)$
	7-1	8	L2	01111111	×	$X \vee Y \vee Z$
4	3-5	32	K5	01001010	=	$(\neg X \wedge Z) \vee (X \wedge Y \wedge \neg Z)$
	3-5	32	K5	01001010		$(\neg X \wedge Z) \vee (X \wedge Y \wedge \neg Z)$
	7-1	7	L1	10111111	×	$X \vee Y \vee \neg Z$
5	3-5	31	K4=H4	01001100	=	$Z \wedge (\neg X \vee \neg Y)$
	2-6	23	S8+O8	00001100		$Z \wedge \neg(X \leftrightarrow Y)$
	7-1	8	L2	01111111	×	$X \vee Y \vee Z$
6	3-5	31	K4=H4	01001100	=	$Z \wedge (\neg X \vee \neg Y)$
	3-5	31	K4=H4	01001100		$Z \wedge (\neg X \vee \neg Y)$
	7-1	7	L1	10111111	×	$X \vee Y \vee \neg Z$
7	3-5	27	K3	01011000	=	$(\neg X \wedge Z) \vee (X \wedge \neg Y \wedge \neg Z)$
	2-6	19	O7+S8	00011000		$\neg X \leftrightarrow Y \leftrightarrow Z$
	7-1	8	L2	01111111	×	$X \vee Y \vee Z$
8	3-5	27	K3	01011000	=	$(\neg X \wedge Z) \vee (X \wedge \neg Y \wedge \neg Z)$
	3-5	27	K3	01011000		$(\neg X \wedge Z) \vee (X \wedge \neg Y \wedge \neg Z)$
	7-1	7	L1	10111111	×	$X \vee Y \vee \neg Z$
9	3-5	23	K2=S2	01101000	=	$\neg X \wedge (Y \vee Z)$
	2-6	15	**S**	00101000		$\neg X \wedge Y$
	7-1	8	L2	01111111	×	$X \vee Y \vee Z$
10	3-5	23	K2=S2	01101000	=	$\neg X \wedge (Y \vee Z)$
	3-5	23	K2=S2	01101000		$\neg X \wedge (Y \vee Z)$
	7-1	7	L1	10111111	×	$X \vee Y \vee \neg Z$
11	3-5	3	K1=T3	11001000	=	$\neg X \wedge (\neg Y \vee Z)$
	2-6	4	T7+S8	10001000		$\neg X \wedge (Y \leftrightarrow Z)$
	7-1	8	L2	01111111	×	$X \vee Y \vee Z$
12	3-5	3	K1=T3	11001000	=	$\neg X \wedge (\neg Y \vee Z)$
	2-6	10	**K**	01001000		$Z \wedge \neg X$

The counterpart of *Ferio* is:

Table 2.42 Derivation of **H** from **M** and **K**

Derivations	Level	#	Symbol	ID		Traditional logical form
	7-1	4	M1	11110111	×	$X \vee \neg Y \vee \neg Z$
1	3-5	33	K6=G2	01001001	=	$Z \wedge (\neg X \vee Y)$
	2-6	7	T7+I8	10000001		$X \leftrightarrow Y \leftrightarrow Z$
	7-1	6	M2	11011111	×	$X \vee \neg Y \vee Z$
2	3-5	33	K6=G2	01001001	=	$Z \wedge (\neg X \vee Y)$
	3-5	33	K6=G2	01001001		$Z \wedge (\neg X \vee Y)$
	7-1	4	M1	11110111	×	$X \vee \neg Y \vee \neg Z$
3	3-5	32	K5	01001010	=	$(\neg X \wedge Z) \vee (X \wedge Y \wedge \neg Z)$
	2-6	12	T8+I7	01000010		$X \leftrightarrow Y \leftrightarrow \neg Z$
	7-1	6	M2	11011111	×	$X \vee \neg Y \vee Z$
4	3-5	32	K5	01001010	=	$(\neg X \wedge Z) \vee (X \wedge Y \wedge \neg Z)$
	3-5	32	K5	01001010		$(\neg X \wedge Z) \vee (X \wedge Y \wedge \neg Z)$
	7-1	4	M1	11110111	×	$X \vee \neg Y \vee \neg Z$
5	3-5	31	K4=H4	01001100	=	$Z \wedge (\neg X \vee \neg Y)$
	2-6	11	**H**	01000100		$Z \wedge \neg Y$
	7-1	6	M2	11011111	×	$X \vee \neg Y \vee Z$
6	3-5	31	K4=H4	01001100	=	$Z \wedge (\neg X \vee \neg Y)$
	3-5	31	K4=H4	01001100		$Z \wedge (\neg X \vee \neg Y)$
	7-1	4	M1	11110111	×	$X \vee \neg Y \vee \neg Z$
7	3-5	27	K3	01011000	=	$(\neg X \wedge Z) \vee (X \wedge \neg Y \wedge \neg Z)$
	2-6	9	T8+O7	01010000		$\neg Y \wedge \neg(X \leftrightarrow Z)$
	7-1	6	M2	11011111	×	$X \vee \neg Y \vee Z$
8	3-5	27	K3	01011000	=	$(\neg X \wedge Z) \vee (X \wedge \neg Y \wedge \neg Z)$
	3-5	27	K3	01011000		$(\neg X \wedge Z) \vee (X \wedge \neg Y \wedge \neg Z)$
	7-1	4	M1	11110111	×	$X \vee \neg Y \vee \neg Z$
9	3-5	23	K2=S2	01101000	=	$\neg X \wedge (Y \vee Z)$
	2-6	8	T8+S7	01100000		$\neg X \wedge \neg(Y \leftrightarrow Z)$
	7-1	6	M2	11011111	×	$X \vee \neg Y \vee Z$
10	3-5	23	K2=S2	01101000	=	$\neg X \wedge (Y \vee Z)$
	2-6	10	**K**	01001000		$Z \wedge \neg X$
	7-1	4	M1	11110111	×	$X \vee \neg Y \vee \neg Z$
11	3-5	3	K1=T3	11001000	=	$\neg X \wedge (\neg Y \vee Z)$
	2-6	1	**T**	11000000		$\neg X \wedge \neg Y$
	7-1	6	M2	11011111	×	$X \vee \neg Y \vee Z$
12	3-5	3	K1=T3	11001000	=	$\neg X \wedge (\neg Y \vee Z)$
	3-5	3	K1=T3	11001000		$\neg X \wedge (\neg Y \vee Z)$

We have exhausted the first group of the first mixed mode. Now, I shall address the second group. The counterpart of *Baroco* is:

Table 2.43 Derivation of **J** from **L** and **H**

Derivations	Level	#	Symbol	ID		Traditional logical form
	7-1	7	L1	10111111	×	$X \vee Y \vee \neg Z$
1	3-5	35	H6=J2	01000101	=	$Z \wedge (X \vee \neg Y)$
	2-6	27	**J**	00000101		$Z \wedge X$
	7-1	8	L2	01111111	×	$X \vee Y \vee Z$
2	3-5	35	H6=J2	01000101	=	$Z \wedge (X \vee \neg Y)$
	3-5	35	H6=J2	01000101		$Z \wedge (X \vee \neg Y)$
	7-1	7	L1	10111111	×	$X \vee Y \vee \neg Z$
3	3-5	34	H5	01000110	=	$(\neg Y \wedge Z) \vee (X \wedge Y \wedge \neg Z)$
	2-6	26	O8+I7	00000110		$X \wedge \neg(Y \leftrightarrow Z)$
	7-1	8	L2	01111111	×	$X \vee Y \vee Z$
4	3-5	34	H5	01000110	=	$(\neg Y \wedge Z) \vee (X \wedge Y \wedge \neg Z)$
	3-5	34	H5	01000110		$(\neg Y \wedge Z) \vee (X \wedge Y \wedge \neg Z)$
	7-1	7	L1	10111111	×	$X \vee Y \vee \neg Z$
5	3-5	31	H4=K4	01001100	=	$Z \wedge (\neg X \vee \neg Y)$
	2-6	23	S8+O8	00001100		$Z \wedge \neg(X \leftrightarrow Y)$
	7-1	8	L2	01111111	×	$X \vee Y \vee Z$
6	3-5	31	H4=K4	01001100	=	$Z \wedge (\neg X \vee \neg Y)$
	3-5	31	H4=K4	01001100		$Z \wedge (\neg X \vee \neg Y)$
	7-1	7	L1	10111111	×	$X \vee Y \vee \neg Z$
7	3-5	28	H3=O2	01010100	=	$\neg Y \wedge (X \vee Z)$
	2-6	20	**O**	00010100		$X \wedge \neg Y$
	7-1	8	L2	01111111	×	$X \vee Y \vee Z$
8	3-5	28	H3=O2	01010100	=	$\neg Y \wedge (X \vee Z)$
	3-5	28	H3=O2	01010100		$\neg Y \wedge (X \vee Z)$
	7-1	7	L1	10111111	×	$X \vee Y \vee \neg Z$
9	3-5	24	H2	01100100	=	$(\neg Y \wedge Z) \vee (\neg X \wedge Y \wedge \neg Z)$
	2-6	16	S7+O8	00100100		$X \leftrightarrow \neg Y \leftrightarrow Z$
	7-1	8	L2	01111111	×	$X \vee Y \vee Z$
10	3-5	24	H2	01100100	=	$(\neg Y \wedge Z) \vee (\neg X \wedge Y \wedge \neg Z)$
	3-5	24	H2	01100100		$(\neg Y \wedge Z) \vee (\neg X \wedge Y \wedge \neg Z)$
	7-1	7	L1	10111111	×	$X \vee Y \vee \neg Z$
11	3-5	4	H1=T4	11000100	=	$\neg Y \wedge (\neg X \vee Z)$
	2-6	5	T7+O8	10000100		$\neg Y \wedge (X \leftrightarrow Z)$
	7-1	8	L2	01111111	×	$X \vee Y \vee Z$
12	3-5	4	H1=T4	11000100	=	$\neg Y \wedge (\neg X \vee Z)$
	2-6	11	**H**	01000100		$Z \wedge \neg Y$

The counterpart of *Festino* is:

Table 2.44 Derivation of **J** from **M** and **G**

Derivations	Level	#	Symbol	ID		Traditional logical form
	7-1	4	M1	11110111	×	$X \vee \neg Y \vee \neg Z$
1	3-5	55	G6=I5	00001011	=	$Y \wedge (X \vee Z)$
	2-6	28	**I**	00000011		$X \wedge Y$
	7-1	6	M2	11011111	×	$X \vee \neg Y \vee Z$
2	3-5	55	G6=I5	00001011	=	$Y \wedge (X \vee Z)$
	3-5	55	G6=I5	00001011		$Y \wedge (X \vee Z)$
	7-1	4	M1	11110111	×	$X \vee \neg Y \vee \neg Z$
3	3-5	54	G5=J5	00001101	=	$Z \wedge (X \vee Y)$
	2-6	27	**J**	00000101		$Z \wedge X$
	7-1	6	M2	11011111	×	$X \vee \neg Y \vee Z$
4	3-5	54	G5=J5	00001101	=	$Z \wedge (X \vee Y)$
	3-5	54	G5=J5	00001101		$Z \wedge (X \vee Y)$
	7-1	4	M1	11110111	×	$X \vee \neg Y \vee \neg Z$
5	3-5	49	G4	00011001	=	$(Y \wedge Z) \vee (X \wedge \neg Y \wedge \neg Z)$
	2-6	22	O7+I8	00010001		$X \wedge (Y \leftrightarrow Z)$
	7-1	6	M2	11011111	×	$X \vee \neg Y \vee Z$
6	3-5	49	G4	00011001	=	$(Y \wedge Z) \vee (X \wedge \neg Y \wedge \neg Z)$
	3-5	49	G4	00011001		$(Y \wedge Z) \vee (X \wedge \neg Y \wedge \neg Z)$
	7-1	4	M1	11110111	×	$X \vee \neg Y \vee \neg Z$
7	3-5	43	G3=S6	00101001	=	$Y \wedge (\neg X \vee Z)$
	2-6	18	S7+I8	00100001		$Y \wedge (X \leftrightarrow Z)$
	7-1	6	M2	11011111	×	$X \vee \neg Y \vee Z$
8	3-5	43	G3=S6	00101001	=	$Y \wedge (\neg X \vee Z)$
	2-6	25	**G**	00001001		$Z \wedge Y$
	7-1	4	M1	11110111	×	$X \vee \neg Y \vee \neg Z$
9	3-5	33	G2=K6	01001001	=	$Z \wedge (\neg X \vee Y)$
	2-6	13	T8+I8	01000001		$Z \wedge (X \leftrightarrow Y)$
	7-1	6	M2	11011111	×	$X \vee \neg Y \vee Z$
10	3-5	33	G2=K6	01001001	=	$Z \wedge (\neg X \vee Y)$
	3-5	33	G2=K6	01001001		$Z \wedge (\neg X \vee Y)$
	7-1	4	M1	11110111	×	$X \vee \neg Y \vee \neg Z$
11	3-5	18	G1	10001001	=	$(Y \wedge Z) \vee (\neg X \wedge \neg Y \wedge \neg Z)$
	2-6	7	T7+I8	10000001		$X \leftrightarrow Y \leftrightarrow Z$
	7-1	6	M2	11011111	×	$X \vee \neg Y \vee Z$
12	3-5	18	G1	10001001	=	$(Y \wedge Z) \vee (\neg X \wedge \neg Y \wedge \neg Z)$
	3-5	18	G1	10001001		$(Y \wedge Z) \vee (\neg X \wedge \neg Y \wedge \neg Z)$

The counterpart of *Camestres* is:

Table 2.45 Derivation of **B** from **L** and **D**

Derivations	Level	#	Symbol	ID		Traditional logical form
	7-1	7	L1=C2	10111111	×	$X \vee Y \vee \neg Z$
1	7-1	1	D1=E1	11111110	=	$\neg X \vee \neg Y \vee \neg Z$
	6-2	16	E1xL1	10111110		$Z \rightarrow \neg(X \leftrightarrow Y)$
	7-1	8	L2	01111111	×	$X \vee Y \vee Z$
2	7-1	1	D1=E1	11111110	=	$\neg X \vee \neg Y \vee \neg Z$
	6-2	22	E1xL2	01111110		$\neg(X \leftrightarrow Y \leftrightarrow Z)$
	7-1	7	L1=C2	10111111	×	$X \vee Y \vee \neg Z$
3	7-1	4	D2=B1=M1	11110111	=	$X \vee \neg Y \vee \neg Z$
	6-2	19	**B**	10110111		$Z \rightarrow X$
	7-1	8	L2	01111111	×	$X \vee Y \vee Z$
4	7-1	4	D2=B1=M1	11110111	=	$X \vee \neg Y \vee \neg Z$
	6-2	25	M1xL2	01110111		$\neg X \rightarrow \neg(Y \leftrightarrow Z)$

The counterpart of *Cesare* is:

Table 2.46 Derivation of **B** from **M** and **C**

Derivations	Level	#	Symbol	ID		Traditional logical form
	7-1	4	M1	11110111	×	$X \vee \neg Y \vee \neg Z$
1	7-1	3	C1=A1	11111011	=	$\neg X \vee Y \vee \neg Z$
	6-2	6	A1xM1	11110011		$Z \rightarrow (X \leftrightarrow Y)$
	7-1	6	M2	11011111	×	$X \vee \neg Y \vee Z$
2	7-1	3	C1=A1	11111011	=	$\neg X \vee Y \vee \neg Z$
	6-2	13	A1xM2	11011011		$\neg(X \leftrightarrow \neg Y \leftrightarrow Z)$
	7-1	4	M1	11110111	×	$X \vee \neg Y \vee \neg Z$
3	7-1	7	C2=B2=L1	10111111	=	$X \vee Y \vee \neg Z$
	6-2	19	**B**	10110111		$Z \rightarrow X$
	7-1	6	M2	11011111	×	$X \vee \neg Y \vee Z$
4	7-1	7	C2=B2=L1	10111111	=	$X \vee Y \vee \neg Z$
	6-2	21	M2xL1	10011111		$\neg X \rightarrow (Y \leftrightarrow Z)$

We have exhausted the second group. I pass now to the third group. The counterpart of *Bocardo* is:

Table 2.47 Derivation of **T** from **H** and **F**

Derivations	Level	#	Symbol	ID		Traditional logical form
	3-5	35	H6=J2	01000101	×	$Z \wedge (X \vee \neg Y)$
1	7-1	1	F1=E1	11111110	=	$\neg X \vee \neg Y \vee \neg Z$
	2-6	11	**H**	01000100		$Z \wedge \neg Y$
	3-5	35	H6=J2	01000101	×	$Z \wedge (X \vee \neg Y)$
2	7-1	3	F2=A1	11111011	=	$\neg X \vee Y \vee \neg Z$
	2-6	13	T8+I8	01000001		$Z \wedge (X \leftrightarrow Y)$
	3-5	34	H5	01000110	×	$(\neg Y \wedge Z) \vee (X \wedge Y \wedge \neg Z)$
3	7-1	1	F1=E1	11111110	=	$\neg X \vee \neg Y \vee \neg Z$
	3-5	34	H5	01000110		$(\neg Y \wedge Z) \vee (X \wedge Y \wedge \neg Z)$
	3-5	34	H5	01000110	×	$(\neg Y \wedge Z) \vee (X \wedge Y \wedge \neg Z)$
4	7-1	3	F2=A1	11111011	=	$\neg X \vee Y \vee \neg Z$
	2-6	12	T8+I7	01000010		$X \leftrightarrow Y \leftrightarrow \neg Z$
	3-5	31	H4=K4	01001100	×	$Z \wedge (\neg X \vee \neg Y)$
5	7-1	1	F1=E1	11111110	=	$\neg X \vee \neg Y \vee \neg Z$
	3-5	31	H4=K4	01001100		$Z \wedge (\neg X \vee \neg Y)$
	3-5	31	H4=K4	01001100	×	$Z \wedge (\neg X \vee \neg Y)$
6	7-1	3	F2=A1	11111011	=	$\neg X \vee Y \vee \neg Z$
	2-6	10	**K**	01001000		$Z \wedge \neg X$
	3-5	28	H3=O2	01010100	×	$\neg Y \wedge (X \vee Z)$
7	7-1	1	F1=E1	11111110	=	$\neg X \vee \neg Y \vee \neg Z$
	3-5	28	H3=O2	01010100		$\neg Y \wedge (X \vee Z)$
	3-5	28	H3=O2	01010100	×	$\neg Y \wedge (X \vee Z)$
8	7-1	3	F2=A1	11111011	=	$\neg X \vee Y \vee \neg Z$
	2-6	9	T8+O7	01010000		$\neg Y \wedge \neg(X \leftrightarrow Z)$
	3-5	24	H2	01100100	×	$(\neg Y \wedge Z) \vee (\neg X \wedge Y \wedge \neg Z)$
9	7-1	1	F1=E1	11111110	=	$\neg X \vee \neg Y \vee \neg Z$
	3-5	24	H2	01100100		$(\neg Y \wedge Z) \vee (\neg X \wedge Y \wedge \neg Z)$
	3-5	24	H2	01100100	×	$(\neg Y \wedge Z) \vee (\neg X \wedge Y \wedge \neg Z)$
10	7-1	3	F2=A1	11111011	=	$\neg X \vee Y \vee \neg Z$
	2-6	8	T8+S7	01100000		$\neg X \wedge \neg(Y \leftrightarrow Z)$
	3-5	4	H1=T4	11000100	×	$\neg Y \wedge (\neg X \vee Z)$
11	7-1	1	F1=E1	11111110	=	$\neg X \vee \neg Y \vee \neg Z$
	3-5	4	H1=T4	11000100		$\neg Y \wedge (\neg X \vee Z)$
	3-5	4	H1=T4	11000100	×	$\neg Y \wedge (\neg X \vee Z)$
12	7-1	3	F2=A1	11111011	=	$\neg X \vee Y \vee \neg Z$
	2-6	1	**T**	11000000		$\neg X \wedge \neg Y$

The counterpart of *Disamis* is:

Table 2.48 Derivation of **S** from **G** and **F**

Derivations	Level	#	Symbol	ID		Traditional logical form
	3-5	55	G6=I5	00001011	×	$Y \wedge (X \vee Z)$
1	7-1	1	F1=E1	11111110	=	$\neg X \vee \neg Y \vee \neg Z$
	2-6	24	S8+I7	00001010		$Y \wedge \neg(X \leftrightarrow Z)$
	3-5	55	G6=I5	00001011	×	$Y \wedge (X \vee Z)$
2	7-1	3	F2=A1	11111011	=	$\neg X \vee Y \vee \neg Z$
	3-5	55	G6=I5	00001011		$Y \wedge (X \vee Z)$
	3-5	54	G5=J5	00001101	×	$Z \wedge (X \vee Y)$
3	7-1	1	F1=E1	11111110	=	$\neg X \vee \neg Y \vee \neg Z$
	2-6	23	S8+O8	00001100		$Z \wedge \neg(X \leftrightarrow Y)$
	3-5	54	G5=J5	00001101	×	$Z \wedge (X \vee Y)$
4	7-1	3	F2=A1	11111011	=	$\neg X \vee Y \vee \neg Z$
	2-6	25	**G**	00001001		$Z \wedge Y$
	3-5	49	G4	00011001	×	$(Y \wedge Z) \vee (X \wedge \neg Y \wedge \neg Z)$
5	7-1	1	F1=E1	11111110	=	$\neg X \vee \neg Y \vee \neg Z$
	2-6	19	O7+S8	00011000		$\neg X \leftrightarrow Y \leftrightarrow Z$
	3-5	49	G4	00011001	×	$(Y \wedge Z) \vee (X \wedge \neg Y \wedge \neg Z)$
6	7-1	3	F2=A1	11111011	=	$\neg X \vee Y \vee \neg Z$
	3-5	49	G4	00011001		$(Y \wedge Z) \vee (X \wedge \neg Y \wedge \neg Z)$
	3-5	43	G3=S6	00101001	×	$Y \wedge (\neg X \vee Z)$
7	7-1	1	F1=E1	11111110	=	$\neg X \vee \neg Y \vee \neg Z$
	2-6	15	**S**	00101000		$\neg X \wedge Y$
	3-5	43	G3=S6	00101001	×	$Y \wedge (\neg X \vee Z)$
8	7-1	3	F2=A1	11111011	=	$\neg X \vee Y \vee \neg Z$
	3-5	43	G3=S6	00101001		$Y \wedge (\neg X \vee Z)$
	3-5	33	G2=K6	01001001	×	$Z \wedge (\neg X \vee Y)$
9	7-1	1	F1=E1	11111110	=	$\neg X \vee \neg Y \vee \neg Z$
	2-6	10	**K**	01001000		$Z \wedge \neg X$
	3-5	33	G2=K6	01001001	×	$Z \wedge (\neg X \vee Y)$
10	7-1	3	F2=A1	11111011	=	$\neg X \vee Y \vee \neg Z$
	3-5	33	G2=K6	01001001		$Z \wedge (\neg X \vee Y)$
	3-5	18	G1	10001001	×	$(Y \wedge Z) \vee (\neg X \wedge \neg Y \wedge \neg Z)$
11	7-1	1	F1=E1	11111110	=	$\neg X \vee \neg Y \vee \neg Z$
	2-6	4	T7+S8	10001000		$\neg X \wedge (Y \leftrightarrow Z)$
	3-5	18	G1	10001001	×	$(Y \wedge Z) \vee (\neg X \wedge \neg Y \wedge \neg Z)$
12	7-1	3	F2=A1	11111011	=	$\neg X \vee Y \vee \neg Z$
	3-5	18	G1	10001001		$(Y \wedge Z) \vee (\neg X \wedge \neg Y \wedge \neg Z)$

The counterpart of *Ferison* is:

Table 2.49 Derivation of **T** from **D** and **K**

Derivations	Level	#	Symbol	ID		Traditional logical form
	7-1	1	D1=E1	11111110	×	$\neg X \vee \neg Y \vee \neg Z$
1	3-5	33	K6=G2	01001001	=	$Z \wedge (\neg X \vee Y)$
	2-6	10	**K**	01001000		$Z \wedge \neg X$
	7-1	4	D2=M1	11110111	×	$X \vee \neg Y \vee \neg Z$
2	3-5	33	K6=G2	01001001	=	$Z \wedge (\neg X \vee Y)$
	2-6	13	T8+I8	01000001		$Z \wedge (X \leftrightarrow Y)$
	7-1	1	D1=E1	11111110	×	$\neg X \vee \neg Y \vee \neg Z$
3	3-5	32	K5	01001010	=	$(\neg X \wedge Z) \vee (X \wedge Y \wedge \neg Z)$
	3-5	32	K5	01001010		$(\neg X \wedge Z) \vee (X \wedge Y \wedge \neg Z)$
	7-1	4	D2=M1	11110111	×	$X \vee \neg Y \vee \neg Z$
4	3-5	32	K5	01001010	=	$(\neg X \wedge Z) \vee (X \wedge Y \wedge \neg Z)$
	2-6	12	T8+I7	01000010		$X \leftrightarrow Y \leftrightarrow \neg Z$
	7-1	1	D1=E1	11111110	×	$\neg X \vee \neg Y \vee \neg Z$
5	3-5	31	K4=H4	01001100	=	$Z \wedge (\neg X \vee \neg Y)$
	3-5	31	K4=H4	01001100		$Z \wedge (\neg X \vee \neg Y)$
	7-1	4	D2=M1	11110111	×	$X \vee \neg Y \vee \neg Z$
6	3-5	31	K4=H4	01001100	=	$Z \wedge (\neg X \vee \neg Y)$
	2-6	11	**H**	01000100		$Z \wedge \neg Y$
	7-1	1	D1=E1	11111110	×	$\neg X \vee \neg Y \vee \neg Z$
7	3-5	27	K3	01011000	=	$(\neg X \wedge Z) \vee (X \wedge \neg Y \wedge \neg Z)$
	3-5	27	K3	01011000		$(\neg X \wedge Z) \vee (X \wedge \neg Y \wedge \neg Z)$
	7-1	4	D2=M1	11110111	×	$X \vee \neg Y \vee \neg Z$
8	3-5	27	K3	01011000	=	$(\neg X \wedge Z) \vee (X \wedge \neg Y \wedge \neg Z)$
	2-6	9	T8+O7	01010000		$\neg Y \wedge \neg(X \leftrightarrow Z)$
	7-1	1	D1=E1	11111110	×	$\neg X \vee \neg Y \vee \neg Z$
9	3-5	23	K2=S2	01101000	=	$\neg X \wedge (Y \vee Z)$
	3-5	23	K2=S2	01101000		$\neg X \wedge (Y \vee Z)$
	7-1	4	D2=M1	11110111	×	$X \vee \neg Y \vee \neg Z$
10	3-5	23	K2=S2	01101000	=	$\neg X \wedge (Y \vee Z)$
	2-6	8	T8+S7	01100000		$\neg X \wedge \neg(Y \leftrightarrow Z)$
	7-1	1	D1=E1	11111110	×	$\neg X \vee \neg Y \vee \neg Z$
11	3-5	3	K1=T3	11001000	=	$\neg X \wedge (\neg Y \vee Z)$
	3-5	3	K1=T3	11001000		$\neg X \wedge (\neg Y \vee Z)$
	7-1	4	D2=M1	11110111	×	$X \vee \neg Y \vee \neg Z$
12	3-5	3	K1=T3	11001000	=	$\neg X \wedge (\neg Y \vee Z)$
	2-6	1	**T**	11000000		$\neg X \wedge \neg Y$

The counterpart of *Datisi* is:

Table 2.50 Derivation of **S** from **C** and **K**

Derivations	Level	#	Symbol	ID		Traditional logical form
	7-1	3	C1=A1	11111011	×	$\neg X \vee Y \vee \neg Z$
1	3-5	33	K6=G2	01001001	=	$Z \wedge (\neg X \vee Y)$
	3-5	33	K6=G2	01001001		$Z \wedge (\neg X \vee Y)$
	7-1	7	C2=L1	10111111	×	$X \vee Y \vee \neg Z$
2	3-5	33	K6=G2	01001001	=	$Z \wedge (\neg X \vee Y)$
	2-6	25	**G**	00001001		$Z \wedge Y$
	7-1	3	C1=A1	11111011	×	$\neg X \vee Y \vee \neg Z$
3	3-5	32	K5	01001010	=	$(\neg X \wedge Z) \vee (X \wedge Y \wedge \neg Z)$
	3-5	32	K5	01001010		$(\neg X \wedge Z) \vee (X \wedge Y \wedge \neg Z)$
	7-1	7	C2=L1	10111111	×	$X \vee Y \vee \neg Z$
4	3-5	32	K5	01001010	=	$(\neg X \wedge Z) \vee (X \wedge Y \wedge \neg Z)$
	2-6	24	S8+I7	00001010		$Y \wedge \neg(X \leftrightarrow Z)$
	7-1	3	C1=A1	11111011	×	$\neg X \vee Y \vee \neg Z$
5	3-5	31	K4=H4	01001100	=	$Z \wedge (\neg X \vee \neg Y)$
	2-6	10	**K**	01001000		$Z \wedge \neg X$
	7-1	7	C2=L1	10111111	×	$X \vee Y \vee \neg Z$
6	3-5	31	K4=H4	01001100	=	$Z \wedge (\neg X \vee \neg Y)$
	2-6	23	S8+O8	00001100		$Z \wedge \neg(X \leftrightarrow Y)$
	7-1	3	C1=A1	11111011	×	$\neg X \vee Y \vee \neg Z$
7	3-5	27	K3	01011000	=	$(\neg X \wedge Z) \vee (X \wedge \neg Y \wedge \neg Z)$
	3-5	27	K3	01011000		$(\neg X \wedge Z) \vee (X \wedge \neg Y \wedge \neg Z)$
	7-1	7	C2=L1	10111111	×	$X \vee Y \vee \neg Z$
8	3-5	27	K3	01011000	=	$(\neg X \wedge Z) \vee (X \wedge \neg Y \wedge \neg Z)$
	2-6	19	O7+S8	00011000		$\neg X \leftrightarrow Y \leftrightarrow Z$
	7-1	3	C1=A1	11111011	×	$\neg X \vee Y \vee \neg Z$
9	3-5	23	K2=S2	01101000	=	$\neg X \wedge (Y \vee Z)$
	3-5	23	K2=S2	01101000		$\neg X \wedge (Y \vee Z)$
	7-1	7	C2=L1	10111111	×	$X \vee Y \vee \neg Z$
10	3-5	23	K2=S2	01101000	=	$\neg X \wedge (Y \vee Z)$
	2-6	15	**S**	00101000		$\neg X \wedge Y$
	7-1	3	C1=A1	11111011	×	$\neg X \vee Y \vee \neg Z$
11	3-5	3	K1=T3	11001000	=	$\neg X \wedge (\neg Y \vee Z)$
	3-5	3	K1=T3	11001000		$\neg X \wedge (\neg Y \vee Z)$
	7-1	7	C2=L1	10111111	×	$X \vee Y \vee \neg Z$
12	3-5	3	K1=T3	11001000	=	$\neg X \wedge (\neg Y \vee Z)$
	2-6	4	T7+S8	10001000		$\neg X \wedge (Y \leftrightarrow Z)$

2.5 Derivation of the Inferences of the Second Mixed Mode through Products

As mentioned, the second mixed mode is characterized by two statements of the classical mode and one out of its extension. The counterpart of *Barbara* is:

Table 2.51 Derivation of **N** from **A** and **Q**

Derivations	Level	#	Symbol	ID		Traditional logical form
	7-1	3	A1	11111011	×	$\neg X \vee Y \vee \neg Z$
1	7-1	6	Q1=M2	11011111	=	$X \vee \neg Y \vee Z$
	6-2	13	A1xM2	11011011		$\neg(X \leftrightarrow \neg Y \leftrightarrow Z)$
	7-1	5	A2=N1	11101111	×	$\neg X \vee Y \vee Z$
2	7-1	6	Q1=M2	11011111	=	$X \vee \neg Y \vee Z$
	6-2	15	A2xM2	11001111		$\neg Z \rightarrow (X \leftrightarrow Y)$
	7-1	3	A1	11111011	×	$\neg X \vee Y \vee \neg Z$
3	7-1	8	Q2=N2=L2	01111111	=	$X \vee Y \vee Z$
	6-2	24	A1xL2	01111011		$\neg Y \rightarrow \neg(X \leftrightarrow Z)$
	7-1	5	A2=N1	11101111	×	$\neg X \vee Y \vee Z$
4	7-1	8	Q2=N2=L2	01111111	=	$X \vee Y \vee Z$
	6-2	26	**N**	01101111		$\neg Z \rightarrow Y$

The counterpart of *Celarent* is:

Table 2.52 Derivation of **P** from **E** and **Q**

Derivations	Level	#	Symbol	ID		Traditional logical form
	7-1	1	E1	11111110	×	$\neg X \vee \neg Y \vee \neg Z$
1	7-1	6	Q1=P2=M2	11011111	=	$X \vee \neg Y \vee Z$
	6-2	11	E1xM2	11011110		$Y \rightarrow \neg(X \leftrightarrow Z)$
	7-1	2	E2=P1	11111101	×	$\neg X \vee \neg Y \vee Z$
2	7-1	6	Q1=P2=M2	11011111	=	$X \vee \neg Y \vee Z$
	6-2	12	**P**	11011101		$\neg Z \rightarrow \neg Y$
	7-1	1	E1	11111110	×	$\neg X \vee \neg Y \vee \neg Z$
3	7-1	8	Q2=N2=L2	01111111	=	$X \vee Y \vee Z$
	6-2	22	E1xL2	01111110		$\neg(X \leftrightarrow Y \leftrightarrow Z)$
	7-1	2	E2=P1	11111101	×	$\neg X \vee \neg Y \vee Z$
4	7-1	8	Q2=N2=L2	01111111	=	$X \vee Y \vee Z$
	6-2	23	E2xL2	01111101		$\neg Z \rightarrow \neg(X \leftrightarrow Y)$

The counterpart of *Darii* is:

Table 2.53 Derivation of **U** from **A** and **W**

Derivations	Level	#	Symbol	ID		Traditional logical form
	7-1	3	A1	11111011	×	$\neg X \vee Y \vee \neg Z$
1	3-5	14	W1=V5	10010010	=	$\neg Z \wedge (X \vee \neg Y)$
	3-5	14	W1=V5	10010010		$\neg Z \wedge (X \vee \neg Y)$
	7-1	5	A2	11101111	×	$\neg X \vee Y \vee Z$
2	3-5	14	W1=V5	10010010	=	$\neg Z \wedge (X \vee \neg Y)$
	2-6	6	T7+I7	10000010		$\neg Z \wedge (X \leftrightarrow Y)$
	7-1	3	A1	11111011	×	$\neg X \vee Y \vee \neg Z$
3	3-5	29	W2	01010010	=	$(\neg Z \wedge X) \vee (\neg X \wedge \neg Y \wedge Z)$
	3-5	29	W2	01010010		$(\neg Z \wedge X) \vee (\neg X \wedge \neg Y \wedge Z)$
	7-1	5	A2	11101111	×	$\neg X \vee Y \vee Z$
4	3-5	29	W2	01010010	=	$(\neg Z \wedge X) \vee (\neg X \wedge \neg Y \wedge Z)$
	2-6	12	T8+I7	01000010		$X \leftrightarrow Y \leftrightarrow \neg Z$
	7-1	3	A1	11111011	×	$\neg X \vee Y \vee \neg Z$
5	3-5	39	W3=U3	00110010	=	$\neg Z \wedge (X \vee Y)$
	3-5	39	W3=U3	00110010		$\neg Z \wedge (X \vee Y)$
	7-1	5	A2	11101111	×	$\neg X \vee Y \vee Z$
6	3-5	39	W3=U3	00110010	=	$\neg Z \wedge (X \vee Y)$
	2-6	17	**U**	00100010		$\neg Z \wedge Y$
	7-1	3	A1	11111011	×	$\neg X \vee Y \vee \neg Z$
7	3-5	48	W4	00011010	=	$(\neg Z \wedge X) \vee (\neg X \wedge Y \wedge Z)$
	3-5	48	W4	00011010		$(\neg Z \wedge X) \vee (\neg X \wedge Y \wedge Z)$
	7-1	5	A2	11101111	×	$\neg X \vee Y \vee Z$
8	3-5	48	W4	00011010	=	$(\neg Z \wedge X) \vee (\neg X \wedge Y \wedge Z)$
	2-6	24	S8+I7	00001010		$Y \wedge \neg(X \leftrightarrow Z)$
	7-1	3	A1	11111011	×	$\neg X \vee Y \vee \neg Z$
9	3-5	50	W5=O5	00010110	=	$X \wedge (\neg Y \vee \neg Z)$
	2-6	21	**W**	00010010		$\neg Z \wedge X$
	7-1	5	A2	11101111	×	$\neg X \vee Y \vee Z$
10	3-5	50	W5=O5	00010110	=	$X \wedge (\neg Y \vee \neg Z)$
	2-6	26	O8+I7	00000110		$X \wedge \neg(Y \leftrightarrow Z)$
	7-1	3	A1	11111011	×	$\neg X \vee Y \vee \neg Z$
11	3-5	52	W6=I4	00010011	=	$X \wedge (Y \vee \neg Z)$
	3-5	52	W6=I4	00010011		$X \wedge (Y \vee \neg Z)$
	7-1	5	A2	11101111	×	$\neg X \vee Y \vee Z$
12	3-5	52	W6=I4	00010011	=	$X \wedge (Y \vee \neg Z)$
	2-6	28	**I**	00000011		$X \wedge Y$

The counterpart of *Ferio* is:

Table 2.54 Derivation of **V** from **E** and **W**

Derivations	Level	#	Symbol	ID		Traditional logical form
	7-1	1	E1	11111110	×	$\neg X \vee \neg Y \vee \neg Z$
1	3-5	14	W1=V5	10010010	=	$\neg Z \wedge (X \vee \neg Y)$
	3-5	14	W1=V5	10010010		$\neg Z \wedge (X \vee \neg Y)$
	7-1	2	E2	11111101	×	$\neg X \vee \neg Y \vee Z$
2	3-5	14	W1=V5	10010010	=	$\neg Z \wedge (X \vee \neg Y)$
	2-6	3	**V**	10010000		$\neg Z \wedge \neg Y$
	7-1	1	E1	11111110	×	$\neg X \vee \neg Y \vee \neg Z$
3	3-5	29	W2	01010010	=	$(\neg Z \wedge X) \vee (\neg X \wedge \neg Y \wedge Z)$
	3-5	29	W2	01010010		$(\neg Z \wedge X) \vee (\neg X \wedge \neg Y \wedge Z)$
	7-1	2	E2	11111101	×	$\neg X \vee \neg Y \vee Z$
4	3-5	29	W2	01010010	=	$(\neg Z \wedge X) \vee (\neg X \wedge \neg Y \wedge Z)$
	2-6	9	T8+O7	01010000		$\neg Y \wedge \neg(X \leftrightarrow Z)$
	7-1	1	E1	11111110	×	$\neg X \vee \neg Y \vee \neg Z$
5	3-5	39	W3=U3	00110010	=	$\neg Z \wedge (X \vee Y)$
	3-5	39	W3=U3	00110010		$\neg Z \wedge (X \vee Y)$
	7-1	2	E2	11111101	×	$\neg X \vee \neg Y \vee Z$
6	3-5	39	W3=U3	00110010	=	$\neg Z \wedge (X \vee Y)$
	2-6	14	S7+O7	00110000		$\neg Z \wedge \neg(X \leftrightarrow Y)$
	7-1	1	E1	11111110	×	$\neg X \vee \neg Y \vee \neg Z$
7	3-5	48	W4	00011010	=	$(\neg Z \wedge X) \vee (\neg X \wedge Y \wedge Z)$
	3-5	48	W4	00011010		$(\neg Z \wedge X) \vee (\neg X \wedge Y \wedge Z)$
	7-1	2	E2	11111101	×	$\neg X \vee \neg Y \vee Z$
8	3-5	48	W4	00011010	=	$(\neg Z \wedge X) \vee (\neg X \wedge Y \wedge Z)$
	2-6	19	O7+S8	00011000		$\neg X \leftrightarrow Y \leftrightarrow Z$
	7-1	1	E1	11111110	×	$\neg X \vee \neg Y \vee \neg Z$
9	3-5	50	W5=O5	00010110	=	$X \wedge (\neg Y \vee \neg Z)$
	3-5	50	W5=O5	00010110	=	$X \wedge (\neg Y \vee \neg Z)$
	7-1	2	E2	11111101	×	$\neg X \vee \neg Y \vee Z$
10	3-5	50	W5=O5	00010110	=	$X \wedge (\neg Y \vee \neg Z)$
	2-6	20	**O**	00010100		$X \wedge \neg Y$
	7-1	1	E1	11111110	×	$\neg X \vee \neg Y \vee \neg Z$
11	3-5	52	W6=I4	00010011	=	$X \wedge (Y \vee \neg Z)$
	2-6	21	**W**	00010010		$\neg Z \wedge X$
	7-1	2	E2	11111101	×	$\neg X \vee \neg Y \vee Z$
12	3-5	52	W6=I4	00010011	=	$X \wedge (Y \vee \neg Z)$
	2-6	22	O7+I8	00010001		$X \wedge (Y \leftrightarrow Z)$

We have completed the first group of the second mixed mode. Now, I shall introduce the second group and present the counterpart of *Baroco*:

Table 2.55 Derivation of **Ω** from **A** and **V**

Derivations	Level	#	Symbol	ID		Traditional logical form
	7-1	3	A1	11111011	×	$\neg X \vee Y \vee \neg Z$
1	3-5	2	V1=T2	11010000	=	$\neg Y \wedge (\neg X \vee \neg Z)$
	3-5	2	V1=T2	11010000		$\neg Y \wedge (\neg X \vee \neg Z)$
	7-1	5	A2	11101111	×	$\neg X \vee Y \vee Z$
2	3-5	2	V1=T2	11010000	=	$\neg Y \wedge (\neg X \vee \neg Z)$
	2-6	1	**T**	11000000		$\neg X \wedge \neg Y$
	7-1	3	A1	11111011	×	$\neg X \vee Y \vee \neg Z$
3	3-5	7	V2=Ω2	10110000	=	$\neg Z \wedge (\neg X \vee \neg Y)$
	3-5	7	V2=Ω2	10110000		$\neg Z \wedge (\neg X \vee \neg Y)$
	7-1	5	A2	11101111	×	$\neg X \vee Y \vee Z$
4	3-5	7	V2=Ω2	10110000	=	$\neg Z \wedge (\neg X \vee \neg Y)$
	2-6	2	**Ω**	10100000		$\neg Z \wedge \neg X$
	7-1	3	A1	11111011	×	$\neg X \vee Y \vee \neg Z$
5	3-5	12	V3	10011000	=	$(\neg Z \wedge \neg Y) \vee (\neg X \wedge Y \wedge Z)$
	3-5	12	V3	10011000		$(\neg Z \wedge \neg Y) \vee (\neg X \wedge Y \wedge Z)$
	7-1	5	A2	11101111	×	$\neg X \vee Y \vee Z$
6	3-5	12	V3	10011000	=	$(\neg Z \wedge \neg Y) \vee (\neg X \wedge Y \wedge Z)$
	2-6	4	T7+S8	10001000		$\neg X \wedge (Y \leftrightarrow Z)$
	7-1	3	A1	11111011	×	$\neg X \vee Y \vee \neg Z$
7	3-5	13	V4=O1	10010100	=	$\neg Y \wedge (X \vee \neg Z)$
	2-6	3	**V**	10010000		$\neg Z \wedge \neg Y$
	7-1	5	A2	11101111	×	$\neg X \vee Y \vee Z$
8	3-5	13	V4=O1	10010100	=	$\neg Y \wedge (X \vee \neg Z)$
	3-5	5	T7+O8	10000100		$\neg Y \wedge (X \leftrightarrow Z)$
	7-1	3	A1	11111011	×	$\neg X \vee Y \vee \neg Z$
9	3-5	14	V5=W1	10010010	=	$\neg Z \wedge (X \vee \neg Y)$
	3-5	14	V5=W1	10010010		$\neg Z \wedge (X \vee \neg Y)$
	7-1	5	A2	11101111	×	$\neg X \vee Y \vee Z$
10	3-5	14	V5=W1	10010010	=	$\neg Z \wedge (X \vee \neg Y)$
	3-5	6	T7+I7	10000010		$\neg Z \wedge (X \leftrightarrow Y)$
	7-1	3	A1	11111011	×	$\neg X \vee Y \vee \neg Z$
11	3-5	15	V6	10010001	=	$(\neg Z \wedge \neg Y) \vee (X \wedge Y \wedge Z)$
	3-5	15	V6	10010001		$(\neg Z \wedge \neg Y) \vee (X \wedge Y \wedge Z)$
	7-1	5	A2	11101111	×	$\neg X \vee Y \vee Z$
12	3-5	15	V6	10010001	=	$(\neg Z \wedge \neg Y) \vee (X \wedge Y \wedge Z)$
	2-6	7	T7+I8	10000001		$X \leftrightarrow Y \leftrightarrow Z$

The counterpart of *Festino* is:

Table 2.56 Derivation of **Ω** from **E** and **U**

Derivations	Level	#	Symbol	ID		Traditional logical form
	7-1	1	E1	11111110	×	$\neg X \vee \neg Y \vee \neg Z$
1	3-5	10	U1=Ω5	10100010	=	$\neg Z \wedge (\neg X \vee Y)$
	3-5	10	U1=Ω5	10100010		$\neg Z \wedge (\neg X \vee Y)$
	7-1	2	E2	11111101	×	$\neg X \vee \neg Y \vee Z$
2	3-5	10	U1=Ω5	10100010	=	$\neg Z \wedge (\neg X \vee Y)$
	2-6	2	**Ω**	10100000		$\neg Z \wedge \neg X$
	7-1	1	E1	11111110	×	$\neg X \vee \neg Y \vee \neg Z$
3	3-5	25	U2	01100010	=	$(\neg Z \wedge Y) \vee (\neg X \wedge \neg Y \wedge Z)$
	3-5	25	U2	01100010		$(\neg Z \wedge Y) \vee (\neg X \wedge \neg Y \wedge Z)$
	7-1	2	E2	11111101	×	$\neg X \vee \neg Y \vee Z$
4	3-5	25	U2	01100010	=	$(\neg Z \wedge Y) \vee (\neg X \wedge \neg Y \wedge Z)$
	2-6	8	T8+S7	01100000		$\neg X \wedge \neg(Y \leftrightarrow Z)$
	7-1	1	E1	11111110	×	$\neg X \vee \neg Y \vee \neg Z$
5	3-5	39	U3=W3	00110010	=	$\neg Z \wedge (X \vee Y)$
	3-5	39	U3=W3	00110010		$\neg Z \wedge (X \vee Y)$
	7-1	2	E2	11111101	×	$\neg X \vee \neg Y \vee Z$
6	3-5	39	U3=W3	00110010	=	$\neg Z \wedge (X \vee Y)$
	2-6	14	S7+O7	00110000		$\neg Z \wedge \neg(X \leftrightarrow Y)$
	7-1	1	E1	11111110	×	$\neg X \vee \neg Y \vee \neg Z$
7	3-5	42	U4=S5	00101010	=	$Y \wedge (\neg X \vee \neg Z)$
	3-5	42	U4=S5	00101010		$Y \wedge (\neg X \vee \neg Z)$
	7-1	2	E2	11111101	×	$\neg X \vee \neg Y \vee Z$
8	3-5	42	U4=S5	00101010	=	$Y \wedge (\neg X \vee \neg Z)$
	2-6 2-6	15	**S**	00101000		$\neg X \wedge Y$
	7-1	1	E1	11111110	×	$\neg X \vee \neg Y \vee \neg Z$
9	3-5	44	U5	00100110	=	$(\neg Z \wedge Y) \vee (X \wedge \neg Y \wedge Z)$
	3-5	44	U5	00100110		$(\neg Z \wedge Y) \vee (X \wedge \neg Y \wedge Z)$
	7-1	2	E2	11111101	×	$\neg X \vee \neg Y \vee Z$
10	3-5	44	U5	00100110	=	$(\neg Z \wedge Y) \vee (X \wedge \neg Y \wedge Z)$
	2-6	16	S7+O8	00100100		$X \leftrightarrow \neg Y \leftrightarrow Z$
	7-1	1	E1	11111110	×	$\neg X \vee \neg Y \vee \neg Z$
11	3-5	46	U6=I3	00100011	=	$Y \wedge (X \vee \neg Z)$
	2-6	17	**U**	00100010		$\neg Z \wedge Y$
	7-1	2	E2	11111101	×	$\neg X \vee \neg Y \vee Z$
12	3-5	46	U6=I3	00100011	=	$Y \wedge (X \vee \neg Z)$
	2-6	18	S7+I8	00100001		$Y \wedge (X \leftrightarrow Z)$

The counterpart of *Camestres* is:

Table 2.57 Derivation of **R** from **A** and **P**

Derivations	Level	#	Symbol	ID		Traditional logical form
	7-1	3	A1	11111011	×	$\neg X \vee Y \vee \neg Z$
1	7-1	2	P1=R1=E2	11111101	=	$\neg X \vee \neg Y \vee Z$
	6-2	3	E2xA1	11111001		$X \rightarrow (Y \leftrightarrow Z)$
	7-1	5	A2	11101111	×	$\neg X \vee Y \vee Z$
2	7-1	2	P1=R1=E2	11111101	=	$\neg X \vee \neg Y \vee Z$
	6-2	8	**R**	11101101		$\neg Z \rightarrow \neg X$
	7-1	3	A1	11111011	×	$\neg X \vee Y \vee \neg Z$
3	7-1	6	P2=M2	11011111	=	$X \vee \neg Y \vee Z$
	6-2	13	A1xM2	11011011		$\neg(X \leftrightarrow \neg Y \leftrightarrow Z)$
	7-1	5	A2	11101111	×	$\neg X \vee Y \vee Z$
4	7-1	6	P2=M2	11011111	=	$X \vee \neg Y \vee Z$
	6-2	15	A2xM2	11001111		$\neg Z \rightarrow (X \leftrightarrow Y)$

The counterpart of *Cesare* is:

Table 2.58 Derivation of **R** from **E** and **N**

Derivations	Level	#	Symbol	ID		Traditional logical form
	7-1	1	E1	11111110	×	$\neg X \vee \neg Y \vee \neg Z$
1	7-1	5	N1=R2=A2	11101111	=	$\neg X \vee Y \vee Z$
	6-2	7	E1xA2	11101110		$X \rightarrow \neg(Y \leftrightarrow Z)$
	7-1	2	E2=R1	11111101	×	$\neg X \vee \neg Y \vee Z$
2	7-1	5	N1=R2=A2	11101111	=	$\neg X \vee Y \vee Z$
	6-2	8	**R**	11101101		$\neg Z \rightarrow \neg X$
	7-1	1	E1	11111110	×	$\neg X \vee \neg Y \vee \neg Z$
3	7-1	8	N2=L2	01111111	=	$X \vee Y \vee Z$
	6-2	22	E1xL2	01111110		$\neg(X \leftrightarrow Y \leftrightarrow Z)$
	7-1	2	E2=R1	11111101	×	$\neg X \vee \neg Y \vee Z$
4	7-1	8	N2=L2	01111111	=	$X \vee Y \vee Z$
	6-2	23	E2xL2	01111101		$\neg Z \rightarrow \neg(X \leftrightarrow Y)$

We have completed the examination of the second group of the second mixed mode. Now we shall deal with the third group. The counterpart of *Bocardo* is:

Table 2.59 Derivation of **O** from **V** and **Q**

Derivations	Level	#	Symbol	ID		Traditional logical form
	3-5	2	V1=T2	11010000	×	$\neg Y \wedge (\neg X \vee \neg Z)$
1	7-1	6	Q1=M2	11011111	=	$X \vee \neg Y \vee Z$
	3-5	2	V1=T2	11010000		$\neg Y \wedge (\neg X \vee \neg Z)$
	3-5	2	V1=T2	11010000	×	$\neg Y \wedge (\neg X \vee \neg Z)$
2	7-1	8	Q2=L2	01111111	=	$X \vee Y \vee Z$
	2-6	9	T8+O7	01010000		$\neg Y \wedge \neg(X \leftrightarrow Z)$
	3-5	7	V2=Ω2	10110000	×	$\neg Z \wedge (\neg X \vee \neg Y)$
3	7-1	6	Q1=M2	11011111	=	$X \vee \neg Y \vee Z$
	2-6	3	**V**	10010000		$\neg Z \wedge \neg Y$
	3-5	7	V2=Ω2	10110000	×	$\neg Z \wedge (\neg X \vee \neg Y)$
4	7-1	8	Q2=L2	01111111	=	$X \vee Y \vee Z$
	2-6	14	S7+O7	00110000		$\neg Z \wedge \neg(X \leftrightarrow Y)$
	3-5	12	V3	10011000	×	$(\neg Z \wedge \neg Y) \vee (\neg X \wedge Y \wedge Z)$
5	7-1	6	Q1=M2	11011111	=	$X \vee \neg Y \vee Z$
	3-5	12	V3	10011000		$(\neg Z \wedge \neg Y) \vee (\neg X \wedge Y \wedge Z)$
	3-5	12	V3	10011000	×	$(\neg Z \wedge \neg Y) \vee (\neg X \wedge Y \wedge Z)$
6	7-1	8	Q2=L2	01111111	=	$X \vee Y \vee Z$
	2-6	19	O7+S8	00011000		$\neg X \leftrightarrow Y \leftrightarrow Z$
	3-5	13	V4=O1	10010100	×	$\neg Y \wedge (X \vee \neg Z)$
7	7-1	6	Q1=M2	11011111	=	$X \vee \neg Y \vee Z$
	3-5	13	V4=O1	10010100		$\neg Y \wedge (X \vee \neg Z)$
	3-5	13	V4=O1	10010100	×	$\neg Y \wedge (X \vee \neg Z)$
8	7-1	8	Q2=L2	01111111	=	$X \vee Y \vee Z$
	2-6	20	**O**	00010100		$X \wedge \neg Y$
	3-5	14	V5=W1	10010010	×	$\neg Z \wedge (X \vee \neg Y)$
9	7-1	6	Q1=M2	11011111	=	$X \vee \neg Y \vee Z$
	3-5	14	V5=W1	10010010		$\neg Z \wedge (X \vee \neg Y)$
	3-5	14	V5=W1	10010010	×	$\neg Z \wedge (X \vee \neg Y)$
10	7-1	8	Q2=L2	01111111	=	$X \vee Y \vee Z$
	2-6	21	**W**	00010010		$\neg Z \wedge X$
	3-5	15	V6	10010001	×	$(\neg Z \wedge \neg Y) \vee (X \wedge Y \wedge Z)$
11	7-1	6	Q1=M2	11011111	=	$X \vee \neg Y \vee Z$
	3-5	15	V6	10010001		$(\neg Z \wedge \neg Y) \vee (X \wedge Y \wedge Z)$
	3-5	15	V6	10010001	×	$(\neg Z \wedge \neg Y) \vee (X \wedge Y \wedge Z)$
12	7-1	8	Q2=L2	01111111	=	$X \vee Y \vee Z$
	2-6	22	O7+I8	00010001		$X \wedge (Y \leftrightarrow Z)$

The counterpart of *Disamis* is:

Table 2.60 Derivation of **I** from **U** and **Q**

Derivations	Level	#	Symbol	ID		Traditional logical form
	3-5	10	U1= Ω5	10100010	×	$\neg Z \wedge (\neg X \vee Y)$
1	7-1	6	Q1=M2	11011111	=	$X \vee \neg Y \vee Z$
	2-6	6	T7+I7	10000010		$\neg Z \wedge (X \leftrightarrow Y)$
	3-5	10	U1= Ω5	10100010	×	$\neg Z \wedge (\neg X \vee Y)$
2	7-1	8	Q2=L2	01111111	=	$X \vee Y \vee Z$
	2-6	17	**U**	00100010		$\neg Z \wedge Y$
	3-5	25	U2	01100010	×	$(\neg Z \wedge Y) \vee (\neg X \wedge \neg Y \wedge Z)$
3	7-1	6	Q1=M2	11011111	=	$X \vee \neg Y \vee Z$
	2-6	13	T8+I8	01000001		$Z \wedge (X \leftrightarrow Y)$
	3-5	25	U2	01100010	×	$(\neg Z \wedge Y) \vee (\neg X \wedge \neg Y \wedge Z)$
4	7-1	8	Q2=L2	01111111	=	$X \vee Y \vee Z$
	3-5	25	U2	01100010		$(\neg Z \wedge Y) \vee (\neg X \wedge \neg Y \wedge Z)$
	3-5	39	U3=W3	00110010	×	$\neg Z \wedge (X \vee Y)$
5	7-1	6	Q1=M2	11011111	=	$X \vee \neg Y \vee Z$
	2-6	21	**W**	00010010		$\neg Z \wedge X$
	3-5	39	U3=W3	00110010	×	$\neg Z \wedge (X \vee Y)$
6	7-1	8	Q2=L2	01111111	=	$X \vee Y \vee Z$
	3-5	39	U3=W3	00110010		$\neg Z \wedge (X \vee Y)$
	3-5	42	U4=S5	00101010	×	$Y \wedge (\neg X \vee \neg Z)$
7	7-1	6	Q1=M2	11011111	=	$X \vee \neg Y \vee Z$
	2-6	24	S8+I7	00001010		$Y \wedge \neg(X \leftrightarrow Z)$
	3-5	42	U4=S5	00101010	×	$Y \wedge (\neg X \vee \neg Z)$
8	7-1	8	Q2=L2	01111111	=	$X \vee Y \vee Z$
	3-5	42	U4=S5	00101010		$Y \wedge (\neg X \vee \neg Z)$
	3-5	44	U5	00100110	×	$(\neg Z \wedge Y) \vee (X \wedge \neg Y \wedge Z)$
9	7-1	6	Q1=M2	11011111	=	$X \vee \neg Y \vee Z$
	2-6	26	O8+I7	00000110		$X \wedge \neg(Y \leftrightarrow Z)$
	3-5	44	U5	00100110	×	$(\neg Z \wedge Y) \vee (X \wedge \neg Y \wedge Z)$
10	7-1	8	Q2=L2	01111111	=	$X \vee Y \vee Z$
	3-5	44	U5	00100110		$(\neg Z \wedge Y) \vee (X \wedge \neg Y \wedge Z)$
	3-5	46	U6=I3	00100011	×	$Y \wedge (X \vee \neg Z)$
11	7-1	6	Q1=M2	11011111	=	$X \vee \neg Y \vee Z$
	2-6	28	**I**	00000011		$X \wedge Y$
	3-5	46	U6=I3	00100011	×	$Y \wedge (X \vee \neg Z)$
12	7-1	8	Q2=L2	01111111	=	$X \vee Y \vee Z$
	3-5	46	U6=I3	00100011		$Y \wedge (X \vee \neg Z)$

The counterpart of *Ferison* is:

Table 2.61 Derivation of **O** from **P** and **W**

Derivations	Level	#	Symbol	ID		Traditional logical form
	7-1	2	P1=E2	11111101	×	$\neg X \vee \neg Y \vee Z$
1	3-5	14	W1=V5	10010010	=	$\neg Z \wedge (X \vee \neg Y)$
	2-6	3	**V**	10010000		$\neg Z \wedge \neg Y$
	7-1	6	P2=M2	11011111	×	$X \vee \neg Y \vee Z$
2	3-5	14	W1=V5	10010010	=	$\neg Z \wedge (X \vee \neg Y)$
	3-5	14	W1=V5	10010010		$\neg Z \wedge (X \vee \neg Y)$
	7-1	2	P1=E2	11111101	×	$\neg X \vee \neg Y \vee Z$
3	3-5	29	W2	01010010	=	$(\neg Z \wedge X) \vee (\neg X \wedge \neg Y \wedge Z)$
	2-6	9	T8+O7	01010000		$\neg Y \wedge \neg(X \leftrightarrow Z)$
	7-1	6	P2=M2	11011111	×	$X \vee \neg Y \vee Z$
4	3-5	29	W2	01010010	=	$(\neg Z \wedge X) \vee (\neg X \wedge \neg Y \wedge Z)$
	3-5	29	W2	01010010		$(\neg Z \wedge X) \vee (\neg X \wedge \neg Y \wedge Z)$
	7-1	2	P1=E2	11111101	×	$\neg X \vee \neg Y \vee Z$
5	3-5	39	W3=U3	00110010	=	$\neg Z \wedge (X \vee Y)$
	2-6	14	S7+O7	00110000		$\neg Z \wedge \neg(X \leftrightarrow Y)$
	7-1	6	P2=M2	11011111	×	$X \vee \neg Y \vee Z$
6	3-5	39	W3=U3	00110010	=	$\neg Z \wedge (X \vee Y)$
	2-6	21	**W**	00010010		$\neg Z \wedge X$
	7-1	2	P1=E2	11111101	×	$\neg X \vee \neg Y \vee Z$
7	3-5	48	W4	00011010	=	$(\neg Z \wedge X) \vee (\neg X \wedge Y \wedge Z)$
	2-6	19	O7+S8	00011000		$\neg X \leftrightarrow Y \leftrightarrow Z$
	7-1	6	P2=M2	11011111	×	$X \vee \neg Y \vee Z$
8	3-5	48	W4	00011010	=	$(\neg Z \wedge X) \vee (\neg X \wedge Y \wedge Z)$
	3-5	48	W4	00011010		$(\neg Z \wedge X) \vee (\neg X \wedge Y \wedge Z)$
	7-1	2	P1=E2	11111101	×	$\neg X \vee \neg Y \vee Z$
9	3-5	50	W5=O5	00010110	=	$X \wedge (\neg Y \vee \neg Z)$
	2-6	20	**O**	00010100		$X \wedge \neg Y$
	7-1	6	P2=M2	11011111	×	$X \vee \neg Y \vee Z$
10	3-5	50	W5=O5	00010110	=	$X \wedge (\neg Y \vee \neg Z)$
	3-5	50	W5=O5	00010110		$X \wedge (\neg Y \vee \neg Z)$
	7-1	2	P1=E2	11111101	×	$\neg X \vee \neg Y \vee Z$
11	3-5	52	W6=I4	00010011	=	$X \wedge (Y \vee \neg Z)$
	2-6	22	O7+I8	00010001		$X \wedge (Y \leftrightarrow Z)$
	7-1	6	P2=M2	11011111	×	$X \vee \neg Y \vee Z$
12	3-5	52	W6=I4	00010011	=	$X \wedge (Y \vee \neg Z)$
	3-5	52	W6=I4	00010011		$X \wedge (Y \vee \neg Z)$

Finally, the counterpart of *Datisi* is:

Table 2.62 Derivation of **I** from **N** and **W**

Derivations	Level	#	Symbol	ID		Traditional logical form
	7-1	5	N1=A2	11101111	×	$\neg X \vee Y \vee Z$
1	3-5	14	W1=V5	10010010	=	$\neg Z \wedge (X \vee \neg Y)$
	2-6	6	T7+I7	10000010		$\neg Z \wedge (X \leftrightarrow Y)$
	7-1	8	N2=L2	01111111	×	$X \vee Y \vee Z$
2	3-5	14	W1=V5	10010010	=	$\neg Z \wedge (X \vee \neg Y)$
	2-6	21	**W**	00010010		$\neg Z \wedge X$
	7-1	5	N1=A2	11101111	×	$\neg X \vee Y \vee Z$
3	3-5	29	W2	01010010	=	$(\neg Z \wedge X) \vee (\neg X \wedge \neg Y \wedge Z)$
	2-6	12	T8+I7	01000010		$X \leftrightarrow Y \leftrightarrow \neg Z$
	7-1	8	N2=L2	01111111	×	$X \vee Y \vee Z$
4	3-5	29	W2	01010010	=	$(\neg Z \wedge X) \vee (\neg X \wedge \neg Y \wedge Z)$
	3-5	29	W2	01010010		$(\neg Z \wedge X) \vee (\neg X \wedge \neg Y \wedge Z)$
	7-1	5	N1=A2	11101111	×	$\neg X \vee Y \vee Z$
5	3-5	39	W3=U3	00110010	=	$\neg Z \wedge (X \vee Y)$
	2-6	17	**U**	00100010		$\neg Z \wedge Y$
	7-1	8	N2=L2	01111111	×	$X \vee Y \vee Z$
6	3-5	39	W3=U3	00110010	=	$\neg Z \wedge (X \vee Y)$
	3-5	39	W3=U3	00110010		$\neg Z \wedge (X \vee Y)$
	7-1	5	N1=A2	11101111	×	$\neg X \vee Y \vee Z$
7	3-5	48	W4	00011010	=	$(\neg Z \wedge X) \vee (\neg X \wedge Y \wedge Z)$
	2-6	24	S8+I7	00001010		$Y \wedge \neg(X \leftrightarrow Z)$
	7-1	8	N2=L2	01111111	×	$X \vee Y \vee Z$
8	3-5	48	W4	00011010	=	$(\neg Z \wedge X) \vee (\neg X \wedge Y \wedge Z)$
	3-5	48	W4	00011010		$(\neg Z \wedge X) \vee (\neg X \wedge Y \wedge Z)$
	7-1	5	N1=A2	11101111	×	$\neg X \vee Y \vee Z$
9	3-5	50	W5=O5	00010110	=	$X \wedge (\neg Y \vee \neg Z)$
	2-6	26	O8+I7	00000110		$X \wedge \neg(Y \leftrightarrow Z)$
	7-1	8	N2=L2	01111111	×	$X \vee Y \vee Z$
10	3-5	50	W5=O5	00010110	=	$X \wedge (\neg Y \vee \neg Z)$
	3-5	50	W5=O5	00010110		$X \wedge (\neg Y \vee \neg Z)$
	7-1	5	N1=A2	11101111	×	$\neg X \vee Y \vee Z$
11	3-5	52	W6=I4	00010011	=	$X \wedge (Y \vee \neg Z)$
	2-6	28	**I**	00000011		$X \wedge Y$
	7-1	8	N2=L2	01111111	×	$X \vee Y \vee Z$
12	3-5	52	W6=I4	00010011	=	$X \wedge (Y \vee \neg Z)$
	3-5	52	W6=I4	00010011		$X \wedge (Y \vee \neg Z)$

First Group							
Classical		1st mixed		2nd mixed		Extension	
A	$X \rightarrow Y$	L	$\neg X \rightarrow Y$	A	$X \rightarrow Y$	L	$\neg X \rightarrow Y$
B	$Z \rightarrow X$	F	$Z \rightarrow \neg X$	Q	$\neg Z \rightarrow X$	R	$\neg Z \rightarrow \neg X$
C	$Z \rightarrow Y$	C	$Z \rightarrow Y$	N	$\neg Z \rightarrow Y$	N	$\neg Z \rightarrow Y$
E	$X \rightarrow \neg Y$	M	$\neg X \rightarrow \neg Y$	E	$X \rightarrow \neg Y$	M	$\neg X \rightarrow \neg Y$
B	$Z \rightarrow X$	F	$Z \rightarrow \neg X$	Q	$\neg Z \rightarrow X$	R	$\neg Z \rightarrow \neg X$
D	$Z \rightarrow \neg Y$	D	$Z \rightarrow \neg Y$	P	$\neg Z \rightarrow \neg Y$	P	$\neg Z \rightarrow \neg Y$
A	$X \rightarrow Y$	L	$\neg X \rightarrow Y$	A	$X \rightarrow Y$	L	$\neg X \rightarrow Y$
J	$Z \wedge X$	K	$Z \wedge \neg X$	W	$\neg Z \wedge X$	Ω	$\neg Z \wedge \neg X$
G	$Z \wedge Y$	G	$Z \wedge Y$	U	$\neg Z \wedge Y$	U	$\neg Z \wedge Y$
E	$X \rightarrow \neg Y$	M	$\neg X \rightarrow \neg Y$	E	$X \rightarrow \neg Y$	M	$\neg X \rightarrow \neg Y$
J	$Z \wedge X$	K	$Z \wedge \neg X$	W	$\neg Z \wedge X$	Ω	$\neg Z \wedge \neg X$
H	$Z \wedge \neg Y$	H	$Z \wedge \neg Y$	V	$\neg Z \wedge \neg Y$	V	$\neg Z \wedge \neg Y$

Second Group							
Classical		1st mixed		2nd mixed		Extension	
A	$X \rightarrow Y$	L	$\neg X \rightarrow Y$	A	$X \rightarrow Y$	L	$\neg X \rightarrow Y$
H	$Z \wedge \neg Y$	H	$Z \wedge \neg Y$	V	$\neg Z \wedge \neg Y$	V	$\neg Z \wedge \neg Y$
K	$Z \wedge \neg X$	J	$Z \wedge X$	Ω	$\neg Z \wedge \neg X$	W	$\neg Z \wedge X$
E	$X \rightarrow \neg Y$	M	$\neg X \rightarrow \neg Y$	E	$X \rightarrow \neg Y$	M	$\neg X \rightarrow \neg Y$
G	$Z \wedge Y$	G	$Z \wedge Y$	U	$\neg Z \wedge Y$	U	$\neg Z \wedge Y$
K	$Z \wedge \neg X$	J	$Z \wedge X$	Ω	$\neg Z \wedge \neg X$	W	$\neg Z \wedge X$
A	$X \rightarrow Y$	L	$\neg X \rightarrow Y$	A	$X \rightarrow Y$	L	$\neg X \rightarrow Y$
D	$Z \rightarrow \neg Y$	D	$Z \rightarrow \neg Y$	P	$\neg Z \rightarrow \neg Y$	P	$\neg Z \rightarrow \neg Y$
F	$Z \rightarrow \neg X$	B	$Z \rightarrow X$	R	$\neg Z \rightarrow \neg X$	Q	$\neg Z \rightarrow X$
E	$X \rightarrow \neg Y$	M	$\neg X \rightarrow \neg Y$	E	$X \rightarrow \neg Y$	M	$\neg X \rightarrow \neg Y$
C	$Z \rightarrow Y$	C	$Z \rightarrow Y$	N	$\neg Z \rightarrow Y$	N	$\neg Z \rightarrow Y$
F	$Z \rightarrow \neg X$	B	$Z \rightarrow X$	R	$\neg Z \rightarrow \neg X$	Q	$\neg Z \rightarrow X$

Third Group							
Classical		1st mixed		2nd mixed		Extension	
H	$Z \wedge \neg Y$	H	$Z \wedge \neg Y$	V	$\neg Z \wedge \neg Y$	V	$\neg Z \wedge \neg Y$
B	$Z \rightarrow X$	F	$Z \rightarrow \neg X$	Q	$\neg Z \rightarrow X$	R	$\neg Z \rightarrow \neg X$
O	$X \wedge \neg Y$	T	$\neg X \wedge \neg Y$	O	$X \wedge \neg Y$	T	$\neg X \wedge \neg Y$
G	$Z \wedge Y$	G	$Z \wedge Y$	U	$\neg Z \wedge Y$	U	$\neg Z \wedge Y$
B	$Z \rightarrow X$	F	$Z \rightarrow \neg X$	Q	$\neg Z \rightarrow X$	R	$\neg Z \rightarrow \neg X$
I	$X \wedge Y$	S	$\neg X \wedge Y$	I	$X \wedge Y$	S	$\neg X \wedge Y$
D	$Z \rightarrow \neg Y$	D	$Z \rightarrow \neg Y$	P	$\neg Z \rightarrow \neg Y$	P	$\neg Z \rightarrow \neg Y$
J	$Z \wedge X$	K	$Z \wedge \neg X$	W	$\neg Z \wedge X$	Ω	$\neg Z \wedge \neg X$
O	$X \wedge \neg Y$	T	$\neg X \wedge \neg Y$	O	$X \wedge \neg Y$	T	$\neg X \wedge \neg Y$
C	$Z \rightarrow Y$	C	$Z \rightarrow Y$	N	$\neg Z \rightarrow Y$	N	$\neg Z \rightarrow Y$
J	$Z \wedge X$	K	$Z \wedge \neg X$	W	$\neg Z \wedge X$	Ω	$\neg Z \wedge \neg X$
I	$X \wedge Y$	S	$\neg X \wedge Y$	I	$X \wedge Y$	S	$\neg X \wedge Y$

Figure 2.2 Summary of all product inferences divided in groups and modes.

We have completed the examination not only of all mixed modes but of all $4 \times 3 \times 4 = 48$ inferences that can be derived through product. It is possible to summarize the results so far obtained as in Fig. 2.2. It may be noted that in all product derivations at least one premise is universal.

Chapter 3

Sums

3.1 Introduction

I have followed so far the way in which inferences are traditionally treated: using products of premises it is equivalent to a conjunction of the latter. As I have mentioned, we can represent this traditional way to deal with inferences, e.g. in the case of the classical *Barbara*, as:

$$(\mathbf{A} \wedge \mathbf{B}) \rightarrow \mathbf{C}\,, \tag{3.1}$$

where the implication means—as customary—sufficient condition: the assertion of the conjunction of the premises **A** and **B** is sufficient condition of **C** to be true. If the inference is correct, this implication is a tautology. However, if we like to use sums instead of products, we need to have something like

$$\neg(\neg\mathbf{A} \vee \neg\mathbf{B}) \rightarrow \mathbf{A}\,, \tag{3.2}$$

which is equivalent to

$$\neg(\mathbf{O} \vee \mathbf{K}) \rightarrow \mathbf{C}\,, \tag{3.3}$$

or also

$$\mathbf{H} \rightarrow (\mathbf{O} \vee \mathbf{K})\,, \tag{3.4}$$

Mechanical Logic in Three-Dimensional Space
Gennaro Auletta

ISBN 978-981-4411-50-9 (Hardcover), 978-981-4411-51-6 (eBook)
www.panstanford.com

since **H** is the contradictory of **C**. However, the last two forms are weird, the former due to the presence of the negation placed before the premises is not sufficiently straight whilst the latter is not very helpful since cannot be computed in a mechanical way: given a statement (in this case **H**) we cannot guess which specific disjunction of statements it can imply (actually too many!). However, as previously mentioned we can interpret a sum-inference in terms of necessary conditions in this way:

$$(\mathbf{O} \vee \mathbf{K}) \leftarrow \mathbf{H}\,, \tag{3.5}$$

where the inversion of the arrow means that the disjunction of the premises is necessary condition of the conclusion. This is fully correct from a logical point of view. In this case, we start from two premises and we obtain a single conclusion, as it is customary. However, what ensures that the conclusion is related through such a logical relation with the premises? The fact that this inference is computed always through operations (in this case through a sum) on statements pertaining to the generating sets of the premises, and the relations that are established through such statements are all equivalences (as before). In the case of the above example (which I shall prove few lines below) we have:

$$(\mathrm{O8} + \mathrm{K7}) = \mathbf{H}\,, \tag{3.6}$$

which indeed implies that

$$(\mathbf{O} \vee \mathbf{K}) \leftarrow \mathbf{H}\,. \tag{3.7}$$

In other words, the equivalence between statements pertaining to the generating sets of the premises is such that, when there is a sum, it implies that the disjunction of the premises is a necessary condition of the conclusion. Obviously the formulation e.g. of *Barbara* in terms of sufficient condition and that in terms of necessary condition are equivalent. However, their forms are very different. In the sufficient-condition (usual) formulation, we say: If

Every X is Y and Every Z is X,
then Every Z is Y

In the necessary-condition formulation, we say: When

Some X is not Y or Some Z is not X,
then we assume that Some Z is not Y

This difference can be made clearer when inferences involving particular statements are considered. For instance, in the sufficient–condition formulation of *Ferio* we can draw a consequence of less generality from one of higher generality (when an assumption of less generality is added), so that: If

$$\text{No } X \text{ is } Y \quad \text{and} \quad \text{Some } Z \text{ is } X,$$
$$\text{then Some } Z \text{ is not } Y$$

However, in the necessary-condition formulation we pass from lesser to higher generality (provided that an universal statement is added) in this way: When

$$\text{Some } X \text{ is } Y \quad \text{or} \quad \text{No } Z \text{ is } X,$$
$$\text{then we assume that Every } Z \text{ is } Y$$

For instance, when

Sometimes light behaves discontinuously or No matter is light,

we can assume that

Every matter behaves discontinuously.

Such an inferential process can be called induction. As far as we compute mechanically, we start from the premises and we assume the conclusion. However, the inductive procedure can be understood as trying to infer (according to the sufficient-condition mode of reasoning) the possible premises given the conclusion. As a matter of fact, the generalization that can be found in this way, is the negation of a possible conclusion to which a deduction could have led: for instance, the conclusion **H** in Eq. (3.7) is the negation of the conclusion **C** of *Barbara* as well as the conclusion

$$\text{Every } Z \text{ is } Y$$

is a generalization that denies the conclusion of the classical *Ferio* (i.e. Some Z is not Y). This is particularly important when we consider that the disjunction of the premises expresses some kind of incertitude, as often occurs in many contexts in which knowledge or practical matters are involved. We know that at least one of the premises is true but we also need a certain decision expressed by the conclusion. In the best case, we can optimize the information contained in the premises. Formally speaking, we can express the last derivation as (where L, D, M stand for light, discontinuity, and matter, respectively)

$$[(L \wedge D) \vee (M \rightarrow \neg L)] \leftarrow (M \rightarrow D). \tag{3.8}$$

In fact, the conclusion is logically compatible with the truth of *both* premises and it expresses the maximum that we can say given these pieces of knowledge. Suppose now that the two premises would be connected by an AND in an ordinary product derivation. We can easily see that in such a case we have the derivation:

$$[(L \wedge D) \wedge (M \rightarrow \neg L)] \rightarrow (D \wedge \neg M), \tag{3.9}$$

where the conclusion is much weaker: there is something discontinuous that is not matter. Summarizing, with this kind of inference we have that when at least one of the premises is true we can assume a certain conclusion that is logically allowed.

We need now to express any of the involved statements as sums of the RGS instead of the LGS. Let us first consider the particular statements expressed as in Table 3.1.

The complete list of the universal statements is obviously longer when dealing with the RGS (in Table 3.2 I only list these statements and recall that any combination of two of them can give rise to the respective bold statement).

Tables 3.1 and 3.2 show all couples of statements involving the three variables capable to generate through sums the main statements considered here (like **A, B, J, K**, and so on).

It may be noted that **A**, **E**, **L** and **M** share no statement in their RGSs (as well as in their LGSs) among them and the same is true for every universal quadruple (displaying contrariety) shown in one of the octagons displayed in Fig. 2.1. On the other hand, any universal bold statement shares one element of its RGS with one out of four other statements that occur pairwise on the other two octagons displayed in Fig. 2.1. For instance, **R** (located in the middle or second octagon) shares a statement with **A** and a statement with **E** (which are located in the central top part of the most internal (first) octagon) as well as shares a statement with **N** and a statement with **P** (which are located in the extreme of the top part of the most external octagon (the third

Table 3.1 Raising generating sets of particular statements

Level	#	Symbol	ID		Traditional logical form
1-7	7	I7	00000010	+	$X \wedge Y \wedge \neg Z$
1-7	8	I8	00000001	=	$X \wedge Y \wedge Z$
2-6	28	**I**	00000011		$X \wedge Y$
1-7	4	O7	00010000	+	$X \wedge \neg Y \wedge \neg Z$
1-7	6	O8	00000100	=	$X \wedge \neg Y \wedge Z$
2-6	20	**O**	00010100		$X \wedge \neg Y$
1-7	3	S7	00100000	+	$\neg X \wedge Y \wedge \neg Z$
1-7	5	S8	00001000	=	$\neg X \wedge Y \wedge Z$
2-6	15	**S**	00101000		$\neg X \wedge Y$
1-7	1	T7	10000000	+	$\neg X \wedge \neg Y \wedge \neg Z$
1-7	2	T8	01000000	=	$\neg X \wedge \neg Y \wedge Z$
2-6	1	**T**	11000000		$\neg X \wedge \neg Y$
1-7	5	G7=S8	00001000	+	$\neg X \wedge Y \wedge Z$
1-7	8	G8=I8	00000001	=	$X \wedge Y \wedge Z$
2-6	25	**G**	00001001		$Z \wedge Y$
1-7	2	H7=T8	01000000	+	$\neg X \wedge \neg Y \wedge Z$
1-7	6	H8=O8	00000100	=	$X \wedge \neg Y \wedge Z$
2-6	11	**H**	01000100		$Z \wedge \neg Y$
1-7	6	J7=O8	00000100	+	$X \wedge \neg Y \wedge Z$
1-7	8	J8=I8	00000001	=	$X \wedge Y \wedge Z$
2-6	27	**J**	00000101		$Z \wedge X$
1-7	2	K7=T8	01000000	+	$\neg X \wedge \neg Y \wedge Z$
1-7	5	K8=S8	00001000	=	$\neg X \wedge Y \wedge Z$
2-6	10	**K**	01001000		$Z \wedge \neg X$
1-7	3	U7=S7	00100000	+	$\neg X \wedge Y \wedge \neg Z$
1-7	7	U8=I7	00000010	=	$X \wedge Y \wedge \neg Z$
2-6	17	**U**	00100010		$\neg Z \wedge Y$
1-7	1	V7=T7	10000000	+	$\neg X \wedge \neg Y \wedge \neg Z$
1-7	4	V8=O7	00010000	=	$X \wedge \neg Y \wedge \neg Z$
2-6	3	**V**	10010000		$\neg Z \wedge \neg Y$
1-7	4	W7=O7	00010000	+	$X \wedge \neg Y \wedge \neg Z$
1-7	7	W8=I7	00000010	=	$X \wedge Y \wedge \neg Z$
2-6	21	**W**	00010010		$\neg Z \wedge X$
1-7	1	Ω7=T7	10000000	+	$\neg X \wedge \neg Y \wedge \neg Z$
1-7	3	Ω8=S7	00100000	=	$\neg X \wedge Y \wedge \neg Z$
2-6	2	**Ω**	10100000		$\neg Z \wedge \neg X$

Table 3.2 Raising generating sets of universal statements

Level	#	Symbol	ID	Traditional logical form
6-2	9	**A**	11101011	$X \rightarrow Y$
5-3	6	A3	11101010	$X \rightarrow (Y \wedge \neg Z)$
	7	A4	11101001	$X \rightarrow (Y \wedge Z)$
	10	A5	11100011	$(\neg X \vee Y) \wedge (X \vee \neg Y \vee \neg Z)$
	19	A6	11001011	$(\neg X \vee Y) \wedge (X \vee \neg Y \vee Z)$
	29	A7	10101011	$\neg Y \rightarrow (\neg X \wedge \neg Z)$
	44	A8	01101011	$\neg Y \rightarrow (\neg X \wedge Z)$
6-2	1	**E**	11111100	$X \rightarrow \neg Y$
5-3	1	E3	11111000	$X \rightarrow (\neg Y \wedge \neg Z)$
	2	E4	11110100	$Y \rightarrow (\neg X \wedge \neg Z)$
	5	E5	11101100	$X \rightarrow (\neg Y \wedge Z)$
	11	E6	11011100	$Y \rightarrow (\neg X \wedge Z)$
	21	E7	10111100	$(\neg X \vee \neg Y) \wedge (X \vee Y \vee \neg Z)$
	36	E8	01111100	$(\neg X \vee \neg Y) \wedge (X \vee Y \vee Z)$
6-2	28	**L**	00111111	$\neg X \rightarrow Y$
5-3	51	L3	00111110	$(X \vee Y) \wedge (\neg X \vee \neg Y \vee \neg Z)$
	52	L4	00111101	$(X \vee Y) \wedge (\neg X \vee \neg Y \vee Z)$
	53	L5	00111011	$\neg Y \rightarrow (X \wedge \neg Z)$
	54	L6	00110111	$\neg X \rightarrow (Y \wedge \neg Z)$
	55	L7	00101111	$\neg Y \rightarrow (X \wedge Z)$
	56	L8	00011111	$\neg X \rightarrow (Y \wedge Z)$
6-2	14	**M**	11010111	$\neg X \rightarrow \neg Y$
5-3	14	M3	11010110	$Y \rightarrow (X \wedge \neg Z)$
	15	M4	11010101	$Y \rightarrow (X \wedge Z)$
	16	M5	11010011	$(X \vee \neg Y) \wedge (\neg X \vee Y \vee \neg Z)$
	20	M6	11000111	$(X \vee \neg Y) \wedge (\neg X \vee Y \vee Z)$
	34	M7	10010111	$\neg X \rightarrow (\neg Y \wedge \neg Z)$
	49	M8	01010111	$\neg X \rightarrow (\neg Y \wedge Z)$
6-2	19	**B**	10110111	$Z \rightarrow X$
5-3	24	B3=D7	10110111	$Z \rightarrow (X \wedge \neg Y)$
	25	B4	10110101	$(X \vee \neg Z) \wedge (\neg X \vee \neg Y \vee Z)$
	26	B5=C5	10110011	$Z \rightarrow (X \wedge Y)$
	30	B6	10100111	$(X \vee \neg Z) \wedge (\neg X \vee Y \vee Z)$
	34	B7=M7	10010111	$\neg X \rightarrow (\neg Y \wedge \neg Z)$
	54	B8=L6	00110111	$\neg X \rightarrow (Y \wedge \neg Z)$

Level	#	Symbol	ID	Traditional logical form
6-2	18	**C**	10111011	$Z \to Y$
5-3	22	C3=F7	10111010	$Z \to (\neg X \wedge Y)$
	23	C4	10111001	$(Y \vee \neg Z) \wedge (\neg X \vee \neg Y \vee Z)$
	26	C5=B5	10110011	$Z \to (X \wedge Y)$
	29	C6=A7	10101011	$\neg Y \to (\neg X \wedge \neg Z)$
	33	C7	10011011	$(Y \vee \neg Z) \wedge (X \vee \neg Y \vee Z)$
	53	C8=L5	00111011	$\neg Y \to (X \wedge \neg Z)$
6-2	4	**D**	11110110	$Z \to \neg Y$
5-3	2	D3=E4	11110100	$Y \to (\neg X \wedge \neg Z)$
	3	D4=F4	11110010	$Z \to (\neg X \wedge \neg Y)$
	8	D5	11100110	$(\neg Y \vee \neg Z) \wedge (\neg X \vee Y \vee Z)$
	14	D6=M3	11010110	$Y \to (X \wedge \neg Z)$
	24	D7=B3	10110111	$Z \to (X \wedge \neg Y)$
	39	D8	01110110	$(\neg Y \vee \neg Z) \wedge (X \vee Y \vee Z)$
6-2	2	**F**	11111010	$Z \to \neg X$
5-3	1	F3=E3	11111000	$X \to (\neg Y \wedge \neg Z)$
	3	F4=D4	11110010	$Z \to (\neg X \wedge \neg Y)$
	6	F5=A3	11101010	$X \to (Y \wedge \neg Z)$
	12	F6	11011010	$(\neg X \vee \neg Z) \wedge (X \vee \neg Y \vee Z)$
	22	F7=C3	10111010	$Z \to (\neg X \wedge Y)$
	37	F8	01111010	$(\neg X \vee \neg Z) \wedge (X \vee Y \vee Z)$
6-2	26	**N**	01101111	$\neg Z \to Y$
5-3	42	N3	01101110	$(Y \vee Z) \wedge (\neg X \vee \neg Y \vee \neg Z)$
	43	N4=R8	01101101	$\neg Z \to (\neg X \wedge Y)$
	44	N5=A8	01101011	$\neg Y \to (\neg X \wedge Z)$
	45	N6	01100111	$(Y \vee Z) \wedge (X \vee \neg Y \vee \neg Z)$
	50	N7=Q7	01001111	$\neg Z \to (X \wedge Y)$
	55	N8=L7	00101111	$\neg Y \to (X \wedge Z)$
6-2	12	**P**	11011101	$\neg Z \to \neg Y$
5-3	11	P3=E6	11011100	$Y \to (\neg X \wedge Z)$
	13	P4	11011001	$(\neg Y \vee Z) \wedge (\neg X \vee Y \vee \neg Z)$
	15	P5=M4	11010101	$Y \to (X \wedge Z)$
	18	P6=R6	11001101	$\neg Z \to (\neg X \wedge \neg Y)$
	32	P7	10011101	$(\neg Y \vee Z) \wedge (X \vee Y \vee \neg Z)$
	47	P8=Q4	01011101	$\neg Z \to (X \wedge \neg Y)$

(Continued)

Table 3.2 (*Continued*)

Level	#	Symbol	ID	Traditional logical form
6-2	27	**Q**	01011111	$\neg Z \to X$
5-3	46	Q3	01011110	$(X \vee Z) \wedge (\neg X \vee \neg Y \vee \neg Z)$
	47	Q4=P8	01011101	$\neg Z \to (X \wedge \neg Y)$
	48	Q5	01011011	$(X \vee Z) \wedge (\neg X \vee Y \vee \neg Z)$
	49	Q6=M8	01010111	$\neg X \to (\neg Y \wedge Z)$
	50	Q7=N7	01001111	$\neg Z \to (X \wedge Y)$
	56	Q8=L8	00011111	$\neg X \to (Y \wedge Z)$
6-2	8	**R**	11101101	$\neg Z \to \neg X$
5-3	5	R3=E5	11101100	$X \to (\neg Y \wedge Z)$
	7	R4=A4	11101001	$X \to (Y \wedge Z)$
	9	R5	11100101	$(\neg X \vee Z) \wedge (X \vee \neg Y \vee \neg Z)$
	18	R6=P6	11001101	$\neg Z \to (\neg X \wedge \neg Y)$
	28	R7	10101101	$(\neg X \vee Z) \wedge (X \vee Y \vee \neg Z)$
	43	R8=N4	01101101	$\neg Z \to (\neg X \wedge Y)$

one)). The same is also true for the particular statements displayed in Table 3.1. For instance, **G** (which is located on the third octagon) shares a statement with **I** and a statement with **S** (which are located on the first octagon) as well a statement with **J** and a statement with **K** (which are located in the second octagon). A similar behavior is true for the LGSs of both universal and particular statements (for a summary have a look at Tables 2.1–2.3. What is remarkable is that whenever a bold statement (whether universal or particular) shares elements with another statements, say of its RGS, it also shares elements of its LGS with the same statement.

An interesting remark is that the shared statements among universal statements have not the form of conjunctions of disjunctions (in a similar way shared statements among particular statements in the product form are not disjunctions of conjunctions). Conversely, all those statements that are not shared have those two forms, respectively. In general, in any raising set of the particular statements we have 2 statements with the form of conjunctions of disjunctions and 4 with the form an implication with a conjunction as a consequent.

Table 3.3 **O** OR **K** as necessary condition of **H**

Derivations	Level	#	Symbol	ID		Traditional logical form
	1-7	4	O7	00010000	+	$X \wedge \neg Y \wedge \neg Z$
1	1-7	2	K7=H7=T8	01000000	=	$\neg X \wedge \neg Y \wedge Z$
	2-6	9	T8+O7	01010000		$\neg Y \wedge \neg(X \leftrightarrow Z)$
	1-7	6	O8=H8	00000100	+	$X \wedge \neg Y \wedge Z$
2	1-7	2	K7=H7=T8	01000000	=	$\neg X \wedge \neg Y \wedge Z$
	2-6	11	**H**	01000100		$Z \wedge \neg Y$
	1-7	4	O7	00010000	+	$X \wedge \neg Y \wedge \neg Z$
3	1-7	5	K8=S8	00001000	=	$\neg X \wedge Y \wedge Z$
	2-6	19	O7+S8	00011000		$\neg X \leftrightarrow Y \leftrightarrow Z$
	1-7	6	O8=H8	00000100	+	$X \wedge \neg Y \wedge Z$
4	1-7	5	K8=S8	00001000	=	$\neg X \wedge Y \wedge Z$
	2-6	23	S8+O8	00001100		$Z \wedge \neg(X \leftrightarrow Y)$

3.2 Classical Inferences through Sums

Let us now start with the first group of classical inferences and therefore by proving *Barbara*: The reader may remark that the conclusion has been precisely obtained by summing the two statements pertaining to the RGS of **H**: indeed they coincide with K7 and O8. Actually, this was already true for all universal derivation through products (see, for instance the derivation of *Barbara* provided in Table 2.12), although the first principle of p. 73 needs to be reformulated as: the conclusion must pertain to the lowest level allowed by the premises. For this reason, we may skip now all inferences through sum that only involve particular statements (they will be anyway summarized in a figure at the end of this exposition) and focus only on derivations involving also universal statements (which correspond to derivations involving particular statements in their product counterpart). The sum-derivation of *Darii* is then:

Table 3.4 **O** OR **F** as necessary condition of **D**

Derivations	Level	#	Symbol	ID		Traditional logical form
	1-7	4	O7	00010000	+	$X \wedge \neg Y \wedge \neg Z$
1	5-3	1	F3=E3	11111000	=	$X \rightarrow (\neg Y \wedge \neg Z)$
	5-3	1	F3=E3	11111000		$X \rightarrow (\neg Y \wedge \neg Z)$

(*Continued*)

Table 3.4 (*Continued*)

Derivations	Level	#	Symbol	ID		Traditional logical form
	1-7	6	08	00000100	+	$X \wedge \neg Y \wedge Z$
2	5-3	1	F3=E3	11111000	=	$X \rightarrow (\neg Y \wedge \neg Z)$
	6-2	1	**E**	11111100		$X \rightarrow \neg Y$
	1-7	4	07	00010000	+	$X \wedge \neg Y \wedge \neg Z$
3	5-3	3	F4=D4	11110010	=	$Z \rightarrow (\neg X \wedge \neg Y)$
	5-3	3	F4=D4	11110010	=	$Z \rightarrow (\neg X \wedge \neg Y)$
	1-7	6	08	00000100	+	$X \wedge \neg Y \wedge Z$
4	5-3	3	F4=D4	11110010	=	$Z \rightarrow (\neg X \wedge \neg Y)$
	6-2	4	**D**	11110110		$Z \rightarrow \neg Y$
	1-7	4	07	00010000	+	$X \wedge \neg Y \wedge \neg Z$
5	5-3	6	F5=A3	11101010	=	$X \rightarrow (Y \wedge \neg Z)$
	6-2	2	**F**	11111010		$Z \rightarrow \neg X$
	1-7	6	08	00000100	+	$X \wedge \neg Y \wedge Z$
6	5-3	6	F5=A3	11101010	=	$X \rightarrow (Y \wedge \neg Z)$
	6-2	7	E1xA2	11101110		$X \rightarrow \neg(Y \leftrightarrow Z)$
	1-7	4	07	00010000	+	$X \wedge \neg Y \wedge \neg Z$
7	5-3	12	F6	11011010	=	$(\neg X \vee \neg Z) \wedge (X \vee \neg Y \vee Z)$
	5-3	12	F6	11011010		$(\neg X \vee \neg Z) \wedge (X \vee \neg Y \vee Z)$
	1-7	6	08	00000100	+	$X \wedge \neg Y \wedge Z$
8	5-3	12	F6	11011010	=	$(\neg X \vee \neg Z) \wedge (X \vee \neg Y \vee Z)$
	6-2	11	E1xM2	11011110		$Y \rightarrow \neg(X \leftrightarrow Z)$
	1-7	4	07	00010000	+	$X \wedge \neg Y \wedge \neg Z$
9	5-3	22	F7=C3	10111010	=	$Z \rightarrow (\neg X \wedge Y)$
	5-3	22	F7=C3	10111010		$Z \rightarrow (\neg X \wedge Y)$
	1-7	6	08	00000100	+	$X \wedge \neg Y \wedge Z$
10	5-3	22	F7=C3	10111010	=	$Z \rightarrow (\neg X \wedge Y)$
	6-2	16	E1xL1	10111110		$Z \rightarrow \neg(X \leftrightarrow Y)$
	1-7	4	07	00010000	+	$X \wedge \neg Y \wedge \neg Z$
11	5-3	37	F8	01111010	=	$(\neg X \vee \neg Z) \wedge (X \vee Y \vee Z)$
	5-3	37	F8	01111010		$(\neg X \vee \neg Z) \wedge (X \vee Y \vee Z)$
	1-7	6	08	00000100	+	$X \wedge \neg Y \wedge Z$
12	5-3	37	F8	01111010	=	$(\neg X \vee \neg Z) \wedge (X \vee Y \vee Z)$
	6-2	22	E1xL2	01111110		$\neg(X \leftrightarrow Y \leftrightarrow Z)$

The reader may note that (similarly to what happens for product derivations) in all cross terms, like E1xL1, there is a recurrent statement. In the above derivations it is E1. This regularity (obviously with each time different statements) is also common to all inferences that correspond to product inferences having a particular premise. We may also note that we have a characteristic inversion between product and sum derivations: in the universal product derivations (like for *Barbara*) the cross terms pertain to Level 6-2 (like A1xM2) while in the derivations involving particular propositions (like *Darii*) the cross terms pertain to Level 2-6 (like S8+O8). In the sum derivations is the opposite.

Derivations involving both universal and particular statements are always more difficult. I shall deal with this problem with the next inference (*Ferio*) since it involves not only statements with different quantity but also with different quality (interchange of affirmative and negative statements). Then, the sum-derivation of *Ferio* is:

Table 3.5 **I** OR **F** as necessary condition of **C**

Derivations	Level	#	Symbol	ID		Traditional logical form
	1-7	7	I7	00000010	+	$X \wedge Y \wedge \neg Z$
1	5-3	1	F3=E3	11111000	=	$X \rightarrow (\neg Y \wedge \neg Z)$
	6-2	2	**F**	11111010		$Z \rightarrow \neg X$
	1-7	8	I8	00000001	+	$X \wedge Y \wedge Z$
2	5-3	1	F3=E3	11111000	=	$X \rightarrow (\neg Y \wedge \neg Z)$
	6-2	3	E2xA1	11111001		$X \rightarrow (Y \leftrightarrow Z)$
	1-7	7	I7	00000010	+	$X \wedge Y \wedge \neg Z$
3	5-3	3	F4=D4	11110010	=	$Z \rightarrow (\neg X \wedge \neg Y)$
	5-3	3	F4=D4	11110010		$Z \rightarrow (\neg X \wedge \neg Y)$
	1-7	8	I8	00000001	+	$X \wedge Y \wedge Z$
4	5-3	3	F4=D4	11110010	=	$Z \rightarrow (\neg X \wedge \neg Y)$
	6-2	6	A1xM1	11110011		$Z \rightarrow (X \leftrightarrow Y)$
	1-7	7	I7	00000010	+	$X \wedge Y \wedge \neg Z$
5	5-3	6	F5=A3	11101010	=	$X \rightarrow (Y \wedge \neg Z)$
	5-3	6	F5=A3	11101010		$X \rightarrow (Y \wedge \neg Z)$
	1-7	8	I8	00000001	+	$X \wedge Y \wedge Z$
6	5-3	6	F5=A3	11101010	=	$X \rightarrow (Y \wedge \neg Z)$
	6-2	9	**A**	11101011		$X \rightarrow Y$

(*Continued*)

Table 3.5 *(Continued)*

Derivations	Level	#	Symbol	ID		Traditional logical form
	1-7	7	I7	00000010	+	$X \wedge Y \wedge \neg Z$
7	5-3	12	F6	11011010	=	$(\neg X \vee \neg Z) \wedge (X \vee \neg Y \vee Z)$
	5-3	12	F6	11011010		$(\neg X \vee \neg Z) \wedge (X \vee \neg Y \vee Z)$
	1-7	8	I8	00000001	+	$X \wedge Y \wedge Z$
8	5-3	12	F6	11011010	=	$(\neg X \vee \neg Z) \wedge (X \vee \neg Y \vee Z)$
	6-2	13	A1xM2	11011011		$\neg(X \leftrightarrow \neg Y \leftrightarrow Z)$
	1-7	7	I7	00000010	+	$X \wedge Y \wedge \neg Z$
9	5-3	22	F7=C3	10111010	=	$Z \rightarrow (\neg X \wedge Y)$
	5-3	22	F7=C3	10111010		$Z \rightarrow (\neg X \wedge Y)$
	1-7	8	I8	00000001	+	$X \wedge Y \wedge Z$
10	5-3	22	F7=C3	10111010	=	$Z \rightarrow (\neg X \wedge Y)$
	6-2	18	**C**	10111011		$Z \rightarrow Y$
	1-7	7	I7	00000010	+	$X \wedge Y \wedge \neg Z$
11	5-3	37	F8	01111010	=	$(\neg X \vee \neg Z) \wedge (X \vee Y \vee Z)$
	5-3	37	F8	01111010		$(\neg X \vee \neg Z) \wedge (X \vee Y \vee Z)$
	1-7	8	I8	00000001	+	$X \wedge Y \wedge Z$
12	5-3	37	F8	01111010	=	$(\neg X \vee \neg Z) \wedge (X \vee Y \vee Z)$
	6-2	24	A1xL2	01111011		$\neg Y \rightarrow \neg(X \leftrightarrow Z)$

We may rewrite Derivation 1 as:

$$(\mathrm{I7} + \mathrm{F3}) = (\mathrm{F7}{=}\mathrm{C3} \ + \ \mathrm{F3})\,, \tag{3.10}$$

where I have again made use of the circumstance that any couple of statements of the RGS of **F** can generate **F**. This allows the context-substitution of I7 by F7 which allows us two write $\mathrm{F7} \leftarrow \mathrm{I7}$ or

$$[Z \rightarrow (\neg X \wedge Y)] \leftarrow (X \wedge Y \wedge \neg Z)\,, \tag{3.11}$$

which is indeed a logical truth. On the other hand we have also the identity A7=C6. This means that we can rewrite Derivation 6 as

$$(\mathrm{I8} \ + \ \mathrm{F5}{=}\mathrm{A3}) = (\mathrm{A7}{=}\mathrm{C6} \ + \ \mathrm{A3}{=}\mathrm{F5})\,, \tag{3.12}$$

which again allows the context-substitution of I8 by A7. Therefore, the desired inference can be written as

$$\begin{aligned}(\mathrm{F5}{=}\mathrm{A3} \ + \ \mathrm{F7}{=}\mathrm{C3}{=}\mathrm{I7} \ + \ \mathrm{I8}{=}\mathrm{A7}{=}\mathrm{C6}) \leftarrow \\ (\mathrm{F7}{=}\mathrm{C3}{=}\mathrm{I7} \ + \ \mathrm{I8}{=}\mathrm{A7}{=}\mathrm{C6})\,,\end{aligned} \tag{3.13}$$

which is a logical truth.

Let us now consider the other inferences. The derivation of *Baroco* is:

Table 3.6 **O** OR **C** as necessary condition of **B**

Derivations	Level	#	Symbol	ID		Traditional logical form
	1-7	4	O7	00010000	+	$X \wedge \neg Y \wedge \neg Z$
1	5-3	22	C3=F7	10111010	=	$Z \rightarrow (\neg X \wedge Y)$
	5-3	22	C3=F7	10111010		$Z \rightarrow (\neg X \wedge Y)$
	1-7	6	O8	00000100	+	$X \wedge \neg Y \wedge Z$
2	5-3	22	C3=F7	10111010	=	$Z \rightarrow (\neg X \wedge Y)$
	6-2	16	E1xL1	10111110		$Z \rightarrow \neg(X \leftrightarrow Y)$
	1-7	4	O7	00010000	+	$X \wedge \neg Y \wedge \neg Z$
3	5-3	23	C4	10111001	=	$(Y \vee \neg Z) \wedge (\neg X \vee \neg Y \vee Z)$
	5-3	23	C4	10111001		$(Y \vee \neg Z) \wedge (\neg X \vee \neg Y \vee Z)$
	1-7	6	O8	00000100	+	$X \wedge \neg Y \wedge Z$
4	5-3	23	C4	10111001	=	$(Y \vee \neg Z) \wedge (\neg X \vee \neg Y \vee Z)$
	6-2	17	E2xL1	10111101		$\neg(X \leftrightarrow Y \leftrightarrow \neg Z)$
	1-7	4	O7	00010000	+	$X \wedge \neg Y \wedge \neg Z$
5	5-3	26	C5=B5	10110011	=	$Z \rightarrow (X \wedge Y)$
	5-3	26	C5=B5	10110011		$Z \rightarrow (X \wedge Y)$
	1-7	6	O8	00000100	+	$X \wedge \neg Y \wedge Z$
6	5-3	26	C5=B5	10110011	=	$Z \rightarrow (X \wedge Y)$
	6-2	19	**B**	10110111		$Z \rightarrow X$
	1-7	4	O7	00010000	+	$X \wedge \neg Y \wedge \neg Z$
7	5-3	29	C6=A7	10101011	=	$\neg Y \rightarrow (\neg X \wedge \neg Z)$
	6-2	18	**C**	10111011		$Z \rightarrow Y$
	1-7	6	O8	00000100	+	$X \wedge \neg Y \wedge Z$
8	5-3	29	C6=A7	10101011	=	$\neg Y \rightarrow (\neg X \wedge \neg Z)$
	6-2	20	A2xL1	10101111		$\neg Y \rightarrow (X \leftrightarrow Z)$
	1-7	4	O7	00010000	+	$X \wedge \neg Y \wedge \neg Z$
9	5-3	33	C7	10011011	=	$(Y \vee \neg Z) \wedge (X \vee \neg Y \vee Z)$
	5-3	33	C7	10011011		$(Y \vee \neg Z) \wedge (X \vee \neg Y \vee Z)$
	1-7	6	O8	00000100	+	$X \wedge \neg Y \wedge Z$
10	5-3	33	C7	10011011	=	$(Y \vee \neg Z) \wedge (X \vee \neg Y \vee Z)$
	6-2	21	M2xL1	10011111		$\neg X \rightarrow (Y \leftrightarrow Z)$
	1-7	4	O7	00010000	+	$X \wedge \neg Y \wedge \neg Z$
11	5-3	53	C8=L5	00111011	=	$\neg Y \rightarrow (X \wedge \neg Z)$
	5-3	53	C8=L5	00111011		$\neg Y \rightarrow (X \wedge \neg Z)$
	1-7	6	O8	00000100	+	$X \wedge \neg Y \wedge Z$
12	5-3	53	C8=L5	00111011	=	$\neg Y \rightarrow (X \wedge \neg Z)$
	6-2	28	**L**	00111111		$\neg X \rightarrow Y$

The derivation of *Festino* is:

Table 3.7 **I** OR **D** as necessary condition of **B**

Derivations	Level	#	Symbol	ID		Traditional logical form
	1-7	7	I7	00000010	+	$X \wedge Y \wedge \neg Z$
1	5-3	2	D3=E4	11110100	=	$Y \rightarrow (\neg X \wedge \neg Z)$
	6-2	4	**D**	11110110		$Z \rightarrow \neg Y$
	1-7	8	I8	00000001	+	$X \wedge Y \wedge Z$
2	5-3	2	D3=E4	11110100	=	$Y \rightarrow (\neg X \wedge \neg Z)$
	6-2	5	E2xM1	11110101		$Y \rightarrow (X \leftrightarrow Z)$
	1-7	7	I7	00000010	+	$X \wedge Y \wedge \neg Z$
3	5-3	3	D4=F4	11110010	=	$Z \rightarrow (\neg X \wedge \neg Y)$
	5-3	3	D4=F4	11110010		$Z \rightarrow (\neg X \wedge \neg Y)$
	1-7	8	I8	00000001	+	$X \wedge Y \wedge Z$
4	5-3	3	D4=F4	11110010	=	$Z \rightarrow (\neg X \wedge \neg Y)$
	6-2	6	A1xM1	11110011		$Z \rightarrow (X \leftrightarrow Y)$
	1-7	7	I7	00000010	+	$X \wedge Y \wedge \neg Z$
5	5-3	8	D5	11100110	=	$(\neg Y \vee \neg Z) \wedge (\neg X \vee Y \vee Z)$
	5-3	8	D5	11100110		$(\neg Y \vee \neg Z) \wedge (\neg X \vee Y \vee Z)$
	1-7	8	I8	00000001	+	$X \wedge Y \wedge Z$
6	5-3	8	D5	11100110	=	$(\neg Y \vee \neg Z) \wedge (\neg X \vee Y \vee Z)$
	6-2	10	M1xA2	11100111		$\neg(\neg X \leftrightarrow Y \leftrightarrow Z)$
	1-7	7	I7	00000010	+	$X \wedge Y \wedge \neg Z$
7	5-3	14	D6=M3	11010110	=	$Y \rightarrow (X \wedge \neg Z)$
	5-3	14	D6=M3	11010110		$Y \rightarrow (X \wedge \neg Z)$
	1-7	8	I8	00000001	+	$X \wedge Y \wedge Z$
8	5-3	14	D6=M3	11010110	=	$Y \rightarrow (X \wedge \neg Z)$
	6-2	14	**M**	11010111		$\neg X \rightarrow \neg Y$
	1-7	7	I7	00000010	+	$X \wedge Y \wedge \neg Z$
9	5-3	24	D7=B3	10110111	=	$Z \rightarrow (X \wedge \neg Y)$
	5-3	24	D7=B3	10110111		$Z \rightarrow (X \wedge \neg Y)$
	1-7	8	I8	00000001	+	$X \wedge Y \wedge Z$
10	5-3	24	D7=B3	10110111	=	$Z \rightarrow (X \wedge \neg Y)$
	6-2	19	**B**	10110111		$Z \rightarrow X$
	1-7	7	I7	00000010	+	$X \wedge Y \wedge \neg Z$
11	5-3	39	D8	01110110	=	$(\neg Y \vee \neg Z) \wedge (X \vee Y \vee Z)$
	5-3	39	D8	01110110		$(\neg Y \vee \neg Z) \wedge (X \vee Y \vee Z)$
	1-7	8	I8	00000001	+	$X \wedge Y \wedge Z$
12	5-3	39	D8	01110110	=	$(\neg Y \vee \neg Z) \wedge (X \vee Y \vee Z)$
	6-2	25	M1xL2	01110111		$\neg X \rightarrow \neg(Y \leftrightarrow Z)$

The derivation of *Bocardo* is:

Table 3.8 **C** OR **K** as necessary condition of **A**

Derivations	Level	#	Symbol	ID		Traditional logical form
	5-3	22	C3=F7	10111010	+	$Z \to (\neg X \wedge Y)$
1	1-7	2	K7=T8	01000000	=	$\neg X \wedge \neg Y \wedge Z$
	6-2	2	**F**	11111010		$Z \to \neg X$
	5-3	22	C3=F7	10111010	+	$Z \to (\neg X \wedge Y)$
2	1-7	5	K8=S8	00001000	=	$\neg X \wedge Y \wedge Z$
	5-3	22	C3=F7	10111010		$Z \to (\neg X \wedge Y)$
	5-3	23	C4	10111001	+	$(Y \vee \neg Z) \wedge (\neg X \vee \neg Y \vee Z)$
3	1-7	2	K7=T8	01000000	=	$\neg X \wedge \neg Y \wedge Z$
	6-2	3	E2xA1	11111001		$X \to (Y \leftrightarrow Z)$
	5-3	23	C4	10111001	+	$(Y \vee \neg Z) \wedge (\neg X \vee \neg Y \vee Z)$
4	1-7	5	K8=S8	00001000	=	$\neg X \wedge Y \wedge Z$
	5-3	23	C4	10111001		$(Y \vee \neg Z) \wedge (\neg X \vee \neg Y \vee Z)$
	5-3	26	C5=B5	10110011	+	$Z \to (X \wedge Y)$
5	1-7	2	K7=T8	01000000	=	$\neg X \wedge \neg Y \wedge Z$
	6-2	6	A1xM1	11110011		$Z \to (X \leftrightarrow Y)$
	5-3	26	C5=B5	10110011	+	$Z \to (X \wedge Y)$
6	1-7	5	K8=S8	00001000	=	$\neg X \wedge Y \wedge Z$
	6-2	18	**C**	10111011		$Z \to Y$
	5-3	29	C6=A7	10101011	+	$\neg Y \to (\neg X \wedge \neg Z)$
7	1-7	2	K7=T8	01000000	=	$\neg X \wedge \neg Y \wedge Z$
	6-2	9	**A**	11101011		$X \to Y$
	5-3	29	C6=A7	10101011	+	$\neg Y \to (\neg X \wedge \neg Z)$
8	1-7	5	K8=S8	00001000	=	$\neg X \wedge Y \wedge Z$
	5-3	29	C6=A7	10101011		$\neg Y \to (\neg X \wedge \neg Z)$
	5-3	33	C7	10011011	+	$(Y \vee \neg Z) \wedge (X \vee \neg Y \vee Z)$
9	1-7	2	K7=T8	01000000	=	$\neg X \wedge \neg Y \wedge Z$
	6-2	13	A1xM2	11011011		$\neg(X \leftrightarrow \neg Y \leftrightarrow Z)$
	5-3	33	C7	10011011	+	$(Y \vee \neg Z) \wedge (X \vee \neg Y \vee Z)$
10	1-7	5	K8=S8	00001000	=	$\neg X \wedge Y \wedge Z$
	5-3	33	C7	10011011		$(Y \vee \neg Z) \wedge (X \vee \neg Y \vee Z)$
	5-3	53	C8=L5	00111011	+	$\neg Y \to (X \wedge \neg Z)$
11	1-7	2	K7=T8	01000000	=	$\neg X \wedge \neg Y \wedge Z$
	6-2	24	A1xL2	01111011		$\neg Y \to \neg(X \leftrightarrow Z)$
	5-3	53	C8=L5	00111011	+	$\neg Y \to (X \wedge \neg Z)$
12	1-7	5	K8=S8	00001000	=	$\neg X \wedge Y \wedge Z$
	5-3	53	C8=L5	00111011		$\neg Y \to (X \wedge \neg Z)$

The derivation of *Disamis* is:

Table 3.9 **D** OR **K** as necessary condition of **E**

Derivations	Level	#	Symbol	ID		Traditional logical form
	5-3	2	D3=E4	11110100	+	$Y \to (\neg X \wedge \neg Z)$
1	1-7	2	K7=T8	01000000	=	$\neg X \wedge \neg Y \wedge Z$
	5-3	2	D3=E4	11110100		$Y \to (\neg X \wedge \neg Z)$
	5-3	2	D3=E4	11110100	+	$Y \to (\neg X \wedge \neg Z)$
2	1-7	5	K8=S8	00001000	=	$\neg X \wedge Y \wedge Z$
	6-2	1	**E**	11111100		$X \to \neg Y$
	5-3	3	D4=F4	11110010	+	$Z \to (\neg X \wedge \neg Y)$
3	1-7	2	K7=T8	01000000	=	$\neg X \wedge \neg Y \wedge Z$
	5-3	3	D4=F4	11110010		$Z \to (\neg X \wedge \neg Y)$
	5-3	3	D4=F4	11110010	+	$Z \to (\neg X \wedge \neg Y)$
4	1-7	5	K8=S8	00001000	=	$\neg X \wedge Y \wedge Z$
	6-2	2	**F**	11111010		$Z \to \neg X$
	5-3	8	D5	11100110	+	$(\neg Y \vee \neg Z) \wedge (\neg X \vee Y \vee Z)$
5	1-7	2	K7=T8	01000000	=	$\neg X \wedge \neg Y \wedge Z$
	5-3	8	D5	11100110		$(\neg Y \vee \neg Z) \wedge (\neg X \vee Y \vee Z)$
	5-3	8	D5	11100110	+	$(\neg Y \vee \neg Z) \wedge (\neg X \vee Y \vee Z)$
6	1-7	5	K8=S8	00001000	=	$\neg X \wedge Y \wedge Z$
	6-2	7	E1xA2	11101110		$X \to \neg(Y \leftrightarrow Z)$
	5-3	14	D6=M3	11010110	+	$Y \to (X \wedge \neg Z)$
7	1-7	2	K7=T8	01000000	=	$\neg X \wedge \neg Y \wedge Z$
	5-3	14	D6=M3	11010110		$Y \to (X \wedge \neg Z)$
	5-3	14	D6=M3	11010110	+	$Y \to (X \wedge \neg Z)$
8	1-7	5	K8=S8	00001000	=	$\neg X \wedge Y \wedge Z$
	6-2	11	E1xM2	11011110		$Y \to \neg(X \leftrightarrow Z)$
	5-3	24	D7=B3	10110111	+	$Z \to (X \wedge \neg Y)$
9	1-7	2	K7=T8	01000000	=	$\neg X \wedge \neg Y \wedge Z$
	6-2	4	**D**	11110110		$Z \to \neg Y$
	5-3	24	D7=B3	10110111	+	$Z \to (X \wedge \neg Y)$
10	1-7	5	K8=S8	00001000	=	$\neg X \wedge Y \wedge Z$
	6-2	16	E1xL1	10111110		$Z \to \neg(X \leftrightarrow Y)$
	5-3	39	D8	01110110	+	$(\neg Y \vee \neg Z) \wedge (X \vee Y \vee Z)$
11	1-7	2	K7=T8	01000000	=	$\neg X \wedge \neg Y \wedge Z$
	5-3	39	D8	01110110		$(\neg Y \vee \neg Z) \wedge (X \vee Y \vee Z)$
	5-3	39	D8	01110110	+	$(\neg Y \vee \neg Z) \wedge (X \vee Y \vee Z)$
12	1-7	5	K8=S8	00001000	=	$\neg X \wedge Y \wedge Z$
	6-2	22	E1xL2	01111110		$\neg(X \leftrightarrow Y \leftrightarrow Z)$

The derivation of *Ferison* is:

Table 3.10 **G** OR **F** as necessary condition of **A**

Derivations	Level	#	Symbol	ID		Traditional logical form
	1-7	5	G7=S8	00001000	+	$\neg X \wedge Y \wedge Z$
1	5-3	1	F3=E3	11111000	=	$X \rightarrow (\neg Y \wedge \neg Z)$
	5-3	1	F3=E3	11111000		$X \rightarrow (\neg Y \wedge \neg Z)$
	1-7	8	G8=I8	00000001	+	$X \wedge Y \wedge Z$
2	5-3	1	F3=E3	11111000	=	$X \rightarrow (\neg Y \wedge \neg Z)$
	6-2	3	E2xA1	11111001		$X \rightarrow (Y \leftrightarrow Z)$
	1-7	5	G7=S8	00001000	+	$\neg X \wedge Y \wedge Z$
3	5-3	3	F4=D4	11110010	=	$Z \rightarrow (\neg X \wedge \neg Y)$
	6-2	2	**F**	11111010		$Z \rightarrow \neg X$
	1-7	8	G8=I8	00000001	+	$X \wedge Y \wedge Z$
4	5-3	3	F4=D4	11110010	=	$Z \rightarrow (\neg X \wedge \neg Y)$
	6-2	6	A1xM1	11110011		$Z \rightarrow (X \leftrightarrow Y)$
	1-7	5	G7=S8	00001000	+	$\neg X \wedge Y \wedge Z$
5	5-3	6	F5=A3	11101010	=	$X \rightarrow (Y \wedge \neg Z)$
	5-3	6	F5=A3	11101010		$X \rightarrow (Y \wedge \neg Z)$
	1-7	8	G8=I8	00000001	+	$X \wedge Y \wedge Z$
6	5-3	6	F5=A3	11101010	=	$X \rightarrow (Y \wedge \neg Z)$
	6-2	9	**A**	11101011		$X \rightarrow Y$
	1-7	5	G7=S8	00001000	+	$\neg X \wedge Y \wedge Z$
7	5-3	12	F6	11011010	=	$(\neg X \vee \neg Z) \wedge (X \vee \neg Y \vee Z)$
	5-3	12	F6	11011010		$(\neg X \vee \neg Z) \wedge (X \vee \neg Y \vee Z)$
	1-7	8	G8=I8	00000001	+	$X \wedge Y \wedge Z$
8	5-3	12	F6	11011010	=	$(\neg X \vee \neg Z) \wedge (X \vee \neg Y \vee Z)$
	6-2	13	A1xM2	11011011		$\neg(X \leftrightarrow \neg Y \leftrightarrow Z)$
	1-7	5	G7=S8	00001000	+	$\neg X \wedge Y \wedge Z$
9	5-3	22	F7=C3	10111010	=	$Z \rightarrow (\neg X \wedge Y)$
	5-3	22	F7=C3	10111010		$Z \rightarrow (\neg X \wedge Y)$
	1-7	8	G8=I8	00000001	+	$X \wedge Y \wedge Z$
10	5-3	22	F7=C3	10111010	=	$Z \rightarrow (\neg X \wedge Y)$
	6-2	18	**C**	10111011		$Z \rightarrow Y$
	1-7	5	G7=S8	00001000	+	$\neg X \wedge Y \wedge Z$
11	5-3	37	F8	01111010	=	$(\neg X \vee \neg Z) \wedge (X \vee Y \vee Z)$
	5-3	37	F8	01111010		$(\neg X \vee \neg Z) \wedge (X \vee Y \vee Z)$
	1-7	8	G8=I8	00000001	+	$X \wedge Y \wedge Z$
12	5-3	37	F8	01111010	=	$(\neg X \vee \neg Z) \wedge (X \vee Y \vee Z)$
	6-2	24	A1xL2	01111011		$\neg Y \rightarrow \neg(X \leftrightarrow Z)$

The derivation of *Datisi* is:

Table 3.11 **H** OR **F** as necessary condition of **E**

Derivations	Level	#	Symbol	ID		Traditional logical form
	1-7	2	H7=T8	01000000	+	$\neg X \wedge \neg Y \wedge Z$
1	5-3	1	F3=E3	11111000	=	$X \rightarrow (\neg Y \wedge \neg Z)$
	5-3	1	F3=E3	11111000		$X \rightarrow (\neg Y \wedge \neg Z)$
	1-7	6	H8=O8	00000100	+	$X \wedge \neg Y \wedge Z$
2	5-3	1	F3=E3	11111000	=	$X \rightarrow (\neg Y \wedge \neg Z)$
	6-2	1	**E**	11111100		$X \rightarrow \neg Y$
	1-7	2	H7=T8	01000000	+	$\neg X \wedge \neg Y \wedge Z$
3	5-3	3	F4=D4	11110010	=	$Z \rightarrow (\neg X \wedge \neg Y)$
	5-3	3	F4=D4	11110010		$Z \rightarrow (\neg X \wedge \neg Y)$
	1-7	6	H8=O8	00000100	+	$X \wedge \neg Y \wedge Z$
4	5-3	3	F4=D4	11110010	=	$Z \rightarrow (\neg X \wedge \neg Y)$
	6-2	4	**D**	11110110		$Z \rightarrow \neg Y$
	1-7	2	H7=T8	01000000	+	$\neg X \wedge \neg Y \wedge Z$
5	5-3	6	F5=A3	11101010	=	$X \rightarrow (Y \wedge \neg Z)$
	5-3	6	F5=A3	11101010		$X \rightarrow (Y \wedge \neg Z)$
	1-7	6	H8=O8	00000100	+	$X \wedge \neg Y \wedge Z$
6	5-3	6	F5=A3	11101010	=	$X \rightarrow (Y \wedge \neg Z)$
	6-2	7	E1xA2	11101110		$X \rightarrow \neg(Y \leftrightarrow Z)$
	1-7	2	H7=T8	01000000	+	$\neg X \wedge \neg Y \wedge Z$
7	5-3	12	F6	11011010	=	$(\neg X \vee \neg Z) \wedge (X \vee \neg Y \vee Z)$
	5-3	12	F6	11011010		$(\neg X \vee \neg Z) \wedge (X \vee \neg Y \vee Z)$
	1-7	6	H8=O8	00000100	+	$X \wedge \neg Y \wedge Z$
8	5-3	12	F6	11011010	=	$(\neg X \vee \neg Z) \wedge (X \vee \neg Y \vee Z)$
	6-2	11	E1xM2	11011110		$Y \rightarrow \neg(X \leftrightarrow Z)$
	1-7	2	H7=T8	01000000	+	$\neg X \wedge \neg Y \wedge Z$
9	5-3	22	F7=C3	10111010	=	$Z \rightarrow (\neg X \wedge Y)$
	6-2	2	**F**	11111010		$Z \rightarrow \neg X$
	1-7	6	H8=O8	00000100	+	$X \wedge \neg Y \wedge Z$
10	5-3	22	F7=C3	10111010	=	$Z \rightarrow (\neg X \wedge Y)$
	6-2	16	E1xL1	10111110		$Z \rightarrow \neg(X \leftrightarrow Y)$
	1-7	2	H7=T8	01000000	+	$\neg X \wedge \neg Y \wedge Z$
11	5-3	37	F8	01111010	=	$(\neg X \vee \neg Z) \wedge (X \vee Y \vee Z)$
	5-3	37	F8	01111010		$(\neg X \vee \neg Z) \wedge (X \vee Y \vee Z)$
	1-7	6	H8=O8	00000100	+	$X \wedge \neg Y \wedge Z$
12	5-3	37	F8	01111010	=	$(\neg X \vee \neg Z) \wedge (X \vee Y \vee Z)$
	6-2	22	E1xL2	01111110		$\neg(X \leftrightarrow Y \leftrightarrow Z)$

We have accomplished the examination of the classical mode through sums.

3.3 Extension of Classical Derivation through Sums

Now, I shall consider the derivations of the first group of the extension of classical derivation. The analogue of *Darii* is:

Table 3.12 **T** OR **Q** as necessary condition of **P**

Derivations	Level	#	Symbol	ID		Traditional logical form
	1-7	1	T7	10000000	+	$\neg X \wedge \neg Y \wedge \neg Z$
1	5-3	46	Q3	01011110	=	$(X \vee Z) \wedge (\neg X \vee \neg Y \vee \neg Z)$
	6-2	11	E1xM2	11011110		$Y \rightarrow \neg(X \leftrightarrow Z)$
	1-7	2	T8	01000000	+	$\neg X \wedge \neg Y \wedge Z$
2	5-3	46	Q3	01011110	=	$(X \vee Z) \wedge (\neg X \vee \neg Y \vee \neg Z)$
	5-3	46	Q3	01011110		$(X \vee Z) \wedge (\neg X \vee \neg Y \vee \neg Z)$
	1-7	1	T7	10000000	+	$\neg X \wedge \neg Y \wedge \neg Z$
3	5-3	47	Q4=P8	01011101	=	$\neg Z \rightarrow (X \wedge \neg Y)$
	6-2	12	**P**	11011101		$\neg Z \rightarrow \neg Y$
	1-7	2	T8	01000000	+	$\neg X \wedge \neg Y \wedge Z$
4	5-3	47	Q4=P8	01011101	=	$\neg Z \rightarrow (X \wedge \neg Y)$
	5-3	47	Q4=P8	01011101		$\neg Z \rightarrow (X \wedge \neg Y)$
	1-7	1	T7	10000000	+	$\neg X \wedge \neg Y \wedge \neg Z$
5	5-3	48	Q5	01011011	=	$(X \vee Z) \wedge (\neg X \vee Y \vee \neg Z)$
	6-2	13	A1xM2	11011011		$\neg(X \leftrightarrow \neg Y \leftrightarrow Z)$
	1-7	2	T8	01000000	+	$\neg X \wedge \neg Y \wedge Z$
6	5-3	48	Q5	01011011	=	$(X \vee Z) \wedge (\neg X \vee Y \vee \neg Z)$
	5-3	48	Q5	01011011		$(X \vee Z) \wedge (\neg X \vee Y \vee \neg Z)$
	1-7	1	T7	10000000	+	$\neg X \wedge \neg Y \wedge \neg Z$
7	5-3	49	Q6=M8	01010111	=	$\neg X \rightarrow (\neg Y \wedge Z)$
	6-2	14	**M**	11010111		$\neg X \rightarrow \neg Y$
	1-7	2	T8	01000000	+	$\neg X \wedge \neg Y \wedge Z$
8	5-3	49	Q6=M8	01010111	=	$\neg X \rightarrow (\neg Y \wedge Z)$
	5-3	49	Q6=M8	01010111		$\neg X \rightarrow (\neg Y \wedge Z)$
	1-7	1	T7	10000000	+	$\neg X \wedge \neg Y \wedge \neg Z$
9	5-3	50	Q7=N7	01001111	=	$\neg Z \rightarrow (X \wedge Y)$
	6-2	15	A2xM2	11001111		$\neg Z \rightarrow (X \leftrightarrow Y)$
	1-7	2	T8	01000000	+	$\neg X \wedge \neg Y \wedge Z$
10	5-3	50	Q7=N7	01001111	=	$\neg Z \rightarrow (X \wedge Y)$
	5-3	50	Q7=N7	01001111		$\neg Z \rightarrow (X \wedge Y)$
	1-7	1	T7	10000000	+	$\neg X \wedge \neg Y \wedge \neg Z$
11	5-3	56	Q8=L8	00011111	=	$\neg X \rightarrow (Y \wedge Z)$
	6-2	21	M2xL1	10011111		$\neg X \rightarrow (Y \leftrightarrow Z)$
	1-7	2	T8	01000000	+	$\neg X \wedge \neg Y \wedge Z$
12	5-3	56	Q8=L8	00011111	=	$\neg X \rightarrow (Y \wedge Z)$
	6-2	27	**Q**	01011111		$\neg Z \rightarrow X$

The derivation of the analogue of *Ferio* is:

Table 3.13 **S** OR **Q** as necessary condition of **N**

Derivations	Level	#	Symbol	ID		Traditional logical form
	1-7	3	S7	00100000	+	$\neg X \wedge Y \wedge \neg Z$
1	5-3	46	Q3	01011110	=	$(X \vee Z) \wedge (\neg X \vee \neg Y \vee \neg Z)$
	6-2	22	E1xL2	01111110		$\neg(X \leftrightarrow Y \leftrightarrow Z)$
	1-7	5	S8	00001000	+	$\neg X \wedge Y \wedge Z$
2	5-3	46	Q3	01011110	=	$(X \vee Z) \wedge (\neg X \vee \neg Y \vee \neg Z)$
	5-3	46	Q3	01011110		$(X \vee Z) \wedge (\neg X \vee \neg Y \vee \neg Z)$
	1-7	3	S7	00100000	+	$\neg X \wedge Y \wedge \neg Z$
3	5-3	47	Q4=P8	01011101	=	$\neg Z \rightarrow (X \wedge \neg Y)$
	6-2	23	E2xL2	01111101		$\neg Z \rightarrow \neg(X \leftrightarrow Y)$
	1-7	5	S8	00001000	+	$\neg X \wedge Y \wedge Z$
4	5-3	47	Q4=P8	01011101	=	$\neg Z \rightarrow (X \wedge \neg Y)$
	5-3	47	Q4=P8	01011101		$\neg Z \rightarrow (X \wedge \neg Y)$
	1-7	3	S7	00100000	+	$\neg X \wedge Y \wedge \neg Z$
5	5-3	48	Q5	01011011	=	$(X \vee Z) \wedge (\neg X \vee Y \vee \neg Z)$
	6-2	24	A1xL2	01111011		$\neg Y \rightarrow \neg(X \leftrightarrow Z)$
	1-7	5	S8	00001000	+	$\neg X \wedge Y \wedge Z$
6	5-3	48	Q5	01011011	=	$(X \vee Z) \wedge (\neg X \vee Y \vee \neg Z)$
	5-3	48	Q5	01011011		$(X \vee Z) \wedge (\neg X \vee Y \vee \neg Z)$
	1-7	3	S7	00100000	+	$\neg X \wedge Y \wedge \neg Z$
7	5-3	49	Q6=M8	01010111	=	$\neg X \rightarrow (\neg Y \wedge Z)$
	6-2	25	M1xL2	01110111		$\neg X \rightarrow \neg(Y \leftrightarrow Z)$
	1-7	5	S8	00001000	+	$\neg X \wedge Y \wedge Z$
8	5-3	49	Q6=M8	01010111	=	$\neg X \rightarrow (\neg Y \wedge Z)$
	6-2	27	**Q**	01011111		$\neg Z \rightarrow X$
	1-7	3	S7	00100000	+	$\neg X \wedge Y \wedge \neg Z$
9	5-3	50	Q7=N7	01001111	=	$\neg Z \rightarrow (X \wedge Y)$
	6-2	26	**N**	01101111		$\neg Z \rightarrow Y$
	1-7	5	S8	00001000	+	$\neg X \wedge Y \wedge Z$
10	5-3	50	Q7=N7	01001111	=	$\neg Z \rightarrow (X \wedge Y)$
	5-3	50	Q7=N7	01001111		$\neg Z \rightarrow (X \wedge Y)$
	1-7	3	S7	00100000	+	$\neg X \wedge Y \wedge \neg Z$
11	5-3	56	Q8=L8	00011111	=	$\neg X \rightarrow (Y \wedge Z)$
	6-2	28	**L**	00111111		$\neg X \rightarrow Y$
	1-7	5	S8	00001000	+	$\neg X \wedge Y \wedge Z$
12	5-3	56	Q8=L8	00011111	=	$\neg X \rightarrow (Y \wedge Z)$
	5-3	56	Q8=L8	00011111		$\neg X \rightarrow (Y \wedge Z)$

The derivation of the analogue of *Baroco* is:

Table 3.14 **T** OR **N** as necessary condition of **R**

Derivations	Level	#	Symbol	ID		Traditional logical form
	1-7	1	T7	10000000	+	$\neg X \wedge \neg Y \wedge \neg Z$
1	5-3	42	N3	01101110	=	$(Y \vee Z) \wedge (\neg X \vee \neg Y \vee \neg Z)$
	6-2	7	E1xA2	11101110		$X \rightarrow \neg(Y \leftrightarrow Z)$
	1-7	2	T8	01000000	+	$\neg X \wedge \neg Y \wedge Z$
2	5-3	42	N3	01101110	=	$(Y \vee Z) \wedge (\neg X \vee \neg Y \vee \neg Z)$
	5-3	42	N3	01101110		$(Y \vee Z) \wedge (\neg X \vee \neg Y \vee \neg Z)$
	1-7	1	T7	10000000	+	$\neg X \wedge \neg Y \wedge \neg Z$
3	5-3	43	N4=R8	01101101	=	$\neg Z \rightarrow (\neg X \wedge Y)$
	6-2	8	**R**	11101101		$\neg Z \rightarrow \neg X$
	1-7	2	T8	01000000	+	$\neg X \wedge \neg Y \wedge Z$
4	5-3	43	N4=R8	01101101	=	$\neg Z \rightarrow (\neg X \wedge Y)$
	5-3	43	N4=R8	01101101		$\neg Z \rightarrow (\neg X \wedge Y)$
	1-7	1	T7	10000000	+	$\neg X \wedge \neg Y \wedge \neg Z$
5	5-3	44	N5=A8	01101011	=	$\neg Y \rightarrow (\neg X \wedge Z)$
	6-2	9	**A**	11101011		$X \rightarrow Y$
	1-7	2	T8	01000000	+	$\neg X \wedge \neg Y \wedge Z$
6	5-3	44	N5=A8	01101011	=	$\neg Y \rightarrow (\neg X \wedge Z)$
	5-3	44	N5=A8	01101011		$\neg Y \rightarrow (\neg X \wedge Z)$
	1-7	1	T7	10000000	+	$\neg X \wedge \neg Y \wedge \neg Z$
7	5-3	45	N6	01100111	=	$(Y \vee Z) \wedge (X \vee \neg Y \vee \neg Z)$
	6-2	10	M1xA2	11100111		$\neg(\neg X \leftrightarrow Y \leftrightarrow Z)$
	1-7	2	T8	01000000	+	$\neg X \wedge \neg Y \wedge Z$
8	5-3	45	N6	01100111	=	$(Y \vee Z) \wedge (X \vee \neg Y \vee \neg Z)$
	5-3	45	N6	01100111		$(Y \vee Z) \wedge (X \vee \neg Y \vee \neg Z)$
	1-7	1	T7	10000000	+	$\neg X \wedge \neg Y \wedge \neg Z$
9	5-3	50	N7=Q7	01001111	=	$\neg Z \rightarrow (X \wedge Y)$
	6-2	15	A2xM2	11001111		$\neg Z \rightarrow (X \leftrightarrow Y)$
	1-7	2	T8	01000000	+	$\neg X \wedge \neg Y \wedge Z$
10	5-3	50	N7=Q7	01001111	=	$\neg Z \rightarrow (X \wedge Y)$
	5-3	50	N7=Q7	01001111		$\neg Z \rightarrow (X \wedge Y)$
	1-7	1	T7	10000000	+	$\neg X \wedge \neg Y \wedge \neg Z$
11	5-3	55	N8=L7	00101111	=	$\neg Y \rightarrow (X \wedge Z)$
	6-2	20	A2xL1	10101111		$\neg Y \rightarrow (X \leftrightarrow Z)$
	1-7	2	T8	01000000	+	$\neg X \wedge \neg Y \wedge Z$
12	5-3	55	N8=L7	00101111	=	$\neg Y \rightarrow (X \wedge Z)$
	6-2	26	**N**	01101111		$\neg Z \rightarrow Y$

The analogue of *Festino* is:

Table 3.15 **S** OR **P** as necessary condition of **R**

Derivations	Level	#	Symbol	ID		Traditional logical form
	1-7	3	S7	00100000	+	$\neg X \wedge Y \wedge \neg Z$
1	5-3	11	P3=E6	11011100	=	$Y \to (\neg X \wedge Z)$
	6-2	1	**E**	11111100		$X \to \neg Y$
	1-7	5	S8	00001000	+	$\neg X \wedge Y \wedge Z$
2	5-3	11	P3=E6	11011100	=	$Y \to (\neg X \wedge Z)$
	5-3	11	P3=E6	11011100		$Y \to (\neg X \wedge Z)$
	1-7	3	S7	00100000	+	$\neg X \wedge Y \wedge \neg Z$
3	5-3	13	P4	11011001	=	$(\neg Y \vee Z) \wedge (\neg X \vee Y \vee \neg Z)$
	6-2	3	E2xA1	11111001		$X \to (Y \leftrightarrow Z)$
	1-7	5	S8	00001000	+	$\neg X \wedge Y \wedge Z$
4	5-3	13	P4	11011001	=	$(\neg Y \vee Z) \wedge (\neg X \vee Y \vee \neg Z)$
	5-3	13	P4	11011001		$(\neg Y \vee Z) \wedge (\neg X \vee Y \vee \neg Z)$
	1-7	3	S7	00100000	+	$\neg X \wedge Y \wedge \neg Z$
5	5-3	15	P5=M4	11010101	=	$Y \to (X \wedge Z)$
	6-2	5	E2xM1	11110101		$Y \to (X \leftrightarrow Z)$
	1-7	5	S8	00001000	+	$\neg X \wedge Y \wedge Z$
6	5-3	15	P5=M4	11010101	=	$Y \to (X \wedge Z)$
	6-2	12	**P**	11011101		$\neg Z \to \neg Y$
	1-7	3	S7	00100000	+	$\neg X \wedge Y \wedge \neg Z$
7	5-3	18	P6=R6	11001101	=	$\neg Z \to (\neg X \wedge \neg Y)$
	6-2	8	**R**	11101101		$\neg Z \to \neg X$
	1-7	5	S8	00001000	+	$\neg X \wedge Y \wedge Z$
8	5-3	18	P6=R6	11001101	=	$\neg Z \to (\neg X \wedge \neg Y)$
	5-3	18	P6=R6	11001101		$\neg Z \to (\neg X \wedge \neg Y)$
	1-7	3	S7	00100000	+	$\neg X \wedge Y \wedge \neg Z$
9	5-3	32	P7	10011101	=	$(\neg Y \vee Z) \wedge (X \vee Y \vee \neg Z)$
	6-2	17	E2xL1	10111101		$\neg(X \leftrightarrow Y \leftrightarrow \neg Z)$
	1-7	5	S8	00001000	+	$\neg X \wedge Y \wedge Z$
10	5-3	32	P7	10011101	=	$(\neg Y \vee Z) \wedge (X \vee Y \vee \neg Z)$
	5-3	32	P7	10011101		$(\neg Y \vee Z) \wedge (X \vee Y \vee \neg Z)$
	1-7	3	S7	00100000	+	$\neg X \wedge Y \wedge \neg Z$
11	5-3	47	P8=Q4	01011101	=	$\neg Z \to (X \wedge \neg Y)$
	6-2	23	E2xL2	01111101		$\neg Z \to \neg(X \leftrightarrow Y)$
	1-7	5	S8	00001000	+	$\neg X \wedge Y \wedge Z$
12	5-3	47	P8=Q4	01011101	=	$\neg Z \to (X \wedge \neg Y)$
	5-3	47	P8=Q4	01011101		$\neg Z \to (X \wedge \neg Y)$

The analogue of *Bocardo* is:

Table 3.16 **N** OR **W** as necessary condition of **L**

Derivations	Level	#	Symbol	ID		Traditional logical form
	5-3	42	N3	01101110	+	$(Y \vee Z) \wedge (\neg X \vee \neg Y \vee \neg Z)$
1	1-7	4	W7=O7	00010000	=	$X \wedge \neg Y \wedge \neg Z$
	6-2	22	E1xL2	01111110		$\neg(X \leftrightarrow Y \leftrightarrow Z)$
	5-3	42	N3	01101110	+	$(Y \vee Z) \wedge (\neg X \vee \neg Y \vee \neg Z)$
2	1-7	7	W8=I7	00000010	=	$X \wedge Y \wedge \neg Z$
	5-3	42	N3	01101110		$(Y \vee Z) \wedge (\neg X \vee \neg Y \vee \neg Z)$
	5-3	43	N4=R8	01101101	+	$\neg Z \rightarrow (\neg X \wedge Y)$
3	1-7	4	W7=O7	00010000	=	$X \wedge \neg Y \wedge \neg Z$
	6-2	23	E2xL2	01111101		$\neg Z \rightarrow \neg(X \leftrightarrow Y)$
	5-3	43	N4=R8	01101101	+	$\neg Z \rightarrow (\neg X \wedge Y)$
4	1-7	7	W8=I7	00000010	=	$X \wedge Y \wedge \neg Z$
	6-2	26	**N**	01101111		$\neg Z \rightarrow Y$
	5-3	44	N5=A8	01101011	+	$\neg Y \rightarrow (\neg X \wedge Z)$
5	1-7	4	W7=O7	00010000	=	$X \wedge \neg Y \wedge \neg Z$
	6-2	24	A1xL2	01111011		$\neg Y \rightarrow \neg(X \leftrightarrow Z)$
	5-3	44	N5=A8	01101011	+	$\neg Y \rightarrow (\neg X \wedge Z)$
6	1-7	7	W8=I7	00000010	=	$X \wedge Y \wedge \neg Z$
	5-3	44	N5=A8	01101011		$\neg Y \rightarrow (\neg X \wedge Z)$
	5-3	45	N6	01100111	+	$(Y \vee Z) \wedge (X \vee \neg Y \vee \neg Z)$
7	1-7	4	W7=O7	00010000	=	$X \wedge \neg Y \wedge \neg Z$
	6-2	25	M1xL2	01110111		$\neg X \rightarrow \neg(Y \leftrightarrow Z)$
	5-3	45	N6	01100111	+	$(Y \vee Z) \wedge (X \vee \neg Y \vee \neg Z)$
8	1-7	7	W8=I7	00000010	=	$X \wedge Y \wedge \neg Z$
	5-3	45	N6	01100111		$(Y \vee Z) \wedge (X \vee \neg Y \vee \neg Z)$
	5-3	50	N7=Q7	01001111	+	$\neg Z \rightarrow (X \wedge Y)$
9	1-7	4	W7=O7	00010000	=	$X \wedge \neg Y \wedge \neg Z$
	6-2	27	**Q**	01011111		$\neg Z \rightarrow X$
	5-3	50	N7=Q7	01001111	+	$\neg Z \rightarrow (X \wedge Y)$
10	1-7	7	W8=I7	00000010	=	$X \wedge Y \wedge \neg Z$
	5-3	50	N7=Q7	01001111		$\neg Z \rightarrow (X \wedge Y)$
	5-3	55	N8=L7	00101111	+	$\neg Y \rightarrow (X \wedge Z)$
11	1-7	4	W7=O7	00010000	=	$X \wedge \neg Y \wedge \neg Z$
	6-2	28	**L**	00111111		$\neg X \rightarrow Y$
	5-3	55	N8=L7	00101111	+	$\neg Y \rightarrow (X \wedge Z)$
12	1-7	7	W8=I7	00000010	=	$X \wedge Y \wedge \neg Z$
	5-3	55	N8=L7	00101111		$\neg Y \rightarrow (X \wedge Z)$

The analogue of *Disamis* is:

Table 3.17 **P** OR **W** as necessary condition of **M**

Derivations	Level	#	Symbol	ID		Traditional logical form
	5-3	11	P3=E6	11011100	+	$Y \to (\neg X \wedge Z)$
1	1-7	4	W7=O7	00010000	=	$X \wedge \neg Y \wedge \neg Z$
	5-3	11	P3=E6	11011100		$Y \to (\neg X \wedge Z)$
	5-3	11	P3=E6	11011100	+	$Y \to (\neg X \wedge Z)$
2	1-7	7	W8=I7	00000010	=	$X \wedge Y \wedge \neg Z$
	6-2	11	E1xM2	11011110		$Y \to \neg(X \leftrightarrow Z)$
	5-3	13	P4	11011001	+	$(\neg Y \vee Z) \wedge (\neg X \vee Y \vee \neg Z)$
3	1-7	4	W7=O7	00010000	=	$X \wedge \neg Y \wedge \neg Z$
	5-3	13	P4	11011001		$(\neg Y \vee Z) \wedge (\neg X \vee Y \vee \neg Z)$
	5-3	13	P4	11011001	+	$(\neg Y \vee Z) \wedge (\neg X \vee Y \vee \neg Z)$
4	1-7	7	W8=I7	00000010	=	$X \wedge Y \wedge \neg Z$
	6-2	13	A1xM2	11011011		$\neg(X \leftrightarrow \neg Y \leftrightarrow Z)$
	5-3	15	P5=M4	11010101	+	$Y \to (X \wedge Z)$
5	1-7	4	W7=O7	00010000	=	$X \wedge \neg Y \wedge \neg Z$
	5-3	15	P5=M4	11010101		$Y \to (X \wedge Z)$
	5-3	15	P5=M4	11010101	+	$Y \to (X \wedge Z)$
6	1-7	7	W8=I7	00000010	=	$X \wedge Y \wedge \neg Z$
	6-2	14	**M**	11010111		$\neg X \to \neg Y$
	5-3	18	P6=R6	11001101	+	$\neg Z \to (\neg X \wedge \neg Y)$
7	1-7	4	W7=O7	00010000	=	$X \wedge \neg Y \wedge \neg Z$
	6-2	12	**P**	11011101		$\neg Z \to \neg Y$
	5-3	18	P6=R6	11001101	+	$\neg Z \to (\neg X \wedge \neg Y)$
8	1-7	7	W8=I7	00000010	=	$X \wedge Y \wedge \neg Z$
	6-2	15	A2xM2	11001111		$\neg Z \to (X \leftrightarrow Y)$
	5-3	32	P7	10011101	+	$(\neg Y \vee Z) \wedge (X \vee Y \vee \neg Z)$
9	1-7	4	W7=O7	00010000	=	$X \wedge \neg Y \wedge \neg Z$
	5-3	32	P7	10011101		$(\neg Y \vee Z) \wedge (X \vee Y \vee \neg Z)$
	5-3	32	P7	10011101	+	$(\neg Y \vee Z) \wedge (X \vee Y \vee \neg Z)$
10	1-7	7	W8=I7	00000010	=	$X \wedge Y \wedge \neg Z$
	6-2	21	M2xL1	10011111		$\neg X \to (Y \leftrightarrow Z)$
	5-3	47	P8=Q4	01011101	+	$\neg Z \to (X \wedge \neg Y)$
11	1-7	4	W7=O7	00010000	=	$X \wedge \neg Y \wedge \neg Z$
	5-3	47	P8=Q4	01011101		$\neg Z \to (X \wedge \neg Y)$
	5-3	47	P8=Q4	01011101	+	$\neg Z \to (X \wedge \neg Y)$
12	1-7	7	W8=I7	00000010	=	$X \wedge Y \wedge \neg Z$
	6-2	27	**Q**	01011111		$\neg Z \to X$

The analogue of *Ferison* is:

Table 3.18 **U** OR **Q** as necessary condition of **L**

Derivations	Level	#	Symbol	ID		Traditional logical form
	1-7	3	U7=S7	00100000	+	$\neg X \wedge Y \wedge \neg Z$
1	5-3	46	Q3	01011110	=	$(X \vee Z) \wedge (\neg X \vee \neg Y \vee \neg Z)$
	6-2	22	E1xL2	01111110		$\neg(X \leftrightarrow Y \leftrightarrow Z)$
	1-7	7	U8=I7	00000010	+	$X \wedge Y \wedge \neg Z$
2	5-3	46	Q3	01011110	=	$(X \vee Z) \wedge (\neg X \vee \neg Y \vee \neg Z)$
	5-3	46	Q3	01011110		$(X \vee Z) \wedge (\neg X \vee \neg Y \vee \neg Z)$
	1-7	3	U7=S7	00100000	+	$\neg X \wedge Y \wedge \neg Z$
3	5-3	47	Q4=P8	01011101	=	$\neg Z \rightarrow (X \wedge \neg Y)$
	6-2	23	E2xL2	01111101		$\neg Z \rightarrow \neg(X \leftrightarrow Y)$
	1-7	7	U8=I7	00000010	+	$X \wedge Y \wedge \neg Z$
4	5-3	47	Q4=P8	01011101	=	$\neg Z \rightarrow (X \wedge \neg Y)$
	6-2	27	**Q**	01011111		$\neg Z \rightarrow X$
	1-7	3	U7=S7	00100000	+	$\neg X \wedge Y \wedge \neg Z$
5	5-3	48	Q5	01011011	=	$(X \vee Z) \wedge (\neg X \vee Y \vee \neg Z)$
	6-2	24	A1xL2	01111011		$\neg Y \rightarrow \neg(X \leftrightarrow Z)$
	1-7	7	U8=I7	00000010	+	$X \wedge Y \wedge \neg Z$
6	5-3	48	Q5	01011011	=	$(X \vee Z) \wedge (\neg X \vee Y \vee \neg Z)$
	5-3	48	Q5	01011011		$(X \vee Z) \wedge (\neg X \vee Y \vee \neg Z)$
	1-7	3	U7=S7	00100000	+	$\neg X \wedge Y \wedge \neg Z$
7	5-3	49	Q6=M8	01010111	=	$\neg X \rightarrow (\neg Y \wedge Z)$
	6-2	25	M1xL2	01110111		$\neg X \rightarrow \neg(Y \leftrightarrow Z)$
	1-7	7	U8=I7	00000010	+	$X \wedge Y \wedge \neg Z$
8	5-3	49	Q6=M8	01010111	=	$\neg X \rightarrow (\neg Y \wedge Z)$
	5-3	49	Q6=M8	01010111		$\neg X \rightarrow (\neg Y \wedge Z)$
	1-7	3	U7=S7	00100000	+	$\neg X \wedge Y \wedge \neg Z$
9	5-3	50	Q7=N7	01001111	=	$\neg Z \rightarrow (X \wedge Y)$
	6-2	26	**N**	01101111		$\neg Z \rightarrow Y$
	1-7	7	U8=I7	00000010	+	$X \wedge Y \wedge \neg Z$
10	5-3	50	Q7=N7	01001111	=	$\neg Z \rightarrow (X \wedge Y)$
	5-3	50	Q7=N7	01001111		$\neg Z \rightarrow (X \wedge Y)$
	1-7	3	U7=S7	00100000	+	$\neg X \wedge Y \wedge \neg Z$
11	5-3	56	Q8=L8	00011111	=	$\neg X \rightarrow (Y \wedge Z)$
	6-2	28	**L**	00111111		$\neg X \rightarrow Y$
	1-7	7	U8=I7	00000010	+	$X \wedge Y \wedge \neg Z$
12	5-3	56	Q8=L8	00011111	=	$\neg X \rightarrow (Y \wedge Z)$
	5-3	56	Q8=L8	00011111		$\neg X \rightarrow (Y \wedge Z)$

The analogue of *Datisi* is:

Table 3.19 **V** OR **Q** as necessary condition of **M**

Derivations	Level	#	Symbol	ID		Traditional logical form
	1-7	1	V7=T7	10000000	+	$\neg X \wedge \neg Y \wedge \neg Z$
1	5-3	46	Q3	01011110	=	$(X \vee Z) \wedge (\neg X \vee \neg Y \vee \neg Z)$
	6-2	11	E1xM2	11011110		$Y \rightarrow \neg(X \leftrightarrow Z)$
	1-7	4	V8=O7	00010000	+	$X \wedge \neg Y \wedge \neg Z$
2	5-3	46	Q3	01011110	=	$(X \vee Z) \wedge (\neg X \vee \neg Y \vee \neg Z)$
	5-3	46	Q3	01011110		$(X \vee Z) \wedge (\neg X \vee \neg Y \vee \neg Z)$
	1-7	1	V7=T7	10000000	+	$\neg X \wedge \neg Y \wedge \neg Z$
3	5-3	47	Q4=P8	01011101	=	$\neg Z \rightarrow (X \wedge \neg Y)$
	6-2	12	**P**	11011101		$\neg Z \rightarrow \neg Y$
	1-7	4	V8=O7	00010000	+	$X \wedge \neg Y \wedge \neg Z$
4	5-3	47	Q4=P8	01011101	=	$\neg Z \rightarrow (X \wedge \neg Y)$
	5-3	47	Q4=P8	01011101		$\neg Z \rightarrow (X \wedge \neg Y)$
	1-7	1	V7=T7	10000000	+	$\neg X \wedge \neg Y \wedge \neg Z$
5	5-3	48	Q5	01011011	=	$(X \vee Z) \wedge (\neg X \vee Y \vee \neg Z)$
	6-2	13	A1xM2	11011011		$\neg(X \leftrightarrow \neg Y \leftrightarrow Z)$
	1-7	4	V8=O7	00010000	+	$X \wedge \neg Y \wedge \neg Z$
6	5-3	48	Q5	01011011	=	$(X \vee Z) \wedge (\neg X \vee Y \vee \neg Z)$
	5-3	48	Q5	01011011		$(X \vee Z) \wedge (\neg X \vee Y \vee \neg Z)$
	1-7	1	V7=T7	10000000	+	$\neg X \wedge \neg Y \wedge \neg Z$
7	5-3	49	Q6=M8	01010111	=	$\neg X \rightarrow (\neg Y \wedge Z)$
	6-2	14	**M**	11010111		$\neg X \rightarrow \neg Y$
	1-7	4	V8=O7	00010000	+	$X \wedge \neg Y \wedge \neg Z$
8	5-3	49	Q6=M8	01010111	=	$\neg X \rightarrow (\neg Y \wedge Z)$
	5-3	49	Q6=M8	01010111		$\neg X \rightarrow (\neg Y \wedge Z)$
	1-7	1	V7=T7	10000000	+	$\neg X \wedge \neg Y \wedge \neg Z$
9	5-3	50	Q7=N7	01001111	=	$\neg Z \rightarrow (X \wedge Y)$
	6-2	15	A2xM2	11001111		$\neg Z \rightarrow (X \leftrightarrow Y)$
	1-7	4	V8=O7	00010000	+	$X \wedge \neg Y \wedge \neg Z$
10	5-3	50	Q7=N7	01001111	=	$\neg Z \rightarrow (X \wedge Y)$
	6-2	27	**Q**	01011111		$\neg Z \rightarrow X$
	1-7	1	V7=T7	10000000	+	$\neg X \wedge \neg Y \wedge \neg Z$
11	5-3	56	Q8=L8	00011111	=	$\neg X \rightarrow (Y \wedge Z)$
	6-2	21	M2xL1	10011111		$\neg X \rightarrow (Y \leftrightarrow Z)$
	1-7	4	V8=O7	00010000	+	$X \wedge \neg Y \wedge \neg Z$
12	5-3	56	Q8=L8	00011111	=	$\neg X \rightarrow (Y \wedge Z)$
	5-3	56	Q8=L8	00011111		$\neg X \rightarrow (Y \wedge Z)$

3.4 First Mixed Mode through Sums

The counterpart of *Darii* is:

Table 3.20 **T** OR **B** as necessary condition of **D**

Derivations	Level	#	Symbol	ID		Traditional logical form
	1-7	1	T7	10000000	+	$\neg X \wedge \neg Y \wedge \neg Z$
1	5-3	24	B3=D7	10110111	=	$Z \rightarrow (X \wedge \neg Y)$
	5-3	24	B3=D7	10110111		$Z \rightarrow (X \wedge \neg Y)$
	1-7	2	T8	01000000	+	$\neg X \wedge \neg Y \wedge Z$
2	5-3	24	B3=D7	10110111	=	$Z \rightarrow (X \wedge \neg Y)$
	6-2	4	**D**	11110110		$Z \rightarrow \neg Y$
	1-7	1	T7	10000000	+	$\neg X \wedge \neg Y \wedge \neg Z$
3	5-3	25	B4	10110101	=	$(X \vee \neg Z) \wedge (\neg X \vee \neg Y \vee Z)$
	5-3	25	B4	10110101		$(X \vee \neg Z) \wedge (\neg X \vee \neg Y \vee Z)$
	1-7	2	T8	01000000	+	$\neg X \wedge \neg Y \wedge Z$
4	5-3	25	B4	10110101	=	$(X \vee \neg Z) \wedge (\neg X \vee \neg Y \vee Z)$
	6-2	5	E2xM1	11110101		$Y \rightarrow (X \leftrightarrow Z)$
	1-7	1	T7	10000000	+	$\neg X \wedge \neg Y \wedge \neg Z$
5	5-3	26	B5=C5	10110011	=	$Z \rightarrow (X \wedge Y)$
	5-3	26	B5=C5	10110011		$Z \rightarrow (X \wedge Y)$
	1-7	2	T8	01000000	+	$\neg X \wedge \neg Y \wedge Z$
6	5-3	26	B5=C5	10110011	=	$Z \rightarrow (X \wedge Y)$
	6-2	6	A1xM1	11110011		$Z \rightarrow (X \leftrightarrow Y)$
	1-7	1	T7	10000000	+	$\neg X \wedge \neg Y \wedge \neg Z$
7	5-3	30	B6	10100111	=	$(X \vee \neg Z) \wedge (\neg X \vee Y \vee Z)$
	5-3	30	B6	10100111		$(X \vee \neg Z) \wedge (\neg X \vee Y \vee Z)$
	1-7	2	T8	01000000	+	$\neg X \wedge \neg Y \wedge Z$
8	5-3	30	B6	10100111	=	$(X \vee \neg Z) \wedge (\neg X \vee Y \vee Z)$
	6-2	10	M1xA2	11100111		$\neg(\neg X \leftrightarrow Y \leftrightarrow Z)$
	1-7	1	T7	10000000	+	$\neg X \wedge \neg Y \wedge \neg Z$
9	5-3	34	B7=M7	10010111	=	$\neg X \rightarrow (\neg Y \wedge \neg Z)$
	5-3	34	B7=M7	10010111		$\neg X \rightarrow (\neg Y \wedge \neg Z)$
	1-7	2	T8	01000000	+	$\neg X \wedge \neg Y \wedge Z$
10	5-3	34	B7=M7	10010111	=	$\neg X \rightarrow (\neg Y \wedge \neg Z)$
	6-2	14	**M**	11010111		$\neg X \rightarrow \neg Y$
	1-7	1	T7	10000000	+	$\neg X \wedge \neg Y \wedge \neg Z$
11	5-3	54	B8=L6	00110111	=	$\neg X \rightarrow (Y \wedge \neg Z)$
	6-2	19	**B**	10110111		$Z \rightarrow X$
	1-7	2	T8	01000000	+	$\neg X \wedge \neg Y \wedge Z$
12	5-3	54	B8=L6	00110111	=	$\neg X \rightarrow (Y \wedge \neg Z)$
	6-2	25	M1xL2	01110111		$\neg X \rightarrow \neg(Y \leftrightarrow Z)$

The counterpart of *Ferio* is:

Table 3.21 **S** OR **B** as necessary condition of **C**

Derivations	Level	#	Symbol	ID		Traditional logical form
	1-7	3	S7	00100000	+	$\neg X \wedge Y \wedge \neg Z$
1	5-3	24	B3=D7	10110111	=	$Z \rightarrow (X \wedge \neg Y)$
	5-3	24	B3=D7	10110111		$Z \rightarrow (X \wedge \neg Y)$
	1-7	5	S8	00001000	+	$\neg X \wedge Y \wedge Z$
2	5-3	24	B3=D7	10110111	=	$Z \rightarrow (X \wedge \neg Y)$
	6-2	16	E1xL1	10111110		$Z \rightarrow \neg(X \leftrightarrow Y)$
	1-7	3	S7	00100000	+	$\neg X \wedge Y \wedge \neg Z$
3	5-3	25	B4	10110101	=	$(X \vee \neg Z) \wedge (\neg X \vee \neg Y \vee Z)$
	5-3	25	B4	10110101		$(X \vee \neg Z) \wedge (\neg X \vee \neg Y \vee Z)$
	1-7	5	S8	00001000	+	$\neg X \wedge Y \wedge Z$
4	5-3	25	B4	10110101	=	$(X \vee \neg Z) \wedge (\neg X \vee \neg Y \vee Z)$
	6-2	17	E2xL1	10111101		$\neg(X \leftrightarrow Y \leftrightarrow \neg Z)$
	1-7	3	S7	00100000	+	$\neg X \wedge Y \wedge \neg Z$
5	5-3	26	B5=C5	10110011	=	$Z \rightarrow (X \wedge Y)$
	5-3	26	B5=C5	10110011		$Z \rightarrow (X \wedge Y)$
	1-7	5	S8	00001000	+	$\neg X \wedge Y \wedge Z$
6	5-3	26	B5=C5	10110011	=	$Z \rightarrow (X \wedge Y)$
	6-2	18	**C**	10111011		$Z \rightarrow Y$
	1-7	3	S7	00100000	+	$\neg X \wedge Y \wedge \neg Z$
7	5-3	30	B6	10100111	=	$(X \vee \neg Z) \wedge (\neg X \vee Y \vee Z)$
	5-3	30	B6	10100111		$(X \vee \neg Z) \wedge (\neg X \vee Y \vee Z)$
	1-7	5	S8	00001000	+	$\neg X \wedge Y \wedge Z$
8	5-3	30	B6	10100111	=	$(X \vee \neg Z) \wedge (\neg X \vee Y \vee Z)$
	6-2	20	A2xL1	10101111		$\neg Y \rightarrow (X \leftrightarrow Z)$
	1-7	3	S7	00100000	+	$\neg X \wedge Y \wedge \neg Z$
9	5-3	34	B7=M7	10010111	=	$\neg X \rightarrow (\neg Y \wedge \neg Z)$
	6-2	19	**B**	10110111		$Z \rightarrow X$
	1-7	5	S8	00001000	+	$\neg X \wedge Y \wedge Z$
10	5-3	34	B7=M7	10010111	=	$\neg X \rightarrow (\neg Y \wedge \neg Z)$
	6-2	21	M2xL1	10011111		$\neg X \rightarrow (Y \leftrightarrow Z)$
	1-7	3	S7	00100000	+	$\neg X \wedge Y \wedge \neg Z$
11	5-3	54	B8=L6	00110111	=	$\neg X \rightarrow (Y \wedge \neg Z)$
	5-3	54	B8=L6	00110111		$\neg X \rightarrow (Y \wedge \neg Z)$
	1-7	5	S8	00001000	+	$\neg X \wedge Y \wedge Z$
12	5-3	54	B8=L6	00110111	=	$\neg X \rightarrow (Y \wedge \neg Z)$
	6-2	28	**L**	00111111		$\neg X \rightarrow Y$

The counterpart of *Baroco* is:

Table 3.22 **T** OR **C** as necessary condition of **F**

Derivations	Level	#	Symbol	ID		Traditional logical form
	1-7	1	T7	10000000	+	$\neg X \wedge \neg Y \wedge \neg Z$
1	5-3	22	C3=F7	10111010	=	$Z \to (\neg X \wedge Y)$
	5-3	22	C3=F7	10111010		$Z \to (\neg X \wedge Y)$
	1-7	2	T8	01000000	+	$\neg X \wedge \neg Y \wedge Z$
2	5-3	22	C3=F7	10111010	=	$Z \to (\neg X \wedge Y)$
	6-2	2	**F**	11111010		$Z \to \neg X$
	1-7	1	T7	10000000	+	$\neg X \wedge \neg Y \wedge \neg Z$
3	5-3	23	C4	10111001	=	$(Y \vee \neg Z) \wedge (\neg X \vee \neg Y \vee Z)$
	5-3	23	C4	10111001		$(Y \vee \neg Z) \wedge (\neg X \vee \neg Y \vee Z)$
	1-7	2	T8	01000000	+	$\neg X \wedge \neg Y \wedge Z$
4	5-3	23	C4	10111001	=	$(Y \vee \neg Z) \wedge (\neg X \vee \neg Y \vee Z)$
	6-2	3	E2xA1	11111001		$X \to (Y \leftrightarrow Z)$
	1-7	1	T7	10000000	+	$\neg X \wedge \neg Y \wedge \neg Z$
5	5-3	26	C5=B5	10110011	=	$Z \to (X \wedge Y)$
	5-3	26	C5=B5	10110011		$Z \to (X \wedge Y)$
	1-7	2	T8	01000000	+	$\neg X \wedge \neg Y \wedge Z$
6	5-3	26	C5=B5	10110011	=	$Z \to (X \wedge Y)$
	6-2	6	A1xM1	11110011		$Z \to (X \leftrightarrow Y)$
	1-7	1	T7	10000000	+	$\neg X \wedge \neg Y \wedge \neg Z$
7	5-3	29	C6=A7	10101011	=	$\neg Y \to (\neg X \wedge \neg Z)$
	5-3	29	C6=A7	10101011		$\neg Y \to (\neg X \wedge \neg Z)$
	1-7	2	T8	01000000	+	$\neg X \wedge \neg Y \wedge Z$
8	5-3	29	C6=A7	10101011	=	$\neg Y \to (\neg X \wedge \neg Z)$
	6-2	9	**A**	11101011		$X \to Y$
	1-7	1	T7	10000000	+	$\neg X \wedge \neg Y \wedge \neg Z$
9	5-3	33	C7	10011011	=	$(Y \vee \neg Z) \wedge (X \vee \neg Y \vee Z)$
	5-3	33	C7	10011011		$(Y \vee \neg Z) \wedge (X \vee \neg Y \vee Z)$
	1-7	2	T8	01000000	+	$\neg X \wedge \neg Y \wedge Z$
10	5-3	33	C7	10011011	=	$(Y \vee \neg Z) \wedge (X \vee \neg Y \vee Z)$
	6-2	13	A1xM2	11011011		$\neg(X \leftrightarrow \neg Y \leftrightarrow Z)$
	1-7	1	T7	10000000	+	$\neg X \wedge \neg Y \wedge \neg Z$
11	5-3	53	C8=L5	00111011	=	$\neg Y \to (X \wedge \neg Z)$
	6-2	18	**C**	10111011		$Z \to Y$
	1-7	2	T8	01000000	+	$\neg X \wedge \neg Y \wedge Z$
12	5-3	53	C8=L5	00111011	=	$\neg Y \to (X \wedge \neg Z)$
	6-2	24	A1xL2	01111011		$\neg Y \to \neg(X \leftrightarrow Z)$

The counterpart of *Festino* is:

Table 3.23 **S** OR **D** as necessary condition of **F**

Derivations	Level	#	Symbol	ID		Traditional logical form
	1-7	3	S7	00100000	+	$\neg X \wedge Y \wedge \neg Z$
1	5-3	2	D3=E4	11110100	=	$Y \to (\neg X \wedge \neg Z)$
	5-3	2	D3=E4	11110100		$Y \to (\neg X \wedge \neg Z)$
	1-7	5	S8	00001000	+	$\neg X \wedge Y \wedge Z$
2	5-3	2	D3=E4	11110100	=	$Y \to (\neg X \wedge \neg Z)$
	6-2	1	**E**	11111100		$X \to \neg Y$
	1-7	3	S7	00100000	+	$\neg X \wedge Y \wedge \neg Z$
3	5-3	3	D4=F4	11110010	=	$Z \to (\neg X \wedge \neg Y)$
	5-3	3	D4=F4	11110010		$Z \to (\neg X \wedge \neg Y)$
	1-7	5	S8	00001000	+	$\neg X \wedge Y \wedge Z$
4	5-3	3	D4=F4	11110010	=	$Z \to (\neg X \wedge \neg Y)$
	6-2	2	**F**	11111010		$Z \to \neg X$
	1-7	3	S7	00100000	+	$\neg X \wedge Y \wedge \neg Z$
5	5-3	8	D5	11100110	=	$(\neg Y \vee \neg Z) \wedge (\neg X \vee Y \vee Z)$
	5-3	8	D5	11100110		$(\neg Y \vee \neg Z) \wedge (\neg X \vee Y \vee Z)$
	1-7	5	S8	00001000	+	$\neg X \wedge Y \wedge Z$
6	5-3	8	D5	11100110	=	$(\neg Y \vee \neg Z) \wedge (\neg X \vee Y \vee Z)$
	6-2	7	E1xA2	11101110		$X \to \neg(Y \leftrightarrow Z)$
	1-7	3	S7	00100000	+	$\neg X \wedge Y \wedge \neg Z$
7	5-3	14	D6=M3	11010110	=	$Y \to (X \wedge \neg Z)$
	6-2	4	**D**	11110110		$Z \to \neg Y$
	1-7	5	S8	00001000	+	$\neg X \wedge Y \wedge Z$
8	5-3	14	D6=M3	11010110	=	$Y \to (X \wedge \neg Z)$
	6-2	11	E1xM2	11011110		$Y \to \neg(X \leftrightarrow Z)$
	1-7	3	S7	00100000	+	$\neg X \wedge Y \wedge \neg Z$
9	5-3	24	D7=B3	10110111	=	$Z \to (X \wedge \neg Y)$
	5-3	24	D7=B3	10110111		$Z \to (X \wedge \neg Y)$
	1-7	5	S8	00001000	+	$\neg X \wedge Y \wedge Z$
10	5-3	24	D7=B3	10110111	=	$Z \to (X \wedge \neg Y)$
	6-2	16	E1xL1	10111110		$Z \to \neg(X \leftrightarrow Y)$
	1-7	3	S7	00100000	+	$\neg X \wedge Y \wedge \neg Z$
11	5-3	39	D8	01110110	=	$(\neg Y \vee \neg Z) \wedge (X \vee Y \vee Z)$
	5-3	39	D8	01110110		$(\neg Y \vee \neg Z) \wedge (X \vee Y \vee Z)$
	1-7	5	S8	00001000	+	$\neg X \wedge Y \wedge Z$
12	5-3	39	D8	01110110	=	$(\neg Y \vee \neg Z) \wedge (X \vee Y \vee Z)$
	6-2	22	E1xL2	01111110		$\neg(X \leftrightarrow Y \leftrightarrow Z)$

The counterpart of *Bocardo* is:

Table 3.24 **C** OR **J** as necessary condition of **L**

Derivations	Level	#	Symbol	ID		Traditional logical form
	5-3	22	C3=F7	10111010	+	$Z \to (\neg X \wedge Y)$
1	1-7	6	J7=O8	00000100	=	$X \wedge \neg Y \wedge Z$
	6-2	16	E1xL1	10111110		$Z \to \neg(X \leftrightarrow Y)$
	5-3	22	C3=F7	10111010	+	$Z \to (\neg X \wedge Y)$
2	1-7	8	J8=I8	00000001	=	$X \wedge Y \wedge Z$
	6-2	18	**C**	10111011		$Z \to Y$
	5-3	23	C4	10111001	+	$(Y \vee \neg Z) \wedge (\neg X \vee \neg Y \vee Z)$
3	1-7	6	J7=O8	00000100	=	$X \wedge \neg Y \wedge Z$
	6-2	17	E2xL1	10111101		$\neg(X \leftrightarrow Y \leftrightarrow \neg Z)$
	5-3	23	C4	10111001	+	$(Y \vee \neg Z) \wedge (\neg X \vee \neg Y \vee Z)$
4	1-7	8	J8=I8	00000001	=	$X \wedge Y \wedge Z$
	5-3	23	C4	10111001		$(Y \vee \neg Z) \wedge (\neg X \vee \neg Y \vee Z)$
	5-3	26	C5=B5	10110011	+	$Z \to (X \wedge Y)$
5	1-7	6	J7=O8	00000100	=	$X \wedge \neg Y \wedge Z$
	6-2	19	**B**	10110111		$Z \to X$
	5-3	26	C5=B5	10110011	+	$Z \to (X \wedge Y)$
6	1-7	8	J8=I8	00000001	=	$X \wedge Y \wedge Z$
	5-3	26	C5=B5	10110011		$Z \to (X \wedge Y)$
	5-3	29	C6=A7	10101011	+	$\neg Y \to (\neg X \wedge \neg Z)$
7	1-7	6	J7=O8	00000100	=	$X \wedge \neg Y \wedge Z$
	6-2	20	A2xL1	10101111		$\neg Y \to (X \leftrightarrow Z)$
	5-3	29	C6=A7	10101011	+	$\neg Y \to (\neg X \wedge \neg Z)$
8	1-7	8	J8=I8	00000001	=	$X \wedge Y \wedge Z$
	5-3	29	C6=A7	10101011		$\neg Y \to (\neg X \wedge \neg Z)$
	5-3	33	C7	10011011	+	$(Y \vee \neg Z) \wedge (X \vee \neg Y \vee Z)$
9	1-7	6	J7=O8	00000100	=	$X \wedge \neg Y \wedge Z$
	6-2	21	M2xL1	10011111		$\neg X \to (Y \leftrightarrow Z)$
	5-3	33	C7	10011011	+	$(Y \vee \neg Z) \wedge (X \vee \neg Y \vee Z)$
10	1-7	8	J8=I8	00000001	=	$X \wedge Y \wedge Z$
	5-3	33	C7	10011011		$(Y \vee \neg Z) \wedge (X \vee \neg Y \vee Z)$
	5-3	53	C8=L5	00111011	+	$\neg Y \to (X \wedge \neg Z)$
11	1-7	6	J7=O8	00000100	=	$X \wedge \neg Y \wedge Z$
	6-2	28	**L**	00111111		$\neg X \to Y$
	5-3	53	C8=L5	00111011	+	$\neg Y \to (X \wedge \neg Z)$
12	1-7	8	J8=I8	00000001	=	$X \wedge Y \wedge Z$
	5-3	53	C8=L5	00111011		$\neg Y \to (X \wedge \neg Z)$

The counterpart of *Disamis* is:

Table 3.25 **D** OR **J** as necessary condition of **M**

Derivations	Level	#	Symbol	ID		Traditional logical form
	5-3	2	D3=E4	11110100	+	$Y \rightarrow (\neg X \wedge \neg Z)$
1	1-7	6	J7=O8	00000100	=	$X \wedge \neg Y \wedge Z$
	5-3	2	D3=E4	11110100		$Y \rightarrow (\neg X \wedge \neg Z)$
	5-3	2	D3=E4	11110100	+	$Y \rightarrow (\neg X \wedge \neg Z)$
2	1-7	8	J8=I8	00000001	=	$X \wedge Y \wedge Z$
	6-2	5	E2xM1	11110101		$Y \rightarrow (X \leftrightarrow Z)$
	5-3	3	D4=F4	11110010	+	$Z \rightarrow (\neg X \wedge \neg Y)$
3	1-7	6	J7=O8	00000100	=	$X \wedge \neg Y \wedge Z$
	6-2	4	**D**	11110110		$Z \rightarrow \neg Y$
	5-3	3	D4=F4	11110010	+	$Z \rightarrow (\neg X \wedge \neg Y)$
4	1-7	8	J8=I8	00000001	=	$X \wedge Y \wedge Z$
	6-2	6	A1xM1	11110011		$Z \rightarrow (X \leftrightarrow Y)$
	5-3	8	D5	11100110	+	$(\neg Y \vee \neg Z) \wedge (\neg X \vee Y \vee Z)$
5	1-7	6	J7=O8	00000100	=	$X \wedge \neg Y \wedge Z$
	5-3	8	D5	11100110		$(\neg Y \vee \neg Z) \wedge (\neg X \vee Y \vee Z)$
	5-3	8	D5	11100110	+	$(\neg Y \vee \neg Z) \wedge (\neg X \vee Y \vee Z)$
6	1-7	8	J8=I8	00000001	=	$X \wedge Y \wedge Z$
	6-2	10	M1xA2	11100111		$\neg(\neg X \leftrightarrow Y \leftrightarrow Z)$
	5-3	14	D6=M3	11010110	+	$Y \rightarrow (X \wedge \neg Z)$
7	1-7	6	J7=O8	00000100	=	$X \wedge \neg Y \wedge Z$
	5-3	14	D6=M3	11010110		$Y \rightarrow (X \wedge \neg Z)$
	5-3	14	D6=M3	11010110	+	$Y \rightarrow (X \wedge \neg Z)$
8	1-7	8	J8=I8	00000001	=	$X \wedge Y \wedge Z$
	6-2	14	**M**	11010111		$\neg X \rightarrow \neg Y$
	5-3	24	D7=B3	10110111	+	$Z \rightarrow (X \wedge \neg Y)$
9	1-7	6	J7=O8	00000100	=	$X \wedge \neg Y \wedge Z$
	5-3	24	D7=B3	10110111		$Z \rightarrow (X \wedge \neg Y)$
	5-3	24	D7=B3	10110111	+	$Z \rightarrow (X \wedge \neg Y)$
10	1-7	8	J8=I8	00000001	=	$X \wedge Y \wedge Z$
	6-2	19	**B**	10110111		$Z \rightarrow X$
	5-3	39	D8	01110110	+	$(\neg Y \vee \neg Z) \wedge (X \vee Y \vee Z)$
11	1-7	6	J7=O8	00000100	=	$X \wedge \neg Y \wedge Z$
	5-3	39	D8	01110110		$(\neg Y \vee \neg Z) \wedge (X \vee Y \vee Z)$
	5-3	39	D8	01110110	+	$(\neg Y \vee \neg Z) \wedge (X \vee Y \vee Z)$
12	1-7	8	J8=I8	00000001	=	$X \wedge Y \wedge Z$
	6-2	25	M1xL2	01110111		$\neg X \rightarrow \neg(Y \leftrightarrow Z)$

The counterpart of *Ferison* is:

Table 3.26 **G** OR **B** as necessary condition of **L**

Derivations	Level	#	Symbol	ID		Traditional logical form
	1-7	5	G7=S8	00001000	+	$\neg X \wedge Y \wedge Z$
1	5-3	24	B3=D7	10110111	=	$Z \rightarrow (X \wedge \neg Y)$
	6-2	16	E1xL1	10111110		$Z \rightarrow \neg(X \leftrightarrow Y)$
	1-7	8	G8=I8	00000001	+	$X \wedge Y \wedge Z$
2	5-3	24	B3=D7	10110111	=	$Z \rightarrow (X \wedge \neg Y)$
	6-2	19	**B**	10110111		$Z \rightarrow X$
	1-7	5	G7=S8	00001000	+	$\neg X \wedge Y \wedge Z$
3	5-3	25	B4	10110101	=	$(X \vee \neg Z) \wedge (\neg X \vee \neg Y \vee Z)$
	6-2	17	E2xL1	10111101		$\neg(X \leftrightarrow Y \leftrightarrow \neg Z)$
	1-7	8	G8=I8	00000001	+	$X \wedge Y \wedge Z$
4	5-3	25	B4	10110101	=	$(X \vee \neg Z) \wedge (\neg X \vee \neg Y \vee Z)$
	5-3	25	B4	10110101		$(X \vee \neg Z) \wedge (\neg X \vee \neg Y \vee Z)$
	1-7	5	G7=S8	00001000	+	$\neg X \wedge Y \wedge Z$
5	5-3	26	B5=C5	10110011	=	$Z \rightarrow (X \wedge Y)$
	6-2	18	**C**	10111011		$Z \rightarrow Y$
	1-7	8	G8=I8	00000001	+	$X \wedge Y \wedge Z$
6	5-3	26	B5=C5	10110011	=	$Z \rightarrow (X \wedge Y)$
	5-3	26	B5=C5	10110011		$Z \rightarrow (X \wedge Y)$
	1-7	5	G7=S8	00001000	+	$\neg X \wedge Y \wedge Z$
7	5-3	30	B6	10100111	=	$(X \vee \neg Z) \wedge (\neg X \vee Y \vee Z)$
	6-2	20	A2xL1	10101111		$\neg Y \rightarrow (X \leftrightarrow Z)$
	1-7	8	G8=I8	00000001	+	$X \wedge Y \wedge Z$
8	5-3	30	B6	10100111	=	$(X \vee \neg Z) \wedge (\neg X \vee Y \vee Z)$
	5-3	30	B6	10100111		$(X \vee \neg Z) \wedge (\neg X \vee Y \vee Z)$
	1-7	5	G7=S8	00001000	+	$\neg X \wedge Y \wedge Z$
9	5-3	34	B7=M7	10010111	=	$\neg X \rightarrow (\neg Y \wedge \neg Z)$
	6-2	21	M2xL1	10011111		$\neg X \rightarrow (Y \leftrightarrow Z)$
	1-7	8	G8=I8	00000001	+	$X \wedge Y \wedge Z$
10	5-3	34	B7=M7	10010111	=	$\neg X \rightarrow (\neg Y \wedge \neg Z)$
	5-3	34	B7=M7	10010111		$\neg X \rightarrow (\neg Y \wedge \neg Z)$
	1-7	5	G7=S8	00001000	+	$\neg X \wedge Y \wedge Z$
11	5-3	54	B8=L6	00110111	=	$\neg X \rightarrow (Y \wedge \neg Z)$
	6-2	28	**L**	00111111		$\neg X \rightarrow Y$
	1-7	8	G8=I8	00000001	+	$X \wedge Y \wedge Z$
12	5-3	54	B8=L6	00110111	=	$\neg X \rightarrow (Y \wedge \neg Z)$
	5-3	54	B8=L6	00110111		$\neg X \rightarrow (Y \wedge \neg Z)$

The counterpart of *Datisi* is:

Table 3.27 **H** OR **B** as necessary condition of **M**

Derivations	Level	#	Symbol	ID		Traditional logical form
	1-7	2	H7=T8	01000000	+	$\neg X \wedge \neg Y \wedge Z$
1	5-3	24	B3=D7	10110111	=	$Z \rightarrow (X \wedge \neg Y)$
	6-2	4	**D**	11110110		$Z \rightarrow \neg Y$
	1-7	6	H8=O8	00000100	+	$X \wedge \neg Y \wedge Z$
2	5-3	24	B3=D7	10110111	=	$Z \rightarrow (X \wedge \neg Y)$
	5-3	24	B3=D7	10110111		$Z \rightarrow (X \wedge \neg Y)$
	1-7	2	H7=T8	01000000	+	$\neg X \wedge \neg Y \wedge Z$
3	5-3	25	B4	10110101	=	$(X \vee \neg Z) \wedge (\neg X \vee \neg Y \vee Z)$
	6-2	5	E2xM1	11110101		$Y \rightarrow (X \leftrightarrow Z)$
	1-7	6	H8=O8	00000100	+	$X \wedge \neg Y \wedge Z$
4	5-3	25	B4	10110101	=	$(X \vee \neg Z) \wedge (\neg X \vee \neg Y \vee Z)$
	5-3	25	B4	10110101		$(X \vee \neg Z) \wedge (\neg X \vee \neg Y \vee Z)$
	1-7	2	H7=T8	01000000	+	$\neg X \wedge \neg Y \wedge Z$
5	5-3	26	B5=C5	10110011	=	$Z \rightarrow (X \wedge Y)$
	6-2	6	A1xM1	11110011		$Z \rightarrow (X \leftrightarrow Y)$
	1-7	6	H8=O8	00000100	+	$X \wedge \neg Y \wedge Z$
6	5-3	26	B5=C5	10110011	=	$Z \rightarrow (X \wedge Y)$
	6-2	19	**B**	10110111		$Z \rightarrow X$
	1-7	2	H7=T8	01000000	+	$\neg X \wedge \neg Y \wedge Z$
7	5-3	30	B6	10100111	=	$(X \vee \neg Z) \wedge (\neg X \vee Y \vee Z)$
	6-2	10	M1xA2	11100111		$\neg(\neg X \leftrightarrow Y \leftrightarrow Z)$
	1-7	6	H8=O8	00000100	+	$X \wedge \neg Y \wedge Z$
8	5-3	30	B6	10100111	=	$(X \vee \neg Z) \wedge (\neg X \vee Y \vee Z)$
	5-3	30	B6	10100111		$(X \vee \neg Z) \wedge (\neg X \vee Y \vee Z)$
	1-7	2	H7=T8	01000000	+	$\neg X \wedge \neg Y \wedge Z$
9	5-3	34	B7=M7	10010111	=	$\neg X \rightarrow (\neg Y \wedge \neg Z)$
	6-2	14	**M**	11010111		$\neg X \rightarrow \neg Y$
	1-7	6	H8=O8	00000100	+	$X \wedge \neg Y \wedge Z$
10	5-3	34	B7=M7	10010111	=	$\neg X \rightarrow (\neg Y \wedge \neg Z)$
	5-3	34	B7=M7	10010111		$\neg X \rightarrow (\neg Y \wedge \neg Z)$
	1-7	2	H7=T8	01000000	+	$\neg X \wedge \neg Y \wedge Z$
11	5-3	54	B8=L6	00110111	=	$\neg X \rightarrow (Y \wedge \neg Z)$
	6-2	25	M1xL2	01110111		$\neg X \rightarrow \neg(Y \leftrightarrow Z)$
	1-7	6	H8=O8	00000100	+	$X \wedge \neg Y \wedge Z$
12	5-3	54	B8=L6	00110111	=	$\neg X \rightarrow (Y \wedge \neg Z)$
	5-3	54	B8=L6	00110111		$\neg X \rightarrow (Y \wedge \neg Z)$

3.5 Second Mixed Mode through Sums

The counterpart of *Darii* is:

Table 3.28 **O** OR **R** as necessary condition of **P**

Derivations	Level	#	Symbol	ID		Traditional logical form
	1-7	4	07	00010000	+	$X \wedge \neg Y \wedge \neg Z$
1	5-3	5	R3=E5	11101100	=	$X \to (\neg Y \wedge Z)$
	6-2	1	**E**	11111100		$X \to \neg Y$
	1-7	6	08	00000100	+	$X \wedge \neg Y \wedge Z$
2	5-3	5	R3=E5	11101100	=	$X \to (\neg Y \wedge Z)$
	5-3	5	R3=E5	11101100		$X \to (\neg Y \wedge Z)$
	1-7	4	07	00010000	+	$X \wedge \neg Y \wedge \neg Z$
3	5-3	7	R4=A4	11101001	=	$X \to (Y \wedge Z)$
	6-2	3	E2xA1	11111001		$X \to (Y \leftrightarrow Z)$
	1-7	6	08	00000100	+	$X \wedge \neg Y \wedge Z$
4	5-3	7	R4=A4	11101001	=	$X \to (Y \wedge Z)$
	6-2	8	**R**	11101101		$\neg Z \to \neg X$
	1-7	4	07	00010000	+	$X \wedge \neg Y \wedge \neg Z$
5	5-3	9	R5	11100101	=	$(\neg X \vee Z) \wedge (X \vee \neg Y \vee \neg Z)$
	6-2	5	E2xM1	11110101		$Y \to (X \leftrightarrow Z)$
	1-7	6	08	00000100	+	$X \wedge \neg Y \wedge Z$
6	5-3	9	R5	11100101	=	$(\neg X \vee Z) \wedge (X \vee \neg Y \vee \neg Z)$
	5-3	9	R5	11100101		$(\neg X \vee Z) \wedge (X \vee \neg Y \vee \neg Z)$
	1-7	4	07	00010000	+	$X \wedge \neg Y \wedge \neg Z$
7	5-3	18	R6=P6	11001101	=	$\neg Z \to (\neg X \wedge \neg Y)$
	6-2	12	**P**	11011101		$\neg Z \to \neg Y$
	1-7	6	08	00000100	+	$X \wedge \neg Y \wedge Z$
8	5-3	18	R6=P6	11001101	=	$\neg Z \to (\neg X \wedge \neg Y)$
	5-3	18	R6=P6	11001101		$\neg Z \to (\neg X \wedge \neg Y)$
	1-7	4	07	00010000	+	$X \wedge \neg Y \wedge \neg Z$
9	5-3	28	R7	10101101	=	$(\neg X \vee Z) \wedge (X \vee Y \vee \neg Z)$
	6-2	17	E2xL1	10111101		$\neg(X \leftrightarrow Y \leftrightarrow \neg Z)$
	1-7	6	08	00000100	+	$X \wedge \neg Y \wedge Z$
10	5-3	28	R7	10101101	=	$(\neg X \vee Z) \wedge (X \vee Y \vee \neg Z)$
	5-3	28	R7	10101101		$(\neg X \vee Z) \wedge (X \vee Y \vee \neg Z)$
	1-7	4	07	00010000	+	$X \wedge \neg Y \wedge \neg Z$
11	5-3	43	R8=N4	01101101	=	$\neg Z \to (\neg X \wedge Y)$
	6-2	23	E2xL2	01111101		$\neg Z \to \neg(X \leftrightarrow Y)$
	1-7	6	08	00000100	+	$X \wedge \neg Y \wedge Z$
12	5-3	43	R8=N4	01101101	=	$\neg Z \to (\neg X \wedge Y)$
	5-3	43	R8=N4	01101101		$\neg Z \to (\neg X \wedge Y)$

The counterpart of *Ferio* is:

Table 3.29 **I** OR **R** as necessary condition of **N**

Derivations	Level	#	Symbol	ID		Traditional logical form
	1-7	7	I7	00000010	+	$X \wedge Y \wedge \neg Z$
1	5-3	5	R3=E5	11101100	=	$X \to (\neg Y \wedge Z)$
	6-2	7	E1xA2	11101110		$X \to \neg(Y \leftrightarrow Z)$
	1-7	8	I8	00000001	+	$X \wedge Y \wedge Z$
2	5-3	5	R3=E5	11101100	=	$X \to (\neg Y \wedge Z)$
	6-2	8	**R**	11101101		$\neg Z \to \neg X$
	1-7	7	I7	00000010	+	$X \wedge Y \wedge \neg Z$
3	5-3	7	R4=A4	11101001	=	$X \to (Y \wedge Z)$
	6-2	9	**A**	11101011		$X \to Y$
	1-7	8	I8	00000001	+	$X \wedge Y \wedge Z$
4	5-3	7	R4=A4	11101001	=	$X \to (Y \wedge Z)$
	5-3	7	R4=A4	11101001		$X \to (Y \wedge Z)$
	1-7	7	I7	00000010	+	$X \wedge Y \wedge \neg Z$
5	5-3	9	R5	11100101	=	$(\neg X \vee Z) \wedge (X \vee \neg Y \vee \neg Z)$
	6-2	10	M1xA2	11100111		$\neg(\neg X \leftrightarrow Y \leftrightarrow Z)$
	1-7	8	I8	00000001	+	$X \wedge Y \wedge Z$
6	5-3	9	R5	11100101	=	$(\neg X \vee Z) \wedge (X \vee \neg Y \vee \neg Z)$
	5-3	9	R5	11100101		$(\neg X \vee Z) \wedge (X \vee \neg Y \vee \neg Z)$
	1-7	7	I7	00000010	+	$X \wedge Y \wedge \neg Z$
7	5-3	18	R6=P6	11001101	=	$\neg Z \to (\neg X \wedge \neg Y)$
	6-2	15	A2xM2	11001111		$\neg Z \to (X \leftrightarrow Y)$
	1-7	8	I8	00000001	+	$X \wedge Y \wedge Z$
8	5-3	18	R6=P6	11001101	=	$\neg Z \to (\neg X \wedge \neg Y)$
	5-3	18	R6=P6	11001101		$\neg Z \to (\neg X \wedge \neg Y)$
	1-7	7	I7	00000010	+	$X \wedge Y \wedge \neg Z$
9	5-3	28	R7	10101101	=	$(\neg X \vee Z) \wedge (X \vee Y \vee \neg Z)$
	6-2	20	A2xL1	10101111		$\neg Y \to (X \leftrightarrow Z)$
	1-7	8	I8	00000001	+	$X \wedge Y \wedge Z$
10	5-3	28	R7	10101101	=	$(\neg X \vee Z) \wedge (X \vee Y \vee \neg Z)$
	5-3	28	R7	10101101		$(\neg X \vee Z) \wedge (X \vee Y \vee \neg Z)$
	1-7	7	I7	00000010	+	$X \wedge Y \wedge \neg Z$
11	5-3	43	R8=N4	01101101	=	$\neg Z \to (\neg X \wedge Y)$
	6-2	26	**N**	01101111		$\neg Z \to Y$
	1-7	8	I8	00000001	+	$X \wedge Y \wedge Z$
12	5-3	43	R8=N4	01101101	=	$\neg Z \to (\neg X \wedge Y)$
	5-3	43	R8=N4	01101101		$\neg Z \to (\neg X \wedge Y)$

The counterpart of *Baroco* is:

Table 3.30 **O** OR **N** as necessary condition of **Q**

Derivations	Level	#	Symbol	ID		Traditional logical form
	1-7	4	O7	00010000	+	$X \wedge \neg Y \wedge \neg Z$
1	5-3	42	N3	01101110	=	$(Y \vee Z) \wedge (\neg X \vee \neg Y \vee \neg Z)$
	6-2	22	E1xL2	01111110		$\neg(X \leftrightarrow Y \leftrightarrow Z)$
	1-7	6	O8	00000100	+	$X \wedge \neg Y \wedge Z$
2	5-3	42	N3	01101110	=	$(Y \vee Z) \wedge (\neg X \vee \neg Y \vee \neg Z)$
	5-3	42	N3	01101110		$(Y \vee Z) \wedge (\neg X \vee \neg Y \vee \neg Z)$
	1-7	4	O7	00010000	+	$X \wedge \neg Y \wedge \neg Z$
3	5-3	43	N4=R8	01101101	=	$\neg Z \rightarrow (\neg X \wedge Y)$
	6-2	23	E2xL2	01111101		$\neg Z \rightarrow \neg(X \leftrightarrow Y)$
	1-7	6	O8	00000100	+	$X \wedge \neg Y \wedge Z$
4	5-3	43	N4=R8	01101101	=	$\neg Z \rightarrow (\neg X \wedge Y)$
	5-3	43	N4=R8	01101101		$\neg Z \rightarrow (\neg X \wedge Y)$
	1-7	4	O7	00010000	+	$X \wedge \neg Y \wedge \neg Z$
5	5-3	44	N5=A8	01101011	=	$\neg Y \rightarrow (\neg X \wedge Z)$
	6-2	24	A1xL2	01111011		$\neg Y \rightarrow \neg(X \leftrightarrow Z)$
	1-7	6	O8	00000100	+	$X \wedge \neg Y \wedge Z$
6	5-3	44	N5=A8	01101011	=	$\neg Y \rightarrow (\neg X \wedge Z)$
	6-2	26	**N**	01101111		$\neg Z \rightarrow Y$
	1-7	4	O7	00010000	+	$X \wedge \neg Y \wedge \neg Z$
7	5-3	45	N6	01100111	=	$(Y \vee Z) \wedge (X \vee \neg Y \vee \neg Z)$
	6-2	25	M1xL2	01110111		$\neg X \rightarrow \neg(Y \leftrightarrow Z)$
	1-7	6	O8	00000100	+	$X \wedge \neg Y \wedge Z$
8	5-3	45	N6	01100111	=	$(Y \vee Z) \wedge (X \vee \neg Y \vee \neg Z)$
	5-3	45	N6	01100111		$(Y \vee Z) \wedge (X \vee \neg Y \vee \neg Z)$
	1-7	4	O7	00010000	+	$X \wedge \neg Y \wedge \neg Z$
9	5-3	50	N7=Q7	01001111	=	$\neg Z \rightarrow (X \wedge Y)$
	6-2	27	**Q**	01011111		$\neg Z \rightarrow X$
	1-7	6	O8	00000100	+	$X \wedge \neg Y \wedge Z$
10	5-3	50	N7=Q7	01001111	=	$\neg Z \rightarrow (X \wedge Y)$
	5-3	50	N7=Q7	01001111		$\neg Z \rightarrow (X \wedge Y)$
	1-7	4	O7	00010000	+	$X \wedge \neg Y \wedge \neg Z$
11	5-3	55	N8=L7	00101111	=	$\neg Y \rightarrow (X \wedge Z)$
	6-2	28	**L**	00111111		$\neg X \rightarrow Y$
	1-7	6	O8	00000100	+	$X \wedge \neg Y \wedge Z$
12	5-3	55	N8=L7	00101111	=	$\neg Y \rightarrow (X \wedge Z)$
	5-3	55	N8=L7	00101111		$\neg Y \rightarrow (X \wedge Z)$

The counterpart of *Festino* is:

Table 3.31 **I** OR **P** as necessary condition of **Q**

Derivations	Level	#	Symbol	ID		Traditional logical form
	1-7	7	I7	00000010	+	$X \wedge Y \wedge \neg Z$
1	5-3	11	P3=E6	11011100	=	$Y \rightarrow (\neg X \wedge Z)$
	6-2	11	E1xM2	11011110		$Y \rightarrow \neg(X \leftrightarrow Z)$
	1-7	8	I8	00000001	+	$X \wedge Y \wedge Z$
2	5-3	11	P3=E6	11011100	=	$Y \rightarrow (\neg X \wedge Z)$
	6-2	12	**P**	11011101		$\neg Z \rightarrow \neg Y$
	1-7	7	I7	00000010	+	$X \wedge Y \wedge \neg Z$
3	5-3	13	P4	11011001	=	$(\neg Y \vee Z) \wedge (\neg X \vee Y \vee \neg Z)$
	6-2	13	A1xM2	11011011		$\neg(X \leftrightarrow \neg Y \leftrightarrow Z)$
	1-7	8	I8	00000001	+	$X \wedge Y \wedge Z$
4	5-3	13	P4	11011001	=	$(\neg Y \vee Z) \wedge (\neg X \vee Y \vee \neg Z)$
	5-3	13	P4	11011001		$(\neg Y \vee Z) \wedge (\neg X \vee Y \vee \neg Z)$
	1-7	7	I7	00000010	+	$X \wedge Y \wedge \neg Z$
5	5-3	15	P5=M4	11010101	=	$Y \rightarrow (X \wedge Z)$
	6-2	14	**M**	11010111		$\neg X \rightarrow \neg Y$
	1-7	8	I8	00000001	+	$X \wedge Y \wedge Z$
6	5-3	15	P5=M4	11010101	=	$Y \rightarrow (X \wedge Z)$
	5-3	15	P5=M4	11010101		$Y \rightarrow (X \wedge Z)$
	1-7	7	I7	00000010	+	$X \wedge Y \wedge \neg Z$
7	5-3	18	P6=R6	11001101	=	$\neg Z \rightarrow (\neg X \wedge \neg Y)$
	6-2	15	A2xM2	11001111		$\neg Z \rightarrow (X \leftrightarrow Y)$
	1-7	8	I8	00000001	+	$X \wedge Y \wedge Z$
8	5-3	18	P6=R6	11001101	=	$\neg Z \rightarrow (\neg X \wedge \neg Y)$
	5-3	18	P6=R6	11001101		$\neg Z \rightarrow (\neg X \wedge \neg Y)$
	1-7	7	I7	00000010	+	$X \wedge Y \wedge \neg Z$
9	5-3	32	P7	10011101	=	$(\neg Y \vee Z) \wedge (X \vee Y \vee \neg Z)$
	6-2	21	M2xL1	10011111		$\neg X \rightarrow (Y \leftrightarrow Z)$
	1-7	8	I8	00000001	+	$X \wedge Y \wedge Z$
10	5-3	32	P7	10011101	=	$(\neg Y \vee Z) \wedge (X \vee Y \vee \neg Z)$
	5-3	32	P7	10011101		$(\neg Y \vee Z) \wedge (X \vee Y \vee \neg Z)$
	1-7	7	I7	00000010	+	$X \wedge Y \wedge \neg Z$
11	5-3	47	P8=Q4	01011101	=	$\neg Z \rightarrow (X \wedge \neg Y)$
	6-2	27	**Q**	01011111		$\neg Z \rightarrow X$
	1-7	8	I8	00000001	+	$X \wedge Y \wedge Z$
12	5-3	47	P8=Q4	01011101	=	$\neg Z \rightarrow (X \wedge \neg Y)$
	5-3	47	P8=Q4	01011101		$\neg Z \rightarrow (X \wedge \neg Y)$

The counterpart of *Bocardo* is:

Table 3.32 **N** OR **Ω** as necessary condition of **A**

Derivations	Level	#	Symbol	ID		Traditional logical form
	5-3	42	N3	01101110	+	$(Y \vee Z) \wedge (\neg X \vee \neg Y \vee \neg Z)$
1	1-7	1	Ω7=T7	10000000	=	$\neg X \wedge \neg Y \wedge \neg Z$
	6-2	7	E1xA2	11101110		$X \rightarrow \neg(Y \leftrightarrow Z)$
	5-3	42	N3	01101110	+	$(Y \vee Z) \wedge (\neg X \vee \neg Y \vee \neg Z)$
2	1-7	3	Ω8=S7	00100000	=	$\neg X \wedge Y \wedge \neg Z$
	5-3	42	N3	01101110		$(Y \vee Z) \wedge (\neg X \vee \neg Y \vee \neg Z)$
	5-3	43	N4=R8	01101101	+	$\neg Z \rightarrow (\neg X \wedge Y)$
3	1-7	1	Ω7=T7	10000000	=	$\neg X \wedge \neg Y \wedge \neg Z$
	6-2	8	**R**	11101101		$\neg Z \rightarrow \neg X$
	5-3	43	N4=R8	01101101	+	$\neg Z \rightarrow (\neg X \wedge Y)$
4	1-7	3	Ω8=S7	00100000	=	$\neg X \wedge Y \wedge \neg Z$
	5-3	43	N4=R8	01101101		$\neg Z \rightarrow (\neg X \wedge Y)$
	5-3	44	N5=A8	01101011	+	$\neg Y \rightarrow (\neg X \wedge Z)$
5	1-7	1	Ω7=T7	10000000	=	$\neg X \wedge \neg Y \wedge \neg Z$
	6-2	9	**A**	11101011		$X \rightarrow Y$
	5-3	44	N5=A8	01101011	+	$\neg Y \rightarrow (\neg X \wedge Z)$
6	1-7	3	Ω8=S7	00100000	=	$\neg X \wedge Y \wedge \neg Z$
	5-3	44	N5=A8	01101011		$\neg Y \rightarrow (\neg X \wedge Z)$
	5-3	45	N6	01100111	+	$(Y \vee Z) \wedge (X \vee \neg Y \vee \neg Z)$
7	1-7	1	Ω7=T7	10000000	=	$\neg X \wedge \neg Y \wedge \neg Z$
	6-2	10	M1xA2	11100111		$\neg(\neg X \leftrightarrow Y \leftrightarrow Z)$
	5-3	45	N6	01100111	+	$(Y \vee Z) \wedge (X \vee \neg Y \vee \neg Z)$
8	1-7	3	Ω8=S7	00100000	=	$\neg X \wedge Y \wedge \neg Z$
	5-3	45	N6	01100111		$(Y \vee Z) \wedge (X \vee \neg Y \vee \neg Z)$
	5-3	50	N7=Q7	01001111	+	$\neg Z \rightarrow (X \wedge Y)$
9	1-7	1	Ω7=T7	10000000	=	$\neg X \wedge \neg Y \wedge \neg Z$
	6-2	15	A2xM2	11001111		$\neg Z \rightarrow (X \leftrightarrow Y)$
	5-3	50	N7=Q7	01001111	+	$\neg Z \rightarrow (X \wedge Y)$
10	1-7	3	Ω8=S7	00100000	=	$\neg X \wedge Y \wedge \neg Z$
	6-2	26	**N**	01101111		$\neg Z \rightarrow Y$
	5-3	55	N8=L7	00101111	+	$\neg Y \rightarrow (X \wedge Z)$
11	1-7	1	Ω7=T7	10000000	=	$\neg X \wedge \neg Y \wedge \neg Z$
	6-2	20	A2xL1	10101111		$\neg Y \rightarrow (X \leftrightarrow Z)$
	5-3	55	N8=L7	00101111	+	$\neg Y \rightarrow (X \wedge Z)$
12	1-7	3	Ω8=S7	00100000	=	$\neg X \wedge Y \wedge \neg Z$
	5-3	55	N8=L7	00101111		$\neg Y \rightarrow (X \wedge Z)$

The derivation of the counterpart of *Disamis* is:

Table 3.33 **P** OR **Ω** as necessary condition of **E**

Derivations	Level	#	Symbol	ID		Traditional logical form
	5-3	11	P3=E6	11011100	+	$Y \to (\neg X \wedge Z)$
1	1-7	1	Ω7=T7	10000000	=	$\neg X \wedge \neg Y \wedge \neg Z$
	5-3	11	P3=E6	11011100		$Y \to (\neg X \wedge Z)$
	5-3	11	P3=E6	11011100	+	$Y \to (\neg X \wedge Z)$
2	1-7	3	Ω8=S7	00100000	=	$\neg X \wedge Y \wedge \neg Z$
	6-2	1	**E**	11111100		$X \to \neg Y$
	5-3	13	P4	11011001	+	$(\neg Y \vee Z) \wedge (\neg X \vee Y \vee \neg Z)$
3	1-7	1	Ω7=T7	10000000	=	$\neg X \wedge \neg Y \wedge \neg Z$
	5-3	13	P4	11011001		$(\neg Y \vee Z) \wedge (\neg X \vee Y \vee \neg Z)$
	5-3	13	P4	11011001	+	$(\neg Y \vee Z) \wedge (\neg X \vee Y \vee \neg Z)$
4	1-7	3	Ω8=S7	00100000	=	$\neg X \wedge Y \wedge \neg Z$
	6-2	3	E2xA1	11111001		$X \to (Y \leftrightarrow Z)$
	5-3	15	P5=M4	11010101	+	$Y \to (X \wedge Z)$
5	1-7	1	Ω7=T7	10000000	=	$\neg X \wedge \neg Y \wedge \neg Z$
	5-3	15	P5=M4	11010101		$Y \to (X \wedge Z)$
	5-3	15	P5=M4	11010101	+	$Y \to (X \wedge Z)$
6	1-7	3	Ω8=S7	00100000	=	$\neg X \wedge Y \wedge \neg Z$
	6-2	5	E2xM1	11110101		$Y \to (X \leftrightarrow Z)$
	5-3	18	P6=R6	11001101	+	$\neg Z \to (\neg X \wedge \neg Y)$
7	1-7	1	Ω7=T7	10000000	=	$\neg X \wedge \neg Y \wedge \neg Z$
	5-3	18	P6=R6	11001101		$\neg Z \to (\neg X \wedge \neg Y)$
	5-3	18	P6=R6	11001101	+	$\neg Z \to (\neg X \wedge \neg Y)$
8	1-7	3	Ω8=S7	00100000	=	$\neg X \wedge Y \wedge \neg Z$
	6-2	8	**R**	11101101		$\neg Z \to \neg X$
	5-3	32	P7	10011101	+	$(\neg Y \vee Z) \wedge (X \vee Y \vee \neg Z)$
9	1-7	1	Ω7=T7	10000000	=	$\neg X \wedge \neg Y \wedge \neg Z$
	5-3	32	P7	10011101		$(\neg Y \vee Z) \wedge (X \vee Y \vee \neg Z)$
	5-3	32	P7	10011101	+	$(\neg Y \vee Z) \wedge (X \vee Y \vee \neg Z)$
10	1-7	3	Ω8=S7	00100000	=	$\neg X \wedge Y \wedge \neg Z$
	6-2	17	E2xL1	10111101		$\neg(X \leftrightarrow Y \leftrightarrow \neg Z)$
	5-3	47	P8=Q4	01011101	+	$\neg Z \to (X \wedge \neg Y)$
11	1-7	1	Ω7=T7	10000000	=	$\neg X \wedge \neg Y \wedge \neg Z$
	6-2	12	**P**	11011101		$\neg Z \to \neg Y$
	5-3	47	P8=Q4	01011101	+	$\neg Z \to (X \wedge \neg Y)$
12	1-7	3	Ω8=S7	00100000	=	$\neg X \wedge Y \wedge \neg Z$
	6-2	23	E2xL2	01111101		$\neg Z \to \neg(X \leftrightarrow Y)$

The counterpart of *Ferison* is:

Table 3.34 **U** OR **R** as necessary condition of **A**

Derivations	Level	#	Symbol	ID		Traditional logical form
	1-7	3	U7=S7	00100000	+	$\neg X \wedge Y \wedge \neg Z$
1	5-3	5	R3=E5	11101100	=	$X \to (\neg Y \wedge Z)$
	5-3	5	R3=E5	11101100		$X \to (\neg Y \wedge Z)$
	1-7	7	U8=I7	00000010	+	$X \wedge Y \wedge \neg Z$
2	5-3	5	R3=E5	11101100	=	$X \to (\neg Y \wedge Z)$
	6-2	7	E1xA2	11101110		$X \to \neg(Y \leftrightarrow Z)$
	1-7	3	U7=S7	00100000	+	$\neg X \wedge Y \wedge \neg Z$
3	5-3	7	R4=A4	11101001	=	$X \to (Y \wedge Z)$
	5-3	7	R4=A4	11101001		$X \to (Y \wedge Z)$
	1-7	7	U8=I7	00000010	+	$X \wedge Y \wedge \neg Z$
4	5-3	7	R4=A4	11101001	=	$X \to (Y \wedge Z)$
	6-2	9	**A**	11101011		$X \to Y$
	1-7	3	U7=S7	00100000	+	$\neg X \wedge Y \wedge \neg Z$
5	5-3	9	R5	11100101	=	$(\neg X \vee Z) \wedge (X \vee \neg Y \vee \neg Z)$
	5-3	9	R5	11100101		$(\neg X \vee Z) \wedge (X \vee \neg Y \vee \neg Z)$
	1-7	7	U8=I7	00000010	+	$X \wedge Y \wedge \neg Z$
6	5-3	9	R5	11100101	=	$(\neg X \vee Z) \wedge (X \vee \neg Y \vee \neg Z)$
	6-2	10	M1xA2	11100111		$\neg(\neg X \leftrightarrow Y \leftrightarrow Z)$
	1-7	3	U7=S7	00100000	+	$\neg X \wedge Y \wedge \neg Z$
7	5-3	18	R6=P6	11001101	=	$\neg Z \to (\neg X \wedge \neg Y)$
	6-2	8	**R**	11101101		$\neg Z \to \neg X$
	1-7	7	U8=I7	00000010	+	$X \wedge Y \wedge \neg Z$
8	5-3	18	R6=P6	11001101	=	$\neg Z \to (\neg X \wedge \neg Y)$
	6-2	15	A2xM2	11001111		$\neg Z \to (X \leftrightarrow Y)$
	1-7	3	U7=S7	00100000	+	$\neg X \wedge Y \wedge \neg Z$
9	5-3	28	R7	10101101	=	$(\neg X \vee Z) \wedge (X \vee Y \vee \neg Z)$
	5-3	28	R7	10101101		$(\neg X \vee Z) \wedge (X \vee Y \vee \neg Z)$
	1-7	7	U8=I7	00000010	+	$X \wedge Y \wedge \neg Z$
10	5-3	28	R7	10101101	=	$(\neg X \vee Z) \wedge (X \vee Y \vee \neg Z)$
	6-2	20	A2xL1	10101111		$\neg Y \to (X \leftrightarrow Z)$
	1-7	3	U7=S7	00100000	+	$\neg X \wedge Y \wedge \neg Z$
11	5-3	43	R8=N4	01101101	=	$\neg Z \to (\neg X \wedge Y)$
	5-3	43	R8=N4	01101101		$\neg Z \to (\neg X \wedge Y)$
	1-7	7	U8=I7	00000010	+	$X \wedge Y \wedge \neg Z$
12	5-3	43	R8=N4	01101101	=	$\neg Z \to (\neg X \wedge Y)$
	6-2	26	**N**	01101111		$\neg Z \to Y$

The counterpart of *Datisi* is:

Table 3.35 **V** OR **R** as necessary condition of **E**

Derivations	Level	#	Symbol	ID		Traditional logical form
	1-7	1	V7=T7	10000000	+	$\neg X \wedge \neg Y \wedge \neg Z$
1	5-3	5	R3=E5	11101100	=	$X \rightarrow (\neg Y \wedge Z)$
	5-3	5	R3=E5	11101100		$X \rightarrow (\neg Y \wedge Z)$
	1-7	4	V8=O7	00010000	+	$X \wedge \neg Y \wedge \neg Z$
2	5-3	5	R3=E5	11101100	=	$X \rightarrow (\neg Y \wedge Z)$
	6-2	1	**E**	11111100		$X \rightarrow \neg Y$
	1-7	1	V7=T7	10000000	+	$\neg X \wedge \neg Y \wedge \neg Z$
3	5-3	7	R4=A4	11101001	=	$X \rightarrow (Y \wedge Z)$
	5-3	7	R4=A4	11101001		$X \rightarrow (Y \wedge Z)$
	1-7	4	V8=O7	00010000	+	$X \wedge \neg Y \wedge \neg Z$
4	5-3	7	R4=A4	11101001	=	$X \rightarrow (Y \wedge Z)$
	6-2	3	E2xA1	11111001		$X \rightarrow (Y \leftrightarrow Z)$
	1-7	1	V7=T7	10000000	+	$\neg X \wedge \neg Y \wedge \neg Z$
5	5-3	9	R5	11100101	=	$(\neg X \vee Z) \wedge (X \vee \neg Y \vee \neg Z)$
	5-3	9	R5	11100101		$(\neg X \vee Z) \wedge (X \vee \neg Y \vee \neg Z)$
	1-7	4	V8=O7	00010000	+	$X \wedge \neg Y \wedge \neg Z$
6	5-3	9	R5	11100101	=	$(\neg X \vee Z) \wedge (X \vee \neg Y \vee \neg Z)$
	6-2	5	E2xM1	11110101		$Y \rightarrow (X \leftrightarrow Z)$
	1-7	1	V7=T7	10000000	+	$\neg X \wedge \neg Y \wedge \neg Z$
7	5-3	18	R6=P6	11001101	=	$\neg Z \rightarrow (\neg X \wedge \neg Y)$
	5-3	18	R6=P6	11001101		$\neg Z \rightarrow (\neg X \wedge \neg Y)$
	1-7	4	V8=O7	00010000	+	$X \wedge \neg Y \wedge \neg Z$
8	5-3	18	R6=P6	11001101	=	$\neg Z \rightarrow (\neg X \wedge \neg Y)$
	6-2	12	**P**	11011101		$\neg Z \rightarrow \neg Y$
	1-7	1	V7=T7	10000000	+	$\neg X \wedge \neg Y \wedge \neg Z$
9	5-3	28	R7	10101101	=	$(\neg X \vee Z) \wedge (X \vee Y \vee \neg Z)$
	5-3	28	R7	10101101		$(\neg X \vee Z) \wedge (X \vee Y \vee \neg Z)$
	1-7	4	V8=O7	00010000	+	$X \wedge \neg Y \wedge \neg Z$
10	5-3	28	R7	10101101	=	$(\neg X \vee Z) \wedge (X \vee Y \vee \neg Z)$
	6-2	17	E2xL1	10111101		$\neg(X \leftrightarrow Y \leftrightarrow \neg Z)$
	1-7	1	V7=T7	10000000	+	$\neg X \wedge \neg Y \wedge \neg Z$
11	5-3	43	R8=N4	01101101	=	$\neg Z \rightarrow (\neg X \wedge Y)$
	6-2	8	**R**	11101101		$\neg Z \rightarrow \neg X$
	1-7	4	V8=O7	00010000	+	$X \wedge \neg Y \wedge \neg Z$
12	5-3	43	R8=N4	01101101	=	$\neg Z \rightarrow (\neg X \wedge Y)$
	6-2	23	E2xL2	01111101		$\neg Z \rightarrow \neg(X \leftrightarrow Y)$

First Group				Second Group				Third Group			
Classical	1st mixed	2nd mixed	Extension	Classical	1st mixed	2nd mixed	Extension	Classical	1st mixed	2nd mixed	Extension
O X ∧ ¬Y	T ¬X ∧ ¬Y	O X ∧ ¬Y	T ¬X ∧ ¬Y	O X ∧ ¬Y	T ¬X ∧ ¬Y	O X ∧ ¬Y	T ¬X ∧ ¬Y	C Z → Y	C Z → Y	N ¬Z → Y	N ¬Z → Y
K Z ∧ ¬X	J X ∧ Z	Ω ¬Z ∧ ¬X	W ¬Z ∧ X	C Z → Y	C Z → Y	N ¬Z → Y	N ¬Z → Y	K Z ∧ ¬X	J X ∧ Z	Ω ¬Z ∧ ¬X	W ¬Z ∧ X
H Z ∧ ¬Y	H Z ∧ ¬Y	V ¬Z ∧ ¬Y	V ¬Z ∧ ¬Y	B Z → X	F Z → ¬X	Q ¬Z → X	R ¬Z → ¬X	A X → Y	L ¬X → Y	A X → Y	L ¬X → Y
I X ∧ Y	S ¬X ∧ Y	I X ∧ Y	S ¬X ∧ Y	I X ∧ Y	S ¬X ∧ Y	I X ∧ Y	S ¬X ∧ Y	D Z → ¬Y	D Z → ¬Y	P ¬Z → ¬Y	P ¬Z → ¬Y
K Z ∧ ¬X	J X ∧ Z	Ω ¬Z ∧ ¬X	W ¬Z ∧ X	D Z → ¬Y	D Z → ¬Y	P ¬Z → ¬Y	P ¬Z → ¬Y	K Z ∧ ¬X	J X ∧ Z	Ω ¬Z ∧ ¬X	W ¬Z ∧ X
G Z ∧ Y	G Z ∧ Y	U ¬Z ∧ Y	U ¬Z ∧ Y	B Z → X	F Z → ¬X	Q ¬Z → X	R ¬Z → ¬X	E X → ¬Y	M ¬X → ¬Y	E X → ¬Y	M ¬X → ¬Y
O X ∧ ¬Y	T ¬X ∧ ¬Y	O X ∧ ¬Y	T ¬X ∧ ¬Y	O X ∧ ¬Y	T ¬X ∧ ¬Y	O X ∧ ¬Y	T ¬X ∧ ¬Y	G Z ∧ Y	G Z ∧ Y	U ¬Z ∧ Y	U ¬Z ∧ Y
F Z → ¬X	B Z → X	R ¬Z → ¬X	Q ¬Z → X	G Z ∧ Y	G Z ∧ Y	U ¬Z ∧ Y	U ¬Z ∧ Y	F Z → ¬X	B Z → X	R ¬Z → ¬X	Q ¬Z → X
D Z → ¬Y	D Z → ¬Y	P ¬Z → ¬Y	P ¬Z → ¬Y	J X ∧ Z	K Z ∧ ¬X	W ¬Z ∧ X	Ω ¬Z ∧ ¬X	A X → Y	L ¬X → Y	A X → Y	L ¬X → Y
I X ∧ Y	S ¬X ∧ Y	I X ∧ Y	S ¬X ∧ Y	I X ∧ Y	S ¬X ∧ Y	I X ∧ Y	S ¬X ∧ Y	H Z ∧ ¬Y	H Z ∧ ¬Y	V ¬Z ∧ ¬Y	V ¬Z ∧ ¬Y
F Z → ¬X	B Z → X	R ¬Z → ¬X	Q ¬Z → X	H Z ∧ ¬Y	H Z ∧ ¬Y	V ¬Z ∧ ¬Y	V ¬Z ∧ ¬Y	F Z → ¬X	B Z → X	R ¬Z → ¬X	Q ¬Z → X
C Z → Y	C Z → Y	N ¬Z → Y	N ¬Z → Y	J Z ∧ X	K Z ∧ ¬X	W ¬Z ∧ X	Ω ¬Z ∧ ¬X	E X → ¬Y	M ¬X → ¬Y	E X → ¬Y	M ¬X → ¬Y

Figure 3.1 Summary of all inferences through sums.

We can summarize all the 48 results obtained through sums as in Fig. 3.1. It may be noted that in all sum derivations at least one premise is particular. We may further remark that when both premises are particular also the conclusion is particular as well as when the two premises are negative also the conclusion is negative. Moreover, this kind of inference is retrograde (given the premises we assume the conclusion or given the conclusion we try to infer the possible premises). Note in particular that with different kinds of premises we assume often the same conclusion (in general there is some exchange in the quantity or quality of the premises), confirming that there is a certain under-determination of the premises relative to the conclusion that we can or wish to assume. Such an under-determination is typical of any inference. However, here it is reinforced as far as we deal with a necessary condition that is actually a disjunction of the premises.

Chapter 4

Subtractions

4.1 Introduction

I shall deal now with a further possibility of inferential reasoning: we can combine two expressions taken from very different generating sets (one from a LGS and the other from a RGS) and obtain nevertheless a valid inference. For instance, let us consider the case of the classical *Barbara*, i.e. $(\mathbf{A} \wedge \mathbf{B}) \rightarrow \mathbf{C}$. We can give rise to an equivalent result in this way:

$$\begin{array}{lll} \mathrm{A1} & 11111011 & - \\ \mathrm{K7} & 01000000 & = \\ \mathbf{C} & 10111011 & \end{array}$$

In other words, we have that (A1 - K7) $=$ **C**. This corresponds to an AND NOT operation (this is also known as non-implication or Boole's *except*: see Fig. 1.3) establishing a sufficient condition (*Hobbes*, 1655):

$$(\mathbf{A} \wedge \neg\mathbf{K}) \rightarrow \mathbf{C}\,. \tag{4.1}$$

In other words, in this case an assumption together with the failure (expressed by the explicit negation) of another assumption is here sufficient condition for obtaining the desired conclusion. We can formulate this e.g. by saying:

Mechanical Logic in Three-Dimensional Space
Gennaro Auletta

ISBN 978-981-4411-50-9 (Hardcover), 978-981-4411-51-6 (eBook)
www.panstanford.com

If Every X is Y and
It is not true that Some Z is not X,
Then it follows that Every Z is Y.

This kind of inference can be considered to a certain extent as a variant of product-like inferences. It has the general form of product inferences but it can also be proved that is equivalent to an inference through sum. Indeed, the last expression is equivalent to

$$(\neg\mathbf{O} \wedge \neg\mathbf{K}) \rightarrow \mathbf{C}\,, \tag{4.2}$$

which is in turn equivalent to

$$\neg(\mathbf{O} \vee \mathbf{K}) \rightarrow \mathbf{C}\,, \tag{4.3}$$

which we can finally express as the *Barbara*-like inference through sum:

$$(\mathbf{O} \vee \mathbf{K}) \leftarrow \mathbf{H}\,. \tag{4.4}$$

I have pointed pout that inference (4.2) is fully equivalent to the classical *Barbara*. Indeed, it is true that, from a formal point of view, the negation of **K** is equivalent to the affirmation of **B**. However, this inference could be interpreted in a different way (we have indeed seen that it has certain connection with the sum-inferences): with the words "It is not true that" we were expressing not simply a negative statement (as the many ones occurring in product inferences), but we were rather affirming the removal of a previous statement (i.e. **K**) due to some kind of results or experience (i.e. a test that results a disproof of that previous statement). In other words, when saying that "It is not true that Some Z is not X" in our current scientific practice but also social-political one this is not necessarily equivalent to"Every Z is not X" but could mean the much weaker statement: Our experience has not so far confirmed that there are Some Zs that are not Xs. We may consider this kind of inference as a weaker form of deduction (like product inferences) or as a milder form of abduction (to be still explored). In other words, when certain experiences suggest that things could stand in a different way as it was previously assumed, we can be tempted to hypothetically consider what would be the (anterograde) consequences if we assume that one of our previous assumptions has been proved to be false through certain tests. It could be considered as an updating of deduction or as the first, very basic form of adductive reasoning.

This difference is clearly expressed by the fact that interchange of lines with product and sum are irrelevant for the result. Instead, interchange of lines with subtraction will lead in general to a completely different result (and the same, as we shall see can be said for division). This means that in subtraction-inferences it does matter which particular premise we are denying.

Summarizing, while product derivations involve LGSs of the premises only and sum derivations RGSs of the premises only, subtraction derivations involve the LGS of the first premise and the RGS of the second premise. I therefore introduce a new operation (the subtraction) defined as in Table 4.1.

Table 4.1 Subtraction

$0-0=0$	$0-1=0$	$1-0=1$	$1-1=0$

The reason for $0-1=0$ is due to the fact that we have only 0 and 1 as domain of the operation. Being 0 the lowest element we cannot subtract any quantity to it (have a look at Table 1.7 for an analogous result for the sum).

4.2 Classical Inferences through Subtractions

The previous examination allows us to perform other 48 derivations. Let us begin with the first group of the classical forms. Apart from *Barbara*, I again skip the universal derivations:

Table 4.2 **A** AND NOT **K** imply **C**

Derivations	Level	#	Symbol	ID		Traditional logical form
	7-1	3	A1	11111011	−	$\neg X \vee Y \vee \neg Z$
1	1-7	2	K7=T8	01000000	=	$\neg X \wedge \neg Y \wedge Z$
	6-2	18	**C**	10111011		$Z \rightarrow Y$
	7-1	5	A2	11101111	−	$\neg X \vee Y \vee Z$
2	1-7	2	K7=T8	01000000	=	$\neg X \wedge \neg Y \wedge Z$
	6-2	20	A2xL1	10101111		$\neg Y \rightarrow (X \leftrightarrow Z)$

(*Continued*)

Table 4.2 (*Continued*)

Derivations	Level	#	Symbol	ID		Traditional logical form
	7-1	3	A1	11111011	–	$\neg X \vee Y \vee \neg Z$
3	1-7	5	K8=S8	00001000	=	$\neg X \wedge Y \wedge Z$
	6-2	6	A1xM1	11110011		$Z \rightarrow (X \leftrightarrow Y)$
	7-1	5	A2	11101111	–	$\neg X \vee Y \vee Z$
4	1-7	5	K8=S8	00001000	=	$\neg X \wedge Y \wedge Z$
	6-2	10	M1xA2	11100111		$\neg(\neg X \leftrightarrow Y \leftrightarrow Z)$

The other classical inferences of the first group in subtraction form are:

Table 4.3 **A** AND NOT **F** imply **G**

Derivations	Level	#	Symbol	ID		Traditional logical form
	7-1	3	A1	11111011	–	$\neg X \vee Y \vee \neg Z$
1	5-3	1	F3=E3	11111000	=	$X \rightarrow (\neg Y \wedge \neg Z)$
	2-6	28	**I**	00000011		$X \wedge Y$
	7-1	5	A2	11101111	–	$\neg X \vee Y \vee Z$
2	5-3	1	F3=E3	11111000	=	$X \rightarrow (\neg Y \wedge \neg Z)$
	3-5	56	J6=I6	00000111		$X \wedge (Y \vee Z)$
	7-1	3	A1	11111011	–	$\neg X \vee Y \vee \neg Z$
3	5-3	3	F4=D4	11110010	=	$Z \rightarrow (\neg X \wedge \neg Y)$
	2-6	25	**G**	00001001		$Z \wedge Y)$
	7-1	5	A2	11101111	–	$\neg X \vee Y \vee Z$
4	5-3	3	F4=D4	11110010	=	$Z \rightarrow (\neg X \wedge \neg Y)$
	3-5	54	J5=G5	00001101		$Z \wedge (X \vee Y)$
	7-1	3	A1	11111011	–	$\neg X \vee Y \vee \neg Z$
5	5-3	6	F5=A3	11101010	=	$X \rightarrow (Y \wedge \neg Z)$
	2-6	22	O7+I8	00010001		$X \wedge (Y \leftrightarrow Z)$
	7-1	5	A2	11101111	–	$\neg X \vee Y \vee Z$
6	5-3	6	F5=A3	11101010	=	$X \rightarrow (Y \wedge \neg Z)$
	2-6	27	**J**	00000101		$Z \wedge X$
	7-1	3	A1	11111011	–	$\neg X \vee Y \vee \neg Z$
7	5-3	12	F6	11011010	=	$(\neg X \vee \neg Z) \wedge (X \vee \neg Y \vee Z)$
	2-6	18	S7+I8	00100001		$Y \wedge (X \leftrightarrow Z)$

Derivations	Level	#	Symbol	ID		Traditional logical form
	7-1	5	A2	11101111	–	$\neg X \vee Y \vee Z$
8	5-3	12	F6	11011010	=	$(\neg X \vee \neg Z) \wedge (X \vee \neg Y \vee Z)$
	3-5	45	J3	00100101		$(X \wedge Z) \vee (\neg X \wedge Y \wedge \neg Z)$
	7-1	3	A1	11111011	–	$\neg X \vee Y \vee \neg Z$
9	5-3	22	F7=C3	10111010	=	$Z \rightarrow (\neg X \wedge Y)$
	2-6	13	T8+I8	01000001		$Z \wedge (X \leftrightarrow Y)$
	7-1	5	A2	11101111	–	$\neg X \vee Y \vee Z$
10	5-3	22	F7=C3	10111010	=	$Z \rightarrow (\neg X \wedge Y)$
	3-5	35	J2=H6	01000101		$Z \wedge (X \vee \neg Y)$
	7-1	3	A1	11111011	–	$\neg X \vee Y \vee \neg Z$
11	5-3	37	F8	01111010	=	$(\neg X \vee \neg Z) \wedge (X \vee Y \vee Z)$
	2-6	7	T7+I8	10000001		$X \leftrightarrow Y \leftrightarrow Z$
	7-1	5	A2	11101111	–	$\neg X \vee Y \vee Z$
12	5-3	37	F8	01111010	=	$(\neg X \vee \neg Z) \wedge (X \vee Y \vee Z)$
	3-5	20	J1	10000101		$(X \wedge Z) \vee (\neg X \wedge \neg Y \wedge \neg Z)$

We may note that the cross terms behave as in the product derivations (statements pertaining to Level 6-2 for *Barbara*-like derivations and statements pertaining to Level 2-6 for *Darii*-like derivations).

Table 4.4 **E** AND NOT **F** imply **H**

Derivations	Level	#	Symbol	ID		Traditional logical form
	7-1	1	E1	11111110	–	$\neg X \vee \neg Y \vee \neg Z$
1	5-3	1	F3=E3	11111000	=	$X \rightarrow (\neg Y \wedge \neg Z)$
	2-6	26	O8+I7	00000110		$X \wedge \neg(Y \leftrightarrow Z)$
	7-1	2	E2	11111101	–	$\neg X \vee \neg Y \vee Z$
2	5-3	1	F3=E3	11111000	=	$X \rightarrow (\neg Y \wedge \neg Z)$
	2-6	27	**J**	00000101		$Z \wedge X$
	7-1	1	E1	11111110	–	$\neg X \vee \neg Y \vee \neg Z$
3	5-3	3	F4=D4	11110010	=	$Z \rightarrow (\neg X \wedge \neg Y)$
	2-6	23	S8+O8	00001100		$Z \wedge \neg(X \leftrightarrow Y)$
	7-1	2	E2	11111101	–	$\neg X \vee \neg Y \vee Z$
4	5-3	3	F4=D4	11110010	=	$Z \rightarrow (\neg X \wedge \neg Y)$
	3-5	54	J5=G5	00001101		$Z \wedge (X \vee Y)$

(*Continued*)

Table 4.4 (*Continued*)

Derivations	Level	#	Symbol	ID		Traditional logical form
	7-1	1	E1	11111110	–	$\neg X \vee \neg Y \vee \neg Z$
5	5-3	6	F5=A3	11101010	=	$X \rightarrow (Y \wedge \neg Z)$
	2-6	20	**O**	00010100		$X \wedge \neg Y$
	7-1	2	E2	11111101	–	$\neg X \vee \neg Y \vee Z$
6	5-3	6	F5=A3	11101010	=	$X \rightarrow (Y \wedge \neg Z)$
	3-5	51	J4=O6	00010101		$X \wedge (\neg Y \vee Z)$
	7-1	1	E1	11111110	–	$\neg X \vee \neg Y \vee \neg Z$
7	5-3	12	F6	11011010	=	$(\neg X \vee \neg Z) \wedge (X \vee \neg Y \vee Z)$
	2-6	16	S7+O8	00100100		$X \leftrightarrow \neg Y \leftrightarrow Z$
	7-1	2	E2	11111101	–	$\neg X \vee \neg Y \vee Z$
8	5-3	12	F6	11011010	=	$(\neg X \vee \neg Z) \wedge (X \vee \neg Y \vee Z)$
	3-5	45	J3	00100101		$(X \wedge Z) \vee (\neg X \wedge Y \wedge \neg Z)$
	7-1	1	E1	11111110	–	$\neg X \vee \neg Y \vee \neg Z$
9	5-3	22	F7=C3	10111010	=	$Z \rightarrow (\neg X \wedge Y)$
	2-6	11	**H**	01000100		$Z \wedge \neg Y$
	7-1	2	E2	11111101	–	$\neg X \vee \neg Y \vee Z$
10	5-3	22	F7=C3	10111010	=	$Z \rightarrow (\neg X \wedge Y)$
	3-5	35	J2=H6	01000101		$Z \wedge (X \vee \neg Y)$
	7-1	1	E1	11111110	–	$\neg X \vee \neg Y \vee \neg Z$
11	5-3	37	F8	01111010	=	$(\neg X \vee \neg Z) \wedge (X \vee Y \vee Z)$
	2-6	5	T7+O8	10000100		$\neg Y \wedge (X \leftrightarrow Z)$
	7-1	2	E2	11111101	–	$\neg X \vee \neg Y \vee Z$
12	5-3	37	F8	01111010	=	$(\neg X \vee \neg Z) \wedge (X \vee Y \vee Z)$
	3-5	20	J1	10000101		$(X \wedge Z) \vee (\neg X \wedge \neg Y \wedge \neg Z)$

The classical inferences of the second group in subtraction form are:

Table 4.5 **A** AND NOT **C** imply **K**

Derivations	Level	#	Symbol	ID		Traditional logical form
	7-1	3	A1	11111011	–	$\neg X \vee Y \vee \neg Z$
1	5-3	22	C3=F7	10111010	=	$Z \rightarrow (\neg X \wedge Y)$
	2-6	13	T8+I8	01000001		$Z \wedge (X \leftrightarrow Y)$
	7-1	5	A2	11101111	–	$\neg X \vee Y \vee Z$
2	5-3	22	C3=F7	10111010	=	$Z \rightarrow (\neg X \wedge Y)$
	3-5	35	H6=J2	01000101		$Z \wedge (X \vee \neg Y)$

Derivations	Level	#	Symbol	ID		Traditional logical form
3	7-1	3	A1	11111011	–	$\neg X \vee Y \vee \neg Z$
	5-3	23	C4	10111001	=	$(Y \vee \neg Z) \wedge (\neg X \vee \neg Y \vee Z)$
	2-6	12	T8+I7	01000010		$X \leftrightarrow Y \leftrightarrow \neg Z$
4	7-1	5	A2	11101111	–	$\neg X \vee Y \vee Z$
	5-3	23	C4	10111001	=	$(Y \vee \neg Z) \wedge (\neg X \vee \neg Y \vee Z)$
	3-5	34	H5	01000110		$(\neg Y \wedge Z) \vee (X \wedge Y \wedge \neg Z)$
5	7-1	3	A1	11111011	–	$\neg X \vee Y \vee \neg Z$
	5-3	26	C5=B5	10110011	=	$Z \rightarrow (X \wedge Y)$
	2-6	10	**K**	01001000		$Z \wedge \neg X$
6	7-1	5	A2	11101111	–	$\neg X \vee Y \vee Z$
	5-3	26	C5=B5	10110011	=	$Z \rightarrow (X \wedge Y)$
	3-5	31	H4=K4	01001100		$Z \wedge (\neg X \vee \neg Y)$
7	7-1	3	A1	11111011	–	$\neg X \vee Y \vee \neg Z$
	5-3 5-3	29	C6=A7	10101011	=	$\neg Y \rightarrow (\neg X \wedge \neg Z)$
	2-6	9	T8+O7	01010000		$\neg Y \wedge \neg(X \leftrightarrow Z)$
8	7-1	5	A2	11101111	–	$\neg X \vee Y \vee Z$
	5-3 5-3	29	C6=A7	10101011	=	$\neg Y \rightarrow (\neg X \wedge \neg Z)$
	2-6	11	**H**	01000100		$Z \wedge \neg Y$
9	7-1	3	A1	11111011	–	$\neg X \vee Y \vee \neg Z$
		33	C7	10011011	=	$(Y \vee \neg Z) \wedge (X \vee \neg Y \vee Z)$
	2-6	8	T8+S7	01100000		$\neg X \wedge \neg(Y \leftrightarrow Z)$
10	7-1	5	A2	11101111	–	$\neg X \vee Y \vee Z$
	5-3	33	C7	10011011	=	$(Y \vee \neg Z) \wedge (X \vee \neg Y \vee Z)$
	3-5	24	H2	01100100		$(\neg Y \wedge Z) \vee (\neg X \wedge Y \wedge \neg Z)$
11	7-1	3	A1	11111011	–	$\neg X \vee Y \vee \neg Z$
	5-3	53	C8=L5	00111011	=	$\neg Y \rightarrow (X \wedge \neg Z)$
	2-6	1	**T**	11000000		$\neg X \wedge \neg Y$
12	7-1	5	A2	11101111	–	$\neg X \vee Y \vee Z$
	5-3	53	C8=L5	00111011	=	$\neg Y \rightarrow (X \wedge \neg Z)$
	3-5	4	H1=T4	11000100		$\neg Y \wedge (\neg X \vee Z)$

Table 4.6 **E** AND NOT **D** imply **K**

Derivations	Level	#	Symbol	ID		Traditional logical form
	7-1	1	E1	11111110	–	$\neg X \vee \neg Y \vee \neg Z$
1	5-3	2	D3=E4	11110100	=	$Y \rightarrow (\neg X \wedge \neg Z)$
	2-6	24	S8+I7	00001010		$Y \wedge \neg(X \leftrightarrow Z)$
	7-1	2	E2	11111101	–	$\neg X \vee \neg Y \vee Z$
2	5-3	2	D3=E4	11110100	=	$Y \rightarrow (\neg X \wedge \neg Z)$
	2-6	25	**G**	00001001		$Z \wedge Y$
	7-1	1	E1	11111110	–	$\neg X \vee \neg Y \vee \neg Z$
3	5-3	3	D4=F4	11110010	=	$Z \rightarrow (\neg X \wedge \neg Y)$
	2-6	23	S8+O8	00001100		$Z \wedge \neg(X \leftrightarrow Y)$
	7-1	2	E2	11111101	–	$\neg X \vee \neg Y \vee Z$
4	5-3	3	D4=F4	11110010	=	$Z \rightarrow (\neg X \wedge \neg Y)$
	3-5	54	G5=J5	00001101		$Z \wedge (X \vee Y)$
	7-1	1	E1	11111110	–	$\neg X \vee \neg Y \vee \neg Z$
5	5-3	8	D5	11100110	=	$(\neg Y \vee \neg Z) \wedge (\neg X \vee Y \vee Z)$
	2-6	19	O7+S8	00011000		$\neg X \leftrightarrow Y \leftrightarrow Z$
	7-1	2	E2	11111101	–	$\neg X \vee \neg Y \vee Z$
6	5-3	8	D5	11100110	=	$(\neg Y \vee \neg Z) \wedge (\neg X \vee Y \vee Z)$
	3-5	49	G4	00011001		$(Y \wedge Z) \vee (X \wedge \neg Y \wedge \neg Z)$
	7-1	1	E1	11111110	–	$\neg X \vee \neg Y \vee \neg Z$
7	5-3	14	D6=M3	11010110	=	$Y \rightarrow (X \wedge \neg Z)$
	2-6	15	**S**	00101000		$\neg X \wedge Y$
	7-1	2	E2	11111101	–	$\neg X \vee \neg Y \vee Z$
8	5-3	14	D6=M3	11010110	=	$Y \rightarrow (X \wedge \neg Z)$
	3-5	43	G3=S6	00101001		$Y \wedge (\neg X \vee Z)$
	7-1	1	E1	11111110	–	$\neg X \vee \neg Y \vee \neg Z$
9	5-3	24	D7=B3	10110111	=	$Z \rightarrow (X \wedge \neg Y)$
	2-6	10	**K**	01001000		$Z \wedge \neg X$
	7-1	2	E2	11111101	–	$\neg X \vee \neg Y \vee Z$
10	5-3	24	D7=B3	10110111	=	$Z \rightarrow (X \wedge \neg Y)$
	3-5	33	G2=K6	01001001		$Z \wedge (\neg X \vee Y)$
	7-1	1	E1	11111110	–	$\neg X \vee \neg Y \vee \neg Z$
11	5-3	39	D8	01110110	=	$(\neg Y \vee \neg Z) \wedge (X \vee Y \vee Z)$
	2-6	4	T7+S8	10001000		$\neg X \wedge (Y \leftrightarrow Z)$
	7-1	2	E2	11111101	–	$\neg X \vee \neg Y \vee Z$
12	5-3	39	D8	01110110	=	$(\neg Y \vee \neg Z) \wedge (X \vee Y \vee Z)$
	3-5	18	G1	10001001		$(Y \wedge Z) \vee (\neg X \wedge \neg Y \wedge \neg Z)$

Finally, the classical inferences of the third group are:

Table 4.7 **H** AND NOT **K** imply **O**

Derivations	Level	#	Symbol	ID		Traditional logical form
	3-5	4	H1=T4	11000100	–	$\neg Y \wedge (\neg X \vee Z)$
1	1-7	2	K7=T8	01000000	=	$\neg X \wedge \neg Y \wedge Z$
	2-6	5	T7+O8	10000100		$\neg Y \wedge (X \leftrightarrow Z)$
	3-5	4	H1=T4	11000100	–	$\neg Y \wedge (\neg X \vee Z)$
2	1-7	5	K8=S8	00001000	=	$\neg X \wedge Y \wedge Z$
	3-5	4	H1=T4	11000100		$\neg Y \wedge (\neg X \vee Z)$
	3-5	24	H2	01100100	–	$(\neg Y \wedge Z) \vee (\neg X \wedge Y \wedge \neg Z)$
3	1-7	2	K7=T8	01000000	=	$\neg X \wedge \neg Y \wedge Z$
	2-6	16	S7+O8	00100100		$X \leftrightarrow \neg Y \leftrightarrow Z$
	3-5	24	H2	01100100	–	$(\neg Y \wedge Z) \vee (\neg X \wedge Y \wedge \neg Z)$
4	1-7	5	K8=S8	00001000	=	$\neg X \wedge Y \wedge Z$
	3-5	24	H2	01100100		$(\neg Y \wedge Z) \vee (\neg X \wedge Y \wedge \neg Z)$
	3-5	28	H3=O2	01010100	–	$\neg Y \wedge (X \vee Z)$
5	1-7	2	K7=T8	01000000	=	$\neg X \wedge \neg Y \wedge Z$
	2-6	20	**O**	00010100		$X \wedge \neg Y$
	3-5	28	H3=O2	01010100	–	$\neg Y \wedge (X \vee Z)$
6	1-7	5	K8=S8	00001000	=	$\neg X \wedge Y \wedge Z$
	3-5	28	H3=O2	01010100		$\neg Y \wedge (X \vee Z)$
	3-5	31	H4=K4	01001100	–	$Z \wedge (\neg X \vee \neg Y)$
7	1-7	2	K7=T8	01000000	=	$\neg X \wedge \neg Y \wedge Z$
	2-6	23	S8+O8	00001100		$Z \wedge \neg(X \leftrightarrow Y)$
	3-5	31	H4=K4	01001100	–	$Z \wedge (\neg X \vee \neg Y)$
8	1-7	5	K8=S8	00001000	=	$\neg X \wedge Y \wedge Z$
	2-6	11	**H**	01000100		$Z \wedge \neg Y$
	3-5	34	H5	01000110	–	$(\neg Y \wedge Z) \vee (X \wedge Y \wedge \neg Z)$
9	1-7	2	K7=T8	01000000	=	$\neg X \wedge \neg Y \wedge Z$
	2-6	26	O8+I7	00000110		$X \wedge \neg(Y \leftrightarrow Z)$
	3-5	34	H5	01000110	–	$(\neg Y \wedge Z) \vee (X \wedge Y \wedge \neg Z)$
10	1-7	5	K8=S8	00001000	=	$\neg X \wedge Y \wedge Z$
	3-5	34	H5	01000110		$(\neg Y \wedge Z) \vee (X \wedge Y \wedge \neg Z)$
	3-5	35	H6=J2	01000101	–	$Z \wedge (X \vee \neg Y)$
11	1-7	2	K7=T8	01000000	=	$\neg X \wedge \neg Y \wedge Z$
	2-6	27	**J**	00000101		$Z \wedge X$
	3-5	35	H6=J2	01000101	–	$Z \wedge (X \vee \neg Y)$
12	1-7	5	K8=S8	00001000	=	$\neg X \wedge Y \wedge Z$
	3-5	35	H6=J2	01000101		$Z \wedge (X \vee \neg Y)$

Table 4.8 **G** AND NOT **K** imply **I**

Derivations	Level	#	Symbol	ID		Traditional logical form
	3-5	18	G1	10001001	–	$(Y \wedge Z) \vee (\neg X \wedge \neg Y \wedge \neg Z)$
1	1-7	2	K7=T8	01000000	=	$\neg X \wedge \neg Y \wedge Z$
	3-5	18	G1	10001001		$(Y \wedge Z) \vee (\neg X \wedge \neg Y \wedge \neg Z)$
	3-5	18	G1	10001001	–	$(Y \wedge Z) \vee (\neg X \wedge \neg Y \wedge \neg Z)$
2	1-7	5	K8=S8	00001000	=	$\neg X \wedge Y \wedge Z$
	2-6	7	T7+I8	10000001		$X \leftrightarrow Y \leftrightarrow Z$
	3-5	33	G2=K6	01001001	–	$Z \wedge (\neg X \vee Y)$
3	1-7	2	K7=T8	01000000	=	$\neg X \wedge \neg Y \wedge Z$
	2-6	25	**G**	00001001		$Z \wedge Y$
	3-5	33	G2=K6	01001001	–	$Z \wedge (\neg X \vee Y)$
4	1-7	5	K8=S8	00001000	=	$\neg X \wedge Y \wedge Z$
	2-6	13	T8+I8	01000001		$Z \wedge (X \leftrightarrow Y)$
	3-5	43	G3=S6	00101001	–	$Y \wedge (\neg X \vee Z)$
5	1-7	2	K7=T8	01000000	=	$\neg X \wedge \neg Y \wedge Z$
	3-5	43	G3=S6	00101001		$Y \wedge (\neg X \vee Z)$
	3-5	43	G3=S6	00101001	–	$Y \wedge (\neg X \vee Z)$
6	1-7	5	K8=S8	00001000	=	$\neg X \wedge Y \wedge Z$
	2-6	18	S7+I8	00100001		$Y \wedge (X \leftrightarrow Z)$
	3-5	49	G4	00011001	–	$(Y \wedge Z) \vee (X \wedge \neg Y \wedge \neg Z)$
7	1-7	2	K7=T8	01000000	=	$\neg X \wedge \neg Y \wedge Z$
	3-5	49	G4	00011001		$(Y \wedge Z) \vee (X \wedge \neg Y \wedge \neg Z)$
	3-5	49	G4	00011001	–	$(Y \wedge Z) \vee (X \wedge \neg Y \wedge \neg Z)$
8	1-7	5	K8=S8	00001000	=	$\neg X \wedge Y \wedge Z$
	2-6	22	O7+I8	00010001		$X \wedge (Y \leftrightarrow Z)$
	3-5	54	G5=J5	00001101	–	$Z \wedge (X \vee Y)$
9	1-7	2	K7=T8	01000000	=	$\neg X \wedge \neg Y \wedge Z$
	3-5	54	G5=J5	00001101		$Z \wedge (X \vee Y)$
	3-5	54	G5=J5	00001101	–	$Z \wedge (X \vee Y)$
10	1-7	5	K8=S8	00001000	=	$\neg X \wedge Y \wedge Z$
	2-6	27	**J**	00000101		$Z \wedge X$
	3-5	55	G6=I5	00001011	–	$Y \wedge (X \vee Z)$
11	1-7	2	K7=T8	01000000	=	$\neg X \wedge \neg Y \wedge Z$
	3-5	55	G6=I5	00001011		$Y \wedge (X \vee Z)$
	3-5	55	G6=I5	00001011	–	$Y \wedge (X \vee Z)$
12	1-7	5	K8=S8	00001000	=	$\neg X \wedge Y \wedge Z$
	2-6	28	**I**	00000011		$X \wedge Y$

Table 4.9 **D** AND NOT **F** imply **O**

Derivations	Level	#	Symbol	ID		Traditional logical form
	7-1	1	D1=E1	11111110	–	$\neg X \vee \neg Y \vee \neg Z$
1	5-3	1	F3=E3	11111000	=	$X \rightarrow (\neg Y \wedge \neg Z)$
	2-6	26	O8+I7	00000110		$X \wedge \neg(Y \leftrightarrow Z)$
	7-1	4	D2=M1	11110111	–	$X \vee \neg Y \vee \neg Z$
2	5-3	1	F3=E3	11111000	=	$X \rightarrow (\neg Y \wedge \neg Z)$
	3-5	56	J6=I6	00000111		$X \wedge (Y \vee Z)$
	7-1	1	D1=E1	11111110	–	$\neg X \vee \neg Y \vee \neg Z$
3	5-3	3	F4=D4	11110010	=	$Z \rightarrow (\neg X \wedge \neg Y)$
	2-6	23	S8+O8	00001100		$Z \wedge \neg(X \leftrightarrow Y)$
	7-1	4	D2=M1	11110111	–	$X \vee \neg Y \vee \neg Z$
4	5-3	3	F4=D4	11110010	=	$Z \rightarrow (\neg X \wedge \neg Y)$
	2-6	27	**J**	00000101		$Z \wedge X$
	7-1	1	D1=E1	11111110	–	$\neg X \vee \neg Y \vee \neg Z$
5	5-3	6	F5=A3	11101010	=	$X \rightarrow (Y \wedge \neg Z)$
	2-6	20	**O**	00010100		$X \wedge \neg Y$
	7-1	4	D2=M1	11110111	–	$X \vee \neg Y \vee \neg Z$
6	5-3	6	F5=A3	11101010	=	$X \rightarrow (Y \wedge \neg Z)$
	3-5	51	J4=O6	00010101		$X \wedge (\neg Y \vee Z)$
	7-1	1	D1=E1	11111110	–	$\neg X \vee \neg Y \vee \neg Z$
7	5-3	12	F6	11011010	=	$(\neg X \vee \neg Z) \wedge (X \vee \neg Y \vee Z)$
	2-6	16	S7+O8	00100100		$X \leftrightarrow \neg Y \leftrightarrow Z$
	7-1	4	D2=M1	11110111	–	$X \vee \neg Y \vee \neg Z$
8	5-3	12	F6	11011010	=	$(\neg X \vee \neg Z) \wedge (X \vee \neg Y \vee Z)$
	3-5	45	J3	00100101		$(X \wedge Z) \vee (\neg X \wedge Y \wedge \neg Z)$
	7-1	1	D1=E1	11111110	–	$\neg X \vee \neg Y \vee \neg Z$
9	5-3	22	F7=C3	10111010	=	$Z \rightarrow (\neg X \wedge Y)$
	2-6	11	**H**	01000100		$Z \wedge \neg Y$
	7-1	4	D2=M1	11110111	–	$X \vee \neg Y \vee \neg Z$
10	5-3	22	F7=C3	10111010	=	$Z \rightarrow (\neg X \wedge Y)$
	3-5	35	J2=H6	01000101		$Z \wedge (X \vee \neg Y)$
	7-1	1	D1=E1	11111110	–	$\neg X \vee \neg Y \vee \neg Z$
11	5-3	37	F8	01111010	=	$(\neg X \vee \neg Z) \wedge (X \vee Y \vee Z)$
	2-6	5	T7+O8	10000100		$\neg Y \wedge (X \leftrightarrow Z)$
	7-1	4	D2=M1	11110111	–	$X \vee \neg Y \vee \neg Z$
12	5-3	37	F8	01111010	=	$(\neg X \vee \neg Z) \wedge (X \vee Y \vee Z)$
	3-5	20	J1	10000101		$(X \wedge Z) \vee (\neg X \wedge \neg Y \wedge \neg Z)$

Table 4.10 **C** AND NOT **F** imply **I**

Derivations	Level	#	Symbol	ID		Traditional logical form
	7-1	3	C1=A1	11111011	−	$\neg X \vee Y \vee \neg Z$
1	5-3	1	F3=E3	11111000	=	$X \rightarrow (\neg Y \wedge \neg Z)$
	2-6	28	**I**	00000011		$X \wedge Y$
	7-1	7	C2=L1	10111111	−	$X \vee Y \vee \neg Z$
2	5-3	1	F3=E3	11111000	=	$X \rightarrow (\neg Y \wedge \neg Z)$
	3-5	56	J6=I6	00000111		$X \wedge (Y \vee Z)$
	7-1	3	C1=A1	11111011	−	$\neg X \vee Y \vee \neg Z$
3	5-3	3	F4=D4	11110010	=	$Z \rightarrow (\neg X \wedge \neg Y)$
	2-6	25	**G**	00001001		$Z \wedge Y$
	7-1	7	C2=L1	10111111	−	$X \vee Y \vee \neg Z$
4	5-3	3	F4=D4	11110010	=	$Z \rightarrow (\neg X \wedge \neg Y)$
	3-5	54	J5=G5	00001101		$Z \wedge (X \vee Y)$
	7-1	3	C1=A1	11111011	−	$\neg X \vee Y \vee \neg Z$
5	5-3	6	F5=A3	11101010	=	$X \rightarrow (Y \wedge \neg Z)$
	2-6	22	O7+I8	00010001		$X \wedge (Y \leftrightarrow Z)$
	7-1	7	C2=L1	10111111	−	$X \vee Y \vee \neg Z$
6	5-3	6	F5=A3	11101010	=	$X \rightarrow (Y \wedge \neg Z)$
	3-5	51	J4=O6	00010101		$X \wedge (\neg Y \vee Z)$
	7-1	3	C1=A1	11111011	−	$\neg X \vee Y \vee \neg Z$
7	5-3	12	F6	11011010	=	$(\neg X \vee \neg Z) \wedge (X \vee \neg Y \vee Z)$
	2-6	18	S7+I8	00100001		$Y \wedge (X \leftrightarrow Z)$
	7-1	7	C2=L1	10111111	−	$X \vee Y \vee \neg Z$
8	5-3	12	F6	11011010	=	$(\neg X \vee \neg Z) \wedge (X \vee \neg Y \vee Z)$
	3-5	45	J3	00100101		$(X \wedge Z) \vee (\neg X \wedge Y \wedge \neg Z)$
	7-1	3	C1=A1	11111011	−	$\neg X \vee Y \vee \neg Z$
9	5-3	22	F7=C3	10111010	=	$Z \rightarrow (\neg X \wedge Y)$
	2-6	13	T8+I8	01000001		$Z \wedge (X \leftrightarrow Y)$
	7-1	7	C2=L1	10111111	−	$X \vee Y \vee \neg Z$
10	5-3	22	F7=C3	10111010	=	$Z \rightarrow (\neg X \wedge Y)$
	2-6	27	**J**	00000101		$Z \wedge X$
	7-1	3	C1=A1	11111011	−	$\neg X \vee Y \vee \neg Z$
11	5-3	37	F8	01111010	=	$(\neg X \vee \neg Z) \wedge (X \vee Y \vee Z)$
	2-6	7	T7+I8	10000001		$X \leftrightarrow Y \leftrightarrow Z$
	7-1	7	C2=L1	10111111	−	$X \vee Y \vee \neg Z$
12	5-3	37	F8	01111010	=	$(\neg X \vee \neg Z) \wedge (X \vee Y \vee Z)$
	3-5	20	J1	10000101		$(X \wedge Z) \vee (\neg X \wedge \neg Y \wedge \neg Z)$

4.3 Extension of Classical Inferences through Subtraction

Again I skip universal inferences and first deal with the first group:

Table 4.11 L AND NOT **Q** imply **U**

Derivations	Level	#	Symbol	ID		Traditional logical form
	7-1	7	L1	10111111	–	$X \vee Y \vee \neg Z$
1	5-3	46	Q3	01011110	=	$(X \vee Z) \wedge (\neg X \vee \neg Y \vee \neg Z)$
	3-5	11	Ω6	10100001		$(\neg Z \wedge \neg X) \vee (X \wedge Y \wedge Z)$
	7-1	8	L2	01111111	–	$X \vee Y \vee Z$
2	5-3	46	Q3	01011110	=	$(X \vee Z) \wedge (\neg X \vee \neg Y \vee \neg Z)$
	2-6	18	S7+I8	00100001		$Y \wedge (X \leftrightarrow Z)$
	7-1	7	L1	10111111	–	$X \vee Y \vee \neg Z$
3	5-3	47	Q4=P8	01011101	=	$\neg Z \rightarrow (X \wedge \neg Y)$
	3-5	10	Ω5=U1	10100010		$\neg Z \wedge (\neg X \vee Y)$
	7-1	8	L2	01111111	–	$X \vee Y \vee Z$
4	5-3	47	Q4=P8	01011101	=	$\neg Z \rightarrow (X \wedge \neg Y)$
	2-6	17	**U**	00100010		$\neg Z \wedge Y$
	7-1	7	L1	10111111	–	$X \vee Y \vee \neg Z$
5	5-3	48	Q5	01011011	=	$(X \vee Z) \wedge (\neg X \vee Y \vee \neg Z)$
	3-5	9	Ω4	10100100		$(\neg Z \wedge \neg X) \vee (X \wedge \neg Y \wedge Z)$
	7-1	8	L2	01111111	–	$X \vee Y \vee Z$
6	5-3	48	Q5	01011011	=	$(X \vee Z) \wedge (\neg X \vee Y \vee \neg Z)$
	2-6	16	S7+O8	00100100		$X \leftrightarrow \neg Y \leftrightarrow Z$
	7-1	7	L1	10111111	–	$X \vee Y \vee \neg Z$
7	5-3	49	Q6=M8	01010111	=	$\neg X \rightarrow (\neg Y \wedge Z)$
	3-5	8	Ω3=S1	10101000		$\neg X \wedge (Y \vee \neg Z)$
	7-1	8	L2	01111111	–	$X \vee Y \vee Z$
8	5-3	49	Q6=M8	01010111	=	$\neg X \rightarrow (\neg Y \wedge Z)$
	2-6	15	**S**	00101000		$\neg X \wedge Y$
	7-1	7	L1	10111111	–	$X \vee Y \vee \neg Z$
9	5-3	50	Q7=N7	01001111	=	$\neg Z \rightarrow (X \wedge Y)$
	3-5	7	Ω2=V2	10110000		$\neg Z \wedge (\neg X \vee \neg Y)$
	7-1	8	L2	01111111	–	$X \vee Y \vee Z$
10	5-3	50	Q7=N7	01001111	=	$\neg Z \rightarrow (X \wedge Y)$
	2-6	14	S7+O7	00110000		$\neg Z \wedge \neg(X \leftrightarrow Y)$
	7-1	7	L1	10111111	–	$X \vee Y \vee \neg Z$
11	5-3	56	Q8=L8	00011111	=	$\neg X \rightarrow (Y \wedge Z)$
	2-6	2	**Ω**	10100000		$\neg Z \wedge \neg X$
	7-1	8	L2	01111111	–	$X \vee Y \vee Z$
12	5-3	56	Q8=L8	00011111	=	$\neg X \rightarrow (Y \wedge Z)$
	2-6	8	T8+S7	01100000		$\neg X \wedge \neg(Y \leftrightarrow Z)$

Table 4.12 **M** AND NOT **Q** imply **V**

Derivations	Level	#	Symbol	ID		Traditional logical form
	7-1	4	M1	11110111	–	$X \vee \neg Y \vee \neg Z$
1	5-3	46	Q3	01011110	=	$(X \vee Z) \wedge (\neg X \vee \neg Y \vee \neg Z)$
	3-5	11	Ω6	10100001		$(\neg Z \wedge \neg X) \vee (X \wedge Y \wedge Z)$
	7-1	6	M2	11011111	–	$X \vee \neg Y \vee Z$
2	5-3	46	Q3	01011110	=	$(X \vee Z) \wedge (\neg X \vee \neg Y \vee \neg Z)$
	2-6	7	T7+I8	10000001		$X \leftrightarrow Y \leftrightarrow Z$
	7-1	4	M1	11110111	–	$X \vee \neg Y \vee \neg Z$
3	5-3	47	Q4=P8	01011101	=	$\neg Z \rightarrow (X \wedge \neg Y)$
	3-5	10	Ω5=U1	10100010		$\neg Z \wedge (\neg X \vee Y)$
	7-1	6	M2	11011111	–	$X \vee \neg Y \vee Z$
4	5-3	47	Q4=P8	01011101	=	$\neg Z \rightarrow (X \wedge \neg Y)$
	2-6	6	T7+I7	10000010		$\neg Z \wedge (X \leftrightarrow Y)$
	7-1	4	M1	11110111	–	$X \vee \neg Y \vee \neg Z$
5	5-3	48	Q5	01011011	=	$(X \vee Z) \wedge (\neg X \vee Y \vee \neg Z)$
	3-5	9	Ω4	10100100		$(\neg Z \wedge \neg X) \vee (X \wedge \neg Y \wedge Z)$
	7-1	6	M2	11011111	–	$X \vee \neg Y \vee Z$
6	5-3	48	Q5	01011011	=	$(X \vee Z) \wedge (\neg X \vee Y \vee \neg Z)$
	2-6	5	T7+O8	10000100		$\neg Y \wedge (X \leftrightarrow Z)$
	7-1	4	M1	11110111	–	$X \vee \neg Y \vee \neg Z$
7	5-3	49	Q6=M8	01010111	=	$\neg X \rightarrow (\neg Y \wedge Z)$
	2-6	2	**Ω**	10100000		$\neg Z \wedge \neg X$
	7-1	6	M2	11011111	–	$X \vee \neg Y \vee Z$
8	5-3	49	Q6=M8	01010111	=	$\neg X \rightarrow (\neg Y \wedge Z)$
	2-6	4	T7+S8	10001000		$\neg X \wedge (Y \leftrightarrow Z)$
	7-1	4	M1	11110111	–	$X \vee \neg Y \vee \neg Z$
9	5-3	50	Q7=N7	01001111	=	$\neg Z \rightarrow (X \wedge Y)$
	3-5	7	Ω2=V2	10110000		$\neg Z \wedge (\neg X \vee \neg Y)$
	7-1	6	M2	11011111	–	$X \vee \neg Y \vee Z$
10	5-3	50	Q7=N7	01001111	=	$\neg Z \rightarrow (X \wedge Y)$
	2-6	3	**V**	10010000		$\neg Z \wedge \neg Y$
	7-1	4	M1	11110111	–	$X \vee \neg Y \vee \neg Z$
11	5-3	56	Q8=L8	00011111	=	$\neg X \rightarrow (Y \wedge Z)$
	3-5	1	Ω1=T1	11100000		$\neg X \wedge (\neg Y \vee \neg Z)$
	7-1	6	M2	11011111	–	$X \vee \neg Y \vee Z$
12	5-3	56	Q8=L8	00011111	=	$\neg X \rightarrow (Y \wedge Z)$
	2-6	1	**T**	11000000		$\neg X \wedge \neg Y$

The second group is given by:

Table 4.13 **L** AND NOT **N** imply **W**

Derivations	Level	#	Symbol	ID		Traditional logical form
	7-1	7	L1	10111111	–	$X \vee Y \vee \neg Z$
1	5-3	42	N3	01101110	=	$(Y \vee Z) \wedge (\neg X \vee \neg Y \vee \neg Z)$
	3-5	15	V6	10010001		$(\neg Z \wedge \neg Y) \vee (X \wedge Y \wedge Z)$
	7-1	8	L2	01111111	–	$X \vee Y \vee Z$
2	5-3	42	N3	01101110	=	$(Y \vee Z) \wedge (\neg X \vee \neg Y \vee \neg Z)$
	2-6	22	O7+I8	00010001		$X \wedge (Y \leftrightarrow Z)$
	7-1	7	L1	10111111	–	$X \vee Y \vee \neg Z$
3	5-3	43	N4=R8	01101101	=	$\neg Z \rightarrow (\neg X \wedge Y)$
	3-5	14	V5=W1	10010010		$\neg Z \wedge (X \vee \neg Y)$
	7-1	8	L2	01111111	–	$X \vee Y \vee Z$
4	5-3	43	N4=R8	01101101	=	$\neg Z \rightarrow (\neg X \wedge Y)$
	3-5 2-6	21	**W**	00010010		$\neg Z \wedge X$
	7-1	7	L1	10111111	–	$X \vee Y \vee \neg Z$
5	5-3	44	N5=A8	01101011	=	$\neg Y \rightarrow (\neg X \wedge Z)$
	3-5	13	V4=O1	10010100		$\neg Y \wedge (X \vee \neg Z)$
	7-1	8	L2	01111111	–	$X \vee Y \vee Z$
6	5-3	44	N5=A8	01101011	=	$\neg Y \rightarrow (\neg X \wedge Z)$
	2-6	20	**O**	00010100		$X \wedge \neg Y$
	7-1	7	L1	10111111	–	$X \vee Y \vee \neg Z$
7	5-3	45	N6	01100111	=	$(Y \vee Z) \wedge (X \vee \neg Y \vee \neg Z)$
	3-5	12	V3	10011000		$(\neg Z \wedge \neg Y) \vee (\neg X \wedge Y \wedge Z)$
	7-1	8	L2	01111111	–	$X \vee Y \vee Z$
8	5-3	45	N6	01100111	=	$(Y \vee Z) \wedge (X \vee \neg Y \vee \neg Z)$
	2-6	19	O7+S8	00011000		$\neg X \leftrightarrow Y \leftrightarrow Z$
	7-1	7	L1	10111111	–	$X \vee Y \vee \neg Z$
9	5-3	50	N7=Q7	01001111	=	$\neg Z \rightarrow (X \wedge Y)$
	3-5	7	V2=Ω2	10110000		$\neg Z \wedge (\neg X \vee \neg Y)$
	7-1	8	L2	01111111	–	$X \vee Y \vee Z$
10	5-3	50	N7=Q7	01001111	=	$\neg Z \rightarrow (X \wedge Y)$
	2-6	14	S7+O7	00110000		$\neg Z \wedge \neg(X \leftrightarrow Y)$
	7-1	7	L1	10111111	–	$X \vee Y \vee \neg Z$
11	5-3	55	N8=L7	00101111	=	$\neg Y \rightarrow (X \wedge Z)$
	2-6	3	**V**	10010000		$\neg Z \wedge \neg Y$
	7-1	8	L2	01111111	–	$X \vee Y \vee Z$
12	5-3	55	N8=L7	00101111	=	$\neg Y \rightarrow (X \wedge Z)$
	2-6	9	T8+O7	01010000		$\neg Y \wedge \neg(X \leftrightarrow Z)$

Table 4.14 **M** AND NOT **P** imply **W**

Derivations	Level	#	Symbol	ID		Traditional logical form
	7-1	4	M1	11110111	–	$X \vee \neg Y \vee \neg Z$
1	5-3	11	P3=E6	11011100	=	$Y \to (\neg X \wedge Z)$
	3-5	46	U6=I3	00100011		$Y \wedge (X \vee \neg Z)$
	7-1	6	M2	11011111	–	$X \vee \neg Y \vee Z$
2	5-3	11	P3=E6	11011100	=	$Y \to (\neg X \wedge Z)$
	2-6	28	**I**	00000011		$X \wedge Y$
	7-1	4	M1	11110111	–	$X \vee \neg Y \vee \neg Z$
3	5-3	13	P4	11011001	=	$(\neg Y \vee Z) \wedge (\neg X \vee Y \vee \neg Z)$
	3-5	44	U5	00100110		$(\neg Z \wedge Y) \vee (X \wedge \neg Y \wedge Z)$
	7-1	6	M2	11011111	–	$X \vee \neg Y \vee Z$
4	5-3	13	P4	11011001	=	$(\neg Y \vee Z) \wedge (\neg X \vee Y \vee \neg Z)$
	2-6	26	O8+I7	00000110		$X \wedge \neg(Y \leftrightarrow Z)$
	7-1	4	M1	11110111	–	$X \vee \neg Y \vee \neg Z$
5	5-3	15	P5=M4	11010101	=	$Y \to (X \wedge Z)$
	2-6	17	**U**	00100010		$\neg Z \wedge Y$
	7-1	6	M2	11011111	–	$X \vee \neg Y \vee Z$
6	5-3	15	P5=M4	11010101	=	$Y \to (X \wedge Z)$
	2-6	24	S8+I7	00001010		$Y \wedge \neg(X \leftrightarrow Z)$
	7-1	4	M1	11110111	–	$X \vee \neg Y \vee \neg Z$
7	5-3	18	P6=R6	11001101	=	$\neg Z \to (\neg X \wedge \neg Y)$
	3-5	39	U3=W3	00110010		$\neg Z \wedge (X \vee Y)$
	7-1	6	M2	11011111	–	$X \vee \neg Y \vee Z$
8	5-3	18	P6=R6	11001101	=	$\neg Z \to (\neg X \wedge \neg Y)$
	2-6	21	**W**	00010010		$\neg Z \wedge X$
	7-1	4	M1	11110111	–	$X \vee \neg Y \vee \neg Z$
9	5-3	32	P7	10011101	=	$(\neg Y \vee Z) \wedge (X \vee Y \vee \neg Z)$
	3-5	25	U2	01100010		$(\neg Z \wedge Y) \vee (\neg X \wedge \neg Y \wedge Z)$
	7-1	6	M2	11011111	–	$X \vee \neg Y \vee Z$
10	5-3	32	P7	10011101	=	$(\neg Y \vee Z) \wedge (X \vee Y \vee \neg Z)$
	2-6	12	T8+I7	01000010		$X \leftrightarrow Y \leftrightarrow \neg Z$
	7-1	4	M1	11110111	–	$X \vee \neg Y \vee \neg Z$
11	5-3	47	P8=Q4	01011101	=	$\neg Z \to (X \wedge \neg Y)$
	3-5	10	U1= Ω25	10100010		$\neg Z \wedge (\neg X \vee Y)$
	7-1	6	M2	11011111	–	$X \vee \neg Y \vee Z$
12	5-3	47	P8=Q4	01011101	=	$\neg Z \to (X \wedge \neg Y)$
	2-6	6	T7+I7	10000010		$\neg Z \wedge (X \leftrightarrow Y)$

The third group encompasses following derivations:

Table 4.15 **V** AND NOT **W** imply **T**

Derivations	Level	#	Symbol	ID		Traditional logical form
	3-5	2	V1=T2	11010000	–	$\neg Y \wedge (\neg X \vee \neg Z)$
1	1-7	4	W7=O7	00010000	=	$X \wedge \neg Y \wedge \neg Z$
	2-6	1	**T**	11000000		$\neg X \wedge \neg Y$
	3-5	2	V1=T2	11010000	–	$\neg Y \wedge (\neg X \vee \neg Z)$
2	1-7	7	W8=I7	00000010	=	$X \wedge Y \wedge \neg Z$
	3-5	2	V1=T2	11010000		$\neg Y \wedge (\neg X \vee \neg Z)$
	3-5	7	V2=Ω2	10110000	–	$\neg Z \wedge (\neg X \vee \neg Y)$
3	1-7	4	W7=O7	00010000	=	$X \wedge \neg Y \wedge \neg Z$
	2-6	2	**Ω**	10100000		$\neg Z \wedge \neg X$
	3-5	7	V2=Ω2	10110000	–	$\neg Z \wedge (\neg X \vee \neg Y)$
4	1-7	7	W8=I7	00000010	=	$X \wedge Y \wedge \neg Z$
	3-5	7	V2=Ω2	10110000		$\neg Z \wedge (\neg X \vee \neg Y)$
	3-5	12	V3	10011000	–	$(\neg Z \wedge \neg Y) \vee (\neg X \wedge Y \wedge Z)$
5	1-7	4	W7=O7	00010000	=	$X \wedge \neg Y \wedge \neg Z$
	2-6	4	T7+S8	10001000		$\neg X \wedge (Y \leftrightarrow Z)$
	3-5	12	V3	10011000	–	$(\neg Z \wedge \neg Y) \vee (\neg X \wedge Y \wedge Z)$
6	1-7	7	W8=I7	00000010	=	$X \wedge Y \wedge \neg Z$
	3-5	12	V3	10011000		$(\neg Z \wedge \neg Y) \vee (\neg X \wedge Y \wedge Z)$
	3-5	13	V4=O1	10010100	–	$\neg Y \wedge (X \vee \neg Z)$
7	1-7	4	W7=O7	00010000	=	$X \wedge \neg Y \wedge \neg Z$
	2-6	5	T7+O8	10000100		$\neg Y \wedge (X \leftrightarrow Z)$
	3-5	13	V4=O1	10010100	–	$\neg Y \wedge (X \vee \neg Z)$
8	1-7	7	W8=I7	00000010	=	$X \wedge Y \wedge \neg Z$
	3-5	13	V4=O1	10010100		$\neg Y \wedge (X \vee \neg Z)$
	3-5	14	V5=W1	10010010	–	$\neg Z \wedge (X \vee \neg Y)$
9	1-7	4	W7=O7	00010000	=	$X \wedge \neg Y \wedge \neg Z$
	2-6	6	T7+I7	10000010		$\neg Z \wedge (X \leftrightarrow Y)$
	3-5	14	V5=W1	10010010	–	$\neg Z \wedge (X \vee \neg Y)$
10	1-7	7	W8=I7	00000010	=	$X \wedge Y \wedge \neg Z$
	2-6	3	**V**	10010000		$\neg Z \wedge \neg Y$
	3-5	15	V6	10010001	–	$(\neg Z \wedge \neg Y) \vee (X \wedge Y \wedge Z)$
11	1-7	4	W7=O7	00010000	=	$X \wedge \neg Y \wedge \neg Z$
	2-6	7	T7+I8	10000001		$X \leftrightarrow Y \leftrightarrow Z$
	3-5	15	V6	10010001	–	$(\neg Z \wedge \neg Y) \vee (X \wedge Y \wedge Z)$
12	1-7	7	W8=I7	00000010	=	$X \wedge Y \wedge \neg Z$
	3-5	15	V6	10010001		$(\neg Z \wedge \neg Y) \vee (X \wedge Y \wedge Z)$

Table 4.16 **U** AND NOT **W** imply **S**

Derivations	Level	#	Symbol	ID		Traditional logical form
	3-5	10	U1= Ω5	10100010	–	$\neg Z \wedge (\neg X \vee Y)$
1	1-7	4	W7=O7	00010000	=	$X \wedge \neg Y \wedge \neg Z$
	3-5	10	U1= Ω5	10100010		$\neg Z \wedge (\neg X \vee Y)$
	3-5	10	U1= Ω5	10100010	–	$\neg Z \wedge (\neg X \vee Y)$
2	1-7	7	W8=I7	00000010	=	$X \wedge Y \wedge \neg Z$
	2-6	2	**Ω**	10100000		$\neg Z \wedge \neg X$
	3-5	25	U2	01100010	–	$(\neg Z \wedge Y) \vee (\neg X \wedge \neg Y \wedge Z)$
3	1-7	4	W7=O7	00010000	=	$X \wedge \neg Y \wedge \neg Z$
	3-5	25	U2	01100010		$(\neg Z \wedge Y) \vee (\neg X \wedge \neg Y \wedge Z)$
	3-5	25	U2	01100010	–	$(\neg Z \wedge Y) \vee (\neg X \wedge \neg Y \wedge Z)$
4	1-7	7	W8=I7	00000010	=	$X \wedge Y \wedge \neg Z$
	2-6	8	T8+S7	01100000		$\neg X \wedge \neg(Y \leftrightarrow Z)$
	3-5	39	U3=W3	00110010	–	$\neg Z \wedge (X \vee Y)$
5	1-7	4	W7=O7	00010000	=	$X \wedge \neg Y \wedge \neg Z$
	2-6	17	**U**	00100010		$\neg Z \wedge Y$
	3-5	39	U3=W3	00110010	–	$\neg Z \wedge (X \vee Y)$
6	1-7	7	W8=I7	00000010	=	$X \wedge Y \wedge \neg Z$
	2-6	14	S7+O7	00110000		$\neg Z \wedge \neg(X \leftrightarrow Y)$
	3-5	42	U4=S5	00101010	–	$Y \wedge (\neg X \vee \neg Z)$
7	1-7	4	W7=O7	00010000	=	$X \wedge \neg Y \wedge \neg Z$
	3-5	42	U4=S5	00101010		$Y \wedge (\neg X \vee \neg Z)$
	3-5	42	U4=S5	00101010	–	$Y \wedge (\neg X \vee \neg Z)$
8	1-7	7	W8=I7	00000010	=	$X \wedge Y \wedge \neg Z$
	2-6	15	**S**	00101000		$\neg X \wedge Y$
	3-5	44	U5	00100110	–	$(\neg Z \wedge Y) \vee (X \wedge \neg Y \wedge Z)$
9	1-7	4	W7=O7	00010000	=	$X \wedge \neg Y \wedge \neg Z$
	3-5	44	U5	00100110		$(\neg Z \wedge Y) \vee (X \wedge \neg Y \wedge Z)$
	3-5	44	U5	00100110	–	$(\neg Z \wedge Y) \vee (X \wedge \neg Y \wedge Z)$
10	1-7	7	W8=I7	00000010	=	$X \wedge Y \wedge \neg Z$
	2-6	16	S7+O8	00100100		$X \leftrightarrow \neg Y \leftrightarrow Z$
	3-5	46	U6=I3	00100011	–	$Y \wedge (X \vee \neg Z)$
11	1-7	4	W7=O7	00010000	=	$X \wedge \neg Y \wedge \neg Z$
	3-5	46	U6=I3	00100011	–	$Y \wedge (X \vee \neg Z)$
	3-5	46	U6=I3	00100011	–	$Y \wedge (X \vee \neg Z)$
12	1-7	7	W8=I7	00000010	=	$X \wedge Y \wedge \neg Z$
	2-6	18	S7+I8	00100001		$Y \wedge (X \leftrightarrow Z)$

Table 4.17 **P** AND NOT **Q** imply **T**

Derivations	Level	#	Symbol	ID		Traditional logical form
	7-1	2	P1=E2	11111101	–	$\neg X \vee \neg Y \vee Z$
1	5-3	46	Q3	01011110	=	$(X \vee Z) \wedge (\neg X \vee \neg Y \vee \neg Z)$
	3-5	11	Ω6	10100001		$(\neg Z \wedge \neg X) \vee (X \wedge Y \wedge Z)$
	7-1	6	P2=M2	11011111	–	$X \vee \neg Y \vee Z$
2	5-3	46	Q3	01011110	=	$(X \vee Z) \wedge (\neg X \vee \neg Y \vee \neg Z)$
	2-6	7	T7+I8	10000001		$X \leftrightarrow Y \leftrightarrow Z$
	7-1	2	P1=E2	11111101	–	$\neg X \vee \neg Y \vee Z$
3	5-3	47	Q4=P8	01011101	=	$\neg Z \rightarrow (X \wedge \neg Y)$
	2-6	2	**Ω**	10100000		$\neg Z \wedge \neg X$
	7-1	6	P2=M2	11011111	–	$X \vee \neg Y \vee Z$
4	5-3	47	Q4=P8	01011101	=	$\neg Z \rightarrow (X \wedge \neg Y)$
	2-6	6	T7+I7	10000010		$\neg Z \wedge (X \leftrightarrow Y)$
	7-1	2	P1=E2	11111101	–	$\neg X \vee \neg Y \vee Z$
5	5-3	48	Q5	01011011	=	$(X \vee Z) \wedge (\neg X \vee Y \vee \neg Z)$
	3-5	9	Ω4	10100100		$(\neg Z \wedge \neg X) \vee (X \wedge \neg Y \wedge Z)$
	7-1	6	P2=M2	11011111	–	$X \vee \neg Y \vee Z$
6	5-3	48	Q5	01011011	=	$(X \vee Z) \wedge (\neg X \vee Y \vee \neg Z)$
	2-6	5	T7+O8	10000100		$\neg Y \wedge (X \leftrightarrow Z)$
	7-1	2	P1=E2	11111101	–	$\neg X \vee \neg Y \vee Z$
7	5-3	49	Q6=M8	01010111	=	$\neg X \rightarrow (\neg Y \wedge Z)$
	3-5	8	Ω3=S1	10101000		$\neg X \wedge (Y \vee \neg Z)$
	7-1	6	P2=M2	11011111	–	$X \vee \neg Y \vee Z$
8	5-3	49	Q6=M8	01010111	=	$\neg X \rightarrow (\neg Y \wedge Z)$
	2-6	4	T7+S8	10001000		$\neg X \wedge (Y \leftrightarrow Z)$
	7-1	2	P1=E2	11111101	–	$\neg X \vee \neg Y \vee Z$
9	5-3	50	Q7=N7	01001111	=	$\neg Z \rightarrow (X \wedge Y)$
	3-5	7	Ω2=V2	10110000		$\neg Z \wedge (\neg X \vee \neg Y)$
	7-1	6	P2=M2	11011111	–	$X \vee \neg Y \vee Z$
10	5-3	50	Q7=N7	01001111	=	$\neg Z \rightarrow (X \wedge Y)$
	2-6	3	**V**	10010000		$\neg Z \wedge \neg Y$
	7-1	2	P1=E2	11111101	–	$\neg X \vee \neg Y \vee Z$
11	5-3	56	Q8=L8	00011111	=	$\neg X \rightarrow (Y \wedge Z)$
	3-5	1	Ω1=T1	11100000		$\neg X \wedge (\neg Y \vee \neg Z)$
	7-1	6	P2=M2	11011111	–	$X \vee \neg Y \vee Z$
12	5-3	56	Q8=L8	00011111	=	$\neg X \rightarrow (Y \wedge Z)$
	2-6	1	**T**	11000000		$\neg X \wedge \neg Y$

Table 4.18 **N** AND NOT **Q** imply **S**

Derivations	Level	#	Symbol	ID		Traditional logical form
	7-1	5	N1=A2	11101111	–	$\neg X \vee Y \vee Z$
1	5-3	46	Q3	01011110	=	$(X \vee Z) \wedge (\neg X \vee \neg Y \vee \neg Z)$
	3-5	11	Ω6	10100001		$(\neg Z \wedge \neg X) \vee (X \wedge Y \wedge Z)$
	7-1	8	N2=L2	01111111	–	$X \vee Y \vee Z$
2	5-3	46	Q3	01011110	=	$(X \vee Z) \wedge (\neg X \vee \neg Y \vee \neg Z)$
	2-6	18	S7+I8	00100001		$Y \wedge (X \leftrightarrow Z)$
	7-1	5	N1=A2	11101111	–	$\neg X \vee Y \vee Z$
3	5-3	47	Q4=P8	01011101	=	$\neg Z \to (X \wedge \neg Y)$
	3-5	10	Ω5=U1	10100010		$\neg Z \wedge (\neg X \vee Y)$
	7-1	8	N2=L2	01111111	–	$X \vee Y \vee Z$
4	5-3	47	Q4=P8	01011101	=	$\neg Z \to (X \wedge \neg Y)$
	2-6	17	**U**	00100010		$\neg Z \wedge Y$
	7-1	5	N1=A2	11101111	–	$\neg X \vee Y \vee Z$
5	5-3	48	Q5	01011011	=	$(X \vee Z) \wedge (\neg X \vee Y \vee \neg Z)$
	3-5	9	Ω4	10100100		$(\neg Z \wedge \neg X) \vee (X \wedge \neg Y \wedge Z)$
	7-1	8	N2=L2	01111111	–	$X \vee Y \vee Z$
6	5-3	48	Q5	01011011	=	$(X \vee Z) \wedge (\neg X \vee Y \vee \neg Z)$
	2-6	16	S7+O8	00100100		$X \leftrightarrow \neg Y \leftrightarrow Z$
	7-1	5	N1=A2	11101111	–	$\neg X \vee Y \vee Z$
7	5-3	49	Q6=M8	01010111	=	$\neg X \to (\neg Y \wedge Z)$
	3-5	8	Ω3=S1	10101000		$\neg X \wedge (Y \vee \neg Z)$
	7-1	8	N2=L2	01111111	–	$X \vee Y \vee Z$
8	5-3	49	Q6=M8	01010111	=	$\neg X \to (\neg Y \wedge Z)$
	2-6	15	**S**	00101000		$\neg X \wedge Y$
	7-1	5	N1=A2	11101111	–	$\neg X \vee Y \vee Z$
9	5-3	50	Q7=N7	01001111	=	$\neg Z \to (X \wedge Y)$
	2-6	2	**Ω**	10100000		$\neg Z \wedge \neg X$
	7-1	8	N2=L2	01111111	–	$X \vee Y \vee Z$
10	5-3	50	Q7=N7	01001111	=	$\neg Z \to (X \wedge Y)$
	2-6	14	S7+O7	00110000		$\neg Z \wedge \neg(X \leftrightarrow Y)$
	7-1	5	N1=A2	11101111	–	$\neg X \vee Y \vee Z$
11	5-3	56	Q8=L8	00011111	=	$\neg X \to (Y \wedge Z)$
	3-5	1	Ω1=T1	11100000		$\neg X \wedge (\neg Y \vee \neg Z)$
	7-1	8	N2=L2	01111111	–	$X \vee Y \vee Z$
12	5-3	56	Q8=L8	00011111	=	$\neg X \to (Y \wedge Z)$
	2-6	8	T8+S7	01100000		$\neg X \wedge \neg(Y \leftrightarrow Z)$

4.4 First Mixed Mode through Subtractions

Again, I skip universal derivations. Those of the first group are:

Table 4.19 **L** AND NOT **B** imply **G**

Derivations	Level	#	Symbol	ID		Traditional logical form
	7-1	7	L1	10111111	–	$X \vee Y \vee \neg Z$
1	5-3	24	B3=D7	10110111	=	$Z \to (X \wedge \neg Y)$
	2-6	25	**G**	00001001		$Z \wedge Y$
	7-1	8	L2	01111111	–	$X \vee Y \vee Z$
2	5-3	24	B3=D7	10110111	=	$Z \to (X \wedge \neg Y)$
	3-5	33	K6=G2	01001001		$Z \wedge (\neg X \vee Y)$
	7-1	7	L1	10111111	–	$X \vee Y \vee \neg Z$
3	5-3	25	B4	10110101	=	$(X \vee \neg Z) \wedge (\neg X \vee \neg Y \vee Z)$
	2-6	24	S8+I7	00001010		$Y \wedge \neg(X \leftrightarrow Z)$
	7-1	8	L2	01111111	–	$X \vee Y \vee Z$
4	5-3	25	B4	10110101	=	$(X \vee \neg Z) \wedge (\neg X \vee \neg Y \vee Z)$
	3-5	32	K5	01001010		$(\neg X \wedge Z) \vee (X \wedge Y \wedge \neg Z)$
	7-1	7	L1	10111111	–	$X \vee Y \vee \neg Z$
5	5-3	26	B5=C5	10110011	=	$Z \to (X \wedge Y)$
	2-6	23	S8+O8	00001100		$Z \wedge \neg(X \leftrightarrow Y)$
	7-1	8	L2	01111111	–	$X \vee Y \vee Z$
6	5-3	26	B5=C5	10110011	=	$Z \to (X \wedge Y)$
	3-5	31	K4=H4	01001100		$Z \wedge (\neg X \vee \neg Y)$
	7-1	7	L1	10111111	–	$X \vee Y \vee \neg Z$
7	5-3	30	B6	10100111	=	$(X \vee \neg Z) \wedge (\neg X \vee Y \vee Z)$
	2-6	19	O7+S8	00011000		$\neg X \leftrightarrow Y \leftrightarrow Z$
	7-1	8	L2	01111111	–	$X \vee Y \vee Z$
8	5-3	30	B6	10100111	=	$(X \vee \neg Z) \wedge (\neg X \vee Y \vee Z)$
	3-5	27	K3	01011000		$(\neg X \wedge Z) \vee (X \wedge \neg Y \wedge \neg Z)$
	7-1	7	L1	10111111	–	$X \vee Y \vee \neg Z$
9	5-3	34	B7=M7	10010111	=	$\neg X \to (\neg Y \wedge \neg Z)$
	2-6	15	**S**	00101000		$\neg X \wedge Y$
	7-1	8	L2	01111111	–	$X \vee Y \vee Z$
10	5-3	34	B7=M7	10010111	=	$\neg X \to (\neg Y \wedge \neg Z)$
	3-5	23	K2=S2	01101000		$\neg X \wedge (Y \vee Z)$
	7-1	7	L1	10111111	–	$X \vee Y \vee \neg Z$
11	5-3	54	B8=L6	00110111	=	$\neg X \to (Y \wedge \neg Z)$
	2-6	4	T7+S8	10001000		$\neg X \wedge (Y \leftrightarrow Z)$
	7-1	8	L2	01111111	–	$X \vee Y \vee Z$
12	5-3	54	B8=L6	00110111	=	$\neg X \to (Y \wedge \neg Z)$
	2-6	10	**K**	01001000		$Z \wedge \neg X$

Table 4.20 **M** AND NOT **B** imply **H**

Derivations	Level	#	Symbol	ID		Traditional logical form
	7-1	4	M1	11110111	–	$X \vee \neg Y \vee \neg Z$
1	5-3	24	B3=D7	10110111	=	$Z \rightarrow (X \wedge \neg Y)$
	2-6	13	T8+I8	01000001		$Z \wedge (X \leftrightarrow Y)$
	7-1	6	M2	11011111	–	$X \vee \neg Y \vee Z$
2	5-3	24	B3=D7	10110111	=	$Z \rightarrow (X \wedge \neg Y)$
	3-5	33	K6=G2	01001001		$Z \wedge (\neg X \vee Y)$
	7-1	4	M1	11110111	–	$X \vee \neg Y \vee \neg Z$
3	5-3	25	B4	10110101	=	$(X \vee \neg Z) \wedge (\neg X \vee \neg Y \vee Z)$
	2-6	12	T8+I7	01000010		$X \leftrightarrow Y \leftrightarrow \neg Z$
	7-1	6	M2	11011111	–	$X \vee \neg Y \vee Z$
4	5-3	25	B4	10110101	=	$(X \vee \neg Z) \wedge (\neg X \vee \neg Y \vee Z)$
	3-5	32	K5	01001010		$(\neg X \wedge Z) \vee (X \wedge Y \wedge \neg Z)$
	7-1	4	M1	11110111	–	$X \vee \neg Y \vee \neg Z$
5	5-3	26	B5=C5	10110011	=	$Z \rightarrow (X \wedge Y)$
	2-6	11	**H**	01000100		$Z \wedge \neg Y$
	7-1	6	M2	11011111	–	$X \vee \neg Y \vee Z$
6	5-3	26	B5=C5	10110011	=	$Z \rightarrow (X \wedge Y)$
	3-5	31	K4=H4	01001100		$Z \wedge (\neg X \vee \neg Y)$
	7-1	4	M1	11110111	–	$X \vee \neg Y \vee \neg Z$
7	5-3	30	B6	10100111	=	$(X \vee \neg Z) \wedge (\neg X \vee Y \vee Z)$
	2-6	9	T8+O7	01010000		$\neg Y \wedge \neg(X \leftrightarrow Z)$
	7-1	6	M2	11011111	–	$X \vee \neg Y \vee Z$
8	5-3	30	B6	10100111	=	$(X \vee \neg Z) \wedge (\neg X \vee Y \vee Z)$
	3-5	27	K3	01011000		$(\neg X \wedge Z) \vee (X \wedge \neg Y \wedge \neg Z)$
	7-1	4	M1	11110111	–	$X \vee \neg Y \vee \neg Z$
9	5-3	34	B7=M7	10010111	=	$\neg X \rightarrow (\neg Y \wedge \neg Z)$
	2-6	8	T8+S7	01100000		$\neg X \wedge \neg(Y \leftrightarrow Z)$
	7-1	6	M2	11011111	–	$X \vee \neg Y \vee Z$
10	5-3	34	B7=M7	10010111	=	$\neg X \rightarrow (\neg Y \wedge \neg Z)$
	2-6	10	**K**	01001000		$Z \wedge \neg X$
	7-1	4	M1	11110111	–	$X \vee \neg Y \vee \neg Z$
11	5-3	54	B8=L6	00110111	=	$\neg X \rightarrow (Y \wedge \neg Z)$
	2-6	1	**T**	11000000		$\neg X \wedge \neg Y$
	7-1	6	M2	11011111	–	$X \vee \neg Y \vee Z$
12	5-3	54	B8=L6	00110111	=	$\neg X \rightarrow (Y \wedge \neg Z)$
	3-5	3	K1=T3	11001000		$\neg X \wedge (\neg Y \vee Z)$

The second group encompasses following derivations:

Table 4.21 **L** AND NOT **C** imply **J**

Derivations	Level	#	Symbol	ID		Traditional logical form
	7-1	7	L1	10111111	–	$X \lor Y \lor \neg Z$
1	5-3	22	C3=F7	10111010	=	$Z \to (\neg X \land Y)$
	2-6	27	**J**	00000101		$Z \land X$
	7-1	8	L2	01111111	–	$X \lor Y \lor Z$
2	5-3	22	C3=F7	10111010	=	$Z \to (\neg X \land Y)$
	3-5	35	H6=J2	01000101		$Z \land (X \lor \neg Y)$
	7-1	7	L1	10111111	–	$X \lor Y \lor \neg Z$
3	5-3	23	C4	10111001	=	$(Y \lor \neg Z) \land (\neg X \lor \neg Y \lor Z)$
	2-6	26	O8+I7	00000110		$X \land \neg(Y \leftrightarrow Z)$
	7-1	8	L2	01111111	–	$X \lor Y \lor Z$
4	5-3	23	C4	10111001	=	$(Y \lor \neg Z) \land (\neg X \lor \neg Y \lor Z)$
	3-5	34	H5	01000110		$(\neg Y \land Z) \lor (X \land Y \land \neg Z)$
	7-1	7	L1	10111111	–	$X \lor Y \lor \neg Z$
5	5-3	26	C5=B5	10110011	=	$Z \to (X \land Y)$
	2-6	23	S8+O8	00001100		$Z \land \neg(X \leftrightarrow Y)$
	7-1	8	L2	01111111	–	$X \lor Y \lor Z$
6	5-3	26	C5=B5	10110011	=	$Z \to (X \land Y)$
	3-5	31	H4=K4	01001100		$Z \land (\neg X \lor \neg Y)$
	7-1	7	L1	10111111	–	$X \lor Y \lor \neg Z$
7	5-3	29	C6=A7	10101011	=	$\neg Y \to (\neg X \land \neg Z)$
	2-6	20	**O**	00010100		$X \land \neg Y$
	7-1	8	L2	01111111	–	$X \lor Y \lor Z$
8	5-3	29	C6=A7	10101011	=	$\neg Y \to (\neg X \land \neg Z)$
	3-5	28	H3=O2	01010100		$\neg Y \land (X \lor Z)$
	7-1	7	L1	10111111	–	$X \lor Y \lor \neg Z$
9	5-3	33	C7	10011011	=	$(Y \lor \neg Z) \land (X \lor \neg Y \lor Z)$
	2-6	16	S7+O8	00100100		$X \leftrightarrow \neg Y \leftrightarrow Z$
	7-1	8	L2	01111111	–	$X \lor Y \lor Z$
10	5-3	33	C7	10011011	=	$(Y \lor \neg Z) \land (X \lor \neg Y \lor Z)$
	3-5	28	H3=O2	01010100		$\neg Y \land (X \lor Z)$
	7-1	7	L1	10111111	–	$X \lor Y \lor \neg Z$
11	5-3	53	C8=L5	00111011	=	$\neg Y \to (X \land \neg Z)$
	2-6	5	T7+O8	10000100		$\neg Y \land (X \leftrightarrow Z)$
	7-1	8	L2	01111111	–	$X \lor Y \lor Z$
12	5-3	53	C8=L5	00111011	=	$\neg Y \to (X \land \neg Z)$
	2-6	11	**H**	01000100		$Z \land \neg Y$

Table 4.22 **M** AND NOT **D** imply **J**

Derivations	Level	#	Symbol	ID		Traditional logical form
	7-1	4	M1	11110111	–	$X \vee \neg Y \vee \neg Z$
1	5-3	2	D3=E4	11110100	=	$Y \rightarrow (\neg X \wedge \neg Z)$
	2-6	28	**I**	00000011		$X \wedge Y$
	7-1	6	M2	11011111	–	$X \vee \neg Y \vee Z$
2	5-3	2	D3=E4	11110100	=	$Y \rightarrow (\neg X \wedge \neg Z)$
	3-5	55	G6=I5	00001011		$Y \wedge (X \vee Z)$
	7-1	4	M1	11110111	–	$X \vee \neg Y \vee \neg Z$
3	5-3	3	D4=F4	11110010	=	$Z \rightarrow (\neg X \wedge \neg Y)$
	2-6	27	**J**	00000101		$Z \wedge X$
	7-1	6	M2	11011111	–	$X \vee \neg Y \vee Z$
4	5-3	3	D4=F4	11110010	=	$Z \rightarrow (\neg X \wedge \neg Y)$
	3-5	54	G5=J5	00001101		$Z \wedge (X \vee Y)$
	7-1	4	M1	11110111	–	$X \vee \neg Y \vee \neg Z$
5	5-3	8	D5	11100110	=	$(\neg Y \vee \neg Z) \wedge (\neg X \vee Y \vee Z)$
	2-6	22	O7+I8	00010001		$X \wedge (Y \leftrightarrow Z)$
	7-1	6	M2	11011111	–	$X \vee \neg Y \vee Z$
6	5-3	8	D5	11100110	=	$(\neg Y \vee \neg Z) \wedge (\neg X \vee Y \vee Z)$
	3-5	49	G4	00011001		$(Y \wedge Z) \vee (X \wedge \neg Y \wedge \neg Z)$
	7-1	4	M1	11110111	–	$X \vee \neg Y \vee \neg Z$
7	5-3	14	D6=M3	11010110	=	$Y \rightarrow (X \wedge \neg Z)$
	2-6	18	S7+I8	00100001		$Y \wedge (X \leftrightarrow Z)$
	7-1	6	M2	11011111	–	$X \vee \neg Y \vee Z$
8	5-3	14	D6=M3	11010110	=	$Y \rightarrow (X \wedge \neg Z)$
	2-6	25	**G**	00001001		$Z \wedge Y$
	7-1	4	M1	11110111	–	$X \vee \neg Y \vee \neg Z$
9	5-3	24	D7=B3	10110111	=	$Z \rightarrow (X \wedge \neg Y)$
	2-6	13	T8+I8	01000001		$Z \wedge (X \leftrightarrow Y)$
	7-1	6	M2	11011111	–	$X \vee \neg Y \vee Z$
10	5-3	24	D7=B3	10110111	=	$Z \rightarrow (X \wedge \neg Y)$
	3-5	33	G2=K6	01001001		$Z \wedge (\neg X \vee Y)$
	7-1	4	M1	11110111	–	$X \vee \neg Y \vee \neg Z$
11	5-3	39	D8	01110110	=	$(\neg Y \vee \neg Z) \wedge (X \vee Y \vee Z)$
	2-6	7	T7+I8	10000001		$X \leftrightarrow Y \leftrightarrow Z$
	7-1	6	M2	11011111	–	$X \vee \neg Y \vee Z$
12	5-3	39	D8	01110110	=	$(\neg Y \vee \neg Z) \wedge (X \vee Y \vee Z)$
	3-5	18	G1	10001001		$(Y \wedge Z) \vee (\neg X \wedge \neg Y \wedge \neg Z)$

The derivations of the third group are:

Table 4.23 **H** AND NOT **J** imply **T**

Derivations	Level	#	Symbol	ID		Traditional logical form
	3-5	4	H1=T4	11000100	–	$\neg Y \wedge (\neg X \vee Z)$
1	1-7	6	J7=O8	00000100	=	$X \wedge \neg Y \wedge Z$
	2-6	1	**T**	11000000		$\neg X \wedge \neg Y$
	3-5	4	H1=T4	11000100	–	$\neg Y \wedge (\neg X \vee Z)$
2	1-7	8	J8=I8	00000001	=	$X \wedge Y \wedge Z$
	3-5	4	H1=T4	11000100		$\neg Y \wedge (\neg X \vee Z)$
	3-5	24	H2	01100100	–	$(\neg Y \wedge Z) \vee (\neg X \wedge Y \wedge \neg Z)$
3	1-7	6	J7=O8	00000100	=	$X \wedge \neg Y \wedge Z$
	2-6	8	T8+S7	01100000		$\neg X \wedge \neg(Y \leftrightarrow Z)$
	3-5	24	H2	01100100	–	$(\neg Y \wedge Z) \vee (\neg X \wedge Y \wedge \neg Z)$
4	1-7	8	J8=I8	00000001	=	$X \wedge Y \wedge Z$
	3-5	24	H2	01100100		$(\neg Y \wedge Z) \vee (\neg X \wedge Y \wedge \neg Z)$
	3-5	28	H3=O2	01010100	–	$\neg Y \wedge (X \vee Z)$
5	1-7	6	J7=O8	00000100	=	$X \wedge \neg Y \wedge Z$
	2-6	9	T8+O7	01010000		$\neg Y \wedge \neg(X \leftrightarrow Z)$
	3-5	28	H3=O2	01010100	–	$\neg Y \wedge (X \vee Z)$
6	1-7	8	J8=I8	00000001	=	$X \wedge Y \wedge Z$
	3-5	28	H3=O2	01010100	–	$\neg Y \wedge (X \vee Z)$
	3-5	31	H4=K4	01001100	–	$Z \wedge (\neg X \vee \neg Y)$
7	1-7	6	J7=O8	00000100	=	$X \wedge \neg Y \wedge Z$
	2-6	10	**K**	01001000		$Z \wedge \neg X$
	3-5	31	H4=K4	01001100	–	$Z \wedge (\neg X \vee \neg Y)$
8	1-7	8	J8=I8	00000001	=	$X \wedge Y \wedge Z$
	3-5	31	H4=K4	01001100	–	$Z \wedge (\neg X \vee \neg Y)$
	3-5	34	H5	01000110	–	$(\neg Y \wedge Z) \vee (X \wedge Y \wedge \neg Z)$
9	1-7	6	J7=O8	00000100	=	$X \wedge \neg Y \wedge Z$
	2-6	12	T8+I7	01000010		$X \leftrightarrow Y \leftrightarrow \neg Z$
	-5	34	H5	01000110	–	$(\neg Y \wedge Z) \vee (X \wedge Y \wedge \neg Z)$
10	1-7	8	J8=I8	00000001	=	$X \wedge Y \wedge Z$
	-5	34	H5	01000110		$(\neg Y \wedge Z) \vee (X \wedge Y \wedge \neg Z)$
	3-5	35	H6=J2	01000101	–	$Z \wedge (X \vee \neg Y)$
11	1-7	6	J7=O8	00000100	=	$X \wedge \neg Y \wedge Z$
	2-6	13	T8+I8	01000001		$Z \wedge (X \leftrightarrow Y)$
	3-5	35	H6=J2	01000101	–	$Z \wedge (X \vee \neg Y)$
12	1-7	8	J8=I8	00000001	=	$X \wedge Y \wedge Z$
	2-6	11	**H**	01000100		$Z \wedge \neg Y$

Table 4.24 **G** AND NOT **J** imply **S**

Derivations	Level	#	Symbol	ID		Traditional logical form
	3-5	18	G1	10001001	–	$(Y \wedge Z) \vee (\neg X \wedge \neg Y \wedge \neg Z)$
1	1-7	6	J7=O8	00000100	=	$X \wedge \neg Y \wedge Z$
	3-5	18	G1	10001001		$(Y \wedge Z) \vee (\neg X \wedge \neg Y \wedge \neg Z)$
	3-5	18	G1	10001001	–	$(Y \wedge Z) \vee (\neg X \wedge \neg Y \wedge \neg Z)$
2	1-7	8	J8=I8	00000001	=	$X \wedge Y \wedge Z$
	2-6	4	T7+S8	10001000		$\neg X \wedge (Y \leftrightarrow Z)$
	3-5	33	G2=K6	01001001	–	$Z \wedge (\neg X \vee Y)$
3	1-7	6	J7=O8	00000100	=	$X \wedge \neg Y \wedge Z$
	3-5	33	G2=K6	01001001		$Z \wedge (\neg X \vee Y)$
	3-5	33	G2=K6	01001001	–	$Z \wedge (\neg X \vee Y)$
4	1-7	8	J8=I8	00000001	=	$X \wedge Y \wedge Z$
	2-6	10	**K**	01001000		$Z \wedge \neg X$
	3-5	43	G3=S6	00101001	–	$Y \wedge (\neg X \vee Z)$
5	1-7	6	J7=O8	00000100	=	$X \wedge \neg Y \wedge Z$
	3-5	43	G3=S6	00101001		$Y \wedge (\neg X \vee Z)$
	3-5	43	G3=S6	00101001	–	$Y \wedge (\neg X \vee Z)$
6	1-7	8	J8=I8	00000001	=	$X \wedge Y \wedge Z$
	2-6	15	**S**	00101000		$\neg X \wedge Y$
	3-5	49	G4	00011001	–	$(Y \wedge Z) \vee (X \wedge \neg Y \wedge \neg Z)$
7	1-7	6	J7=O8	00000100	=	$X \wedge \neg Y \wedge Z$
	3-5	49	G4	00011001		$(Y \wedge Z) \vee (X \wedge \neg Y \wedge \neg Z)$
	3-5	49	G4	00011001	–	$(Y \wedge Z) \vee (X \wedge \neg Y \wedge \neg Z)$
8	1-7	8	J8=I8	00000001	=	$X \wedge Y \wedge Z$
	2-6	19	O7+S8	00011000		$\neg X \leftrightarrow Y \leftrightarrow Z$
	3-5	54	G5=J5	00001101	–	$Z \wedge (X \vee Y)$
9	1-7	6	J7=O8	00000100	=	$X \wedge \neg Y \wedge Z$
	2-6	25	**G**	00001001		$Z \wedge Y$
	3-5	54	G5=J5	00001101	–	$Z \wedge (X \vee Y)$
10	1-7	8	J8=I8	00000001	=	$X \wedge Y \wedge Z$
	2-6	23	S8+O8	00001100		$Z \wedge \neg(X \leftrightarrow Y)$
	3-5	55	G6=I5	00001011	–	$Y \wedge (X \vee Z)$
11	1-7	6	J7=O8	00000100	=	$X \wedge \neg Y \wedge Z$
	3-5	55	G6=I5	00001011		$Y \wedge (X \vee Z)$
	3-5	55	G6=I5	00001011	–	$Y \wedge (X \vee Z)$
12	11-7	8	J8=I8	00000001	=	$X \wedge Y \wedge Z$
	2-6	24	S8+I7	00001010		$Y \wedge \neg(X \leftrightarrow Z)$

Table 4.25 **D** AND NOT **B** imply **T**

Derivations	Level	#	Symbol	ID		Traditional logical form
	7-1	1	D1=E1	11111110	–	$\neg X \vee \neg Y \vee \neg Z$
1	5-3	24	B3=D7	10110111	=	$Z \rightarrow (X \wedge \neg Y)$
	2-6	10	**K**	01001000		$Z \wedge \neg X$
	7-1	4	D2=M1	11110111	–	$X \vee \neg Y \vee \neg Z$
2	5-3	24	B3=D7	10110111	=	$Z \rightarrow (X \wedge \neg Y)$
	2-6	13	T8+I8	01000001		$Z \wedge (X \leftrightarrow Y)$
	7-1	1	D1=E1	11111110	–	$\neg X \vee \neg Y \vee \neg Z$
3	5-3	25	B4	10110101	=	$(X \vee \neg Z) \wedge (\neg X \vee \neg Y \vee Z)$
	3-5	32	K5	01001010		$(\neg X \wedge Z) \vee (X \wedge Y \wedge \neg Z)$
	7-1	4	D2=M1	11110111	–	$X \vee \neg Y \vee \neg Z$
4	5-3	25	B4	10110101	=	$(X \vee \neg Z) \wedge (\neg X \vee \neg Y \vee Z)$
	2-6	12	T8+I7	01000010		$X \leftrightarrow Y \leftrightarrow \neg Z$
	7-1	1	D1=E1	11111110	–	$\neg X \vee \neg Y \vee \neg Z$
5	5-3	26	B5=C5	10110011	=	$Z \rightarrow (X \wedge Y)$
	3-5	31	K4=H4	01001100		$Z \wedge (\neg X \vee \neg Y)$
	7-1	4	D2=M1	11110111	–	$X \vee \neg Y \vee \neg Z$
6	5-3	26	B5=C5	10110011	=	$Z \rightarrow (X \wedge Y)$
	2-6	11	**H**	01000100		$Z \wedge \neg Y$
	7-1	1	D1=E1	11111110	–	$\neg X \vee \neg Y \vee \neg Z$
7	5-3	30	B6	10100111	=	$(X \vee \neg Z) \wedge (\neg X \vee Y \vee Z)$
	3-5	27	K3	01011000		$(\neg X \wedge Z) \vee (X \wedge \neg Y \wedge \neg Z)$
	7-1	4	D2=M1	11110111	–	$X \vee \neg Y \vee \neg Z$
8	5-3	30	B6	10100111	=	$(X \vee \neg Z) \wedge (\neg X \vee Y \vee Z)$
	2-6	9	T8+O7	01010000		$\neg Y \wedge \neg(X \leftrightarrow Z)$
	7-1	1	D1=E1	11111110	–	$\neg X \vee \neg Y \vee \neg Z$
9	5-3	34	B7=M7	10010111	=	$\neg X \rightarrow (\neg Y \wedge \neg Z)$
	3-5	23	K2=S2	01101000		$\neg X \wedge (Y \vee Z)$
	7-1	4	D2=M1	11110111	–	$X \vee \neg Y \vee \neg Z$
10	5-3	34	B7=M7	10010111	=	$\neg X \rightarrow (\neg Y \wedge \neg Z)$
	2-6	8	T8+S7	01100000		$\neg X \wedge \neg(Y \leftrightarrow Z)$
	7-1	1	D1=E1	11111110	–	$\neg X \vee \neg Y \vee \neg Z$
11	5-3	54	B8=L6	00110111	=	$\neg X \rightarrow (Y \wedge \neg Z)$
	3-5	3	K1=T3	11001000		$\neg X \wedge (\neg Y \vee Z)$
	7-1	4	D2=M1	11110111	–	$X \vee \neg Y \vee \neg Z$
12	5-3	54	B8=L6	00110111	=	$\neg X \rightarrow (Y \wedge \neg Z)$
	2-6	1	**T**	11000000		$\neg X \wedge \neg Y$

Table 4.26 **C** AND NOT **B** imply **S**

Derivations	Level	#	Symbol	ID		Traditional logical form
	7-1	3	C1=A1	11111011	–	$\neg X \vee Y \vee \neg Z$
1	5-3	24	B3=D7	10110111	=	$Z \rightarrow (X \wedge \neg Y)$
	3-5	33	K6=G2	01001001		$Z \wedge (\neg X \vee Y)$
	7-1	7	C2=L1	10111111	–	$X \vee Y \vee \neg Z$
2	5-3	24	B3=D7	10110111	=	$Z \rightarrow (X \wedge \neg Y)$
	2-6	25	**G**	00001001		$Z \wedge Y$
	7-1	3	C1=A1	11111011	–	$\neg X \vee Y \vee \neg Z$
3	5-3	25	B4	10110101	=	$(X \vee \neg Z) \wedge (\neg X \vee \neg Y \vee Z)$
	3-5	32	K5	01001010		$(\neg X \wedge Z) \vee (X \wedge Y \wedge \neg Z)$
	7-1	7	C2=L1	10111111	–	$X \vee Y \vee \neg Z$
4	5-3	25	B4	10110101	=	$(X \vee \neg Z) \wedge (\neg X \vee \neg Y \vee Z)$
	2-6	24	S8+I7	00001010		$Y \wedge \neg(X \leftrightarrow Z)$
	7-1	3	C1=A1	11111011	–	$\neg X \vee Y \vee \neg Z$
5	5-3	26	B5=C5	10110011	=	$Z \rightarrow (X \wedge Y)$
	2-6	10	**K**	01001000		$Z \wedge \neg X$
	7-1	7	C2=L1	10111111	–	$X \vee Y \vee \neg Z$
6	5-3	26	B5=C5	10110011	=	$Z \rightarrow (X \wedge Y)$
	2-6	23	S8+O8	00001100		$Z \wedge \neg(X \leftrightarrow Y)$
	7-1	3	C1=A1	11111011	–	$\neg X \vee Y \vee \neg Z$
7	5-3	30	B6	10100111	=	$(X \vee \neg Z) \wedge (\neg X \vee Y \vee Z)$
	3-5	27	K3	01011000		$(\neg X \wedge Z) \vee (X \wedge \neg Y \wedge \neg Z)$
	7-1	7	C2=L1	10111111	–	$X \vee Y \vee \neg Z$
8	5-3	30	B6	10100111	=	$(X \vee \neg Z) \wedge (\neg X \vee Y \vee Z)$
	2-6	19	O7+S8	00011000		$\neg X \leftrightarrow Y \leftrightarrow Z$
	7-1	3	C1=A1	11111011	–	$\neg X \vee Y \vee \neg Z$
9	5-3	34	B7=M7	10010111	=	$\neg X \rightarrow (\neg Y \wedge \neg Z)$
	3-5	23	K2=S2	01101000		$\neg X \wedge (Y \vee Z)$
	7-1	7	C2=L1	10111111	–	$X \vee Y \vee \neg Z$
10	5-3	34	B7=M7	10010111	=	$\neg X \rightarrow (\neg Y \wedge \neg Z)$
	2-6	15	**S**	00101000		$\neg X \wedge Y$
	7-1	3	C1=A1	11111011	–	$\neg X \vee Y \vee \neg Z$
11	5-3	54	B8=L6	00110111	=	$\neg X \rightarrow (Y \wedge \neg Z)$
	3-5	3	K1=T3	11001000		$\neg X \wedge (\neg Y \vee Z)$
	7-1	7	C2=L1	10111111	–	$X \vee Y \vee \neg Z$
12	5-3	54	B8=L6	00110111	=	$\neg X \rightarrow (Y \wedge \neg Z)$
	2-6	4	T7+S8	10001000		$\neg X \wedge (Y \leftrightarrow Z)$

4.5 Second Mixed Mode through Subtractions

The first group of the second mixed mode encompasses following derivations:

Table 4.27 A AND NOT **R** imply **U**

Derivations	Level	#	Symbol	ID		Traditional logical form
	7-1	3	A1	11111011	–	$\neg X \vee Y \vee \neg Z$
1	5-3	5	R3=E5	11101100	=	$X \rightarrow (\neg Y \wedge Z)$
	3-5	52	W6=I4	00010011		$X \wedge (Y \vee \neg Z)$
	7-1	5	A2	11101111	–	$\neg X \vee Y \vee Z$
2	5-3	5	R3=E5	11101100	=	$X \rightarrow (\neg Y \wedge Z)$
	2-6	28	**I**	00000011		$X \wedge Y$
	7-1	3	A1	11111011	–	$\neg X \vee Y \vee \neg Z$
3	5-3	7	R4=A4	11101001	=	$X \rightarrow (Y \wedge Z)$
	2-6	21	**W**	00010010		$\neg Z \wedge X$
	7-1	5	A2	11101111	–	$\neg X \vee Y \vee Z$
4	5-3	7	R4=A4	11101001	=	$X \rightarrow (Y \wedge Z)$
	2-6	26	O8+I7	00000110		$X \wedge \neg(Y \leftrightarrow Z)$
	7-1	3	A1	11111011	–	$\neg X \vee Y \vee \neg Z$
5	5-3	9	R5	11100101	=	$(\neg X \vee Z) \wedge (X \vee \neg Y \vee \neg Z)$
	3-5	48	W4	00011010		$(\neg Z \wedge X) \vee (\neg X \wedge Y \wedge Z)$
	7-1	5	A2	11101111	–	$\neg X \vee Y \vee Z$
6	5-3	9	R5	11100101	=	$(\neg X \vee Z) \wedge (X \vee \neg Y \vee \neg Z)$
	2-6	24	S8+I7	00001010		$Y \wedge \neg(X \leftrightarrow Z)$
	7-1	3	A1	11111011	–	$\neg X \vee Y \vee \neg Z$
7	5-3	18	R6=P6	11001101	=	$\neg Z \rightarrow (\neg X \wedge \neg Y)$
	3-5	39	W3=U3	00110010		$\neg Z \wedge (X \vee Y)$
	7-1	5	A2	11101111	–	$\neg X \vee Y \vee Z$
8	5-3	18	R6=P6	11001101	=	$\neg Z \rightarrow (\neg X \wedge \neg Y)$
	2-6	17	**U**	00100010		$\neg Z \wedge Y$
	7-1	3	A1	11111011	–	$\neg X \vee Y \vee \neg Z$
9	5-3	28	R7	10101101	=	$(\neg X \vee Z) \wedge (X \vee Y \vee \neg Z)$
	3-5	29	W2	01010010		$(\neg Z \wedge X) \vee (\neg X \wedge \neg Y \wedge Z)$
	7-1	5	A2	11101111	–	$\neg X \vee Y \vee Z$
10	5-3	28	R7	10101101	=	$(\neg X \vee Z) \wedge (X \vee Y \vee \neg Z)$
	2-6	12	T8+I7	01000010		$X \leftrightarrow Y \leftrightarrow \neg Z$
	7-1	3	A1	11111011	–	$\neg X \vee Y \vee \neg Z$
11	5-3	43	R8=N4	01101101	=	$\neg Z \rightarrow (\neg X \wedge Y)$
	3-5	14	W1=V5	10010010		$\neg Z \wedge (X \vee \neg Y)$
	7-1	5	A2	11101111	–	$\neg X \vee Y \vee Z$
12	5-3	43	R8=N4	01101101	=	$\neg Z \rightarrow (\neg X \wedge Y)$
	2-6	6	T7+I7	10000010		$\neg Z \wedge (X \leftrightarrow Y)$

Table 4.28 **E** AND NOT **R** imply **V**

Derivations	Level	#	Symbol	ID		Traditional logical form
	7-1	1	E1	11111110	–	$\neg X \vee \neg Y \vee \neg Z$
1	5-3	5	R3=E5	11101100	=	$X \to (\neg Y \wedge Z)$
	2-6	21	**W**	00010010		$\neg Z \wedge X$
	7-1	2	E2	11111101	–	$\neg X \vee \neg Y \vee Z$
2	5-3	5	R3=E5	11101100	=	$X \to (\neg Y \wedge Z)$
	2-6	22	O7+I8	00010001		$X \wedge (Y \leftrightarrow Z)$
	7-1	1	E1	11111110	–	$\neg X \vee \neg Y \vee \neg Z$
3	5-3	7	R4=A4	11101001	=	$X \to (Y \wedge Z)$
	3-5	50	W5=O5	00010110		$X \wedge (\neg Y \vee \neg Z)$
	7-1	2	E2	11111101	–	$\neg X \vee \neg Y \vee Z$
4	5-3	7	R4=A4	11101001	=	$X \to (Y \wedge Z)$
	2-6	20	**O**	00010100		$X \wedge \neg Y$
	7-1	1	E1	11111110	–	$\neg X \vee \neg Y \vee \neg Z$
5	5-3	9	R5	11100101	=	$(\neg X \vee Z) \wedge (X \vee \neg Y \vee \neg Z)$
	3-5	48	W4	00011010		$(\neg Z \wedge X) \vee (\neg X \wedge Y \wedge Z)$
	7-1	2	E2	11111101	–	$\neg X \vee \neg Y \vee Z$
6	5-3	9	R5	11100101	=	$(\neg X \vee Z) \wedge (X \vee \neg Y \vee \neg Z)$
	2-6	19	O7+S8	00011000		$\neg X \leftrightarrow Y \leftrightarrow Z$
	7-1	1	E1	11111110	–	$\neg X \vee \neg Y \vee \neg Z$
7	5-3	18	R6=P6	11001101	=	$\neg Z \to (\neg X \wedge \neg Y)$
	3-5	39	W3=U3	00110010		$\neg Z \wedge (X \vee Y)$
	7-1	2	E2	11111101	–	$\neg X \vee \neg Y \vee Z$
8	5-3	18	R6=P6	11001101	=	$\neg Z \to (\neg X \wedge \neg Y)$
	2-6	14	S7+O7	00110000		$\neg Z \wedge \neg(X \leftrightarrow Y)$
	7-1	1	E1	11111110	–	$\neg X \vee \neg Y \vee \neg Z$
9	5-3	28	R7	10101101	=	$(\neg X \vee Z) \wedge (X \vee Y \vee \neg Z)$
	3-5	29	W2	01010010		$(\neg Z \wedge X) \vee (\neg X \wedge \neg Y \wedge Z)$
	7-1	2	E2	11111101	–	$\neg X \vee \neg Y \vee Z$
10	5-3	28	R7	10101101	=	$(\neg X \vee Z) \wedge (X \vee Y \vee \neg Z)$
	2-6	9	T8+O7	01010000		$\neg Y \wedge \neg(X \leftrightarrow Z)$
	7-1	1	E1	11111110	–	$\neg X \vee \neg Y \vee \neg Z$
11	5-3	43	R8=N4	01101101	=	$\neg Z \to (\neg X \wedge Y)$
	3-5	14	W1=V5	10010010		$\neg Z \wedge (X \vee \neg Y)$
	7-1	2	E2	11111101	–	$\neg X \vee \neg Y \vee Z$
12	5-3	43	R8=N4	01101101	=	$\neg Z \to (\neg X \wedge Y)$
	2-6	3	**V**	10010000		$\neg Z \wedge \neg Y$

The derivations of the second group are:

Table 4.29 A AND NOT **N** imply **Ω**

Derivations	Level	#	Symbol	ID		Traditional logical form
	7-1	3	A1	11111011	–	$\neg X \vee Y \vee \neg Z$
1	5-3	42	N3	01101110	=	$(Y \vee Z) \wedge (\neg X \vee \neg Y \vee \neg Z)$
	3-5	15	V6	10010001		$(\neg Z \wedge \neg Y) \vee (X \wedge Y \wedge Z)$
	7-1	5	A2	11101111	–	$\neg X \vee Y \vee Z$
2	5-3	42	N3	01101110	=	$(Y \vee Z) \wedge (\neg X \vee \neg Y \vee \neg Z)$
	2-6	7	T7+I8	10000001		$X \leftrightarrow Y \leftrightarrow Z$
	7-1	3	A1	11111011	–	$\neg X \vee Y \vee \neg Z$
3	5-3	43	N4=R8	01101101	=	$\neg Z \rightarrow (\neg X \wedge Y)$
	3-5	14	V5=W1	10010010		$\neg Z \wedge (X \vee \neg Y)$
	7-1	5	A2	11101111	–	$\neg X \vee Y \vee Z$
4	5-3	43	N4=R8	01101101	=	$\neg Z \rightarrow (\neg X \wedge Y)$
	2-6	6	T7+I7	10000010		$\neg Z \wedge (X \leftrightarrow Y)$
	7-1	3	A1	11111011	–	$\neg X \vee Y \vee \neg Z$
5	5-3	44	N5=A8	01101011	=	$\neg Y \rightarrow (\neg X \wedge Z)$
	2-6	3	**V**	10010000		$\neg Z \wedge \neg Y$
	7-1	5	A2	11101111	–	$\neg X \vee Y \vee Z$
6	5-3	44	N5=A8	01101011	=	$\neg Y \rightarrow (\neg X \wedge Z)$
	2-6	5	T7+O8	10000100		$\neg Y \wedge (X \leftrightarrow Z)$
	7-1	3	A1	11111011	–	$\neg X \vee Y \vee \neg Z$
7	5-3	45	N6	01100111	=	$(Y \vee Z) \wedge (X \vee \neg Y \vee \neg Z)$
	3-5	12	V3	10011000		$(\neg Z \wedge \neg Y) \vee (\neg X \wedge Y \wedge Z)$
	7-1	5	A2	11101111	–	$\neg X \vee Y \vee Z$
8	5-3	45	N6	01100111	=	$(Y \vee Z) \wedge (X \vee \neg Y \vee \neg Z)$
	2-6	4	T7+S8	10001000		$\neg X \wedge (Y \leftrightarrow Z)$
	7-1	3	A1	11111011	–	$\neg X \vee Y \vee \neg Z$
9	5-3	50	N7=Q7	01001111	=	$\neg Z \rightarrow (X \wedge Y)$
	3-5	7	V2=Ω2	10110000		$\neg Z \wedge (\neg X \vee \neg Y)$
	7-1	5	A2	11101111	–	$\neg X \vee Y \vee Z$
10	5-3	50	N7=Q7	01001111	=	$\neg Z \rightarrow (X \wedge Y)$
	2-6	2	**Ω**	10100000		$\neg Z \wedge \neg X$
	7-1	3	A1	11111011	–	$\neg X \vee Y \vee \neg Z$
11	5-3	55	N8=L7	00101111	=	$\neg Y \rightarrow (X \wedge Z)$
	3-5	2	V1=T2	11010000		$\neg Y \wedge (\neg X \vee \neg Z)$
	7-1	5	A2	11101111	–	$\neg X \vee Y \vee Z$
12	5-3	55	N8=L7	00101111	=	$\neg Y \rightarrow (X \wedge Z)$
	2-6	1	**T**	11000000		$\neg X \wedge \neg Y$

Table 4.30 **E** AND NOT **P** imply **Ω**

Derivations	Level	#	Symbol	ID		Traditional logical form
	7-1	1	E1	11111110	–	$\neg X \vee \neg Y \vee \neg Z$
1	5-3	11	P3=E6	11011100	=	$Y \to (\neg X \wedge Z)$
	2-6	17	**U**	00100010		$\neg Z \wedge Y$
	7-1	2	E2	11111101	–	$\neg X \vee \neg Y \vee Z$
2	5-3	11	P3=E6	11011100	=	$Y \to (\neg X \wedge Z)$
	2-6	18	S7+I8	00100001		$Y \wedge (X \leftrightarrow Z)$
	7-1	1	E1	11111110	–	$\neg X \vee \neg Y \vee \neg Z$
3	5-3	13	P4	11011001	=	$(\neg Y \vee Z) \wedge (\neg X \vee Y \vee \neg Z)$
	3-5	44	U5	00100110		$(\neg Z \wedge Y) \vee (X \wedge \neg Y \wedge Z)$
	7-1	2	E2	11111101	–	$\neg X \vee \neg Y \vee Z$
4	5-3	13	P4	11011001	=	$(\neg Y \vee Z) \wedge (\neg X \vee Y \vee \neg Z)$
	2-6	16	S7+O8	00100100		$X \leftrightarrow \neg Y \leftrightarrow Z$
	7-1	1	E1	11111110	–	$\neg X \vee \neg Y \vee \neg Z$
5	5-3	15	P5=M4	11010101	=	$Y \to (X \wedge Z)$
	3-5	42	U4=S5	00101010		$Y \wedge (\neg X \vee \neg Z)$
	7-1	2	E2	11111101	–	$\neg X \vee \neg Y \vee Z$
6	5-3	15	P5=M4	11010101	=	$Y \to (X \wedge Z)$
	2-6	15	**S**	00101000		$\neg X \wedge Y$
	7-1	1	E1	11111110	–	$\neg X \vee \neg Y \vee \neg Z$
7	5-3	18	P6=R6	11001101	=	$\neg Z \to (\neg X \wedge \neg Y)$
	3-5	39	U3=W3	00110010		$\neg Z \wedge (X \vee Y)$
	7-1	2	E2	11111101	–	$\neg X \vee \neg Y \vee Z$
8	5-3	18	P6=R6	11001101	=	$\neg Z \to (\neg X \wedge \neg Y)$
	2-6	14	S7+O7	00110000		$\neg Z \wedge \neg(X \leftrightarrow Y)$
	7-1	1	E1	11111110	–	$\neg X \vee \neg Y \vee \neg Z$
9	5-3	32	P7	10011101	=	$(\neg Y \vee Z) \wedge (X \vee Y \vee \neg Z)$
	3-5	25	U2	01100010		$(\neg Z \wedge Y) \vee (\neg X \wedge \neg Y \wedge Z)$
	7-1	2	E2	11111101	–	$\neg X \vee \neg Y \vee Z$
10	5-3	32	P7	10011101	=	$(\neg Y \vee Z) \wedge (X \vee Y \vee \neg Z)$
	2-6	8	T8+S7	01100000		$\neg X \wedge \neg(Y \leftrightarrow Z)$
	7-1	1	E1	11111110	–	$\neg X \vee \neg Y \vee \neg Z$
11	5-3	47	P8=Q4	01011101	=	$\neg Z \to (X \wedge \neg Y)$
	3-5	10	U1= Ω5	10100010		$\neg Z \wedge (\neg X \vee Y)$
	7-1	2	E2	11111101	–	$\neg X \vee \neg Y \vee Z$
12	5-3	47	P8=Q4	01011101	=	$\neg Z \to (X \wedge \neg Y)$
	2-6	2	**Ω**	10100000		$\neg Z \wedge \neg X$

The derivations of the third group are:

Table 4.31 **V** AND NOT **Ω** imply **O**

Derivations	Level	#	Symbol	ID		Traditional logical form
	3-5	2	V1=T2	11010000	–	$\neg Y \wedge (\neg X \vee \neg Z)$
1	1-7	1	Ω7=T7	10000000	=	$\neg X \wedge \neg Y \wedge \neg Z$
	2-6	9	T8+O7	01010000		$\neg Y \wedge \neg(X \leftrightarrow Z)$
	3-5	2	V1=T2	11010000	–	$\neg Y \wedge (\neg X \vee \neg Z)$
2	1-7	3	Ω8=S7	00100000	=	$\neg X \wedge Y \wedge \neg Z$
	3-5	2	V1=T2	11010000		$\neg Y \wedge (\neg X \vee \neg Z)$
	3-5	7	V2=Ω2	10110000	–	$\neg Z \wedge (\neg X \vee \neg Y)$
3	1-7	1	Ω7=T7	10000000	=	$\neg X \wedge \neg Y \wedge \neg Z$
	2-6	14	S7+O7	00110000		$\neg Z \wedge \neg(X \leftrightarrow Y)$
	3-5	7	V2=Ω2	10110000	–	$\neg Z \wedge (\neg X \vee \neg Y)$
4	1-7	3	Ω8=S7	00100000	=	$\neg X \wedge Y \wedge \neg Z$
	2-6	3	**V**	10010000		$\neg Z \wedge \neg Y$
	3-5	12	V3	10011000	–	$(\neg Z \wedge \neg Y) \vee (\neg X \wedge Y \wedge Z)$
5	1-7	1	Ω7=T7	10000000	=	$\neg X \wedge \neg Y \wedge \neg Z$
	2-6	19	O7+S8	00011000		$\neg X \leftrightarrow Y \leftrightarrow Z$
	3-5	12	V3	10011000	–	$(\neg Z \wedge \neg Y) \vee (\neg X \wedge Y \wedge Z)$
6	1-7	3	Ω8=S7	00100000	=	$\neg X \wedge Y \wedge \neg Z$
	3-5	12	V3	10011000		$(\neg Z \wedge \neg Y) \vee (\neg X \wedge Y \wedge Z)$
	3-5	13	V4=O1	10010100	–	$\neg Y \wedge (X \vee \neg Z)$
7	1-7	1	Ω7=T7	10000000	=	$\neg X \wedge \neg Y \wedge \neg Z$
	2-6	20	**O**	00010100		$X \wedge \neg Y$
	3-5	13	V4=O1	10010100	–	$\neg Y \wedge (X \vee \neg Z)$
8	1-7	3	Ω8=S7	00100000	=	$\neg X \wedge Y \wedge \neg Z$
	3-5	13	V4=O1	10010100		$\neg Y \wedge (X \vee \neg Z)$
	3-5	14	V5=W1	10010010	–	$\neg Z \wedge (X \vee \neg Y)$
9	1-7	1	Ω7=T7	10000000	=	$\neg X \wedge \neg Y \wedge \neg Z$
	2-6	21	**W**	00010010		$\neg Z \wedge X$
	3-5	14	V5=W1	10010010	–	$\neg Z \wedge (X \vee \neg Y)$
10	1-7	3	Ω8=S7	00100000	=	$\neg X \wedge Y \wedge \neg Z$
	3-5	14	V5=W1	10010010		$\neg Z \wedge (X \vee \neg Y)$
	3-5	15	V6	10010001	–	$(\neg Z \wedge \neg Y) \vee (X \wedge Y \wedge Z)$
11	1-7	1	Ω7=T7	10000000	=	$\neg X \wedge \neg Y \wedge \neg Z$
	2-6	22	O7+I8	00010001		$X \wedge (Y \leftrightarrow Z)$
	3-5	15	V6	10010001	–	$(\neg Z \wedge \neg Y) \vee (X \wedge Y \wedge Z)$
12	1-7	3	Ω8=S7	00100000	=	$\neg X \wedge Y \wedge \neg Z$
	3-5	15	V6	10010001		$(\neg Z \wedge \neg Y) \vee (X \wedge Y \wedge Z)$

Table 4.32 **U** AND NOT **Ω** imply **I**

Derivations	Level	#	Symbol	ID		Traditional logical form
	3-5	10	U1= Ω5	10100010	–	$\neg Z \wedge (\neg X \vee Y)$
1	1-7	1	Ω7=T7	10000000	=	$\neg X \wedge \neg Y \wedge \neg Z$
	2-6	17	**U**	00100010		$\neg Z \wedge Y$
	3-5	10	U1= Ω5	10100010	–	$\neg Z \wedge (\neg X \vee Y)$
2	1-7	3	Ω8=S7	00100000	=	$\neg X \wedge Y \wedge \neg Z$
	2-6	6	T7+I7	10000010		$\neg Z \wedge (X \leftrightarrow Y)$
	3-5	25	U2	01100010	–	$(\neg Z \wedge Y) \vee (\neg X \wedge \neg Y \wedge Z)$
3	1-7	1	Ω7=T7	10000000	=	$\neg X \wedge \neg Y \wedge \neg Z$
	3-5	25	U2	01100010		$(\neg Z \wedge Y) \vee (\neg X \wedge \neg Y \wedge Z)$
	3-5	25	U2	01100010	–	$(\neg Z \wedge Y) \vee (\neg X \wedge \neg Y \wedge Z)$
4	1-7	3	Ω8=S7	00100000	=	$\neg X \wedge Y \wedge \neg Z$
	2-6	12	T8+I7	01000010		$X \leftrightarrow Y \leftrightarrow \neg Z$
	3-5	39	U3=W3	00110010	–	$\neg Z \wedge (X \vee Y)$
5	1-7	1	Ω7=T7	10000000	=	$\neg X \wedge \neg Y \wedge \neg Z$
	3-5	39	U3=W3	00110010		$\neg Z \wedge (X \vee Y)$
	3-5	39	U3=W3	00110010	–	$\neg Z \wedge (X \vee Y)$
6	1-7	3	Ω8=S7	00100000	=	$\neg X \wedge Y \wedge \neg Z$
	2-6	21	**W**	00010010		$\neg Z \wedge X$
	3-5	42	U4=S5	00101010	–	$Y \wedge (\neg X \vee \neg Z)$
7	1-7	1	Ω7=T7	10000000	=	$\neg X \wedge \neg Y \wedge \neg Z$
	3-5	42	U4=S5	00101010		$Y \wedge (\neg X \vee \neg Z)$
	3-5	42	U4=S5	00101010	–	$Y \wedge (\neg X \vee \neg Z)$
8	1-7	3	Ω8=S7	00100000	=	$\neg X \wedge Y \wedge \neg Z$
	2-6	24	S8+I7	00001010		$Y \wedge \neg(X \leftrightarrow Z)$
	3-5	44	U5	00100110	–	$(\neg Z \wedge Y) \vee (X \wedge \neg Y \wedge Z)$
9	1-7	1	Ω7=T7	10000000	=	$\neg X \wedge \neg Y \wedge \neg Z$
	3-5	44	U5	00100110		$(\neg Z \wedge Y) \vee (X \wedge \neg Y \wedge Z)$
	3-5	44	U5	00100110	–	$(\neg Z \wedge Y) \vee (X \wedge \neg Y \wedge Z)$
10	1-7	3	Ω8=S7	00100000	=	$\neg X \wedge Y \wedge \neg Z$
	2-6	26	O8+I7	00000110		$X \wedge \neg(Y \leftrightarrow Z)$
	3-5	46	U6=I3	00100011	–	$Y \wedge (X \vee \neg Z)$
11	1-7	1	Ω7=T7	10000000	=	$\neg X \wedge \neg Y \wedge \neg Z$
	3-5	46	U6=I3	00100011		$Y \wedge (X \vee \neg Z)$
	3-5	46	U6=I3	00100011	–	$Y \wedge (X \vee \neg Z)$
12	1-7	3	Ω8=S7	00100000	=	$\neg X \wedge Y \wedge \neg Z$
	2-6	28	**I**	00000011		$X \wedge Y$

Table 4.33 **P** AND NOT **R** imply **O**

Derivations	Level	#	Symbol	ID		Traditional logical form
	7-1	2	P1=E2	11111101	–	$\neg X \vee \neg Y \vee Z$
1	5-3	5	R3=E5	11101100	=	$X \rightarrow (\neg Y \wedge Z)$
	2-6	22	07+I8	00010001		$X \wedge (Y \leftrightarrow Z)$
	7-1	6	P2=M2	11011111	–	$X \vee \neg Y \vee Z$
2	5-3	5	R3=E5	11101100	=	$X \rightarrow (\neg Y \wedge Z)$
	3-5	52	W6=I4	00010011		$X \wedge (Y \vee \neg Z)$
	7-1	2	P1=E2	11111101	–	$\neg X \vee \neg Y \vee Z$
3	5-3	7	R4=A4	11101001	=	$X \rightarrow (Y \wedge Z)$
	2-6	20	**O**	00010100		$X \wedge \neg Y$
	7-1	6	P2=M2	11011111	–	$X \vee \neg Y \vee Z$
4	5-3	7	R4=A4	11101001	=	$X \rightarrow (Y \wedge Z)$
	3-5	50	W5=05	00010110		$X \wedge (\neg Y \vee \neg Z)$
	7-1	2	P1=E2	11111101	–	$\neg X \vee \neg Y \vee Z$
5	5-3	9	R5	11100101	=	$(\neg X \vee Z) \wedge (X \vee \neg Y \vee \neg Z)$
	2-6	19	07+S8	00011000		$\neg X \leftrightarrow Y \leftrightarrow Z$
	7-1	6	P2=M2	11011111	–	$X \vee \neg Y \vee Z$
6	5-3	9	R5	11100101	=	$(\neg X \vee Z) \wedge (X \vee \neg Y \vee \neg Z)$
	3-5	48	W4	00011010		$(\neg Z \wedge X) \vee (\neg X \wedge Y \wedge Z)$
	7-1	2	P1=E2	11111101	–	$\neg X \vee \neg Y \vee Z$
7	5-3	18	R6=P6	11001101	=	$\neg Z \rightarrow (\neg X \wedge \neg Y)$
	2-6	14	S7+07	00110000		$\neg Z \wedge \neg(X \leftrightarrow Y)$
	7-1	6	P2=M2	11011111	–	$X \vee \neg Y \vee Z$
8	5-3	18	R6=P6	11001101	=	$\neg Z \rightarrow (\neg X \wedge \neg Y)$
	2-6	21	**W**	00010010		$\neg Z \wedge X$
	7-1	2	P1=E2	11111101	–	$\neg X \vee \neg Y \vee Z$
9	5-3	28	R7	10101101	=	$(\neg X \vee Z) \wedge (X \vee Y \vee \neg Z)$
	2-6	9	T8+07	01010000		$\neg Y \wedge \neg(X \leftrightarrow Z)$
	7-1	6	P2=M2	11011111	–	$X \vee \neg Y \vee Z$
10	5-3	28	R7	10101101	=	$(\neg X \vee Z) \wedge (X \vee Y \vee \neg Z)$
	3-5	29	W2	01010010		$(\neg Z \wedge X) \vee (\neg X \wedge \neg Y \wedge Z)$
	7-1	2	P1=E2	11111101	–	$\neg X \vee \neg Y \vee Z$
11	5-3	43	R8=N4	01101101	=	$\neg Z \rightarrow (\neg X \wedge Y)$
	2-6	3	**V**	10010000		$\neg Z \wedge \neg Y$
	7-1	6	P2=M2	11011111	–	$X \vee \neg Y \vee Z$
12	5-3	43	R8=N4	01101101	=	$\neg Z \rightarrow (\neg X \wedge Y)$
	3-5	14	W1=V5	10010010		$\neg Z \wedge (X \vee \neg Y)$

Table 4.34 **N** AND NOT **R** imply **I**

Derivations	Level	#	Symbol	ID		Traditional logical form
	7-1	5	N1=A2	11101111	–	$\neg X \vee Y \vee Z$
1	5-3	5	R3=E5	11101100	=	$X \rightarrow (\neg Y \wedge Z)$
	2-6	28	**I**	00000011		$X \wedge Y$
	7-1	8	N2=L2	01111111	–	$X \vee Y \vee Z$
2	5-3	5	R3=E5	11101100	=	$X \rightarrow (\neg Y \wedge Z)$
	3-5	52	W6=I4	00010011		$X \wedge (Y \vee \neg Z)$
	7-1	5	N1=A2	11101111	–	$\neg X \vee Y \vee Z$
3	5-3	7	R4=A4	11101001	=	$X \rightarrow (Y \wedge Z)$
	2-6	26	O8+I7	00000110		$X \wedge \neg(Y \leftrightarrow Z)$
	7-1	8	N2=L2	01111111	–	$X \vee Y \vee Z$
4	5-3	7	R4=A4	11101001	=	$X \rightarrow (Y \wedge Z)$
	3-5	50	W5=O5	00010110		$X \wedge (\neg Y \vee \neg Z)$
	7-1	5	N1=A2	11101111	–	$\neg X \vee Y \vee Z$
5	5-3	9	R5	11100101	=	$(\neg X \vee Z) \wedge (X \vee \neg Y \vee \neg Z)$
	2-6	24	S8+I7	00001010		$Y \wedge \neg(X \leftrightarrow Z)$
	7-1	8	N2=L2	01111111	–	$X \vee Y \vee Z$
6	5-3	9	R5	11100101	=	$(\neg X \vee Z) \wedge (X \vee \neg Y \vee \neg Z)$
	3-5	48	W4	00011010		$(\neg Z \wedge X) \vee (\neg X \wedge Y \wedge Z)$
	7-1	5	N1=A2	11101111	–	$\neg X \vee Y \vee Z$
7	5-3	18	R6=P6	11001101	=	$\neg Z \rightarrow (\neg X \wedge \neg Y)$
	2-6	17	**U**	00100010		$\neg Z \wedge Y$
	7-1	8	N2=L2	01111111	–	$X \vee Y \vee Z$
8	5-3	18	R6=P6	11001101	=	$\neg Z \rightarrow (\neg X \wedge \neg Y)$
	3-5	39	W3=U3	00110010		$\neg Z \wedge (X \vee Y)$
	7-1	5	N1=A2	11101111	–	$\neg X \vee Y \vee Z$
9	5-3	28	R7	10101101	=	$(\neg X \vee Z) \wedge (X \vee Y \vee \neg Z)$
	2-6	12	T8+I7	01000010		$X \leftrightarrow Y \leftrightarrow \neg Z$
	7-1	8	N2=L2	01111111	–	$X \vee Y \vee Z$
10	5-3	28	R7	10101101	=	$(\neg X \vee Z) \wedge (X \vee Y \vee \neg Z)$
	3-5	29	W2	01010010		$(\neg Z \wedge X) \vee (\neg X \wedge \neg Y \wedge Z)$
	7-1	5	N1=A2	11101111	–	$\neg X \vee Y \vee Z$
11	5-3	43	R8=N4	01101101	=	$\neg Z \rightarrow (\neg X \wedge Y)$
	2-6	6	T7+I7	10000010		$\neg Z \wedge (X \leftrightarrow Y)$
	7-1	8	N2=L2	01111111	–	$X \vee Y \vee Z$
12	5-3	43	R8=N4	01101101	=	$\neg Z \rightarrow (\neg X \wedge Y)$
	2-6	21	**W**	00010010		$\neg Z \wedge X$

First Group				Second Group				Third Group			
Classical	1st mixed	2nd mixed	Extension	Classical	1st mixed	2nd mixed	Extension	Classical	1st mixed	2nd mixed	Extension
A $X \rightarrow Y$	L $\neg X \rightarrow Y$	A $X \rightarrow Y$	L $\neg X \rightarrow Y$	A $X \rightarrow Y$	L $\neg X \rightarrow Y$	A $X \rightarrow Y$	L $\neg X \rightarrow Y$	H $Z \wedge \neg Y$	H $Z \wedge \neg Y$	V $\neg Z \wedge \neg Y$	V $\neg Z \wedge \neg Y$
B $Z \rightarrow X$	F $Z \rightarrow \neg X$	Q $\neg Z \rightarrow X$	R $\neg Z \rightarrow \neg X$	H $Z \wedge \neg Y$	H $Z \wedge \neg Y$	V $\neg Z \wedge \neg Y$	V $\neg Z \wedge \neg Y$	B $Z \rightarrow X$	F $Z \rightarrow \neg X$	Q $\neg Z \rightarrow X$	R $\neg Z \rightarrow \neg X$
C $Z \rightarrow Y$	C $Z \rightarrow Y$	N $\neg Z \rightarrow Y$	N $\neg Z \rightarrow Y$	K $Z \wedge \neg X$	J $Z \wedge X$	Ω $\neg Z \wedge \neg X$	W $\neg Z \wedge X$	O $X \wedge \neg Y$	T $\neg X \wedge \neg Y$	O $X \wedge \neg Y$	T $\neg X \wedge \neg Y$
E $X \rightarrow \neg Y$	M $\neg X \rightarrow \neg Y$	E $X \rightarrow \neg Y$	M $\neg X \rightarrow \neg Y$	E $X \rightarrow \neg Y$	M $\neg X \rightarrow \neg Y$	E $X \rightarrow \neg Y$	M $\neg X \rightarrow \neg Y$	G $Z \wedge Y$	G $Z \wedge Y$	U $\neg Z \wedge Y$	U $\neg Z \wedge Y$
B $Z \rightarrow X$	F $Z \rightarrow \neg X$	Q $\neg Z \rightarrow X$	R $\neg Z \rightarrow \neg X$	G $Z \wedge Y$	G $Z \wedge Y$	U $\neg Z \wedge Y$	U $\neg Z \wedge Y$	B $Z \rightarrow X$	F $Z \rightarrow \neg X$	Q $\neg Z \rightarrow X$	R $\neg Z \rightarrow \neg X$
D $Z \rightarrow \neg Y$	D $Z \rightarrow \neg Y$	P $\neg Z \rightarrow \neg Y$	P $\neg Z \rightarrow \neg Y$	K $Z \wedge \neg X$	J $Z \wedge X$	Ω $\neg Z \wedge \neg X$	W $\neg Z \wedge X$	I $X \wedge Y$	S $\neg X \wedge Y$	I $X \wedge Y$	S $\neg X \wedge Y$
A $X \rightarrow Y$	L $\neg X \rightarrow Y$	A $X \rightarrow Y$	L $\neg X \rightarrow Y$	A $X \rightarrow Y$	L $\neg X \rightarrow Y$	A $X \rightarrow Y$	L $\neg X \rightarrow Y$	D $Z \rightarrow \neg Y$	D $Z \rightarrow \neg Y$	P $\neg Z \rightarrow \neg Y$	P $\neg Z \rightarrow \neg Y$
J $Z \wedge X$	K $Z \wedge \neg X$	W $\neg Z \wedge X$	Ω $\neg Z \wedge \neg X$	D $Z \rightarrow \neg Y$	D $Z \rightarrow \neg Y$	P $\neg Z \rightarrow \neg Y$	P $\neg Z \rightarrow \neg Y$	J $Z \wedge X$	K $Z \wedge \neg X$	W $\neg Z \wedge X$	Ω $\neg Z \wedge \neg X$
G $Z \wedge Y$	G $Z \wedge Y$	U $\neg Z \wedge Y$	U $\neg Z \wedge Y$	F $Z \rightarrow \neg X$	B $Z \rightarrow X$	R $\neg Z \rightarrow \neg X$	Q $\neg Z \rightarrow X$	O $X \wedge \neg Y$	T $\neg X \wedge \neg Y$	O $X \wedge \neg Y$	T $\neg X \wedge \neg Y$
E $X \rightarrow \neg Y$	M $\neg X \rightarrow \neg Y$	E $X \rightarrow \neg Y$	M $\neg X \rightarrow \neg Y$	E $X \rightarrow \neg Y$	M $\neg X \rightarrow \neg Y$	E $X \rightarrow \neg Y$	M $\neg X \rightarrow \neg Y$	C $Z \rightarrow Y$	C $Z \rightarrow Y$	N $\neg Z \rightarrow Y$	N $\neg Z \rightarrow Y$
J $Z \wedge X$	K $Z \wedge \neg X$	W $\neg Z \wedge X$	Ω $\neg Z \wedge \neg X$	C $Z \rightarrow Y$	C $Z \rightarrow Y$	N $\neg Z \rightarrow Y$	N $\neg Z \rightarrow Y$	J $Z \wedge X$	K $Z \wedge \neg X$	W $\neg Z \wedge X$	Ω $\neg Z \wedge \neg X$
H $Z \wedge \neg Y$	H $Z \wedge \neg Y$	V $\neg Z \wedge \neg Y$	V $\neg Z \wedge \neg Y$	F $Z \rightarrow \neg X$	B $Z \rightarrow X$	R $\neg Z \rightarrow \neg X$	Q $\neg Z \rightarrow X$	I $X \wedge Y$	S $\neg X \wedge Y$	I $X \wedge Y$	S $\neg X \wedge Y$

Figure 4.1 Summary of all subtraction inferences.

I have summarized all derivations through subtractions in Fig. 4.1 (recall the second assumption needs to be subtracted to the first one). It may be noted that all subtraction inferences have at least one premise that is negative e and one premise that is particular (they may coincide).

Chapter 5

Divisions

5.1 Introduction

In this chapter I shall introduce the final form of derivation that takes recourse to divisions. We can again start with *Barbara* in its classical product form:

$$(\mathbf{A} \wedge \mathbf{B}) \rightarrow \mathbf{C} . \tag{5.1}$$

Due to the previous considerations about sum and subtraction inferences, it is easy to see that such a derivation is fully equivalent to

$$(\mathbf{O} \vee \neg\mathbf{B}) \leftarrow \mathbf{H} . \tag{5.2}$$

Indeed, *Barbara* is equivalent to

$$(\neg\mathbf{O} \wedge \mathbf{B}) \rightarrow \mathbf{C} , \tag{5.3}$$

which in turn is equivalent to:

$$\neg(\mathbf{O} \vee \neg\mathbf{B}) \rightarrow \neg\mathbf{H} , \tag{5.4}$$

which is finally equivalent to the desired expression (5.2), since inversion of the arrow drops the negation of both the antecedent and the consequent being a kind of Transposition. The expression $\mathbf{O} \vee \neg\mathbf{B}$

Mechanical Logic in Three-Dimensional Space
Gennaro Auletta

ISBN 978-981-4411-50-9 (Hardcover), 978-981-4411-51-6 (eBook)
www.panstanford.com

could be called an OR NOT operation or also converse implication (see Fig. 1.3). This allows us to write the inference (5.2) as:

$$(\mathbf{O} \leftarrow \mathbf{B}) \leftarrow \mathbf{H}\,, \tag{5.5}$$

which means that **O** being necessary condition of **B** is necessary condition of **H**. Although this formulation possesses a certain beauty, I prefer in the following to make use of the OR NOT formulation. In such a case, to affirm that the premises

$$\text{Every } X \text{ is } Y \text{ and Every } Z \text{ is } X$$

are sufficient conditions of

$$\text{Every } Z \text{ is } Y$$

is fully equivalent to affirm that the premise

$$\text{Some } X \text{ is not } Y \quad \text{or} \quad \text{It is not true that Every } Z \text{ is } X$$

is necessary condition of

$$\text{Not Every } Z \text{ is } Y.$$

In other words,

$$\begin{gathered}\text{When some } X \text{ is not } Y \qquad \text{or} \qquad \text{It is not true that Every } Z \text{ is } X,\\ \text{We can assume that it is not true that Every } Z \text{ is } Y.\end{gathered}$$

Perhaps an easier example is the following:

$$\begin{gathered}\text{When some } X \text{ is } Y \qquad \text{or} \qquad \text{It is not true that Every } Z \text{ is } X,\\ \text{We can assume that Some } Z \text{ is } Y.\end{gathered}$$

We have seen (in Chapter 3) that we could associate the kind of inference determined by the sum to the inference called induction (or statistical generalization). We can now associate the kind of inference determined by the division to the inference mode called abduction, i.e. the inference trying to guess new properties able to single out new classes of objects (Peirce, 1878; Auletta, 2009). Indeed, by denying that Every Z is X, we are affirming that we have badly chosen the set of objects Z, and we were induced into such an error by believing that those objects are characterized by the feature X. Therefore, in the conclusion we now assume that at least some of those objects are characterized by the property Y. Note that both induction (sum) and abduction (division) are retrograde modes of reasoning while both product and subtraction are anterograde (although I have stressed that subtraction is kind of weak abduction or deduction: see Section 4.1). When we speak of retrograde inference, in general we start from the conclusion and try to guess which kinds of premises may imply it. However, when reasoning mechanically, it is more convenient to start from the premises and assume a conclusion, as shown here. In both cases the inference has a retrograde form.

Perhaps, we can understand better that abduction is a kind of retrograde generalization by considering the following inference:

When some X is not Y or It is not true that Some Z is not Y,
We can assume that Every Z is X.

This inference is fully valid, as we shall discover, and has the form

$$(\mathbf{O} \vee \neg\mathbf{H}) \leftarrow \mathbf{B}\,. \tag{5.6}$$

This inference tells us that we had previously assumed that there are certain objects Z that do not have the property Y, what now turn out to be false or also some objects X do not have the property Y. In such a case we can assume that all objects Z are characterized by the feature X.

It is important to remark that, although inspired by him, I am not following Peirce's approach to the problem in details. Although being the discoverer of the form that we now call abduction, he was led to the distinction among deduction, abduction and induction by starting with the Aristotelian three figures (he proved that the so-called forth one is not really a figure) (Peirce, 1865). It is also true that he tried later to overcome this initial standpoint (Peirce, 1901). What I am trying to show is that this distinction is much more caught in its specificity when considering different forms of inferences.

It is obviously true that all the kind of inferences I am considering are full equivalent from a formal point of view and can be understood as a "variation on the same theme". Nevertheless, it may be interesting to make use of different forms in different contexts. As already stressed, in empirical sciences or practical contexts to make use of one or the other can make a very important difference to the extent to which we can be led or not to further assumptions or experiences.

The new operation introduced here can be defines as follows:

Table 5.1 Division

$0:0=1$	$0:1=0$	$1:0=1$	$1:1=1$

In other words, we deal always with a set of two basic numbers (0 and 1). Do not worry about the arithmetic significance of 0 and 1 (in which case the operation 0 : 0 would not be allowed), since here they play here a logical role (although we make use for the sake of notation of counterparts of arithmetic operations). In such a case, 0 has the meaning of the lowest set element and 1 of the highest set element. Clarified this point, we can make also use of arithmetic notions. Then, it is clear that to divide these numbers by themselves gives always the unity , as it indeed occurs for any ordinary number apart from the arithmetic 0 (which, however, has this particular property due to both its position in the numbers series and its specific arithmetic significance). On the other hand, we cannot divide the lowest element further, so that any division of this number by other numbers gives again itself. Finally, to divide the highest element by the lowest elements gives again the highest element.

I also summarize all the four operations in a single table (see Tables 1.7 and 4.1):

Table 5.2 All operations

Sum	$0+0=0$	$0+1=1$	$1+0=1$	$1+1=1$
Product	$0\times 0=0$	$0\times 1=0$	$1\times 0=0$	$1\times 1=1$
Subtraction	$0-0=0$	$0-1=0$	$1-0=1$	$1-1=0$
Division	$0:0=1$	$0:1=0$	$1:0=1$	$1:1=1$

I recall that sum and product remain unchanged when interchanging the two lines involved in the operation (the premises) while this is not the case with subtraction and division. This confirms the fact that the denial of a possible premise is not fully equivalent to the countervalent statement of such a premise. One may also worry that these operations have no inverse (division is not the inverse of product and subtraction not of sum). However, we need to take again into account that we deal here with logical operations and not with arithmetic ones, and the former are irreversible. To convince ourselves we need to start with product (whose definition is quite straightforward). If I do following operation:

$$\begin{array}{ll} 00110000 & \times \\ 01010000 & = \\ 00010000 & \end{array}$$

it is easy to see that there can be no possible operation that, combining the last line (the result) with e.g. the first line, could bring us back to the second line:

$$\begin{array}{ll} 00010000 & ? \\ 00110000 & = \\ 01010000 & \end{array}$$

The reason is quite clear: such an operation would give both 0 and 1 as a result of the combination of two zeros (the first two columns). Similar considerations are true for subtraction and sum. Indeed,

$$\begin{array}{ll} 01011011 & - \\ 11010101 & = \\ 00001010 & \end{array}$$

For the same reason mentioned before, that there can be no possible operation that, combining the last line (the result) with the second line, could bring us back to the first line. The case of sum is a little different but does not change this essential result.

5.2 Classical Derivations through Divisions

When we like to pass now to inferences by making use of the division, we must take into account that we need to combine a RGS for the first premise and a LGS for the second premise (at the opposite of what happened with subtraction). This makes the derivations a little bit difficult. Let us now start with classical derivations, and in particular with the first group:

Table 5.3 **O** OR NOT **B** as necessary condition of **H**

Derivations	Level	#	Symbol	ID		Traditional logical form
	1-7	4	O7	00010000	:	$X \wedge \neg Y \wedge \neg Z$
1	7-1	4	B1=M1	11110111	=	$X \vee \neg Y \vee \neg Z$
	2-6	19	O7+S8	00011000		$\neg X \leftrightarrow Y \leftrightarrow Z$
	1-7	6	O8=H8	00000100	:	$X \wedge \neg Y \wedge Z$
2	7-1	4	B1=M1	11110111	=	$X \vee \neg Y \vee \neg Z$
	2-6	11	**H**	01000100		$Z \wedge \neg Y$
	1-7	4	O7	00010000	:	$X \wedge \neg Y \wedge \neg Z$
3	7-1	7	B2=L1	10111111	=	$X \vee Y \vee \neg Z$
	2-6	9	T8+O7	01010000		$\neg Y \wedge \neg(X \leftrightarrow Z)$
	1-7	6	O8=H8	00000100	:	$X \wedge \neg Y \wedge Z$
4	7-1	7	B2=L1	10111111	=	$X \vee Y \vee \neg Z$
	2-6	23	S8+O8	00001100		$Z \wedge \neg(X \leftrightarrow Y)$

In the following I shall skip universal derivations:

Table 5.4 **O** OR NOT **J** as necessary condition of **D**

Derivations	Level	#	Symbol	ID		Traditional logical form
	1-7	4	O7	00010000	:	$X \wedge \neg Y \wedge \neg Z$
1	3-5	20	J1	10000101	=	$(X \wedge Z) \vee (\neg X \wedge \neg Y \wedge \neg Z)$
	5-3	37	F8	01111010		$(\neg X \vee \neg Z) \wedge (X \vee Y \vee Z)$
	1-7	6	O8	00000100	:	$X \wedge \neg Y \wedge Z$
2	3-5	20	J1	10000101	=	$(X \wedge Z) \vee (\neg X \wedge \neg Y \wedge \neg Z)$
	6-2	22	E1xL2	01111110		$\neg(X \leftrightarrow Y \leftrightarrow Z)$
	1-7	4	O7	00010000	:	$X \wedge \neg Y \wedge \neg Z$
3	3-5	35	J2=H6	01000101	=	$Z \wedge (X \vee \neg Y)$
	5-3	22	F7=C3	10111010		$Z \rightarrow (\neg X \wedge Y)$
	1-7	6	O8	00000100	:	$X \wedge \neg Y \wedge Z$
4	3-5	35	J2=H6	01000101	=	$Z \wedge (X \vee \neg Y)$
	6-2	16	E1xL1	10111110		$Z \rightarrow \neg(X \leftrightarrow Y)$
	1-7	4	O7	00010000	:	$X \wedge \neg Y \wedge \neg Z$
5	3-5	45	J3	00100101	=	$(X \wedge Z) \vee (\neg X \wedge Y \wedge \neg Z)$
	5-3	12	F6	11011010		$(\neg X \vee \neg Z) \wedge (X \vee \neg Y \vee Z)$
	1-7	6	O8	00000100	:	$X \wedge \neg Y \wedge Z$
6	3-5	45	J3	00100101	=	$(X \wedge Z) \vee (\neg X \wedge Y \wedge \neg Z)$
	6-2	11	E1xM2	11011110		$Y \rightarrow \neg(X \leftrightarrow Z)$

Derivations	Level	#	Symbol	ID		Traditional logical form
	1-7	4	07	00010000	:	$X \wedge \neg Y \wedge \neg Z$
7	3-5	51	J4=06	00010101	=	$X \wedge (\neg Y \vee Z)$
	6-2	2	**F**	11111010		$Z \rightarrow \neg X$
	1-7	6	08	00000100	:	$X \wedge \neg Y \wedge Z$
8	3-5	51	J4=06	00010101	=	$X \wedge (\neg Y \vee Z)$
	6-2	7	E1xA2	11101110		$X \rightarrow \neg(Y \leftrightarrow Z)$
	1-7	4	07	00010000	:	$X \wedge \neg Y \wedge \neg Z$
9	3-5	54	J5=G5	00001101	=	$Z \wedge (X \vee Y)$
	5-3	3	F4=D4	11110010		$Z \rightarrow (\neg X \wedge \neg Y)$
	1-7	6	08	00000100	:	$X \wedge \neg Y \wedge Z$
10	3-5	54	J5=G5	00001101	=	$Z \wedge (X \vee Y)$
	6-2	4	**D**	11110110		$Z \rightarrow \neg Y$
	1-7	4	07	00010000	:	$X \wedge \neg Y \wedge \neg Z$
11	3-5	56	J6=I6	00000111	=	$X \wedge (Y \vee Z)$
	5-3	1	F3=E3	11111000		$X \rightarrow (\neg Y \wedge \neg Z)$
	1-7	6	08	00000100	:	$X \wedge \neg Y \wedge Z$
12	3-5	56	J6=I6	00000111	=	$X \wedge (Y \vee Z)$
	6-2	1	**E**	11111100		$X \rightarrow \neg Y$

Note that the cross terms (like E1xA2 or S8+08) behave as in the sum derivations (those pertaining to Level 2-6 occur in universal derivations while those pertaining to Level 6-2 in the other ones). It may be also noted that 5 elements of the RGS of **F** are derived while **F** itself is the result of Derivation 7. This is typical also of this kind of "particular" derivations.

Table 5.5 **I** OR NOT **J** as necessary condition of **C**

Derivations	Level	#	Symbol	ID		Traditional logical form
	1-7	7	I7	00000010	:	$X \wedge Y \wedge \neg Z$
1	3-5	20	J1	10000101	=	$(X \wedge Z) \vee (\neg X \wedge \neg Y \wedge \neg Z)$
	5-3	37	F8	01111010		$(\neg X \vee \neg Z) \wedge (X \vee Y \vee Z)$
	1-7	8	I8	00000001	:	$X \wedge Y \wedge Z$
2	3-5	20	J1	10000101	=	$(X \wedge Z) \vee (\neg X \wedge \neg Y \wedge \neg Z)$
	6-2	24	A1xL2	01111011		$\neg Y \rightarrow \neg(X \leftrightarrow Z)$

(*Continued*)

Table 5.5 *(Continued)*

Derivations	Level	#	Symbol	ID		Traditional logical form
3	1-7	7	I7	00000010	:	$X \wedge Y \wedge \neg Z$
	3-5	35	J2=H6	01000101	=	$Z \wedge (X \vee \neg Y)$
	5-3	22	F7=C3	10111010		$Z \rightarrow (\neg X \wedge Y)$
4	1-7	8	I8	00000001	:	$X \wedge Y \wedge Z$
	3-5	35	J2=H6	01000101	=	$Z \wedge (X \vee \neg Y)$
	6-2	18	**C**	10111011		$Z \rightarrow Y$
5	1-7	7	I7	00000010	:	$X \wedge Y \wedge \neg Z$
	3-5	45	J3	00100101	=	$(X \wedge Z) \vee (\neg X \wedge Y \wedge \neg Z)$
	5-3	12	F6	11011010		$(\neg X \vee \neg Z) \wedge (X \vee \neg Y \vee Z)$
6	1-7	8	I8	00000001	:	$X \wedge Y \wedge Z$
	3-5	45	J3	00100101	=	$(X \wedge Z) \vee (\neg X \wedge Y \wedge \neg Z)$
	6-2	13	A1xM2	11011011		$\neg(X \leftrightarrow \neg Y \leftrightarrow Z)$
7	1-7	7	I7	00000010	:	$X \wedge Y \wedge \neg Z$
	3-5	51	J4=O6	00010101	=	$X \wedge (\neg Y \vee Z)$
	5-3	6	F5=A3	11101010		$X \rightarrow (Y \wedge \neg Z)$
8	1-7	8	I8	00000001	:	$X \wedge Y \wedge Z$
	3-5	51	J4=O6	00010101	=	$X \wedge (\neg Y \vee Z)$
	6-2	9	**A**	11101011		$X \rightarrow Y$
9	1-7	7	I7	00000010	:	$X \wedge Y \wedge \neg Z$
	3-5	54	J5=G5	00001101	=	$Z \wedge (X \vee Y)$
	5-3	3	F4=D4	11110010		$Z \rightarrow (\neg X \wedge \neg Y)$
10	1-7	8	I8	00000001	:	$X \wedge Y \wedge Z$
	3-5	54	J5=G5	00001101	=	$Z \wedge (X \vee Y)$
	6-2	6	A1xM1	11110011		$Z \rightarrow (X \leftrightarrow Y)$
11	1-7	7	I7	00000010	:	$X \wedge Y \wedge \neg Z$
	3-5	56	J6=I6	00000111	=	$X \wedge (Y \vee Z)$
	6-2	2	**F**	11111010		$Z \rightarrow \neg X$
12	1-7	8	I8	00000001	:	$X \wedge Y \wedge Z$
	3-5	56	J6=I6	00000111	=	$X \wedge (Y \vee Z)$
	6-2	3	E2xA1	11111001		$X \rightarrow (Y \leftrightarrow Z)$

The second group of classical inferences is:

Table 5.6 **O** OR NOT **H** as necessary condition of **B**

Derivations	Level	#	Symbol	ID		Traditional logical form
	1-7	4	O7	00010000	:	$X \wedge \neg Y \wedge \neg Z$
1	3-5	4	H1=T4	11000100	=	$\neg Y \wedge (\neg X \vee Z)$
	5-3	53	C8=L5	00111011		$\neg Y \rightarrow (X \wedge \neg Z)$
	1-7	6	O8	00000100	:	$X \wedge \neg Y \wedge Z$
2	3-5	4	H1=T4	11000100	=	$\neg Y \wedge (\neg X \vee Z)$
	6-2	28	**L**	00111111		$\neg X \rightarrow Y$
	1-7	4	O7	00010000	:	$X \wedge \neg Y \wedge \neg Z$
3	3-5	24	H2	01100100	=	$(\neg Y \wedge Z) \vee (\neg X \wedge Y \wedge \neg Z)$
	5-3	33	C7	10011011		$(Y \vee \neg Z) \wedge (X \vee \neg Y \vee Z)$
	1-7	6	O8	00000100	:	$X \wedge \neg Y \wedge Z$
4	3-5	24	H2	01100100	=	$(\neg Y \wedge Z) \vee (\neg X \wedge Y \wedge \neg Z)$
	6-2	21	M2xL1	10011111		$\neg X \rightarrow (Y \leftrightarrow Z)$
	1-7	4	O7	00010000	:	$X \wedge \neg Y \wedge \neg Z$
5	3-5	28	H3=O2	01010100	=	$\neg Y \wedge (X \vee Z)$
	6-2	18	**C**	10111011		$Z \rightarrow Y$
	1-7	6	O8	00000100	:	$X \wedge \neg Y \wedge Z$
6	3-5	28	H3=O2	01010100	=	$\neg Y \wedge (X \vee Z)$
	6-2	20	A2xL1	10101111		$\neg Y \rightarrow (X \leftrightarrow Z)$
	1-7	4	O7	00010000	:	$X \wedge \neg Y \wedge \neg Z$
7	3-5	31	H4=K4	01001100	=	$Z \wedge (\neg X \vee \neg Y)$
	5-3	26	C5=B5	10110011		$Z \rightarrow (X \wedge Y)$
	1-7	6	O8	00000100	:	$X \wedge \neg Y \wedge Z$
8	3-5	31	H4=K4	01001100	=	$Z \wedge (\neg X \vee \neg Y)$
	6-2	19	**B**	10110111		$Z \rightarrow X$
	1-7	4	O7	00010000	:	$X \wedge \neg Y \wedge \neg Z$
9	3-5	34	H5	01000110	=	$(\neg Y \wedge Z) \vee (X \wedge Y \wedge \neg Z)$
	5-3	23	C4	10111001		$(Y \vee \neg Z) \wedge (\neg X \vee \neg Y \vee Z)$
	1-7	6	O8	00000100	:	$X \wedge \neg Y \wedge Z$
10	3-5	34	H5	01000110	=	$(\neg Y \wedge Z) \vee (X \wedge Y \wedge \neg Z)$
	6-2	17	E2xL1	10111101		$\neg(X \leftrightarrow Y \leftrightarrow \neg Z)$
	1-7	4	O7	00010000	:	$X \wedge \neg Y \wedge \neg Z$
11	3-5	35	H6=J2	01000101	=	$Z \wedge (X \vee \neg Y)$
	5-3	22	C3=F7	10111010		$Z \rightarrow (\neg X \wedge Y)$
	1-7	6	O8	00000100	:	$X \wedge \neg Y \wedge Z$
12	3-5	35	H6=J2	01000101	=	$Z \wedge (X \vee \neg Y)$
	6-2	16	E1xL1	10111110		$Z \rightarrow \neg(X \leftrightarrow Y)$

Table 5.7 **I** OR NOT **G** as necessary condition of **B**

Derivations	Level	#	Symbol	ID		Traditional logical form
	1-7	7	I7	00000010	:	$X \wedge Y \wedge \neg Z$
1	3-5	18	G1	10001001	=	$(Y \wedge Z) \vee (\neg X \wedge \neg Y \wedge \neg Z)$
	5-3	39	D8	01110110		$(\neg Y \vee \neg Z) \wedge (X \vee Y \vee Z)$
	1-7	8	I8	00000001	:	$X \wedge Y \wedge Z$
2	3-5	18	G1	10001001	=	$(Y \wedge Z) \vee (\neg X \wedge \neg Y \wedge \neg Z)$
	6-2	25	M1xL2	01110111		$\neg X \rightarrow \neg(Y \leftrightarrow Z)$
	1-7	7	I7	00000010	:	$X \wedge Y \wedge \neg Z$
3	3-5	33	G2=K6	01001001	=	$Z \wedge (\neg X \vee Y)$
	5-3	24	D7=B3	10110110		$Z \rightarrow (X \wedge \neg Y)$
	1-7	8	I8	00000001	:	$X \wedge Y \wedge Z$
4	3-5	33	G2=K6	01001001	=	$Z \wedge (\neg X \vee Y)$
	6-2	19	**B**	10110111		$Z \rightarrow X$
	1-7	7	I7	00000010	:	$X \wedge Y \wedge \neg Z$
5	3-5	43	G3=S6	00101001	=	$Y \wedge (\neg X \vee Z)$
	5-3	14	D6=M3	11010110		$Y \rightarrow (X \wedge \neg Z)$
	1-7	8	I8	00000001	:	$X \wedge Y \wedge Z$
6	3-5	43	G3=S6	00101001	=	$Y \wedge (\neg X \vee Z)$
	6-2	14	**M**	11010111		$\neg X \rightarrow \neg Y$
	1-7	7	I7	00000010	:	$X \wedge Y \wedge \neg Z$
7	3-5	49	G4	00011001	=	$(Y \wedge Z) \vee (X \wedge \neg Y \wedge \neg Z)$
	5-3	8	D5	11100110		$(\neg Y \vee \neg Z) \wedge (\neg X \vee Y \vee Z)$
	1-7	8	I8	00000001	:	$X \wedge Y \wedge Z$
8	3-5	49	G4	00011001	=	$(Y \wedge Z) \vee (X \wedge \neg Y \wedge \neg Z)$
	6-2	10	M1xA2	11100111		$\neg(\neg X \leftrightarrow Y \leftrightarrow Z)$
	1-7	7	I7	00000010	:	$X \wedge Y \wedge \neg Z$
9	3-5	54	G5=J5	00001101	=	$Z \wedge (X \vee Y)$
	5-3	3	D4=F4	11110010		$Z \rightarrow (\neg X \wedge \neg Y)$
	1-7	8	I8	00000001	:	$X \wedge Y \wedge Z$
10	3-5	54	G5=J5	00001101	=	$Z \wedge (X \vee Y)$
	6-2	6	A1xM1	11110011		$Z \rightarrow (X \leftrightarrow Y)$
	1-7	7	I7	00000010	:	$X \wedge Y \wedge \neg Z$
11	3-5	55	G6=I5	00001011	=	$Y \wedge (X \vee Z)$
	6-2	4	**D**	11110110		$Z \rightarrow \neg Y$
	1-7	8	I8	00000001	:	$X \wedge Y \wedge Z$
12	3-5	55	G6=I5	00001011	=	$Y \wedge (X \vee Z)$
	6-2	5	E2xM1	11110101		$Y \rightarrow (X \leftrightarrow Z)$

The third group of classical inferences is:

Table 5.8 **C** OR NOT **B** as necessary condition of **A**

Derivations	Level	#	Symbol	ID		Traditional logical form
	5-3	22	C3=F7	10111010	:	$Z \to (\neg X \wedge Y)$
1	7-1	4	B1=M1	11110111	=	$X \vee \neg Y \vee \neg Z$
	5-3	22	C3=F7	10111010		$Z \to (\neg X \wedge Y)$
	5-3	22	C3=F7	10111010	:	$Z \to (\neg X \wedge Y)$
2	7-1	7	B2=L1	10111111	=	$X \vee Y \vee \neg Z$
	6-2	2	**F**	11111010		$Z \to \neg X$
	5-3	23	C4	10111001	:	$(Y \vee \neg Z) \wedge (\neg X \vee \neg Y \vee Z)$
3	7-1	4	B1=M1	11110111	=	$X \vee \neg Y \vee \neg Z$
	5-3	23	C4	10111001		$(Y \vee \neg Z) \wedge (\neg X \vee \neg Y \vee Z)$
	5-3	23	C4	10111001	:	$(Y \vee \neg Z) \wedge (\neg X \vee \neg Y \vee Z)$
4	7-1	7	B2=L1	10111111	=	$X \vee Y \vee \neg Z$
	6-2	3	E2xA1	11111001		$X \to (Y \leftrightarrow Z)$
	5-3	26	C5=B5	10110011	:	$Z \to (X \wedge Y)$
5	7-1	4	B1=M1	11110111	=	$X \vee \neg Y \vee \neg Z$
	6-2	18	**C**	10111011		$Z \to Y$
	5-3	26	C5=B5	10110011	:	$Z \to (X \wedge Y)$
6	7-1	7	B2=L1	10111111	=	$X \vee Y \vee \neg Z$
	6-2	6	A1xM1	11110011		$Z \to (X \leftrightarrow Y)$
	5-3	29	C6=A7	10101011	:	$\neg Y \to (\neg X \wedge \neg Z)$
7	7-1	4	B1=M1	11110111	=	$X \vee \neg Y \vee \neg Z$
	5-3	29	C6=A7	10101011		$\neg Y \to (\neg X \wedge \neg Z)$
	5-3	29	C6=A7	10101011	:	$\neg Y \to (\neg X \wedge \neg Z)$
8	7-1	7	B2=L1	10111111	=	$X \vee Y \vee \neg Z$
	6-2	9	**A**	11101011		$X \to Y$
	5-3	33	C7	10011011	:	$(Y \vee \neg Z) \wedge (X \vee \neg Y \vee Z)$
9	7-1	4	B1=M1	11110111	=	$X \vee \neg Y \vee \neg Z$
	5-3	33	C7	10011011		$(Y \vee \neg Z) \wedge (X \vee \neg Y \vee Z)$
	5-3	33	C7	10011011	:	$(Y \vee \neg Z) \wedge (X \vee \neg Y \vee Z)$
10	7-1	7	B2=L1	10111111	=	$X \vee Y \vee \neg Z$
	6-2	13	A1xM2	11011011		$\neg(X \leftrightarrow \neg Y \leftrightarrow Z)$
	5-3	53	C8=L5	00111011	:	$\neg Y \to (X \wedge \neg Z)$
11	7-1	4	B1=M1	11110111	=	$X \vee \neg Y \vee \neg Z$
	5-3	53	C8=L5	00111011		$\neg Y \to (X \wedge \neg Z)$
	5-3	53	C8=L5	00111011	:	$\neg Y \to (X \wedge \neg Z)$
12	7-1	7	B2=L1	10111111	=	$X \vee Y \vee \neg Z$
	6-2	24	A1xL2	01111011		$\neg Y \to \neg(X \leftrightarrow Z)$

Table 5.9 **D** OR NOT **B** as necessary condition of **E**

Derivations	Level	#	Symbol	ID		Traditional logical form
	5-3	2	D3=E4	11110100	:	$Y \to (\neg X \wedge \neg Z)$
1	7-1	4	B1=M1	11110111	=	$X \vee \neg Y \vee \neg Z$
	6-2	1	**E**	11111100		$X \to \neg Y$
	5-3	2	D3=E4	11110100	:	$Y \to (\neg X \wedge \neg Z)$
2	7-1	7	B2=L1	10111111	=	$X \vee Y \vee \neg Z$
	5-3	2	D3=E4	11110100		$Y \to (\neg X \wedge \neg Z)$
	5-3	3	D4=F4	11110010	:	$Z \to (\neg X \wedge \neg Y)$
3	7-1	4	B1=M1	11110111	=	$X \vee \neg Y \vee \neg Z$
	6-2	2	**F**	11111010		$Z \to \neg X$
	5-3	3	D4=F4	11110010	:	$Z \to (\neg X \wedge \neg Y)$
4	7-1	7	B2=L1	10111111	=	$X \vee Y \vee \neg Z$
	5-3	3	D4=F4	11110010		$Z \to (\neg X \wedge \neg Y)$
	5-3	8	D5	11100110	:	$(\neg Y \vee \neg Z) \wedge (\neg X \vee Y \vee Z)$
5	7-1	4	B1=M1	11110111	=	$X \vee \neg Y \vee \neg Z$
	6-2	7	E1xA2	11101110		$X \to \neg(Y \leftrightarrow Z)$
	5-3	8	D5	11100110	:	$(\neg Y \vee \neg Z) \wedge (\neg X \vee Y \vee Z)$
6	7-1	7	B2=L1	10111111	=	$X \vee Y \vee \neg Z$
	5-3	8	D5	11100110		$(\neg Y \vee \neg Z) \wedge (\neg X \vee Y \vee Z)$
	5-3	14	D6=M3	11010110	:	$Y \to (X \wedge \neg Z)$
7	7-1	4	B1=M1	11110111	=	$X \vee \neg Y \vee \neg Z$
	6-2	11	E1xM2	11011110		$Y \to \neg(X \leftrightarrow Z)$
	5-3	14	D6=M3	11010110	:	$Y \to (X \wedge \neg Z)$
8	7-1	7	B2=L1	10111111	=	$X \vee Y \vee \neg Z$
	5-3	14	D6=M3	11010110		$Y \to (X \wedge \neg Z)$
	5-3	24	D7=B3	10110111	:	$Z \to (X \wedge \neg Y)$
9	7-1	4	B1=M1	11110111	=	$X \vee \neg Y \vee \neg Z$
	6-2	16	E1xL1	10111110		$Z \to \neg(X \leftrightarrow Y)$
	5-3	24	D7=B3	10110111	:	$Z \to (X \wedge \neg Y)$
10	7-1	7	B2=L1	10111111	=	$X \vee Y \vee \neg Z$
	6-2	4	**D**	11110110		$Z \to \neg Y$
	5-3	39	D8	01110110	:	$(\neg Y \vee \neg Z) \wedge (X \vee Y \vee Z)$
11	7-1	4	B1=M1	11110111	=	$X \vee \neg Y \vee \neg Z$
	6-2	22	E1xL2	01111110		$\neg(X \leftrightarrow Y \leftrightarrow Z)$
	5-3	39	D8	01110110	:	$(\neg Y \vee \neg Z) \wedge (X \vee Y \vee Z)$
12	7-1	7	B2=L1	10111111	=	$X \vee Y \vee \neg Z$
	5-3	39	D8	01110110		$(\neg Y \vee \neg Z) \wedge (X \vee Y \vee Z)$

Table 5.10 **G** OR NOT **J** as necessary condition of **A**

Derivations	Level	#	Symbol	ID		Traditional logical form
	1-7	5	G7=S8	00001000	:	$\neg X \wedge Y \wedge Z$
1	3-5	20	J1	10000101	=	$(X \wedge Z) \vee (\neg X \wedge \neg Y \wedge \neg Z)$
	5-3	37	F8	01111010		$(\neg X \vee \neg Z) \wedge (X \vee Y \vee Z)$
	1-7	8	G8=I8	00000001	:	$X \wedge Y \wedge Z$
2	3-5	20	J1	10000101	=	$(X \wedge Z) \vee (\neg X \wedge \neg Y \wedge \neg Z)$
	6-2	24	A1xL2	01111011		$\neg Y \rightarrow \neg(X \leftrightarrow Z)$
	1-7	5	G7=S8	00001000	:	$\neg X \wedge Y \wedge Z$
3	3-5	35	J2=H6	01000101	=	$Z \wedge (X \vee \neg Y)$
	5-3	22	F7=C3	10111010		$Z \rightarrow (\neg X \wedge Y)$
	1-7	8	G8=I8	00000001	:	$X \wedge Y \wedge Z$
4	3-5	35	J2=H6	01000101		$Z \wedge (X \vee \neg Y)$
	6-2	18	**C**	10111011		$Z \rightarrow Y$
	1-7	5	G7=S8	00001000	:	$\neg X \wedge Y \wedge Z$
5	3-5	45	J3	00100101	=	$(X \wedge Z) \vee (\neg X \wedge Y \wedge \neg Z)$
	5-3	12	F6	11011010		$(\neg X \vee \neg Z) \wedge (X \vee \neg Y \vee Z)$
	1-7	8	G8=I8	00000001	:	$X \wedge Y \wedge Z$
6	3-5	45	J3	00100101	=	$(X \wedge Z) \vee (\neg X \wedge Y \wedge \neg Z)$
	6-2	13	A1xM2	11011011		$\neg(X \leftrightarrow \neg Y \leftrightarrow Z)$
	1-7	5	G7=S8	00001000	:	$\neg X \wedge Y \wedge Z$
7	3-5	51	J4=O6	00010101	=	$X \wedge (\neg Y \vee Z)$
	5-3	6	F5=A3	11101010		$X \rightarrow (Y \wedge \neg Z)$
	1-7	8	G8=I8	00000001	:	$X \wedge Y \wedge Z$
8	3-5	51	J4=O6	00010101	=	$X \wedge (\neg Y \vee Z)$
	6-2	9	**A**	11101011		$X \rightarrow Y$
	1-7	5	G7=S8	00001000	:	$\neg X \wedge Y \wedge Z$
9	3-5	54	J5=G5	00001101	=	$Z \wedge (X \vee Y)$
	6-2	2	**F**	11111010		$Z \rightarrow \neg X$
	1-7	8	G8=I8	00000001	:	$X \wedge Y \wedge Z$
10	3-5	54	J5=G5	00001101	=	$Z \wedge (X \vee Y)$
	6-2	6	A1xM1	11110011		$Z \rightarrow (X \leftrightarrow Y)$
	1-7	5	G7=S8	00001000	:	$\neg X \wedge Y \wedge Z$
11	3-5	56	J6=I6	00000111	=	$X \wedge (Y \vee Z)$
	5-3	1	F3=E3	11111000		$X \rightarrow (\neg Y \wedge \neg Z)$
	1-7	8	G8=I8	00000001	:	$X \wedge Y \wedge Z$
12	3-5	56	J6=I6	00000111	=	$X \wedge (Y \vee Z)$
	6-2	3	E2xA1	11111001		$X \rightarrow (Y \leftrightarrow Z)$

Table 5.11 **H** OR NOT **J** as necessary condition of **E**

Derivations	Level	#	Symbol	ID		Traditional logical form
	1-7	2	H7=T8	01000000	:	$\neg X \wedge \neg Y \wedge Z$
1	3-5	20	J1	10000101	=	$(X \wedge Z) \vee (\neg X \wedge \neg Y \wedge \neg Z)$
	5-3	37	F8	01111010		$(\neg X \vee \neg Z) \wedge (X \vee Y \vee Z)$
	1-7	6	H8=O8	00000100	:	$X \wedge \neg Y \wedge Z$
2	3-5	20	J1	10000101	=	$(X \wedge Z) \vee (\neg X \wedge \neg Y \wedge \neg Z)$
	6-2	22	E1xL2	01111110		$\neg(X \leftrightarrow Y \leftrightarrow Z)$
	1-7	2	H7=T8	01000000	:	$\neg X \wedge \neg Y \wedge Z$
3	3-5	35	J2=H6	01000101	=	$Z \wedge (X \vee \neg Y)$
	6-2	2	**F**	11111010		$Z \rightarrow \neg X$
	1-7	6	H8=O8	00000100	:	$X \wedge \neg Y \wedge Z$
4	3-5	35	J2=H6	01000101	=	$Z \wedge (X \vee \neg Y)$
	6-2	16	E1xL1	10111110		$Z \rightarrow \neg(X \leftrightarrow Y)$
	1-7	2	H7=T8	01000000	:	$\neg X \wedge \neg Y \wedge Z$
5	3-5	45	J3	00100101	=	$(X \wedge Z) \vee (\neg X \wedge Y \wedge \neg Z)$
	5-3	12	F6	11011010		$(\neg X \vee \neg Z) \wedge (X \vee \neg Y \vee Z)$
	1-7	6	H8=O8	00000100	:	$X \wedge \neg Y \wedge Z$
6	3-5	45	J3	00100101	=	$(X \wedge Z) \vee (\neg X \wedge Y \wedge \neg Z)$
	6-2	11	E1xM2	11011110		$Y \rightarrow \neg(X \leftrightarrow Z)$
	1-7	2	H7=T8	01000000	:	$\neg X \wedge \neg Y \wedge Z$
7	3-5	51	J4=O6	00010101	=	$X \wedge (\neg Y \vee Z)$
	5-3	6	F5=A3	11101010		$X \rightarrow (Y \wedge \neg Z)$
	1-7	6	H8=O8	00000100	:	$X \wedge \neg Y \wedge Z$
8	3-5	51	J4=O6	00010101	=	$X \wedge (\neg Y \vee Z)$
	6-2	7	E1xA2	11101110		$X \rightarrow \neg(Y \leftrightarrow Z)$
	1-7	2	H7=T8	01000000	:	$\neg X \wedge \neg Y \wedge Z$
9	3-5	54	J5=G5	00001101	=	$Z \wedge (X \vee Y)$
	5-3	3	F4=D4	11110010		$Z \rightarrow (\neg X \wedge \neg Y)$
	1-7	6	H8=O8	00000100	:	$X \wedge \neg Y \wedge Z$
10	3-5	54	J5=G5	00001101	=	$Z \wedge (X \vee Y)$
	6-2	4	**D**	11110110		$Z \rightarrow \neg Y$
	1-7	2	H7=T8	01000000	:	$\neg X \wedge \neg Y \wedge Z$
11	3-5	56	J6=I6	00000111	=	$X \wedge (Y \vee Z)$
	5-3	1	F3=E3	11111000		$X \rightarrow (\neg Y \wedge \neg Z)$
	1-7	6	H8=O8	00000100	:	$X \wedge \neg Y \wedge Z$
12	3-5	56	J6=I6	00000111	=	$X \wedge (Y \vee Z)$
	6-2	1	**E**	11111100		$X \rightarrow \neg Y$

5.3 Extension of Classical Derivations through Divisions

As always I skip derivations of universal kind. The first group is represented by following inferences:

Table 5.12 **T** OR NOT **Ω** as necessary condition of **P**

Derivations	Level	#	Symbol	ID		Traditional logical form
	1-7	1	T7	10000000	:	$\neg X \wedge \neg Y \wedge \neg Z$
1	3-5	1	Ω1=T1	11100000	=	$\neg X \wedge (\neg Y \vee \neg Z)$
	6-2	21	M2xL1	10011111		$\neg X \to (Y \leftrightarrow Z)$
	1-7	2	T8	01000000	:	$\neg X \wedge \neg Y \wedge Z$
2	3-5	1	Ω1=T1	11100000	=	$\neg X \wedge (\neg Y \vee \neg Z)$
	6-2	27	**Q**	01011111		$\neg Z \to X$
	1-7	1	T7	10000000	:	$\neg X \wedge \neg Y \wedge \neg Z$
3	3-5	7	Ω2=V2	10110000	=	$\neg Z \wedge (\neg X \vee \neg Y)$
	6-2	15	A2xM2	11001111		$\neg Z \to (X \leftrightarrow Y)$
	1-7	2	T8	01000000	:	$\neg X \wedge \neg Y \wedge Z$
4	3-5	7	Ω2=V2	10110000	=	$\neg Z \wedge (\neg X \vee \neg Y)$
	5-3	50	Q7=N7	01001111		$\neg Z \to (X \wedge Y)$
	1-7	1	T7	10000000	:	$\neg X \wedge \neg Y \wedge \neg Z$
5	3-5	8	Ω3=S1	10101000	=	$\neg X \wedge (Y \vee \neg Z)$
	6-2	14	**M**	11010111		$\neg X \to \neg Y$
	1-7	2	T8	01000000	:	$\neg X \wedge \neg Y \wedge Z$
6	3-5	8	Ω3=S1	10101000	=	$\neg X \wedge (Y \vee \neg Z)$
	5-3	49	Q6=M8	01010111		$\neg X \to (\neg Y \wedge Z)$
	1-7	1	T7	10000000	:	$\neg X \wedge \neg Y \wedge \neg Z$
7	3-5	9	Ω4	10100100	=	$(\neg Z \wedge \neg X) \vee (X \wedge \neg Y \wedge Z)$
	6-2	13	A1xM2	11011011		$\neg(X \leftrightarrow \neg Y \leftrightarrow Z)$
	1-7	2	T8	01000000	:	$\neg X \wedge \neg Y \wedge Z$
8	3-5	9	Ω4	10100100	=	$(\neg Z \wedge \neg X) \vee (X \wedge \neg Y \wedge Z)$
	5-3	48	Q5	01011011		$(X \vee Z) \wedge (\neg X \vee Y \vee \neg Z)$
	1-7	1	T7	10000000	:	$\neg X \wedge \neg Y \wedge \neg Z$
9	3-5	10	Ω5=U1	10100010	=	$\neg Z \wedge (\neg X \vee Y)$
	6-2	12	**P**	11011101		$\neg Z \to \neg Y$
	1-7	2	T8	01000000	:	$\neg X \wedge \neg Y \wedge Z$
10	3-5	10	Ω5=U1	10100010	=	$\neg Z \wedge (\neg X \vee Y)$
	5-3	47	Q4=P8	01011101		$\neg Z \to (X \wedge \neg Y)$
	1-7	1	T7	10000000	:	$\neg X \wedge \neg Y \wedge \neg Z$
11	3-5	11	Ω6	10100001	=	$(\neg Z \wedge \neg X) \vee (X \wedge Y \wedge Z)$
	6-2	11	E1xM2	11011110		$Y \to \neg(X \leftrightarrow Z)$
	1-7	2	T8	01000000	:	$\neg X \wedge \neg Y \wedge Z$
12	3-5	11	Ω6	10100001	=	$(\neg Z \wedge \neg X) \vee (X \wedge Y \wedge Z)$
	5-3	46	Q3	01011110		$(X \vee Z) \wedge (\neg X \vee \neg Y \vee \neg Z)$

Table 5.13 **S** OR NOT **Ω** as necessary condition of **N**

Derivations	Level	#	Symbol	ID		Traditional logical form
	1-7	3	S7	00100000	:	$\neg X \wedge Y \wedge \neg Z$
1	3-5	1	Ω1=T1	11100000	=	$\neg X \wedge (\neg Y \vee \neg Z)$
	6-2	28	**L**	00111111		$\neg X \rightarrow Y$
	1-7	5	S8	00001000	:	$\neg X \wedge Y \wedge Z$
2	3-5	1	Ω1=T1	11100000	=	$\neg X \wedge (\neg Y \vee \neg Z)$
	5-3	56	Q8=L8	00011111		$\neg X \rightarrow (Y \wedge Z)$
	1-7	3	S7	00100000	:	$\neg X \wedge Y \wedge \neg Z$
3	3-5	7	Ω2=V2	10110000	=	$\neg Z \wedge (\neg X \vee \neg Y)$
	6-2	26	**N**	01101111		$\neg Z \rightarrow Y$
	1-7	5	S8	00001000	:	$\neg X \wedge Y \wedge Z$
4	3-5	7	Ω2=V2	10110000	=	$\neg Z \wedge (\neg X \vee \neg Y)$
	5-3	50	Q7=N7	01001111		$\neg Z \rightarrow (X \wedge Y)$
	1-7	3	S7	00100000	:	$\neg X \wedge Y \wedge \neg Z$
5	3-5	8	Ω3=S1	10101000	=	$\neg X \wedge (Y \vee \neg Z)$
	6-2	25	M1xL2	01110111		$\neg X \rightarrow \neg(Y \leftrightarrow Z)$
	1-7	5	S8	00001000	:	$\neg X \wedge Y \wedge Z$
6	3-5	8	Ω3=S1	10101000	=	$\neg X \wedge (Y \vee \neg Z)$
	6-2	27	**Q**	01011111		$\neg Z \rightarrow X$
	1-7	3	S7	00100000	:	$\neg X \wedge Y \wedge \neg Z$
7	3-5	9	Ω4	10100100	=	$(\neg Z \wedge \neg X) \vee (X \wedge \neg Y \wedge Z)$
	6-2	24	A1xL2	01111011		$\neg Y \rightarrow \neg(X \leftrightarrow Z)$
	1-7	5	S8	00001000	:	$\neg X \wedge Y \wedge Z$
8	3-5	9	Ω4	10100100	=	$(\neg Z \wedge \neg X) \vee (X \wedge \neg Y \wedge Z)$
	5-3	48	Q5	01011011		$(X \vee Z) \wedge (\neg X \vee Y \vee \neg Z)$
	1-7	3	S7	00100000	:	$\neg X \wedge Y \wedge \neg Z$
9	3-5	10	Ω5=U1	10100010	=	$\neg Z \wedge (\neg X \vee Y)$
	6-2	23	E2xL2	01111101		$\neg Z \rightarrow \neg(X \leftrightarrow Y)$
	1-7	5	S8	00001000	:	$\neg X \wedge Y \wedge Z$
10	3-5	10	Ω5=U1	10100010	=	$\neg Z \wedge (\neg X \vee Y)$
	5-3	47	Q4=P8	01011101		$\neg Z \rightarrow (X \wedge \neg Y)$
	1-7	3	S7	00100000	:	$\neg X \wedge Y \wedge \neg Z$
11	3-5	11	Ω6	10100001	=	$(\neg Z \wedge \neg X) \vee (X \wedge Y \wedge Z)$
	6-2	22	E1xL2	01111110		$\neg(X \leftrightarrow Y \leftrightarrow Z)$
	1-7	5	S8	00001000	:	$\neg X \wedge Y \wedge Z$
12	3-5	11	Ω6	10100001	=	$(\neg Z \wedge \neg X) \vee (X \wedge Y \wedge Z)$
	5-3	46	Q3	01011110		$(X \vee Z) \wedge (\neg X \vee \neg Y \vee \neg Z)$

The inferences of the second group are:

Table 5.14 **T** OR NOT **V** as necessary condition of **R**

Derivations	Level	#	Symbol	ID		Traditional logical form
	1-7	1	T7	10000000	:	$\neg X \wedge \neg Y \wedge \neg Z$
1	3-5	2	V1=T2	11010000	=	$\neg Y \wedge (\neg X \vee \neg Z)$
	6-2	20	A2xL1	10101111		$\neg Y \rightarrow (X \leftrightarrow Z)$
	1-7	2	T8	01000000	:	$\neg X \wedge \neg Y \wedge Z$
2	3-5	2	V1=T2	11010000	=	$\neg Y \wedge (\neg X \vee \neg Z)$
	6-2	26	**N**	01101111		$\neg Z \rightarrow Y$
	1-7	1	T7	10000000	:	$\neg X \wedge \neg Y \wedge \neg Z$
3	3-5	7	V2=Ω2	10110000	=	$\neg Z \wedge (\neg X \vee \neg Y)$
	6-2	15	A2xM2	11001111		$\neg Z \rightarrow (X \leftrightarrow Y)$
	1-7	2	T8	01000000	:	$\neg X \wedge \neg Y \wedge Z$
4	3-5	7	V2=Ω2	10110000	=	$\neg Z \wedge (\neg X \vee \neg Y)$
	5-3	50	N7=Q7	01001111		$\neg Z \rightarrow (X \wedge Y)$
	1-7	1	T7	10000000	:	$\neg X \wedge \neg Y \wedge \neg Z$
5	3-5	12	V3	10011000	=	$(\neg Z \wedge \neg Y) \vee (\neg X \wedge Y \wedge Z)$
	6-2	10	M1xA2	11100111		$\neg(\neg X \leftrightarrow Y \leftrightarrow Z)$
	1-7	2	T8	01000000	:	$\neg X \wedge \neg Y \wedge Z$
6	3-5	12	V3	10011000	=	$(\neg Z \wedge \neg Y) \vee (\neg X \wedge Y \wedge Z)$
	5-3	45	N6	01100111		$(Y \vee Z) \wedge (X \vee \neg Y \vee \neg Z)$
	1-7	1	T7	10000000	:	$\neg X \wedge \neg Y \wedge \neg Z$
7	3-5	13	V4=O1	10010100	=	$\neg Y \wedge (X \vee \neg Z)$
	6-2	9	**A**	11101011		$X \rightarrow Y$
	1-7	2	T8	01000000	:	$\neg X \wedge \neg Y \wedge Z$
8	3-5	13	V4=O1	10010100	=	$\neg Y \wedge (X \vee \neg Z)$
	5-3	44	N5=A8	01101011		$\neg Y \rightarrow (\neg X \wedge Z)$
	1-7	1	T7	10000000	:	$\neg X \wedge \neg Y \wedge \neg Z$
9	3-5	14	V5=W1	10010010	=	$\neg Z \wedge (X \vee \neg Y)$
	6-2	8	**R**	11101101		$\neg Z \rightarrow \neg X$
	1-7	2	T8	01000000	:	$\neg X \wedge \neg Y \wedge Z$
10	3-5	14	V5=W1	10010010	=	$\neg Z \wedge (X \vee \neg Y)$
	5-3	43	N4=R8	01101101		$\neg Z \rightarrow (\neg X \wedge Y)$
	1-7	1	T7	10000000	:	$\neg X \wedge \neg Y \wedge \neg Z$
11	3-5	15	V6	10010001	=	$(\neg Z \wedge \neg Y) \vee (X \wedge Y \wedge Z)$
	6-2	7	E1xA2	11101110		$X \rightarrow \neg(Y \leftrightarrow Z)$
	1-7	2	T8	01000000	:	$\neg X \wedge \neg Y \wedge Z$
12	3-5	15	V6	10010001	=	$(\neg Z \wedge \neg Y) \vee (X \wedge Y \wedge Z)$
	5-3	42	N3	01101110		$(Y \vee Z) \wedge (\neg X \vee \neg Y \vee \neg Z)$

Table 5.15 **S** OR NOT **U** as necessary condition of **R**

Derivations	Level	#	Symbol	ID		Traditional logical form
	1-7	3	S7	00100000	:	$\neg X \wedge Y \wedge \neg Z$
1	3-5	10	U1= Ω5	10100010	=	$\neg Z \wedge (\neg X \vee Y)$
	6-2	23	E2xL2	01111101		$\neg Z \rightarrow \neg(X \leftrightarrow Y)$
	1-7	5	S8	00001000	:	$\neg X \wedge Y \wedge Z$
2	3-5	10	U1= Ω5	10100010	=	$\neg Z \wedge (\neg X \vee Y)$
	5-3	46	P8=Q4	01011101		$(X \vee Z) \wedge (\neg X \vee \neg Y \vee \neg Z)$
	1-7	3	S7	00100000	:	$\neg X \wedge Y \wedge \neg Z$
3	3-5	25	U2	01100010	=	$(\neg Z \wedge Y) \vee (\neg X \wedge \neg Y \wedge Z)$
	6-2	17	E2xL1	10111101		$\neg(X \leftrightarrow Y \leftrightarrow \neg Z)$
	1-7	5	S8	00001000	:	$\neg X \wedge Y \wedge Z$
4	3-5	25	U2	01100010	=	$(\neg Z \wedge Y) \vee (\neg X \wedge \neg Y \wedge Z)$
	5-3	32	P7	10011101		$(\neg Y \vee Z) \wedge (X \vee Y \vee \neg Z)$
	1-7	3	S7	00100000	:	$\neg X \wedge Y \wedge \neg Z$
5	3-5	39	U3=W3	00110010	=	$\neg Z \wedge (X \vee Y)$
	6-2	8	**R**	11101101		$\neg Z \rightarrow \neg X$
	1-7	5	S8	00001000	:	$\neg X \wedge Y \wedge Z$
6	3-5	39	U3=W3	00110010	=	$\neg Z \wedge (X \vee Y)$
	5-3	18	P6=R6	11001101		$\neg Z \rightarrow (\neg X \wedge \neg Y)$
	1-7	3	S7	00100000	:	$\neg X \wedge Y \wedge \neg Z$
7	3-5	42	U4=S5	00101010	=	$Y \wedge (\neg X \vee \neg Z)$
	6-2	5	E2xM1	11110101		$Y \rightarrow (X \leftrightarrow Z)$
	1-7	5	S8	00001000	:	$\neg X \wedge Y \wedge Z$
8	3-5	42	U4=S5	00101010	=	$Y \wedge (\neg X \vee \neg Z)$
	6-2	12	**P**	11011101		$\neg Z \rightarrow \neg Y$
	1-7	3	S7	00100000	:	$\neg X \wedge Y \wedge \neg Z$
9	3-5	44	U5	00100110	=	$(\neg Z \wedge Y) \vee (X \wedge \neg Y \wedge Z)$
	6-2	3	E2xA1	11111001		$X \rightarrow (Y \leftrightarrow Z)$
	1-7	5	S8	00001000	:	$\neg X \wedge Y \wedge Z$
10	3-5	44	U5	00100110	=	$(\neg Z \wedge Y) \vee (X \wedge \neg Y \wedge Z)$
	5-3	13	P4	11011001		$(\neg Y \vee Z) \wedge (\neg X \vee Y \vee \neg Z)$
	1-7	3	S7	00100000	:	$\neg X \wedge Y \wedge \neg Z$
11	3-5	46	U6=I3	00100011	=	$Y \wedge (X \vee \neg Z)$
	6-2	1	**E**	11111100		$X \rightarrow \neg Y$
	1-7	5	S8	00001000	:	$\neg X \wedge Y \wedge Z$
12	3-5	46	U6=I3	00100011	=	$Y \wedge (X \vee \neg Z)$
	5-3	11	P3=E6	11011100		$Y \rightarrow (\neg X \wedge Z)$

The inferences of the third group are:

Table 5.16 **N** OR NOT **R** as necessary condition of **L**

Derivations	Level	#	Symbol	ID		Traditional logical form
	5-3	42	N3	01101110	:	$(Y \vee Z) \wedge (\neg X \vee \neg Y \vee \neg Z)$
1	7-1	2	R1=E2	11111101	=	$\neg X \vee \neg Y \vee Z$
	5-3	42	N3	01101110		$(Y \vee Z) \wedge (\neg X \vee \neg Y \vee \neg Z)$
	5-3	42	N3	01101110	:	$(Y \vee Z) \wedge (\neg X \vee \neg Y \vee \neg Z)$
2	7-1	5	R2=A2	11101111	=	$\neg X \vee Y \vee Z$
	6-2	22	E1xL2	01111110		$\neg(X \leftrightarrow Y \leftrightarrow Z)$
	5-3	43	N4=R8	01101101	:	$\neg Z \rightarrow (\neg X \wedge Y)$
3	7-1	2	R1=E2	11111101	=	$\neg X \vee \neg Y \vee Z$
	6-2	26	**N**	01101111		$\neg Z \rightarrow Y$
	5-3	43	N4=R8	01101101	:	$\neg Z \rightarrow (\neg X \wedge Y)$
4	7-1	5	R2=A2	11101111	=	$\neg X \vee Y \vee Z$
	6-2	23	E2xL2	01111101		$\neg Z \rightarrow \neg(X \leftrightarrow Y)$
	5-3	44	N5=A8	01101011	:	$\neg Y \rightarrow (\neg X \wedge Z)$
5	7-1	2	R1=E2	11111101	=	$\neg X \vee \neg Y \vee Z$
	5-3	44	N5=A8	01101011		$\neg Y \rightarrow (\neg X \wedge Z)$
	5-3	44	N5=A8	01101011	:	$\neg Y \rightarrow (\neg X \wedge Z)$
6	7-1	5	R2=A2	11101111	=	$\neg X \vee Y \vee Z$
	6-2	24	A1xL2	01111011		$\neg Y \rightarrow \neg(X \leftrightarrow Z)$
	5-3	45	N6	01100111	:	$(Y \vee Z) \wedge (X \vee \neg Y \vee \neg Z)$
7	7-1	2	R1=E2	11111101	=	$\neg X \vee \neg Y \vee Z$
	5-3	45	N6	01100111		$(Y \vee Z) \wedge (X \vee \neg Y \vee \neg Z)$
	5-3	45	N6	01100111	:	$(Y \vee Z) \wedge (X \vee \neg Y \vee \neg Z)$
8	7-1	5	R2=A2	11101111	=	$\neg X \vee Y \vee Z$
	6-2	25	M1xL2	01110111		$\neg X \rightarrow \neg(Y \leftrightarrow Z)$
	5-3	50	N7=Q7	01001111	:	$\neg Z \rightarrow (X \wedge Y)$
9	7-1	2	R1=E2	11111101	=	$\neg X \vee \neg Y \vee Z$
	5-3	50	N7=Q7	01001111		$\neg Z \rightarrow (X \wedge Y)$
	5-3	50	N7=Q7	01001111	:	$\neg Z \rightarrow (X \wedge Y)$
10	7-1	5	R2=A2	11101111	=	$\neg X \vee Y \vee Z$
	6-2	27	**Q**	01011111		$\neg Z \rightarrow X$
	5-3	55	N8=L7	00101111	:	$\neg Y \rightarrow (X \wedge Z)$
11	7-1	2	R1=E2	11111101	=	$\neg X \vee \neg Y \vee Z$
	5-3	55	N8=L7	00101111		$\neg Y \rightarrow (X \wedge Z)$
	5-3	55	N8=L7	00101111	:	$\neg Y \rightarrow (X \wedge Z)$
12	7-1	5	R2=A2	11101111	=	$\neg X \vee Y \vee Z$
	6-2	28	**L**	00111111		$\neg X \rightarrow Y$

Table 5.17 **P** OR NOT **R** as necessary condition of **M**

Derivations	Level	#	Symbol	ID		Traditional logical form
	5-3	11	P3=E6	11011100	:	$Y \to (\neg X \land Z)$
1	7-1	2	R1=E2	11111101	=	$\neg X \lor \neg Y \lor Z$
	6-2	11	E1xM2	11011110		$Y \to \neg(X \leftrightarrow Z)$
	5-3	11	P3=E6	11011100	:	$Y \to (\neg X \land Z)$
2	7-1	5	R2=A2	11101111	=	$\neg X \lor Y \lor Z$
	5-3	11	P3=E6	11011100		$Y \to (\neg X \land Z)$
	5-3	13	P4	11011001	:	$(\neg Y \lor Z) \land (\neg X \lor Y \lor \neg Z)$
3	7-1	2	R1=E2	11111101	=	$\neg X \lor \neg Y \lor Z$
	6-2	13	A1xM2	11011011		$\neg(X \leftrightarrow \neg Y \leftrightarrow Z)$
	5-3	13	P4	11011001	:	$(\neg Y \lor Z) \land (\neg X \lor Y \lor \neg Z)$
4	7-1	5	R2=A2	11101111	=	$\neg X \lor Y \lor Z$
	5-3	13	P4	11011001		$(\neg Y \lor Z) \land (\neg X \lor Y \lor \neg Z)$
	5-3	15	P5=M4	11010101	:	$Y \to (X \land Z)$
5	7-1	2	R1=E2	11111101	=	$\neg X \lor \neg Y \lor Z$
	6-2	14	**M**	11010111		$\neg X \to \neg Y$
	5-3	15	P5=M4	11010101	:	$Y \to (X \land Z)$
6	7-1	5	R2=A2	11101111	=	$\neg X \lor Y \lor Z$
	5-3	15	P5=M4	11010101		$Y \to (X \land Z)$
	5-3	18	P6=R6	11001101	:	$\neg Z \to (\neg X \land \neg Y)$
7	7-1	2	R1=E2	11111101	=	$\neg X \lor \neg Y \lor Z$
	6-2	15	A2xM2	11001111		$\neg Z \to (X \leftrightarrow Y)$
	5-3	18	P6=R6	11001101	:	$\neg Z \to (\neg X \land \neg Y)$
8	7-1	5	R2=A2	11101111	=	$\neg X \lor Y \lor Z$
	6-2	12	**P**	11011101		$\neg Z \to \neg Y$
	5-3	32	P7	10011101	:	$(\neg Y \lor Z) \land (X \lor Y \lor \neg Z)$
9	7-1	2	R1=E2	11111101	=	$\neg X \lor \neg Y \lor Z$
	6-2	21	M2xL1	10011111		$\neg X \to (Y \leftrightarrow Z)$
	5-3	32	P7	10011101	:	$(\neg Y \lor Z) \land (X \lor Y \lor \neg Z)$
10	7-1	5	R2=A2	11101111	=	$\neg X \lor Y \lor Z$
	5-3	32	P7	10011101		$(\neg Y \lor Z) \land (X \lor Y \lor \neg Z)$
	5-3	47	P8=Q4	01011101	:	$\neg Z \to (X \land \neg Y)$
11	7-1	2	R1=E2	11111101	=	$\neg X \lor \neg Y \lor Z$
	6-2	27	**Q**	01011111		$\neg Z \to X$
	5-3	47	P8=Q4	01011101	:	$\neg Z \to (X \land \neg Y)$
12	7-1	5	R2=A2	11101111	=	$\neg X \lor Y \lor Z$
	5-3	47	P8=Q4	01011101		$\neg Z \to (X \land \neg Y)$

Table 5.18 **U** OR NOT **Ω** as necessary condition of **L**

Derivations	Level	#	Symbol	ID		Traditional logical form
	1-7	3	U7=S7	00100000	:	$\neg X \wedge Y \wedge \neg Z$
1	3-5	1	Ω1=T1	11100000	=	$\neg X \wedge (\neg Y \vee \neg Z)$
	6-2	28	**L**	00111111		$\neg X \rightarrow Y$
	1-7	7	U8=I7	00000010	:	$X \wedge Y \wedge \neg Z$
2	3-5	1	Ω1=T1	11100000	=	$\neg X \wedge (\neg Y \vee \neg Z)$
	5-3	56	Q8=L8	00011111		$\neg X \rightarrow (Y \wedge Z)$
	1-7	3	U7=S7	00100000	:	$\neg X \wedge Y \wedge \neg Z$
3	3-5	7	Ω2=V2	10110000	=	$\neg Z \wedge (\neg X \vee \neg Y)$
	6-2	26	**N**	01101111		$\neg Z \rightarrow Y$
	1-7	7	U8=I7	00000010	:	$X \wedge Y \wedge \neg Z$
4	3-5	7	Ω2=V2	10110000	=	$\neg Z \wedge (\neg X \vee \neg Y)$
	5-3	50	Q7=N7	01001111		$\neg Z \rightarrow (X \wedge Y)$
	1-7	3	U7=S7	00100000	:	$\neg X \wedge Y \wedge \neg Z$
5	3-5	8	Ω3=S1	10101000	=	$\neg X \wedge (Y \vee \neg Z)$
	6-2	25	M1xL2	01110111		$\neg X \rightarrow \neg(Y \leftrightarrow Z)$
	1-7	7	U8=I7	00000010	:	$X \wedge Y \wedge \neg Z$
6	3-5	8	Ω3=S1	10101000	=	$\neg X \wedge (Y \vee \neg Z)$
	5-3	49	Q6=M8	01010111		$\neg X \rightarrow (\neg Y \wedge Z)$
	1-7	3	U7=S7	00100000	:	$\neg X \wedge Y \wedge \neg Z$
7	3-5	9	Ω4	10100100	=	$(\neg Z \wedge \neg X) \vee (X \wedge \neg Y \wedge Z)$
	6-2	24	A1xL2	01111011		$\neg Y \rightarrow \neg(X \leftrightarrow Z)$
	1-7	7	U8=I7	00000010	:	$X \wedge Y \wedge \neg Z$
8	3-5	9	Ω4	10100100	=	$(\neg Z \wedge \neg X) \vee (X \wedge \neg Y \wedge Z)$
	5-3	48	Q5	01011011		$(X \vee Z) \wedge (\neg X \vee Y \vee \neg Z)$
	1-7	3	U7=S7	00100000	:	$\neg X \wedge Y \wedge \neg Z$
9	3-5	10	Ω5=U1	10100010	=	$\neg Z \wedge (\neg X \vee Y)$
	6-2	23	E2xL2	01111101		$\neg Z \rightarrow \neg(X \leftrightarrow Y)$
	1-7	7	U8=I7	00000010	:	$X \wedge Y \wedge \neg Z$
10	3-5	10	Ω5=U1	10100010	=	$\neg Z \wedge (\neg X \vee Y)$
	6-2	27	**Q**	01011111		$\neg Z \rightarrow X$
	1-7	3	U7=S7	00100000	:	$\neg X \wedge Y \wedge \neg Z$
11	3-5	11	Ω6	10100001	=	$(\neg Z \wedge \neg X) \vee (X \wedge Y \wedge Z)$
	6-2	22	E1xL2	01111110		$\neg(X \leftrightarrow Y \leftrightarrow Z)$
	1-7	7	U8=I7	00000010	:	$X \wedge Y \wedge \neg Z$
12	3-5	11	Ω6	10100001	=	$(\neg Z \wedge \neg X) \vee (X \wedge Y \wedge Z)$
	5-3	46	Q3	01011110		$(X \vee Z) \wedge (\neg X \vee \neg Y \vee \neg Z)$

Table 5.19 **V** OR NOT **Ω** as necessary condition of **M**

Derivations	Level	#	Symbol	ID		Traditional logical form
	1-7	1	V7=T7	10000000	:	$\neg X \wedge \neg Y \wedge \neg Z$
1	3-5	1	Ω1=T1	11100000	=	$\neg X \wedge (\neg Y \vee \neg Z)$
	6-2	21	M2xL1	10011111		$\neg X \rightarrow (Y \leftrightarrow Z)$
	1-7	4	V8=O7	00010000	:	$X \wedge \neg Y \wedge \neg Z$
2	3-5	1	Ω1=T1	11100000	=	$\neg X \wedge (\neg Y \vee \neg Z)$
	5-3	56	Q8=L8	00011111		$\neg X \rightarrow (Y \wedge Z)$
	1-7	1	V7=T7	10000000	:	$\neg X \wedge \neg Y \wedge \neg Z$
3	3-5	7	Ω2=V2	10110000	=	$\neg Z \wedge (\neg X \vee \neg Y)$
	6-2	15	A2xM2	11001111		$\neg Z \rightarrow (X \leftrightarrow Y)$
	1-7	4	V8=O7	00010000	:	$X \wedge \neg Y \wedge \neg Z$
4	3-5	7	Ω2=V2	10110000	=	$\neg Z \wedge (\neg X \vee \neg Y)$
	6-2	27	**Q**	01011111		$\neg Z \rightarrow X$
	1-7	1	V7=T7	10000000	:	$\neg X \wedge \neg Y \wedge \neg Z$
5	3-5	8	Ω3=S1	10101000	=	$\neg X \wedge (Y \vee \neg Z)$
	6-2	14	**M**	11010111		$\neg X \rightarrow \neg Y$
	1-7	4	V8=O7	00010000	:	$X \wedge \neg Y \wedge \neg Z$
6	3-5	8	Ω3=S1	10101000	=	$\neg X \wedge (Y \vee \neg Z)$
	5-3	49	Q6=M8	01010111		$\neg X \rightarrow (\neg Y \wedge Z)$
	1-7	1	V7=T7	10000000	:	$\neg X \wedge \neg Y \wedge \neg Z$
7	3-5	9	Ω4	10100100	=	$(\neg Z \wedge \neg X) \vee (X \wedge \neg Y \wedge Z)$
	6-2	13	A1xM2	11011011		$\neg(X \leftrightarrow \neg Y \leftrightarrow Z)$
	1-7	4	V8=O7	00010000	:	$X \wedge \neg Y \wedge \neg Z$
8	3-5	9	Ω4	10100100	=	$(\neg Z \wedge \neg X) \vee (X \wedge \neg Y \wedge Z)$
	5-3	48	Q5	01011011		$(X \vee Z) \wedge (\neg X \vee Y \vee \neg Z)$
	1-7	1	V7=T7	10000000	:	$\neg X \wedge \neg Y \wedge \neg Z$
9	3-5	10	Ω5=U1	10100010	=	$\neg Z \wedge (\neg X \vee Y)$
	6-2	12	**P**	11011101		$\neg Z \rightarrow \neg Y$
	1-7	4	V8=O7	00010000	:	$X \wedge \neg Y \wedge \neg Z$
10	3-5	10	Ω5=U1	10100010	=	$\neg Z \wedge (\neg X \vee Y)$
	5-3	47	Q4=P8	01011101		$\neg Z \rightarrow (X \wedge \neg Y)$
	1-7	1	V7=T7	10000000	:	$\neg X \wedge \neg Y \wedge \neg Z$
11	3-5	11	Ω6	10100001	=	$(\neg Z \wedge \neg X) \vee (X \wedge Y \wedge Z)$
	6-2	11	E1xM2	11011110		$Y \rightarrow \neg(X \leftrightarrow Z)$
	1-7	4	V8=O7	00010000	:	$X \wedge \neg Y \wedge \neg Z$
12	3-5	11	Ω6	10100001	=	$(\neg Z \wedge \neg X) \vee (X \wedge Y \wedge Z)$
	5-3	46	Q3	01011110		$(X \vee Z) \wedge (\neg X \vee \neg Y \vee \neg Z)$

5.4 Inferences of the First Mixed Mode through Divisions

The inferences of the first group are:

Table 5.20 **T** OR NOT **K** as necessary condition of **D**

Derivations	Level	#	Symbol	ID		Traditional logical form
	1-7	1	T7	10000000	:	$\neg X \wedge \neg Y \wedge \neg Z$
1	3-5	3	K1=T3	11001000	=	$\neg X \wedge (\neg Y \vee Z)$
	6-2	19	**B**	10110111		$Z \rightarrow X$
	1-7	2	T8	01000000	:	$\neg X \wedge \neg Y \wedge Z$
2	3-5	3	K1=T3	11001000	=	$\neg X \wedge (\neg Y \vee Z)$
	6-2	25	M1xL2	01110111		$\neg X \rightarrow \neg(Y \leftrightarrow Z)$
	1-7	1	T7	10000000	:	$\neg X \wedge \neg Y \wedge \neg Z$
3	3-5	23	K2=S2	01101000	=	$\neg X \wedge (Y \vee Z)$
	5-3	34	B7=M7	10010111		$\neg X \rightarrow (\neg Y \wedge \neg Z)$
	1-7	2	T8	01000000	:	$\neg X \wedge \neg Y \wedge Z$
4	3-5	23	K2=S2	01101000	=	$\neg X \wedge (Y \vee Z)$
	6-2	14	**M**	11010111		$\neg X \rightarrow \neg Y$
	1-7	1	T7	10000000	:	$\neg X \wedge \neg Y \wedge \neg Z$
5	3-5	27	K3	01011000	=	$(\neg X \wedge Z) \vee (X \wedge \neg Y \wedge \neg Z)$
	5-3	30	B6	10100111		$(X \vee \neg Z) \wedge (\neg X \vee Y \vee Z)$
	1-7	2	T8	01000000	:	$\neg X \wedge \neg Y \wedge Z$
6	3-5	27	K3	01011000	=	$(\neg X \wedge Z) \vee (X \wedge \neg Y \wedge \neg Z)$
	6-2	10	M1xA2	11100111		$\neg(\neg X \leftrightarrow Y \leftrightarrow Z)$
	1-7	1	T7	10000000	:	$\neg X \wedge \neg Y \wedge \neg Z$
7	3-5	31	K4=H4	01001100	=	$Z \wedge (\neg X \vee \neg Y)$
	5-3	26	B5=C5	10110011		$Z \rightarrow (X \wedge Y)$
	1-7	2	T8	01000000	:	$\neg X \wedge \neg Y \wedge Z$
8	3-5	31	K4=H4	01001100	=	$Z \wedge (\neg X \vee \neg Y)$
	6-2	6	A1xM1	11110011		$Z \rightarrow (X \leftrightarrow Y)$
	1-7	1	T7	10000000	:	$\neg X \wedge \neg Y \wedge \neg Z$
9	3-5	32	K5	01001010	=	$(\neg X \wedge Z) \vee (X \wedge Y \wedge \neg Z)$
	5-3	25	B4	10110101		$(X \vee \neg Z) \wedge (\neg X \vee \neg Y \vee Z)$
	1-7	2	T8	01000000	:	$\neg X \wedge \neg Y \wedge Z$
10	3-5	32	K5	01001010	=	$(\neg X \wedge Z) \vee (X \wedge Y \wedge \neg Z)$
	6-2	5	E2xM1	11110101		$Y \rightarrow (X \leftrightarrow Z)$
	1-7	1	T7	10000000	:	$\neg X \wedge \neg Y \wedge \neg Z$
11	3-5	33	K6=G2	01001001	=	$Z \wedge (\neg X \vee Y)$
	5-3	24	B3=D7	10110110		$Z \rightarrow (X \wedge \neg Y)$
	1-7	2	T8	01000000	:	$\neg X \wedge \neg Y \wedge Z$
12	3-5	33	K6=G2	01001001	=	$Z \wedge (\neg X \vee Y)$
	6-2	4	**D**	11110110		$Z \rightarrow \neg Y$

Table 5.21 **S** OR NOT **K** as necessary condition of **C**

Derivations	Level	#	Symbol	ID		Traditional logical form
	1-7	3	S7	00100000	:	$\neg X \wedge Y \wedge \neg Z$
1	3-5	3	K1=T3	11001000	=	$\neg X \wedge (\neg Y \vee Z)$
	5-3	54	B8=L6	00110111		$\neg X \rightarrow (Y \wedge \neg Z)$
	1-7	5	S8	00001000	:	$\neg X \wedge Y \wedge Z$
2	3-5	3	K1=T3	11001000	=	$\neg X \wedge (\neg Y \vee Z)$
	6-2	28	**L**	00111111		$\neg X \rightarrow Y$
	1-7	3	S7	00100000	:	$\neg X \wedge Y \wedge \neg Z$
3	3-5	23	K2=S2	01101000	=	$\neg X \wedge (Y \vee Z)$
	6-2	19	**B**	10110111		$Z \rightarrow X$
	1-7	5	S8	00001000	:	$\neg X \wedge Y \wedge Z$
4	3-5	23	K2=S2	01101000	=	$\neg X \wedge (Y \vee Z)$
	6-2	21	M2xL1	10011111		$\neg X \rightarrow (Y \leftrightarrow Z)$
	1-7	3	S7	00100000	:	$\neg X \wedge Y \wedge \neg Z$
5	3-5	27	K3	01011000	=	$(\neg X \wedge Z) \vee (X \wedge \neg Y \wedge \neg Z)$
	5-3	30	B6	10100111		$(X \vee \neg Z) \wedge (\neg X \vee Y \vee Z)$
	1-7	5	S8	00001000	:	$\neg X \wedge Y \wedge Z$
6	3-5	27	K3	01011000	=	$(\neg X \wedge Z) \vee (X \wedge \neg Y \wedge \neg Z)$
	6-2	20	A2xL1	10101111		$\neg Y \rightarrow (X \leftrightarrow Z)$
	1-7	3	S7	00100000	:	$\neg X \wedge Y \wedge \neg Z$
7	3-5	31	K4=H4	01001100	=	$Z \wedge (\neg X \vee \neg Y)$
	5-3	26	B5=C5	10110011		$Z \rightarrow (X \wedge Y)$
	1-7	5	S8	00001000	:	$\neg X \wedge Y \wedge Z$
8	3-5	31	K4=H4	01001100	=	$Z \wedge (\neg X \vee \neg Y)$
	6-2	18	**C**	10111011		$Z \rightarrow Y$
	1-7	3	S7	00100000	:	$\neg X \wedge Y \wedge \neg Z$
9	3-5	32	K5	01001010	=	$(\neg X \wedge Z) \vee (X \wedge Y \wedge \neg Z)$
	5-3	25	B4	10110101		$(X \vee \neg Z) \wedge (\neg X \vee \neg Y \vee Z)$
	1-7	5	S8	00001000	:	$\neg X \wedge Y \wedge Z$
10	3-5	32	K5	01001010	=	$(\neg X \wedge Z) \vee (X \wedge Y \wedge \neg Z)$
	6-2	17	E2xL1	10111101		$\neg(X \leftrightarrow Y \leftrightarrow \neg Z)$
	1-7	3	S7	00100000	:	$\neg X \wedge Y \wedge \neg Z$
11	3-5	33	K6=G2	01001001	=	$Z \wedge (\neg X \vee Y)$
	5-3	24	B3=D7	10110110		$Z \rightarrow (X \wedge \neg Y)$
	1-7	5	S8	00001000	:	$\neg X \wedge Y \wedge Z$
12	3-5	33	K6=G2	01001001	=	$Z \wedge (\neg X \vee Y)$
	6-2	16	E1xL1	10111110		$Z \rightarrow \neg(X \leftrightarrow Y)$

The inferences of the second group are:

Table 5.22 **T** OR NOT **H** as necessary condition of **F**

Derivations	Level	#	Symbol	ID		Traditional logical form
	1-7	1	T7	10000000	:	$\neg X \wedge \neg Y \wedge \neg Z$
1	3-5	4	H1=T4	11000100	=	$\neg Y \wedge (\neg X \vee Z)$
	6-2	18	**C**	10111011		$Z \rightarrow Y$
	1-7	2	T8	01000000	:	$\neg X \wedge \neg Y \wedge Z$
2	3-5	4	H1=T4	11000100	=	$\neg Y \wedge (\neg X \vee Z)$
	6-2	24	A1xL2	01111011		$\neg Y \rightarrow \neg(X \leftrightarrow Z)$
	1-7	1	T7	10000000	:	$\neg X \wedge \neg Y \wedge \neg Z$
3	3-5	24	H2	01100100	=	$(\neg Y \wedge Z) \vee (\neg X \wedge Y \wedge \neg Z)$
	5-3	33	C7	10011011		$(Y \vee \neg Z) \wedge (X \vee \neg Y \vee Z)$
	1-7	2	T8	01000000	:	$\neg X \wedge \neg Y \wedge Z$
4	3-5	24	H2	01100100	=	$(\neg Y \wedge Z) \vee (\neg X \wedge Y \wedge \neg Z)$
	6-2	13	A1xM2	11011011		$\neg(X \leftrightarrow \neg Y \leftrightarrow Z)$
	1-7	1	T7	10000000	:	$\neg X \wedge \neg Y \wedge \neg Z$
5	3-5	28	H3=O2	01010100	=	$\neg Y \wedge (X \vee Z)$
	5-3	29	C6=A7	10101011		$\neg Y \rightarrow (\neg X \wedge \neg Z)$
	1-7	2	T8	01000000	:	$\neg X \wedge \neg Y \wedge Z$
6	3-5	28	H3=O2	01010100	=	$\neg Y \wedge (X \vee Z)$
	6-2	9	**A**	11101011		$X \rightarrow Y$
	1-7	1	T7	10000000	:	$\neg X \wedge \neg Y \wedge \neg Z$
7	3-5	31	H4=K4	01001100	=	$Z \wedge (\neg X \vee \neg Y)$
	5-3	26	C5=B5	10110011		$Z \rightarrow (X \wedge Y)$
	1-7	2	T8	01000000	:	$\neg X \wedge \neg Y \wedge Z$
8	3-5	31	H4=K4	01001100	=	$Z \wedge (\neg X \vee \neg Y)$
	6-2	6	A1xM1	11110011		$Z \rightarrow (X \leftrightarrow Y)$
	1-7	1	T7	10000000	:	$\neg X \wedge \neg Y \wedge \neg Z$
9	3-5	34	H5	01000110	=	$(\neg Y \wedge Z) \vee (X \wedge Y \wedge \neg Z)$
	5-3	23	C4	10111001		$(Y \vee \neg Z) \wedge (\neg X \vee \neg Y \vee Z)$
	1-7	2	T8	01000000	:	$\neg X \wedge \neg Y \wedge Z$
10	3-5	34	H5	01000110	=	$(\neg Y \wedge Z) \vee (X \wedge Y \wedge \neg Z)$
	6-2	3	E2xA1	11111001		$X \rightarrow (Y \leftrightarrow Z)$
	1-7	1	T7	10000000	:	$\neg X \wedge \neg Y \wedge \neg Z$
11	3-5	35	H6=J2	01000101	=	$Z \wedge (X \vee \neg Y)$
	5-3	22	C3=F7	10111010		$Z \rightarrow (\neg X \wedge Y)$
	1-7	2	T8	01000000	:	$\neg X \wedge \neg Y \wedge Z$
12	3-5	35	H6=J2	01000101	=	$Z \wedge (X \vee \neg Y)$
	6-2	2	**F**	11111010		$Z \rightarrow \neg X$

Table 5.23 **S** OR NOT **G** as necessary condition of **F**

Derivations	Level	#	Symbol	ID		Traditional logical form
	1-7	3	S7	00100000	:	$\neg X \wedge Y \wedge \neg Z$
1	3-5	18	G1	10001001	=	$(Y \wedge Z) \vee (\neg X \wedge \neg Y \wedge \neg Z)$
	5-3	39	D8	01110110		$(\neg Y \vee \neg Z) \wedge (X \vee Y \vee Z)$
	1-7	5	S8	00001000	:	$\neg X \wedge Y \wedge Z$
2	3-5	18	G1	10001001	=	$(Y \wedge Z) \vee (\neg X \wedge \neg Y \wedge \neg Z)$
	6-2	22	E1xL2	01111110		$\neg(X \leftrightarrow Y \leftrightarrow Z)$
	1-7	3	S7	00100000	:	$\neg X \wedge Y \wedge \neg Z$
3	3-5	33	G2=K6	01001001	=	$Z \wedge (\neg X \vee Y)$
	5-3	24	D7=B3	10110110		$Z \rightarrow (X \wedge \neg Y)$
	1-7	5	S8	00001000	:	$\neg X \wedge Y \wedge Z$
4	3-5	33	G2=K6	01001001	=	$Z \wedge (\neg X \vee Y)$
	6-2	16	E1xL1	10111110		$Z \rightarrow \neg(X \leftrightarrow Y)$
	1-7	3	S7	00100000	:	$\neg X \wedge Y \wedge \neg Z$
5	3-5	43	G3=S6	00101001	=	$Y \wedge (\neg X \vee Z)$
	6-2	4	**D**	11110110		$Z \rightarrow \neg Y$
	1-7	5	S8	00001000	:	$\neg X \wedge Y \wedge Z$
6	3-5	43	G3=S6	00101001	=	$Y \wedge (\neg X \vee Z)$
	6-2	11	E1xM2	11011110		$Y \rightarrow \neg(X \leftrightarrow Z)$
	1-7	3	S7	00100000	:	$\neg X \wedge Y \wedge \neg Z$
7	3-5	49	G4	00011001	=	$(Y \wedge Z) \vee (X \wedge \neg Y \wedge \neg Z)$
	5-3	8	D5	11100110		$(\neg Y \vee \neg Z) \wedge (\neg X \vee Y \vee Z)$
	1-7	5	S8	00001000	:	$\neg X \wedge Y \wedge Z$
8	3-5	49	G4	00011001	=	$(Y \wedge Z) \vee (X \wedge \neg Y \wedge \neg Z)$
	6-2	7	E1xA2	11101110		$X \rightarrow \neg(Y \leftrightarrow Z)$
	1-7	3	S7	00100000	:	$\neg X \wedge Y \wedge \neg Z$
9	3-5	54	G5=J5	00001101	=	$Z \wedge (X \vee Y)$
	5-3	3	D4=F4	11110010		$Z \rightarrow (\neg X \wedge \neg Y)$
	1-7	5	S8	00001000	:	$\neg X \wedge Y \wedge Z$
10	3-5	54	G5=J5	00001101	=	$Z \wedge (X \vee Y)$
	6-2	2	**F**	11111010		$Z \rightarrow \neg X$
	1-7	3	S7	00100000	:	$\neg X \wedge Y \wedge \neg Z$
11	3-5	55	G6=I5	00001011	=	$Y \wedge (X \vee Z)$
	5-3	2	D3=E4	11110100		$Y \rightarrow (\neg X \wedge \neg Z)$
	1-7	5	S8	00001000	:	$\neg X \wedge Y \wedge Z$
12	3-5	55	G6=I5	00001011	=	$Y \wedge (X \vee Z)$
	6-2	1	**E**	11111100		$X \rightarrow \neg Y$

The inferences of the third group are:

Table 5.24 **C** OR NOT **F** as necessary condition of **L**

Derivations	Level	#	Symbol	ID		Traditional logical form
	5-3	22	C3=F7	10111010	:	$Z \to (\neg X \wedge Y)$
1	7-1	1	F1=E1	11111110	=	$\neg X \vee \neg Y \vee \neg Z$
	6-2	18	**C**	10111011		$Z \to Y$
	5-3	22	C3=F7	10111010	:	$Z \to (\neg X \wedge Y)$
2	7-1	3	F2=A1	11111011	=	$\neg X \vee Y \vee \neg Z$
	6-2	16	E1xL1	10111110		$Z \to \neg(X \leftrightarrow Y)$
	5-3	23	C4	10111001	:	$(Y \vee \neg Z) \wedge (\neg X \vee \neg Y \vee Z)$
3	7-1	1	F1=E1	11111110	=	$\neg X \vee \neg Y \vee \neg Z$
	5-3	23	C4	10111001		$(Y \vee \neg Z) \wedge (\neg X \vee \neg Y \vee Z)$
	5-3	23	C4	10111001	:	$(Y \vee \neg Z) \wedge (\neg X \vee \neg Y \vee Z)$
4	7-1	3	F2=A1	11111011	=	$\neg X \vee Y \vee \neg Z$
	6-2	17	E2xL1	10111101		$\neg(X \leftrightarrow Y \leftrightarrow \neg Z)$
	5-3	26	C5=B5	10110011	:	$Z \to (X \wedge Y)$
5	7-1	1	F1=E1	11111110	=	$\neg X \vee \neg Y \vee \neg Z$
	5-3	26	C5=B5	10110011		$Z \to (X \wedge Y)$
	5-3	26	C5=B5	10110011	:	$Z \to (X \wedge Y)$
6	7-1	3	F2=A1	11111011	=	$\neg X \vee Y \vee \neg Z$
	6-2	19	**B**	10110111		$Z \to X$
	5-3	29	C6=A7	10101011	:	$\neg Y \to (\neg X \wedge \neg Z)$
7	7-1	1	F1=E1	11111110	=	$\neg X \vee \neg Y \vee \neg Z$
	5-3	29	C6=A7	10101011		$\neg Y \to (\neg X \wedge \neg Z)$
	5-3	29	C6=A7	10101011	:	$\neg Y \to (\neg X \wedge \neg Z)$
8	7-1	3	F2=A1	11111011	=	$\neg X \vee Y \vee \neg Z$
	6-2	20	A2xL1	10101111		$\neg Y \to (X \leftrightarrow Z)$
	5-3	33	C7	10011011	:	$(Y \vee \neg Z) \wedge (X \vee \neg Y \vee Z)$
9	7-1	1	F1=E1	11111110	=	$\neg X \vee \neg Y \vee \neg Z$
	5-3	33	C7	10011011		$(Y \vee \neg Z) \wedge (X \vee \neg Y \vee Z)$
	5-3	33	C7	10011011	:	$(Y \vee \neg Z) \wedge (X \vee \neg Y \vee Z)$
10	7-1	3	F2=A1	11111011	=	$\neg X \vee Y \vee \neg Z$
	6-2	21	M2xL1	10011111		$\neg X \to (Y \leftrightarrow Z)$
	5-3	53	C8=L5	00111011	:	$\neg Y \to (X \wedge \neg Z)$
11	7-1	1	F1=E1	11111110	=	$\neg X \vee \neg Y \vee \neg Z$
	5-3	53	C8=L5	00111011		$\neg Y \to (X \wedge \neg Z)$
	5-3	53	C8=L5	00111011	:	$\neg Y \to (X \wedge \neg Z)$
12	7-1	3	F2=A1	11111011	=	$\neg X \vee Y \vee \neg Z$
	6-2	28	**L**	00111111		$\neg X \to Y$

Table 5.25 **D** OR NOT **F** as necessary condition of **M**

Derivations	Level	#	Symbol	ID		Traditional logical form
	5-3	2	D3=E4	11110100	:	$Y \to (\neg X \wedge \neg Z)$
1	7-1	1	F1=E1	11111110	=	$\neg X \vee \neg Y \vee \neg Z$
	6-2	5	E2xM1	11110101		$Y \to (X \leftrightarrow Z)$
	5-3	2	D3=E4	11110100	:	$Y \to (\neg X \wedge \neg Z)$
2	7-1	3	F2=A1	11111011	=	$\neg X \vee Y \vee \neg Z$
	5-3	2	D3=E4	11110100		$Y \to (\neg X \wedge \neg Z)$
	5-3	3	D4=F4	11110010	:	$Z \to (\neg X \wedge \neg Y)$
3	7-1	1	F1=E1	11111110	=	$\neg X \vee \neg Y \vee \neg Z$
	6-2	6	A1xM1	11110011		$Z \to (X \leftrightarrow Y)$
	5-3	3	D4=F4	11110010	:	$Z \to (\neg X \wedge \neg Y)$
4	7-1	3	F2=A1	11111011	=	$\neg X \vee Y \vee \neg Z$
	6-2	4	**D**	11110110		$Z \to \neg Y$
	5-3	8	D5	11100110	:	$(\neg Y \vee \neg Z) \wedge (\neg X \vee Y \vee Z)$
5	7-1	1	F1=E1	11111110	=	$\neg X \vee \neg Y \vee \neg Z$
	6-2	10	M1xA2	11100111		$\neg(\neg X \leftrightarrow Y \leftrightarrow Z)$
	5-3	8	D5	11100110	:	$(\neg Y \vee \neg Z) \wedge (\neg X \vee Y \vee Z)$
6	7-1	3	F2=A1	11111011	=	$\neg X \vee Y \vee \neg Z$
	5-3	8	D5	11100110		$(\neg Y \vee \neg Z) \wedge (\neg X \vee Y \vee Z)$
	5-3	14	D6=M3	11010110	:	$Y \to (X \wedge \neg Z)$
7	7-1	1	F1=E1	11111110	=	$\neg X \vee \neg Y \vee \neg Z$
	6-2	14	**M**	11010111		$\neg X \to \neg Y$
	5-3	14	D6=M3	11010110	:	$Y \to (X \wedge \neg Z)$
8	7-1	3	F2=A1	11111011	=	$\neg X \vee Y \vee \neg Z$
	5-3	14	D6=M3	11010110		$Y \to (X \wedge \neg Z)$
	5-3	24	D7=B3	10110110	:	$Z \to (X \wedge \neg Y)$
9	7-1	1	F1=E1	11111110	=	$\neg X \vee \neg Y \vee \neg Z$
	6-2	19	**B**	10110111		$Z \to X$
	5-3	24	D7=B3	10110110	:	$Z \to (X \wedge \neg Y)$
10	7-1	3	F2=A1	11111011	=	$\neg X \vee Y \vee \neg Z$
	5-3	24	D7=B3	10110110		$Z \to (X \wedge \neg Y)$
	5-3	39	D8	01110110	:	$(\neg Y \vee \neg Z) \wedge (X \vee Y \vee Z)$
11	7-1	1	F1=E1	11111110	=	$\neg X \vee \neg Y \vee \neg Z$
	6-2	25	M1xL2	01110111		$\neg X \to \neg(Y \leftrightarrow Z)$
	5-3	39	D8	01110110	:	$(\neg Y \vee \neg Z) \wedge (X \vee Y \vee Z)$
12	7-1	3	F2=A1	11111011	=	$\neg X \vee Y \vee \neg Z$
	5-3	39	D8	01110110		$(\neg Y \vee \neg Z) \wedge (X \vee Y \vee Z)$

Table 5.26 **G** OR NOT **K** as necessary condition of **L**

Derivations	Level	#	Symbol	ID		Traditional logical form
	1-7	5	G7=S8	00001000	:	$\neg X \wedge Y \wedge Z$
1	3-5	3	K1=T3	11001000	=	$\neg X \wedge (\neg Y \vee Z)$
	6-2	28	**L**	00111111		$\neg X \rightarrow Y$
	1-7	8	G8=I8	00000001	:	$X \wedge Y \wedge Z$
2	3-5	3	K1=T3	11001000	=	$\neg X \wedge (\neg Y \vee Z)$
	5-3	54	B8=L6	00110111		$\neg X \rightarrow (Y \wedge \neg Z)$
	1-7	5	G7=S8	00001000	:	$\neg X \wedge Y \wedge Z$
3	3-5	23	K2=S2	01101000	=	$\neg X \wedge (Y \vee Z)$
	6-2	21	M2xL1	10011111		$\neg X \rightarrow (Y \leftrightarrow Z)$
	1-7	8	G8=I8	00000001	:	$X \wedge Y \wedge Z$
4	3-5	23	K2=S2	01101000	=	$\neg X \wedge (Y \vee Z)$
	5-3	34	B7=M7	10010111		$\neg X \rightarrow (\neg Y \wedge \neg Z)$
	1-7	5	G7=S8	00001000	:	$\neg X \wedge Y \wedge Z$
5	3-5	27	K3	01011000	=	$(\neg X \wedge Z) \vee (X \wedge \neg Y \wedge \neg Z)$
	6-2	20	A2xL1	10101111		$\neg Y \rightarrow (X \leftrightarrow Z)$
	1-7	8	G8=I8	00000001	:	$X \wedge Y \wedge Z$
6	3-5	27	K3	01011000	=	$(\neg X \wedge Z) \vee (X \wedge \neg Y \wedge \neg Z)$
	5-3	30	B6	10100111		$(X \vee \neg Z) \wedge (\neg X \vee Y \vee Z)$
	1-7	5	G7=S8	00001000	:	$\neg X \wedge Y \wedge Z$
7	3-5	31	K4=H4	01001100	=	$Z \wedge (\neg X \vee \neg Y)$
	6-2	18	**C**	10111011		$Z \rightarrow Y$
	1-7	8	G8=I8	00000001	:	$X \wedge Y \wedge Z$
8	3-5	31	K4=H4	01001100	=	$Z \wedge (\neg X \vee \neg Y)$
	5-3	26	B5=C5	10110011		$Z \rightarrow (X \wedge Y)$
	1-7	5	G7=S8	00001000	:	$\neg X \wedge Y \wedge Z$
9	3-5	32	K5	01001010	=	$(\neg X \wedge Z) \vee (X \wedge Y \wedge \neg Z)$
	6-2	17	E2xL1	10111101		$\neg(X \leftrightarrow Y \leftrightarrow \neg Z)$
	1-7	8	G8=I8	00000001	:	$X \wedge Y \wedge Z$
10	3-5	32	K5	01001010	=	$(\neg X \wedge Z) \vee (X \wedge Y \wedge \neg Z)$
	5-3	25	B4	10110101		$(X \vee \neg Z) \wedge (\neg X \vee \neg Y \vee Z)$
	1-7	5	G7=S8	00001000	:	$\neg X \wedge Y \wedge Z$
11	3-5	33	K6=G2	01001001	=	$Z \wedge (\neg X \vee Y)$
	6-2	16	E1xL1	10111110		$Z \rightarrow \neg(X \leftrightarrow Y)$
	1-7	8	G8=I8	00000001	:	$X \wedge Y \wedge Z$
12	3-5	33	K6=G2	01001001	=	$Z \wedge (\neg X \vee Y)$
	6-2	19	**B**	10110111		$Z \rightarrow X$

Table 5.27 **H** OR NOT **K** as necessary condition of **M**

Derivations	Level	#	Symbol	ID		Traditional logical form
	1-7	2	H7=T8	01000000	:	$\neg X \wedge \neg Y \wedge Z$
1	3-5	3	K1=T3	11001000	=	$\neg X \wedge (\neg Y \vee Z)$
	6-2	25	M1xL2	01110111		$\neg X \rightarrow \neg(Y \leftrightarrow Z)$
	1-7	6	H8=O8	00000100	:	$X \wedge \neg Y \wedge Z$
2	3-5	3	K1=T3	11001000	=	$\neg X \wedge (\neg Y \vee Z)$
	5-3	54	B8=L6	00110111		$\neg X \rightarrow (Y \wedge \neg Z)$
	1-7	2	H7=T8	01000000	:	$\neg X \wedge \neg Y \wedge Z$
3	3-5	23	K2=S2	01101000	=	$\neg X \wedge (Y \vee Z)$
	6-2	14	**M**	11010111		$\neg X \rightarrow \neg Y$
	1-7	6	H8=O8	00000100	:	$X \wedge \neg Y \wedge Z$
4	3-5	23	K2=S2	01101000	=	$\neg X \wedge (Y \vee Z)$
	5-3	34	B7=M7	10010111		$\neg X \rightarrow (\neg Y \wedge \neg Z)$
	1-7	2	H7=T8	01000000	:	$\neg X \wedge \neg Y \wedge Z$
5	3-5	27	K3	01011000	=	$(\neg X \wedge Z) \vee (X \wedge \neg Y \wedge \neg Z)$
	6-2	10	M1xA2	11100111		$\neg(\neg X \leftrightarrow Y \leftrightarrow Z)$
	1-7	6	H8=O8	00000100	:	$X \wedge \neg Y \wedge Z$
6	3-5	27	K3	01011000	=	$(\neg X \wedge Z) \vee (X \wedge \neg Y \wedge \neg Z)$
	5-3	30	B6	10100111		$(X \vee \neg Z) \wedge (\neg X \vee Y \vee Z)$
	1-7	2	H7=T8	01000000	:	$\neg X \wedge \neg Y \wedge Z$
7	3-5	31	K4=H4	01001100	=	$Z \wedge (\neg X \vee \neg Y)$
	6-2	6	A1xM1	11110011		$Z \rightarrow (X \leftrightarrow Y)$
	1-7	6	H8=O8	00000100	:	$X \wedge \neg Y \wedge Z$
8	3-5	31	K4=H4	01001100	=	$Z \wedge (\neg X \vee \neg Y)$
	6-2	19	**B**	10110111		$Z \rightarrow X$
	1-7	2	H7=T8	01000000	:	$\neg X \wedge \neg Y \wedge Z$
9	3-5	32	K5	01001010	=	$(\neg X \wedge Z) \vee (X \wedge Y \wedge \neg Z)$
	6-2	5	E2xM1	11110101		$Y \rightarrow (X \leftrightarrow Z)$
	1-7	6	H8=O8	00000100	:	$X \wedge \neg Y \wedge Z$
10	3-5	32	K5	01001010	=	$(\neg X \wedge Z) \vee (X \wedge Y \wedge \neg Z)$
	5-3	25	B4	10110101		$(X \vee \neg Z) \wedge (\neg X \vee \neg Y \vee Z)$
	1-7	2	H7=T8	01000000	:	$\neg X \wedge \neg Y \wedge Z$
11	3-5	33	K6=G2	01001001	=	$Z \wedge (\neg X \vee Y)$
	6-2	4	**D**	11110110		$Z \rightarrow \neg Y$
	1-7	6	H8=O8	00000100	:	$X \wedge \neg Y \wedge Z$
12	3-5	33	K6=G2	01001001	=	$Z \wedge (\neg X \vee Y)$
	5-3	24	B3=D7	10110110		$Z \rightarrow (X \wedge \neg Y)$

5.5 Inferences of the Second Mixed Mode through Divisions

Let us consider the derivation of the first group:

Table 5.28 **O** OR NOT **W** as necessary condition of **P**

Derivations	Level	#	Symbol	ID		Traditional logical form
	1-7	4	O7	00010000	:	$X \wedge \neg Y \wedge \neg Z$
1	3-5	14	W1=V5	10010010	=	$\neg Z \wedge (X \vee \neg Y)$
	6-2	23	E2xL2	01111101		$\neg Z \rightarrow \neg(X \leftrightarrow Y)$
	1-7	6	O8	00000100	:	$X \wedge \neg Y \wedge Z$
2	3-5	14	W1=V5	10010010	=	$\neg Z \wedge (X \vee \neg Y)$
	5-3	43	R8=N4	01101101		$\neg Z \rightarrow (\neg X \wedge Y)$
	1-7	4	O7	00010000	:	$X \wedge \neg Y \wedge \neg Z$
3	3-5	29	W2	01010010	=	$(\neg Z \wedge X) \vee (\neg X \wedge \neg Y \wedge Z)$
	6-2	17	E2xL1	10111101		$\neg(X \leftrightarrow Y \leftrightarrow \neg Z)$
	1-7	6	O8	00000100	:	$X \wedge \neg Y \wedge Z$
4	3-5	29	W2	01010010	=	$(\neg Z \wedge X) \vee (\neg X \wedge \neg Y \wedge Z)$
	5-3	28	R7	10101101		$(\neg X \vee Z) \wedge (X \vee Y \vee \neg Z)$
	1-7	4	O7	00010000	:	$X \wedge \neg Y \wedge \neg Z$
5	3-5	39	W3=U3	00110010	=	$\neg Z \wedge (X \vee Y)$
	6-2	12	**P**	11011101		$\neg Z \rightarrow \neg Y$
	1-7	6	O8	00000100	:	$X \wedge \neg Y \wedge Z$
6	3-5	39	W3=U3	00110010	=	$\neg Z \wedge (X \vee Y)$
	5-3	18	R6=P6	11001101		$\neg Z \rightarrow (\neg X \wedge \neg Y)$
	1-7	4	O7	00010000	:	$X \wedge \neg Y \wedge \neg Z$
7	3-5	48	W4	00011010	=	$(\neg Z \wedge X) \vee (\neg X \wedge Y \wedge Z)$
	6-2	5	E2xM1	11110101		$Y \rightarrow (X \leftrightarrow Z)$
	1-7	6	O8	00000100	:	$X \wedge \neg Y \wedge Z$
8	3-5	48	W4	00011010	=	$(\neg Z \wedge X) \vee (\neg X \wedge Y \wedge Z)$
	5-3	9	R5	11100101		$(\neg X \vee Z) \wedge (X \vee \neg Y \vee \neg Z)$
	1-7	4	O7	00010000	:	$X \wedge \neg Y \wedge \neg Z$
9	3-5	50	W5=O5	00010110	=	$X \wedge (\neg Y \vee \neg Z)$
	6-2	3	E2xA1	11111001		$X \rightarrow (Y \leftrightarrow Z)$
	1-7	6	O8	00000100	:	$X \wedge \neg Y \wedge Z$
10	3-5	50	W5=O5	00010110	=	$X \wedge (\neg Y \vee \neg Z)$
	6-2	8	**R**	11101101		$\neg Z \rightarrow \neg X$
	1-7	4	O7	00010000	:	$X \wedge \neg Y \wedge \neg Z$
11	3-5	52	W6=I4	00010011	=	$X \wedge (Y \vee \neg Z)$
	6-2	1	**E**	11111100		$X \rightarrow \neg Y$
	1-7	6	O8	00000100	:	$X \wedge \neg Y \wedge Z$
12	3-5	52	W6=I4	00010011	=	$X \wedge (Y \vee \neg Z)$
	5-3	5	R3=E5	11101100		$X \rightarrow (\neg Y \wedge Z)$

Table 5.29 **I** OR NOT **W** as necessary condition of **N**

Derivations	Level	#	Symbol	ID		Traditional logical form
	1-7	7	I7	00000010	:	$X \wedge Y \wedge \neg Z$
1	3-5	14	W1=V5	10010010	=	$\neg Z \wedge (X \vee \neg Y)$
	6-2	26	**N**	01101111		$\neg Z \rightarrow Y$
	1-7	8	I8	00000001	:	$X \wedge Y \wedge Z$
2	3-5	14	W1=V5	10010010	=	$\neg Z \wedge (X \vee \neg Y)$
	5-3	43	R8=N4	01101101		$\neg Z \rightarrow (\neg X \wedge Y)$
	1-7	7	I7	00000010	:	$X \wedge Y \wedge \neg Z$
3	3-5	29	W2	01010010	=	$(\neg Z \wedge X) \vee (\neg X \wedge \neg Y \wedge Z)$
	6-2	20	A2xL1	10101111		$\neg Y \rightarrow (X \leftrightarrow Z)$
	1-7	8	I8	00000001	:	$X \wedge Y \wedge Z$
4	3-5	29	W2	01010010	=	$(\neg Z \wedge X) \vee (\neg X \wedge \neg Y \wedge Z)$
	5-3	28	R7	10101101		$(\neg X \vee Z) \wedge (X \vee Y \vee \neg Z)$
	1-7	7	I7	00000010	:	$X \wedge Y \wedge \neg Z$
5	3-5	39	W3=U3	00110010	=	$\neg Z \wedge (X \vee Y)$
	6-2	15	A2xM2	11001111		$\neg Z \rightarrow (X \leftrightarrow Y)$
	1-7	8	I8	00000001	:	$X \wedge Y \wedge Z$
6	3-5	39	W3=U3	00110010	=	$\neg Z \wedge (X \vee Y)$
	5-3	18	R6=P6	11001101		$\neg Z \rightarrow (\neg X \wedge \neg Y)$
	1-7	7	I7	00000010	:	$X \wedge Y \wedge \neg Z$
7	3-5	48	W4	00011010	=	$(\neg Z \wedge X) \vee (\neg X \wedge Y \wedge Z)$
	6-2	10	M1xA2	11100111		$\neg(\neg X \leftrightarrow Y \leftrightarrow Z)$
	1-7	8	I8	00000001	:	$X \wedge Y \wedge Z$
8	3-5	48	W4	00011010	=	$(\neg Z \wedge X) \vee (\neg X \wedge Y \wedge Z)$
	5-3	9	R5	11100101		$(\neg X \vee Z) \wedge (X \vee \neg Y \vee \neg Z)$
	1-7	7	I7	00000010	:	$X \wedge Y \wedge \neg Z$
9	3-5	50	W5=O5	00010110	=	$X \wedge (\neg Y \vee \neg Z)$
	6-2	9	**A**	11101011		$X \rightarrow Y$
	1-7	8	I8	00000001	:	$X \wedge Y \wedge Z$
10	3-5	50	W5=O5	00010110	=	$X \wedge (\neg Y \vee \neg Z)$
	5-3	7	R4=A4	11101001		$X \rightarrow (Y \wedge Z)$
	1-7	7	I7	00000010	:	$X \wedge Y \wedge \neg Z$
11	3-5	52	W6=I4	00010011	=	$X \wedge (Y \vee \neg Z)$
	6-2	7	E1xA2	11101110		$X \rightarrow \neg(Y \leftrightarrow Z)$
	1-7	8	I8	00000001	:	$X \wedge Y \wedge Z$
12	3-5	52	W6=I4	00010011	=	$X \wedge (Y \vee \neg Z)$
	6-2	8	**R**	11101101		$\neg Z \rightarrow \neg X$

The inferences of the second group are:

Table 5.30 **O** OR NOT **V** as necessary condition of **Q**

Derivations	Level	#	Symbol	ID		Traditional logical form
	1-7	4	O7	00010000	:	$X \wedge \neg Y \wedge \neg Z$
1	3-5	2	V1=T2	11010000	=	$\neg Y \wedge (\neg X \vee \neg Z)$
	6-2	28	**L**	00111111		$\neg X \rightarrow Y$
	1-7	6	O8	00000100	:	$X \wedge \neg Y \wedge Z$
2	3-5	2	V1=T2	11010000	=	$\neg Y \wedge (\neg X \vee \neg Z)$
	5-3	55	N8=L7	00101111		$\neg Y \rightarrow (X \wedge Z)$
	1-7	4	O7	00010000	:	$X \wedge \neg Y \wedge \neg Z$
3	3-5	7	V2=Ω2	10110000	=	$\neg Z \wedge (\neg X \vee \neg Y)$
	6-2	27	**Q**	01011111		$\neg Z \rightarrow X$
	1-7	6	O8	00000100	:	$X \wedge \neg Y \wedge Z$
4	3-5	7	V2=Ω2	10110000	=	$\neg Z \wedge (\neg X \vee \neg Y)$
	5-3	50	N7=Q7	01001111		$\neg Z \rightarrow (X \wedge Y)$
	1-7	4	O7	00010000	:	$X \wedge \neg Y \wedge \neg Z$
5	3-5	12	V3	10011000	=	$(\neg Z \wedge \neg Y) \vee (\neg X \wedge Y \wedge Z)$
	6-2	25	M1xL2	01110111		$\neg X \rightarrow \neg(Y \leftrightarrow Z)$
	1-7	6	O8	00000100	:	$X \wedge \neg Y \wedge Z$
6	3-5	12	V3	10011000	=	$(\neg Z \wedge \neg Y) \vee (\neg X \wedge Y \wedge Z)$
	5-3	45	N6	01100111		$(Y \vee Z) \wedge (X \vee \neg Y \vee \neg Z)$
	1-7	4	O7	00010000	:	$X \wedge \neg Y \wedge \neg Z$
7	3-5	13	V4=O1	10010100	=	$\neg Y \wedge (X \vee \neg Z)$
	6-2	24	A1xL2	01111011		$\neg Y \rightarrow \neg(X \leftrightarrow Z)$
	1-7	6	O8	00000100	:	$X \wedge \neg Y \wedge Z$
8	3-5	13	V4=O1	10010100	=	$\neg Y \wedge (X \vee \neg Z)$
	6-2	26	**N**	01101111		$\neg Z \rightarrow Y$
	1-7	4	O7	00010000	:	$X \wedge \neg Y \wedge \neg Z$
9	3-5	14	V5=W1	10010010	=	$\neg Z \wedge (X \vee \neg Y)$
	6-2	23	E2xL2	01111101		$\neg Z \rightarrow \neg(X \leftrightarrow Y)$
	1-7	6	O8	00000100	:	$X \wedge \neg Y \wedge Z$
10	3-5	14	V5=W1	10010010	=	$\neg Z \wedge (X \vee \neg Y)$
	5-3	43	N4=R8	01101101		$\neg Z \rightarrow (\neg X \wedge Y)$
	1-7	4	O7	00010000	:	$X \wedge \neg Y \wedge \neg Z$
11	3-5	15	V6	10010001	=	$(\neg Z \wedge \neg Y) \vee (X \wedge Y \wedge Z)$
	6-2	22	E1xL2	01111110		$\neg(X \leftrightarrow Y \leftrightarrow Z)$
	1-7	6	O8	00000100	:	$X \wedge \neg Y \wedge Z$
12	3-5	15	V6	10010001	=	$(\neg Z \wedge \neg Y) \vee (X \wedge Y \wedge Z)$
	5-3	42	N3	01101110		$(Y \vee Z) \wedge (\neg X \vee \neg Y \vee \neg Z)$

Table 5.31 **I** OR NOT **U** as necessary condition of **Q**

Derivations	Level	#	Symbol	ID		Traditional logical form
	1-7	7	I7	00000010	:	$X \wedge Y \wedge \neg Z$
1	3-5	10	U1= Ω5	10100010	=	$\neg Z \wedge (\neg X \vee Y)$
	6-2	27	**Q**	01011111		$\neg Z \rightarrow X$
	1-7	8	I8	00000001	:	$X \wedge Y \wedge Z$
2	3-5	10	U1= Ω5	10100010	=	$\neg Z \wedge (\neg X \vee Y)$
	5-3	47	P8=Q4	01011101		$\neg Z \rightarrow (X \wedge \neg Y)$
	1-7	7	I7	00000010	:	$X \wedge Y \wedge \neg Z$
3	3-5	25	U2	01100010		$(\neg Z \wedge Y) \vee (\neg X \wedge \neg Y \wedge Z)$
	6-2	21	M2xL1	10011111		$\neg X \rightarrow (Y \leftrightarrow Z)$
	1-7	8	I8	00000001	:	$X \wedge Y \wedge Z$
4	3-5	25	U2	01100010		$(\neg Z \wedge Y) \vee (\neg X \wedge \neg Y \wedge Z)$
	5-3	32	P7	10011101		$(\neg Y \vee Z) \wedge (X \vee Y \vee \neg Z)$
	1-7	7	I7	00000010	:	$X \wedge Y \wedge \neg Z$
5	3-5	39	U3=W3	00110010		$\neg Z \wedge (X \vee Y)$
	6-2	15	A2xM2	11001111		$\neg Z \rightarrow (X \leftrightarrow Y)$
	1-7	8	I8	00000001	:	$X \wedge Y \wedge Z$
6	3-5	39	U3=W3	00110010		$\neg Z \wedge (X \vee Y)$
	5-3	18	P6=R6	11001101		$\neg Z \rightarrow (\neg X \wedge \neg Y)$
	1-7	7	I7	00000010	:	$X \wedge Y \wedge \neg Z$
7	3-5	42	U4=S5	00101010	=	$Y \wedge (\neg X \vee \neg Z)$
	6-2	14	**M**	11010111		$\neg X \rightarrow \neg Y$
	1-7	8	I8	00000001	:	$X \wedge Y \wedge Z$
8	3-5	42	U4=S5	00101010	=	$Y \wedge (\neg X \vee \neg Z)$
	5-3	15	P5=M4	11010101		$Y \rightarrow (X \wedge Z)$
	1-7	7	I7	00000010	:	$X \wedge Y \wedge \neg Z$
9	3-5	44	U5	00100110	=	$(\neg Z \wedge Y) \vee (X \wedge \neg Y \wedge Z)$
	6-2	13	A1xM2	11011011		$\neg(X \leftrightarrow \neg Y \leftrightarrow Z)$
	1-7	8	I8	00000001	:	$X \wedge Y \wedge Z$
10	3-5	44	U5	00100110	=	$(\neg Z \wedge Y) \vee (X \wedge \neg Y \wedge Z)$
	5-3	15	P5=M4	11010101		$Y \rightarrow (X \wedge Z)$
	1-7	7	I7	00000010	:	$X \wedge Y \wedge \neg Z$
11	3-5	46	U6=I3	00100011	=	$Y \wedge (X \vee \neg Z)$
	6-2	11	E1xM2	11011110		$Y \rightarrow \neg(X \leftrightarrow Z)$
	1-7	8	I8	00000001	:	$X \wedge Y \wedge Z$
12	3-5	46	U6=I3	00100011	=	$Y \wedge (X \vee \neg Z)$
	6-2	12	**P**	11011101		$\neg Z \rightarrow \neg Y$

The third group encompasses following inferences:

Table 5.32 **N** OR NOT **Q** as necessary condition of **A**

Derivations	Level	#	Symbol	ID		Traditional logical form
	5-3	42	N3	01101110	:	$(Y \vee Z) \wedge (\neg X \vee \neg Y \vee \neg Z)$
1	7-1	6	Q1=M2	11011111	=	$X \vee \neg Y \vee Z$
	5-3	42	N3	01101110		$(Y \vee Z) \wedge (\neg X \vee \neg Y \vee \neg Z)$
	5-3	42	N3	01101110	:	$(Y \vee Z) \wedge (\neg X \vee \neg Y \vee \neg Z)$
2	7-1	8	Q2=L2	01111111	=	$X \vee Y \vee Z$
	6-2	7	E1xA2	11101110		$X \rightarrow \neg(Y \leftrightarrow Z)$
	5-3	43	N4=R8	01101101	:	$\neg Z \rightarrow (\neg X \wedge Y)$
3	7-1	6	Q1=M2	11011111	=	$X \vee \neg Y \vee Z$
	5-3	43	N4=R8	01101101		Z
	5-3	43	N4=R8	01101101	:	$\neg Z \rightarrow (\neg X \wedge Y)$
4	7-1	8	Q2=L2	01111111	=	$X \vee Y \vee Z$
	6-2	8	**R**	11101101		$\neg Z \rightarrow \neg X$
	5-3	44	N5=A8	01101011	:	$\neg Y \rightarrow (\neg X \wedge Z)$
5	7-1	6	Q1=M2	11011111	=	$X \vee \neg Y \vee Z$
	5-3	44	N5=A8	01101011		$\neg Y \rightarrow (\neg X \wedge Z)$
	5-3	44	N5=A8	01101011	:	$\neg Y \rightarrow (\neg X \wedge Z)$
6	7-1	8	Q2=L2	01111111	=	$X \vee Y \vee Z$
	6-2	9	**A**	11101011		$X \rightarrow Y$
	5-3	45	N6	01100111	:	$(Y \vee Z) \wedge (X \vee \neg Y \vee \neg Z)$
7	7-1	6	Q1=M2	11011111	=	$X \vee \neg Y \vee Z$
	5-3	45	N6	01100111		$(Y \vee Z) \wedge (X \vee \neg Y \vee \neg Z)$
	5-3	45	N6	01100111	:	$(Y \vee Z) \wedge (X \vee \neg Y \vee \neg Z)$
8	7-1	8	Q2=L2	01111111	=	$X \vee Y \vee Z$
	6-2	10	M1xA2	11100111		$\neg(\neg X \leftrightarrow Y \leftrightarrow Z)$
	5-3	50	N7=Q7	01001111	:	$\neg Z \rightarrow (X \wedge Y)$
9	7-1	6	Q1=M2	11011111	=	$X \vee \neg Y \vee Z$
	6-2	26	**N**	01101111		$\neg Z \rightarrow Y$
	5-3	50	N7=Q7	01001111	:	$\neg Z \rightarrow (X \wedge Y)$
10	7-1	8	Q2=L2	01111111	=	$X \vee Y \vee Z$
	6-2	15	A2xM2	11001111		$\neg Z \rightarrow (X \leftrightarrow Y)$
	5-3	55	N8=L7	00101111	:	$\neg Y \rightarrow (X \wedge Z)$
11	7-1	6	Q1=M2	11011111	=	$X \vee \neg Y \vee Z$
	5-3	55	N8=L7	00101111		$\neg Y \rightarrow (X \wedge Z)$
	5-3	55	N8=L7	00101111	:	$\neg Y \rightarrow (X \wedge Z)$
12	7-1	8	Q2=L2	01111111	=	$X \vee Y \vee Z$
	6-2	20	A2xL1	10101111		$\neg Y \rightarrow (X \leftrightarrow Z)$

Table 5.33 **P** OR NOT **Q** as necessary condition of **E**

Derivations	Level	#	Symbol	ID		Traditional logical form
	5-3	11	P3=E6	11011100	:	$Y \to (\neg X \wedge Z)$
1	7-1	6	Q1=M2	11011111	=	$X \vee \neg Y \vee Z$
	6-2	1	**E**	11111100		$X \to \neg Y$
	5-3	11	P3=E6	11011100	:	$Y \to (\neg X \wedge Z)$
2	7-1	8	Q2=L2	01111111	=	$X \vee Y \vee Z$
	5-3	11	P3=E6	11011100		$Y \to (\neg X \wedge Z)$
	5-3	13	P4	11011001	:	$(\neg Y \vee Z) \wedge (\neg X \vee Y \vee \neg Z)$
3	7-1	6	Q1=M2	11011111	=	$X \vee \neg Y \vee Z$
	6-2	3	E2xA1	11111001		$X \to (Y \leftrightarrow Z)$
	5-3	13	P4	11011001	:	$(\neg Y \vee Z) \wedge (\neg X \vee Y \vee \neg Z)$
4	7-1	8	Q2=L2	01111111	=	$X \vee Y \vee Z$
	5-3	13	P4	11011001		$(\neg Y \vee Z) \wedge (\neg X \vee Y \vee \neg Z)$
	5-3	15	P5=M4	11010101	:	$Y \to (X \wedge Z)$
5	7-1	6	Q1=M2	11011111	=	$X \vee \neg Y \vee Z$
	6-2	5	E2xM1	11110101		$Y \to (X \leftrightarrow Z)$
	5-3	15	P5=M4	11010101	:	$Y \to (X \wedge Z)$
6	7-1	8	Q2=L2	01111111	=	$X \vee Y \vee Z$
	5-3	15	P5=M4	11010101	:	$Y \to (X \wedge Z)$
	5-3	18	P6=R6	11001101	:	$\neg Z \to (\neg X \wedge \neg Y)$
7	7-1	6	Q1=M2	11011111	=	$X \vee \neg Y \vee Z$
	6-2	8	**R**	11101101		$\neg Z \to \neg X$
	5-3	18	P6=R6	11001101	:	$\neg Z \to (\neg X \wedge \neg Y)$
8	7-1	8	Q2=L2	01111111	=	$X \vee Y \vee Z$
	5-3	18	P6=R6	11001101		$\neg Z \to (\neg X \wedge \neg Y)$
	5-3	32	P7	10011101	:	$(\neg Y \vee Z) \wedge (X \vee Y \vee \neg Z)$
9	7-1	6	Q1=M2	11011111	=	$X \vee \neg Y \vee Z$
	6-2	17	E2xL1	10111101		$\neg(X \leftrightarrow Y \leftrightarrow \neg Z)$
	5-3	32	P7	10011101	:	$(\neg Y \vee Z) \wedge (X \vee Y \vee \neg Z)$
10	7-1	8	Q2=L2	01111111	=	$X \vee Y \vee Z$
	5-3	32	P7	10011101		$(\neg Y \vee Z) \wedge (X \vee Y \vee \neg Z)$
	5-3	47	P8=Q4	01011101	:	$\neg Z \to (X \wedge \neg Y)$
11	7-1	6	Q1=M2	11011111	=	$X \vee \neg Y \vee Z$
	6-2	23	E2xL2	01111101		$\neg Z \to \neg(X \leftrightarrow Y)$
	5-3	47	P8=Q4	01011101	:	$\neg Z \to (X \wedge \neg Y)$
12	7-1	8	Q2=L2	01111111	=	$X \vee Y \vee Z$
	6-2	12	**P**	11011101		$\neg Z \to \neg Y$

Table 5.34 **U** OR NOT **W** as necessary condition of **A**

Derivations	Level	#	Symbol	ID		Traditional logical form
	1-7	3	U7=S7	00100000	:	$\neg X \wedge Y \wedge \neg Z$
1	3-5	14	W1=V5	10010010	=	$\neg Z \wedge (X \vee \neg Y)$
	5-3	43	R8=N4	01101101		$\neg Z \rightarrow (\neg X \wedge Y)$
	1-7	7	U8=I7	00000010	:	$X \wedge Y \wedge \neg Z$
2	3-5	14	W1=V5	10010010	=	$\neg Z \wedge (X \vee \neg Y)$
	6-2	26	**N**	01101111		$\neg Z \rightarrow Y$
	1-7	3	U7=S7	00100000	:	$\neg X \wedge Y \wedge \neg Z$
3	3-5	29	W2	01010010	=	$(\neg Z \wedge X) \vee (\neg X \wedge \neg Y \wedge Z)$
	5-3	28	R7	10101101		$(\neg X \vee Z) \wedge (X \vee Y \vee \neg Z)$
	1-7	7	U8=I7	00000010	:	$X \wedge Y \wedge \neg Z$
4	3-5	29	W2	01010010	=	$(\neg Z \wedge X) \vee (\neg X \wedge \neg Y \wedge Z)$
	6-2	20	A2xL1	10101111		$\neg Y \rightarrow (X \leftrightarrow Z)$
	1-7	3	U7=S7	00100000	:	$\neg X \wedge Y \wedge \neg Z$
5	3-5	39	W3=U3	00110010	=	$\neg Z \wedge (X \vee Y)$
	6-2	8	**R**	11101101		$\neg Z \rightarrow \neg X$
	1-7	7	U8=I7	00000010	:	$X \wedge Y \wedge \neg Z$
6	3-5	39	W3=U3	00110010	=	$\neg Z \wedge (X \vee Y)$
	6-2	15	A2xM2	11001111		$\neg Z \rightarrow (X \leftrightarrow Y)$
	1-7	3	U7=S7	00100000	:	$\neg X \wedge Y \wedge \neg Z$
7	3-5	48	W4	00011010	=	$(\neg Z \wedge X) \vee (\neg X \wedge Y \wedge Z)$
	5-3	9	R5	11100101		$(\neg X \vee Z) \wedge (X \vee \neg Y \vee \neg Z)$
	1-7	7	U8=I7	00000010	:	$X \wedge Y \wedge \neg Z$
8	3-5	48	W4	00011010	=	$(\neg Z \wedge X) \vee (\neg X \wedge Y \wedge Z)$
	6-2	10	M1xA2	11100111		$\neg(\neg X \leftrightarrow Y \leftrightarrow Z)$
	1-7	3	U7=S7	00100000	:	$\neg X \wedge Y \wedge \neg Z$
9	3-5	50	W5=O5	00010110	=	$X \wedge (\neg Y \vee \neg Z)$
	5-3	7	R4=A4	11101001		$X \rightarrow (Y \wedge Z)$
	1-7	7	U8=I7	00000010	:	$X \wedge Y \wedge \neg Z$
10	3-5	50	W5=O5	00010110	=	$X \wedge (\neg Y \vee \neg Z)$
	6-2	9	**A**	11101011		$X \rightarrow Y$
	1-7	3	U7=S7	00100000	:	$\neg X \wedge Y \wedge \neg Z$
11	3-5	52	W6=I4	00010011	=	$X \wedge (Y \vee \neg Z)$
	5-3	5	R3=E5	11101100		$X \rightarrow (\neg Y \wedge Z)$
	1-7	7	U8=I7	00000010	:	$X \wedge Y \wedge \neg Z$
12	3-5	52	W6=I4	00010011	=	$X \wedge (Y \vee \neg Z)$
	6-2	7	E1xA2	11101110		$X \rightarrow \neg(Y \leftrightarrow Z)$

Table 5.35 **V** OR NOT **W** as necessary condition of **E**

Derivations	Level	#	Symbol	ID		Traditional logical form
	1-7	1	V7=T7	10000000	:	$\neg X \wedge \neg Y \wedge \neg Z$
1	3-5	14	W1=V5	10010010	=	$\neg Z \wedge (X \vee \neg Y)$
	6-2	8	**R**	11101101		$\neg Z \rightarrow \neg X$
	1-7	4	V8=O7	00010000	:	$X \wedge \neg Y \wedge \neg Z$
2	3-5	14	W1=V5	10010010	=	$\neg Z \wedge (X \vee \neg Y)$
	6-2	23	E2xL2	01111101		$\neg Z \rightarrow \neg(X \leftrightarrow Y)$
	1-7	1	V7=T7	10000000	:	$\neg X \wedge \neg Y \wedge \neg Z$
3	3-5	29	W2	01010010	=	$(\neg Z \wedge X) \vee (\neg X \wedge \neg Y \wedge Z)$
	5-3	28	R7	10101101		$(\neg X \vee Z) \wedge (X \vee Y \vee \neg Z)$
	1-7	4	V8=O7	00010000	:	$X \wedge \neg Y \wedge \neg Z$
4	3-5	29	W2	01010010	=	$(\neg Z \wedge X) \vee (\neg X \wedge \neg Y \wedge Z)$
	6-2	17	E2xL1	10111101		$\neg(X \leftrightarrow Y \leftrightarrow \neg Z)$
	1-7	1	V7=T7	10000000	:	$\neg X \wedge \neg Y \wedge \neg Z$
5	3-5	39	W3=U3	00110010	=	$\neg Z \wedge (X \vee Y)$
	5-3	18	R6=P6	11001101		$\neg Z \rightarrow (\neg X \wedge \neg Y)$
	1-7	4	V8=O7	00010000	:	$X \wedge \neg Y \wedge \neg Z$
6	3-5	39	W3=U3	00110010	=	$\neg Z \wedge (X \vee Y)$
	6-2	12	**P**	11011101		$\neg Z \rightarrow \neg Y$
	1-7	1	V7=T7	10000000	:	$\neg X \wedge \neg Y \wedge \neg Z$
7	3-5	48	W4	00011010	=	$(\neg Z \wedge X) \vee (\neg X \wedge Y \wedge Z)$
	5-3	9	R5	11100101		$(\neg X \vee Z) \wedge (X \vee \neg Y \vee \neg Z)$
	1-7	4	V8=O7	00010000	:	$X \wedge \neg Y \wedge \neg Z$
8	3-5	48	W4	00011010	=	$(\neg Z \wedge X) \vee (\neg X \wedge Y \wedge Z)$
	6-2	5	E2xM1	11110101		$Y \rightarrow (X \leftrightarrow Z)$
	1-7	1	V7=T7	10000000	:	$\neg X \wedge \neg Y \wedge \neg Z$
9	3-5	50	W5=O5	00010110	=	$X \wedge (\neg Y \vee \neg Z)$
	5-3	7	R4=A4	11101001		$X \rightarrow (Y \wedge Z)$
	1-7	4	V8=O7	00010000	:	$X \wedge \neg Y \wedge \neg Z$
10	3-5	50	W5=O5	00010110	=	$X \wedge (\neg Y \vee \neg Z)$
	6-2	3	E2xA1	11111001		$X \rightarrow (Y \leftrightarrow Z)$
	1-7	1	V7=T7	10000000	:	$\neg X \wedge \neg Y \wedge \neg Z$
11	3-5	52	W6=I4	00010011	=	$X \wedge (Y \vee \neg Z)$
	5-3	5	R3=E5	11101100		$X \rightarrow (\neg Y \wedge Z)$
	1-7	4	V8=O7	00010000	:	$X \wedge \neg Y \wedge \neg Z$
12	3-5	52	W6=I4	00010011	=	$X \wedge (Y \vee \neg Z)$
	6-2	1	**E**	11111100		$X \rightarrow \neg Y$

First Group				Second Group				Third Group			
Classical	1st mixed	2nd mixed	Extension	Classical	1st mixed	2nd mixed	Extension	Classical	1st mixed	2nd mixed	Extension
O X ∧ ¬Y	T ¬X ∧ ¬Y	O X ∧ ¬Y	T ¬X ∧ ¬Y	O X ∧ ¬Y	T ¬X ∧ ¬Y	O X ∧ ¬Y	T ¬X ∧ ¬Y	C Z → Y	C Z → Y	N ¬Z → Y	N ¬Z → Y
B Z → X	F Z → ¬X	Q ¬Z → X	R ¬Z → ¬X	H Z ∧ ¬Y	H Z ∧ ¬Y	V ¬Z ∧ ¬Y	V ¬Z ∧ ¬Y	B Z → X	F Z → ¬X	Q ¬Z → X	R ¬Z → ¬X
H Z ∧ ¬Y	H Z ∧ ¬Y	V ¬Z ∧ ¬Y	V ¬Z ∧ ¬Y	B Z → X	F Z → ¬X	Q ¬Z → X	R ¬Z → ¬X	A X → Y	L ¬X → Y	A X → Y	L ¬X → Y
I X ∧ Y	S ¬X ∧ Y	I X ∧ Y	S ¬X ∧ Y	I X ∧ Y	S ¬X ∧ Y	I X ∧ Y	S ¬X ∧ Y	D Z → ¬Y	D Z → ¬Y	P ¬Z → ¬Y	P ¬Z → ¬Y
B Z → X	F Z → ¬X	Q ¬Z → X	R ¬Z → ¬X	G Z ∧ Y	G Z ∧ Y	U ¬Z ∧ Y	U ¬Z ∧ Y	B Z → X	F Z → ¬X	Q ¬Z → X	R ¬Z → ¬X
G Z ∧ Y	G Z ∧ Y	U ¬Z ∧ Y	U ¬Z ∧ Y	B Z → X	F Z → ¬X	Q ¬Z → X	R ¬Z → ¬X	E X → ¬Y	M ¬X → ¬Y	E X → ¬Y	M ¬X → ¬Y
O X ∧ ¬Y	T ¬X ∧ ¬Y	O X ∧ ¬Y	T ¬X ∧ ¬Y	O X ∧ ¬Y	T ¬X ∧ ¬Y	O X ∧ ¬Y	T ¬X ∧ ¬Y	G Z ∧ Y	G Z ∧ Y	U ¬Z ∧ Y	U ¬Z ∧ Y
J X ∧ Z	K Z ∧ ¬X	W ¬Z ∧ X	Ω ¬Z ∧ ¬X	D Z → ¬Y	D Z → ¬Y	P ¬Z → ¬Y	P ¬Z → ¬Y	J X ∧ Z	K Z ∧ ¬X	W ¬Z ∧ X	Ω ¬Z ∧ ¬X
D Z → ¬Y	D Z → ¬Y	P ¬Z → ¬Y	P ¬Z → ¬Y	J X ∧ Z	K Z ∧ ¬X	W ¬Z ∧ X	Ω ¬Z ∧ ¬X	A X → Y	L ¬X → Y	A X → Y	L ¬X → Y
I X ∧ Y	S ¬X ∧ Y	I X ∧ Y	S ¬X ∧ Y	I X ∧ Y	S ¬X ∧ Y	I X ∧ Y	S ¬X ∧ Y	H Z ∧ ¬Y	H Z ∧ ¬Y	V ¬Z ∧ ¬Y	V ¬Z ∧ ¬Y
J X ∧ Z	K Z ∧ ¬X	W ¬Z ∧ X	Ω ¬Z ∧ ¬X	C Z → Y	C Z → Y	N ¬Z → Y	N ¬Z → Y	J X ∧ Z	K Z ∧ ¬X	W ¬Z ∧ X	Ω ¬Z ∧ ¬X
C Z → Y	C Z → Y	N ¬Z → Y	N ¬Z → Y	J Z ∧ X	K Z ∧ ¬X	W ¬Z ∧ X	Ω ¬Z ∧ ¬X	E X → ¬Y	M ¬X → ¬Y	E X → ¬Y	M ¬X → ¬Y

Figure 5.1 Summary of all division inferences.

All inferences through divisions may be summarized as in Fig. 5.1. It may be noted that all division inference have at least one premise that is negative. Moreover, I recall that this kind of inference is retrograde (given the premises we assume the conclusion). In other words, we assume a statement that could be able to explain the circumstance that at least one of the two premises is false. Note also that when two premises are one particular and the other the negation of an universal, also the conclusion is particular.

Part III

Generalizations and Applications

Chapter 6

Assessment of All the Previous Inferences

6.1 General Considerations

It may be helpful to summarize in some tables all contradictory (countervalent) pairs that I have used in my previous derivations. I consider in the following the main statements involved and their generating sets (for product and sum inferences the data are extracted from Tables 2.4–2.11, 2.24–2.25 and 3.1–3.2).

The level is ascertained by counting the zeros and ones. The first four columns of Table 6.1 represent universal statements whilst the second four columns particular statements. On each row there are two contradictory (countervalent) statements. The first two statements of the generating set of an universal statement (e.g. A1 and A2 of **A**) constitute the LGS whilst the latter 6 (e.g. A3–A8) the RGS. In the case of particular statement, the first two statements of the generating set (e.g. O7 and O8 of **O**) constitute the RGS whilst the latter six (e.g. O1–O6) the LGS. There are also other countervalent

Mechanical Logic in Three-Dimensional Space
Gennaro Auletta

ISBN 978-981-4411-50-9 (Hardcover), 978-981-4411-51-6 (eBook)
www.panstanford.com

Table 6.1 Summary of the main countervalent pairs

#	Symbol	ID	Logical form	#	Symbol	ID	Logical form
9	**A**	11101011	$X \rightarrow Y$	20	**O**	00010100	$X \wedge \neg Y$
3	A1	11111011	$\neg X \vee Y \vee \neg Z$	6	O8	00000100	$X \wedge \neg Y \wedge Z$
5	A2	11101111	$\neg X \vee Y \vee Z$	4	O7	00010000	$X \wedge \neg Y \wedge \neg Z$
6	A3	11101010	$X \rightarrow (Y \wedge \neg Z)$	51	O6	00010101	$X \wedge (\neg Y \vee Z)$
7	A4	11101001	$X \rightarrow (Y \wedge Z)$	50	O5	00010110	$X \wedge (\neg Y \vee \neg Z)$
10	A5	11100011	$(\neg X \vee Y) \wedge (X \vee \neg Y \vee \neg Z)$	47	O4	00011100	$(X \wedge \neg Y) \vee (\neg X \wedge Y \wedge Z)$
19	A6	11001011	$(\neg X \vee Y) \wedge (X \vee \neg Y \vee Z)$	38	O3	00110100	$(X \wedge \neg Y) \vee (\neg X \wedge Y \wedge \neg Z)$
29	A7	10101011	$\neg Y \rightarrow (\neg X \wedge \neg Z)$	28	O2	01010100	$\neg Y \wedge (X \vee Z)$
44	A8	01101011	$\neg Y \rightarrow (\neg X \wedge Z)$	13	O1	10010100	$\neg Y \wedge (X \vee \neg Z)$
1	**E**	11111100	$X \rightarrow \neg Y$	28	**I**	00000011	$X \wedge Y$
1	E1	11111110	$\neg X \vee \neg Y \vee \neg Z$	8	I8	00000001	$X \wedge Y \wedge Z$
2	E2	11111101	$\neg X \vee \neg Y \vee Z$	7	I7	00000010	$X \wedge Y \wedge \neg Z$
1	E3	11111000	$X \rightarrow (\neg Y \wedge \neg Z)$	56	I6	00000111	$X \wedge (Y \vee Z)$
2	E4	11110100	$Y \rightarrow (\neg X \wedge \neg Z)$	55	I5	00001011	$Y \wedge (X \vee Z)$
5	E5	11101100	$X \rightarrow (\neg Y \wedge Z)$	52	I4	00010011	$X \wedge (Y \vee \neg Z)$
11	E6	11011100	$Y \rightarrow (\neg X \wedge Z)$	46	I3	00100011	$Y \wedge (X \vee \neg Z)$
21	E7	10111100	$(\neg X \vee \neg Y) \wedge (X \vee Y \vee \neg Z)$	36	I2	01000011	$(X \wedge Y) \vee (\neg X \wedge \neg Y \wedge Z)$
36	E8	01111100	$(\neg X \vee \neg Y) \wedge (X \vee Y \vee Z)$	21	I1	10000011	$(X \wedge Y) \vee (\neg X \wedge \neg Y \wedge \neg Z)$
28	**L**	00111111	$\neg X \rightarrow Y$	1	**T**	11000000	$\neg X \wedge \neg Y$
7	L1	10111111	$X \vee Y \vee \neg Z$	2	T8	01000000	$\neg X \wedge \neg Y \wedge Z$
8	L2	01111111	$X \vee Y \vee Z$	1	T7	10000000	$\neg X \wedge \neg Y \wedge \neg Z$

51	L3	00111110	$(X \vee Y) \wedge (\neg X \vee \neg Y \vee \neg Z)$	6	T6	11000001	$(\neg X \wedge \neg Y) \vee (X \wedge Y \wedge Z)$
52	L4	00111101	$(X \vee Y) \wedge (\neg X \vee \neg Y \vee Z)$	5	T5	11000010	$(\neg X \wedge \neg Y) \vee (X \wedge Y \wedge \neg Z)$
53	L5	00111011	$\neg Y \rightarrow (X \wedge \neg Z)$	4	T4	11000100	$\neg Y \wedge (\neg X \vee Z)$
54	L6	00110111	$\neg X \rightarrow (Y \wedge \neg Z)$	3	T3	11001000	$\neg X \wedge (\neg Y \vee Z)$
55	L7	00101111	$\neg Y \rightarrow (X \wedge Z)$	2	T2	11010000	$\neg Y \wedge (\neg X \vee \neg Z)$
56	L8	00011111	$\neg X \rightarrow (Y \wedge Z)$	1	T1	11100000	$\neg X \wedge (\neg Y \vee \neg Z)$
14	**M**	11010111	$\neg X \rightarrow \neg Y$	15	**S**	00101000	$\neg X \wedge Y$
4	M1	11110111	$X \vee \neg Y \vee \neg Z$	5	S8	00001000	$\neg X \wedge Y \wedge Z$
6	M2	11011111	$X \vee \neg Y \vee Z$	3	S7	00100000	$\neg X \wedge Y \wedge \neg Z$
14	M3	11010110	$Y \rightarrow (X \wedge \neg Z)$	43	S6	00101001	$Y \wedge (\neg X \vee Z)$
15	M4	11010101	$Y \rightarrow (X \wedge Z)$	42	S5	00101010	$Y \wedge (\neg X \vee \neg Z)$
16	M5	11010011	$(X \vee \neg Y) \wedge (\neg X \vee Y \vee \neg Z)$	41	S4	00101100	$(\neg X \wedge Y) \vee (X \wedge \neg Y \wedge Z)$
20	M6	11000111	$(X \vee \neg Y) \wedge (\neg X \vee Y \vee Z)$	37	S3	00111000	$(\neg X \wedge Y) \vee (X \wedge \neg Y \wedge \neg Z)$
34	M7	10010111	$\neg X \rightarrow (\neg Y \wedge \neg Z)$	23	S2	01101000	$\neg X \wedge (Y \vee Z)$
49	M8	01010111	$\neg X \rightarrow (\neg Y \wedge Z)$	8	S1	10101000	$\neg X \wedge (Y \vee \neg Z)$
19	**B**	10110111	$Z \rightarrow X$	10	**K**	01001000	$Z \wedge \neg X$
4	B1=M1	11110111	$X \vee \neg Y \vee \neg Z$	5	K8=S8	00001000	$\neg X \wedge Y \wedge Z$
7	B2=L1	10111111	$X \vee Y \vee \neg Z$	2	K7=T8	01000000	$\neg X \wedge \neg Y \wedge Z$
24	B3=D7	10110110	$Z \rightarrow (X \wedge \neg Y)$	33	K6=G2	01001001	$Z \wedge (\neg X \vee Y)$
25	B4	10110101	$(X \vee \neg Z) \wedge (\neg X \vee \neg Y \vee Z)$	32	K5	01001010	$(\neg X \wedge Z) \vee (X \wedge Y \wedge \neg Z)$
26	B5=C5	10110011	$Z \rightarrow (X \wedge Y)$	31	K4=H4	01001100	$Z \wedge (\neg X \vee \neg Y)$
30	B6	10100111	$(X \vee \neg Z) \wedge (\neg X \vee Y \vee Z)$	27	K3	01011000	$(\neg X \wedge Z) \vee (X \wedge \neg Y \wedge \neg Z)$

(Continued)

Table 6.1 (*Continued*)

#	Symbol	ID	Logical form	#	Symbol	ID	Logical form
34	B7=M7	10010111	$\neg X \to (\neg Y \wedge \neg Z)$	23	K2=S2	01101000	$\neg X \wedge (Y \vee Z)$
54	B8=L6	00110111	$\neg X \to (Y \wedge \neg Z)$	3	K1=T3	11001000	$\neg X \wedge (\neg Y \vee Z)$
18	**C**	10111011	$Z \to Y$	11	**H**	01000100	$Z \wedge \neg Y$
3	C1=A1	11111011	$\neg X \vee Y \vee \neg Z$	6	H8=O8	00000100	$X \wedge \neg Y \wedge Z$
7	C2=L1	10111111	$X \vee Y \vee \neg Z$	2	H7=T8	01000000	$\neg X \wedge \neg Y \wedge Z$
22	C3=F7	10111010	$Z \to (\neg X \wedge Y)$	35	H6=J2	01000101	$Z \wedge (X \vee \neg Y)$
23	C4	10111001	$(Y \vee \neg Z) \wedge (\neg X \vee \neg Y \vee Z)$	34	H5	01000110	$(\neg Y \wedge Z) \vee (X \wedge Y \wedge \neg Z)$
26	C5=B5	10110011	$Z \to (X \wedge Y)$	31	H4=K4	01001100	$Z \wedge (\neg X \vee \neg Y)$
29	C6=A7	10101011	$\neg Y \to (\neg X \wedge \neg Z)$	28	H3=O2	01010100	$\neg Y \wedge (X \vee Z)$
33	C7	10011011	$(Y \vee \neg Z) \wedge (X \vee \neg Y \vee Z)$	24	H2	01100100	$(\neg Y \wedge Z) \vee (\neg X \wedge Y \wedge \neg Z)$
53	C8=L5	00111011	$\neg Y \to (X \wedge \neg Z)$	4	H1=T4	11000100	$\neg Y \wedge (\neg X \vee Z)$
4	**D**	11110110	$Z \to \neg Y$	25	**G**	00001001	$Z \wedge Y$
1	D1=E1	11111110	$\neg X \vee \neg Y \vee \neg Z$	8	G8=I8	00000001	$X \wedge Y \wedge Z$
4	D2=M1	11110111	$X \vee \neg Y \vee \neg Z$	5	G7=S8	00001000	$\neg X \wedge Y \wedge Z$
2	D3=E4	11110100	$Y \to (\neg X \wedge \neg Z)$	55	G6=I5	00001011	$Y \wedge (X \vee Z)$
3	D4=F4	11110010	$Z \to (\neg X \wedge \neg Y)$	54	G5=J5	00001101	$Z \wedge (X \vee Y)$
8	D5	11100110	$(\neg Y \vee \neg Z) \wedge (\neg X \vee Y \vee Z)$	49	G4	00011001	$(Y \wedge Z) \vee (X \wedge \neg Y \wedge \neg Z)$
14	D6=M3	11010110	$Y \to (X \wedge \neg Z)$	43	G3=S6	00101001	$Y \wedge (\neg X \vee Z)$
24	D7=B3	10110110	$Z \to (X \wedge \neg Y)$	33	G2=K6	01001001	$Z \wedge (\neg X \vee Y)$
39	D8	01110110	$(\neg Y \vee \neg Z) \wedge (X \vee Y \vee Z)$	18	G1	10001001	$(Y \wedge Z) \vee (\neg X \wedge \neg Y \wedge \neg Z)$

2	**F**	11111010	$Z \to \neg X$	27	**J**	00000101	$Z \wedge X$
1	F1=E1	11111110	$\neg X \vee \neg Y \vee \neg Z$	8	J8=I8	00000001	$X \wedge Y \wedge Z$
3	F2=A1	11111011	$\neg X \vee Y \vee \neg Z$	6	J7=O8	00000100	$X \wedge \neg Y \wedge Z$
1	F3=E3	11111000	$X \to (\neg Y \wedge \neg Z)$	56	J6=I6	00000111	$X \wedge (Y \vee Z)$
3	F4=D4	11110010	$Z \to (\neg X \wedge \neg Y)$	54	J5=G5	00001101	$Z \wedge (X \vee Y)$
6	F5=A3	11101010	$X \to (Y \wedge \neg Z)$	51	J4=O6	00010101	$X \wedge (\neg Y \vee Z)$
12	F6	11011010	$(\neg X \vee \neg Z) \wedge (X \vee \neg Y \vee Z)$	45	J3	00100101	$(X \wedge Z) \vee (\neg X \wedge Y \wedge \neg Z)$
22	F7=C3	10111010	$Z \to (\neg X \wedge Y)$	35	J2=H6	01000101	$Z \wedge (X \vee \neg Y)$
37	F8	01111010	$(\neg X \vee \neg Z) \wedge (X \vee Y \vee Z)$	20	J1	10000101	$(X \wedge Z) \vee (\neg X \wedge \neg Y \wedge \neg Z)$
26	**N**	01101111	$\neg Z \to Y$	3	**V**	10010000	$\neg Z \wedge \neg Y$
5	N1=A2	11101111	$\neg X \vee Y \vee Z$	4	V8=O7	00010000	$X \wedge \neg Y \wedge \neg Z$
8	N2=L2	01111111	$X \vee Y \vee Z$	1	V7=T7	10000000	$\neg X \wedge \neg Y \wedge \neg Z$
42	N3	01101110	$(Y \vee Z) \wedge (\neg X \vee \neg Y \vee \neg Z)$	15	V6	10010001	$(\neg Z \wedge \neg Y) \vee (X \wedge Y \wedge Z)$
43	N4=R8	01101101	$\neg Z \to (\neg X \wedge Y)$	14	V5=W1	10010010	$\neg Z \wedge (X \vee \neg Y)$
44	N5=A8	01101011	$\neg Y \to (\neg X \wedge Z)$	13	V4=O1	10010100	$\neg Y \wedge (X \vee \neg Z)$
45	N6	01100111	$(Y \vee Z) \wedge (X \vee \neg Y \vee \neg Z)$	12	V3	10011000	$(\neg Z \wedge \neg Y) \vee (\neg X \wedge Y \wedge Z)$
50	N7=Q7	01001111	$\neg Z \to (X \wedge Y)$	7	V2=Ω2	10110000	$\neg Z \wedge (\neg X \vee \neg Y)$
55	N8=L7	00101111	$\neg Y \to (X \wedge Z)$	2	V1=T2	11010000	$\neg Y \wedge (\neg X \vee \neg Z)$
12	**P**	11011101	$\neg Z \to \neg Y$	17	**U**	00100010	$\neg Z \wedge Y$
2	P1=E2	11111101	$\neg X \vee \neg Y \vee Z$	7	U8=I7	00000010	$X \wedge Y \wedge \neg Z$
6	P2=M2	11011111	$X \vee \neg Y \vee Z$	3	U7=S7	00100000	$\neg X \wedge Y \wedge \neg Z$
11	P3=E6	11011100	$Y \to (\neg X \wedge Z)$	46	U6=I3	00100011	$Y \wedge (X \vee \neg Z)$
13	P4	11011001	$(\neg Y \vee Z) \wedge (\neg X \vee Y \vee \neg Z)$	44	U5	00100110	$(\neg Z \wedge Y) \vee (X \wedge \neg Y \wedge Z)$
15	P5=M4	11010101	$Y \to (X \wedge Z)$	42	U4=S5	00101010	$Y \wedge (\neg X \vee \neg Z)$

(Continued)

Table 6.1 *(Continued)*

#	Symbol	ID	Logical form	#	Symbol	ID	Logical form
18	P6=R6	11001101	$\neg Z \to (\neg X \wedge \neg Y)$	39	U3=W3	00110010	$\neg Z \wedge (X \vee Y)$
32	P7	10011101	$(\neg Y \vee Z) \wedge (X \vee Y \vee \neg Z)$	25	U2	01100010	$(\neg Z \wedge Y) \vee (\neg X \wedge \neg Y \wedge Z)$
47	P8=Q4	01011101	$\neg Z \to (X \wedge \neg Y)$	10	U1=Ω5	10100010	$\neg Z \wedge (\neg X \vee Y)$
27	**Q**	01011111	$\neg Z \to X$	2	**Ω**	10100000	$\neg Z \wedge \neg X$
6	Q1=M2	11011111	$X \vee \neg Y \vee Z$	3	Ω8=S7	00100000	$\neg X \wedge Y \wedge \neg Z$
8	Q2=L2	01111111	$X \vee Y \vee Z$	1	Ω7=T7	10000000	$\neg X \wedge \neg Y \wedge \neg Z$
46	Q3	01011110	$(X \vee Z) \wedge (\neg X \vee \neg Y \vee \neg Z)$	11	Ω6	10100001	$(\neg Z \wedge \neg X) \vee (X \wedge Y \wedge Z)$
47	Q4=P8	01011101	$\neg Z \to (X \wedge \neg Y)$	10	Ω5=U1	10100010	$\neg Z \wedge (\neg X \vee Y)$
48	Q5	01011011	$(X \vee Z) \wedge (\neg X \vee Y \vee \neg Z)$	9	Ω4	10100100	$(\neg Z \wedge \neg X) \vee (X \wedge \neg Y \wedge Z)$
49	Q6=M8	01010111	$\neg X \to (\neg Y \wedge Z)$	8	Ω3=S1	10101000	$\neg X \wedge (Y \vee \neg Z)$
50	Q7=N7	01001111	$\neg Z \to (X \wedge Y)$	7	Ω2=V2	10110000	$\neg Z \wedge (\neg X \vee \neg Y)$
56	Q8=L8	00011111	$\neg X \to (Y \wedge Z)$	1	Ω1=T1	11100000	$\neg X \wedge (\neg Y \vee \neg Z)$
8	**R**	11101101	$\neg Z \to \neg X$	21	**W**	00010010	$\neg Z \wedge X$
2	R1=E2	11111101	$\neg X \vee \neg Y \vee Z$	7	W8=I7	00000010	$X \wedge Y \wedge \neg Z$
5	R2=A2	11101111	$\neg X \vee Y \vee Z$	4	W7=O7	00010000	$X \wedge \neg Y \wedge \neg Z$
5	R3=E5	11101100	$X \to (\neg Y \wedge Z)$	52	W6=I4	00010011	$X \wedge (Y \vee \neg Z)$
7	R4=A4	11101001	$X \to (Y \wedge Z)$	50	W5=O5	00010110	$X \wedge (\neg Y \vee \neg Z)$
9	R5	11100101	$(\neg X \vee Z) \wedge (X \vee \neg Y \vee \neg Z)$	48	W4	00011010	$(\neg Z \wedge X) \vee (\neg X \wedge Y \wedge Z)$
18	R6=P6	11001101	$\neg Z \to (\neg X \wedge \neg Y)$	39	W3=U3	00110010	$\neg Z \wedge (X \vee Y)$
28	R7	10101101	$(\neg X \vee Z) \wedge (X \vee Y \vee \neg Z)$	29	W2	01010010	$(\neg Z \wedge X) \vee (\neg X \wedge \neg Y \wedge Z)$
43	R8=N4	01101101	$\neg Z \to (\neg X \wedge Y)$	14	W1=V5	10010010	$\neg Z \wedge (X \vee \neg Y)$

statements that have been considered in the previous derivations. I summarize them in the following table:

Table 6.2 Other countervalent pairs

#	Symbol	ID	Logical form	#	Symbol	ID	Logical form
3	E2xA1	11111001	$X \to (Y \leftrightarrow Z)$	26	O8+I7	00000110	$X \wedge \neg(Y \leftrightarrow Z)$
5	E2xM1	11110101	$Y \to (X \leftrightarrow Z)$	24	S8+I7	00001010	$Y \wedge \neg(X \leftrightarrow Z)$
6	A1xM1	11110011	$Z \to (X \leftrightarrow Y)$	23	S8+O8	00001100	$Z \wedge \neg(X \leftrightarrow Y)$
7	E1xA2	11101110	$X \to \neg(Y \leftrightarrow Z)$	22	O7+I8	00010001	$X \wedge (Y \leftrightarrow Z)$
10	M1xA2	11100111	$\neg(\neg X \leftrightarrow Y \leftrightarrow Z)$	19	O7+S8	00011000	$\neg X \leftrightarrow Y \leftrightarrow Z$
11	E1xM2	11011110	$Y \to \neg(X \leftrightarrow Z)$	18	S7+I8	00100001	$Y \wedge (X \leftrightarrow Z)$
13	A1xM2	11011011	$\neg(X \leftrightarrow \neg Y \leftrightarrow Z)$	16	S7+O8	00100100	$X \leftrightarrow \neg Y \leftrightarrow Z$
15	A2xM2	11001111	$\neg Z \to (X \leftrightarrow Y)$	14	S7+O7	00110000	$\neg Z \wedge \neg(X \leftrightarrow Y)$
16	E1xL1	10111110	$Z \to \neg(X \leftrightarrow Y)$	13	T8+I8	01000001	$Z \wedge (X \leftrightarrow Y)$
17	E2xL1	10111101	$\neg(X \leftrightarrow Y \leftrightarrow \neg Z)$	12	T8+I7	01000010	$X \leftrightarrow Y \leftrightarrow \neg Z$
20	A2xL1	10101111	$\neg Y \to (X \leftrightarrow Z)$	9	T8+O7	01010000	$\neg Y \wedge \neg(X \leftrightarrow Z)$
21	M2xL1	10011111	$\neg X \to (Y \leftrightarrow Z)$	8	T8+S7	01100000	$\neg X \wedge \neg(Y \leftrightarrow Z)$
22	E1xL2	01111110	$\neg(X \leftrightarrow Y \leftrightarrow Z)$	7	T7+I8	10000001	$X \leftrightarrow Y \leftrightarrow Z$
23	E2xL2	01111101	$\neg Z \to \neg(X \leftrightarrow Y)$	6	T7+I7	10000010	$\neg Z \wedge (X \leftrightarrow Y)$
24	A1xL2	01111011	$\neg Y \to \neg(X \leftrightarrow Z)$	5	T7+O8	10000100	$\neg Y \wedge (X \leftrightarrow Z)$
25	M1xL2	01110111	$\neg X \to \neg(Y \leftrightarrow Z)$	4	T7+S8	10001000	$\neg X \wedge (Y \leftrightarrow Z)$

Obviously, all statements on the first four columns pertain to Level 6-2 whilst all statements on the second four columns to Level 2-6. Again, each row represents contradictory pairs.

It may be further noticed that the 8 statements of Level 5-3 that are not involved in any of the previous derivations are formed by a combination of two statements of their LGS whose pool is constituted by the 16 statements of Table 6.2, whilst all 48 statements of Level 5-3 that are involved in the previous derivation are a combination of two out of 3 statements of the 12 of Level 6-2 present in Table 6.1. In particular, these 8 statements are:

Table 6.3 Statements of Level 5-3 not involved in the derivations

#	LGS
4	{3, 5, 6}
17	{7, 11, 15}
27	{7, 16, 20}

(Continued)

Table 6.3 (*Continued*)

#	LGS
31	{11, 16, 21}
35	{15, 20, 21}
38	{3, 23, 24}
40	{5, 23, 25}
41	{6, 24, 25}

Similarly, the 8 statements of Level 3-5 that are not involved in any of the previous derivations are formed by a combination of two statements of their RGS whose pool is constituted by the 16 statements of Table 6.2, whilst all 48 statements of Level 3-5 that are involved in the previous derivation are a combination of two out of 3 statements of the 12 Level 2-6 present in Table 6.1. In particular, these 8 statements are:

Table 6.4 Statements of Level 3-5 not involved in the derivations

#	RGS
16	{4, 5, 23}
17	{4, 6, 24}
19	{5, 6, 26}
22	{8, 9, 14}
26	{8, 13, 18}
30	{9, 13, 22}
40	{14, 18, 22}
53	{23, 24, 26}

Although not so straight as the statements listed in statements of Table 6.1, those of Table 6.2 are very important for charactering and classifying the previous derivations. In the case of derivations involving only universal statements, we have the presence of three statements included in Table 6.2. For instance, in the case of the classical *Barbara* (Table 2.12), i.e. the derivation of **C** from **A** and **B**, we have: A1xM1, M1xA2 and A2xL1. It is interesting to note that all of them include either A1 or A2 whilst two of them have the component M1 which is identical to one of the two statements of the LGS of the premise **B** (B1=M1) whilst L1 occurring in the statement A2xL1 is equal to both (i) the other statement of the LGS of **B** (B2=L1) and (ii) the statement of the LGS of the conclusion **C** that is involved in this derivation (L1=C2). Note that both B1 and B2 are involved. Moreover, the other statement of the LGS of the conclusion (i.e. C1) is equal to one of the statements of the LGS of the other premise (C1=A1). Summarizing, we have:

$$(\mathrm{A1}=\mathrm{C1})\times(\mathrm{M1}=\mathrm{B1}) \quad (\mathrm{M1}=\mathrm{B1})\times\mathrm{A2} \quad \mathrm{A2}\times(\mathrm{L1}=\mathrm{B2}=\mathrm{C2})$$

Derivations involving both universal and particular statements are far richer. Let us recall that in each derivation of this kind, three grounding (bold) statements are deduced (see also the comments at the derivation of *Ferio*, Table 2.15). For instance, let us take the example of *Darii* in its classical form (Table 2.14). This derivations tells us that **A** and **J** are sufficient condition of **G**, that is, $(\mathbf{A} \wedge \mathbf{J}) \rightarrow \mathbf{G}$, where the implication needs to be a tautology to be a valid inference. The three deduced statements in the table are **G**, **J**, and **I**. Statement **G** is the correct result of the derivation; **J** is already a premise and so is not relevant; **I** is very interesting since if we have a look at the statements involved in such a derivation having the form displayed in Table 6.2 we shall discover that they are the following four (all of them pertaining to Level 2-6):

O7+I8 (# 22), S7+I8 (# 18), T8+I8 (# 13), T7+I8 (# 7).

What these statements have in common is I8 (which is one of the two statements constituting the RGS of **I**). Now, the relevant point is that we have following equality: J8=G8=I8. This is crucial, since this equality connects thee statements: one premise of *Darii* (namely **J**), having the same quantity of the conclusion, the conclusion **G** and the third statement occurring in the derivation (**I**). This equality is what characterizes the general form of this derivation and indeed can be found in any similar derivation. For this reason, let us call it the *formal identity* (FI in short). However, in the same derivation there is also another interesting equality, namely G5=J5. This is precisely the equality that allows the inference since it establishes the equivalence

$$(\mathrm{A1} \times (\mathrm{J5}=\mathrm{G5})) = \mathbf{G}$$

that allows the inference $(\mathbf{A} \wedge \mathbf{J}) \rightarrow \mathbf{G}$. Let us call the latter equivalence the *grounding equivalence* (GE in short) and an equality of the kind G5=J5 *inferential identity* (II in short). The inferential identity owes its name to the fact that makes the specific difference of *Darii* relative to any other statements having the general form characterized by the FI. We may extend these expressions also to derivation involving only universal statements. On these bases let us now consider all inferences so far.

6.2 Product Inferences

Thanks to the previous analysis, we can summarize all the 48 derivations (tautologies) involving products as in the following table:

Table 6.5 Summary of all product inferences

Group		Table	Inference	GE	FI	II
	1	2.12	(**A** ∧ **B**) → **C**	(A1 x B2) = **C**	L1=B2=C2	A1=C1
	2	2.13	(**E** ∧ **B**) → **D**	(E1 x B1) = **D**	M1=B1=D2	E1=D1
	3	2.14	(**A** ∧ **J**) → **G**	(A1 x J5) = **G**	I8=G8=J8	J5=G5
	4	2.15	(**E** ∧ **J**) → **H**	(E1 x J2) = **H**	O8=H8=J7	J2=H6
	5	2.39	(**L** ∧ **F**) → **C**	(L1 x F2) = **C**	A1=C1=F2	L1=C2
	6	2.40	(**M** ∧ **F**) → **D**	(M1 x F1) = **D**	E1=D1=F1	M1=D2
	7	2.41	(**L** ∧ **K**) → **G**	(L1 x K6) = **G**	S8=G7=K8	G2=K6
1st group	8	2.42	(**M** ∧ **K**) → **H**	(M1 x K4) = **H**	T8=H7=K7	H4=K4
	9	2.51	(**A** ∧ **Q**) → **N**	(A2 x Q2) = **N**	L2=N2=Q2	A2=N1
	10	2.52	(**E** ∧ **Q**) → **P**	(E2 x Q1) = **P**	M2=P2=Q1	E2=P1
	11	2.53	(**A** ∧ **W**) → **U**	(A1 x W3) = **U**	7=U8=W8	U3=W3
	12	2.54	(**E** ∧ **W**) → **V**	(E2 x W1) = **V**	O7=V8=W	V5=W1
	13	2.27	(**L** ∧ **R**) → **N**	(L2 x R2) = **N**	A2=N1=R2	L2=N2
	14	2.28	(**M** ∧ **R**) → **P**	(M2 x R1) = **P**	E2=P1=R1	M2=P2
	15	2.29	(**L** ∧ **Ω**) → **U**	(L1 x Ω5) = **U**	S7=U7=Ω8	U1=Ω5
	16	2.30	(**M** ∧ **Ω**) → **V**	(M2 x Ω2) = **V**	T7=V7=Ω7	V2=Ω2
	17	2.16	(**A** ∧ **H**) → **K**	(A1 x H4) = **K**	T8=H7=K7	H4=K4
	18	2.17	(**E** ∧ **G**) → **K**	(E1 x G2) = **K**	S8=G7=K8	G2=K6
	19	2.18	(**A** ∧ **D**) → **F**	(A1 x D1) = **F**	E1=D1=F1	A1=F2
	20	2.19	(**E** ∧ **C**) → **F**	(E1 x C1) = **F**	A1=C1=F2	E1=F1
	21	2.43	(**L** ∧ **H**) → **J**	(L1 x H6) = **J**	O8=H8=J7	H6=J2
	22	2.44	(**M** ∧ **G**) → **J**	(M1 x G5) = **J**	I8=G8=J8	G5=J5
	23	2.45	(**L** ∧ **D**) → **B**	(L1 x D2) = **B**	M1=B1=D2	L1=B2
2nd group	24	2.46	(**M** ∧ **C**) → **B**	(M1 x C2) = **B**	L1=B2=C2	M1=B1
	25	2.55	(**A** ∧ **V**) → **Ω**	(A2 x V2) = **Ω**	T7=V7=Ω7	V2=Ω2

Group		Table	Inference	GE	FI	II
	26	2.56	$(\mathbf{E} \wedge \mathbf{U}) \rightarrow \mathbf{\Omega}$	(E2 x U1) = **Ω**	S7=U7=Ω8	U1=Ω5
	27	2.57	$(\mathbf{A} \wedge \mathbf{P}) \rightarrow \mathbf{R}$	(A2 x P1) = **R**	E2=P1=R1	A2=R1
	28	2.58	$(\mathbf{E} \wedge \mathbf{N}) \rightarrow \mathbf{R}$	(E2 x N1) = **R**	A2=N1=R2	E2=R1
	29	2.31	$(\mathbf{L} \wedge \mathbf{V}) \rightarrow \mathbf{W}$	(L2 x V5) =**W**	O7=V8=W7	V5=W1
	30	2.32	$(\mathbf{M} \wedge \mathbf{U}) \rightarrow \mathbf{W}$	(M2 x U3) =**W**	I7=U8=W8	U3=W3
	31	2.33	$(\mathbf{L} \wedge \mathbf{P}) \rightarrow \mathbf{Q}$	(L2 x P2) = **Q**	M2=P2=Q1	L2=Q2
	32	2.34	$(\mathbf{M} \wedge \mathbf{N}) \rightarrow \mathbf{Q}$	(M2 x N2) = **Q**	L2=N2=Q2	M2=Q1
3rd group	33	2.20	$(\mathbf{H} \wedge \mathbf{B}) \rightarrow \mathbf{O}$	(H3 x B2) = **O**	O8=H8=J7	O2=H3
	34	2.21	$(\mathbf{G} \wedge \mathbf{B}) \rightarrow \mathbf{I}$	(G6 x B1) = **I**	8=G8=J8	I5=G6
	35	2.22	$(\mathbf{D} \wedge \mathbf{J}) \rightarrow \mathbf{O}$	(D1 x J4) = **O**	O8=H8=J7	O6=J4
	36	2.23	$(\mathbf{C} \wedge \mathbf{J}) \rightarrow \mathbf{I}$	(C1 x J6) = **I**	8=G8=J8	I6=J6
	37	2.47	$(\mathbf{H} \wedge \mathbf{F}) \rightarrow \mathbf{T}$	(H1 x F2) = **T**	T8=H7=K7	T4=H1
	38	2.48	$(\mathbf{G} \wedge \mathbf{F}) \rightarrow \mathbf{S}$	(G3 x F1) = **S**	S8=G7=K8	S6=G3
	39	2.49	$(\mathbf{D} \wedge \mathbf{K}) \rightarrow \mathbf{T}$	(D2 x K1) = **T**	T8=H7=K7	T3=K1
	40	2.50	$(\mathbf{C} \wedge \mathbf{K}) \rightarrow \mathbf{S}$	(C2 x K2) = **S**	S8=G7=K8	S2=K2
	41	2.59	$(\mathbf{V} \wedge \mathbf{Q}) \rightarrow \mathbf{O}$	(V4 x Q2) = **S**	O7=V8=W7	O1=V4
	42	2.60	$(\mathbf{U} \wedge \mathbf{Q}) \rightarrow \mathbf{I}$	(U6 x Q1) = **I**	7=U8=W8	I3=U6
	43	2.61	$(\mathbf{P} \wedge \mathbf{W}) \rightarrow \mathbf{O}$	(P1 x W5) = **S**	O7=V8=W7	O5=W5
	44	2.62	$(\mathbf{N} \wedge \mathbf{W}) \rightarrow \mathbf{I}$	(N1 x W6) = **I**	7=U8=W8	I4=W6
	45	2.35	$(\mathbf{V} \wedge \mathbf{R}) \rightarrow \mathbf{T}$	(V1 x R2) = **T**	T7=V7=Ω7	T2=V1
	46	2.36	$(\mathbf{U} \wedge \mathbf{R}) \rightarrow \mathbf{S}$	(U4 x R1) = **S**	S7=U7=Ω8	S5=U4
	47	2.37	$(\mathbf{P} \wedge \mathbf{\Omega}) \rightarrow \mathbf{T}$	(P2 x Ω1) = **T**	T7=V7=Ω7	T1=Ω1
	48	2.38	$(\mathbf{N} \wedge \mathbf{\Omega}) \rightarrow \mathbf{S}$	(N2 x Ω3) = **V**	S7=U7=Ω8	S1=Ω3

It can be easily remarked that following universal derivations have the same FI (an therefore are equivalent derivations):

- 1, 24;
- 2, 23;
- 5, 20;
- 6, 19;
- 9, 32;
- 10, 31;
- 13, 28;
- 14, 27.

Similarly, the following particular derivations (having a particular premise) have the same FI:

- 3, 22, 34, 36;
- 4, 21, 33, 35;

- 7, 18, 38, 40;
- 8, 17, 37, 39;
- 11, 29, 41, 43;
- 15, 26, 46, 48;
- 16, 25, 45, 47.

We can further observe that there are 16 universal FI (also characterized by the FI A1=C1=F2) and 32 particular ones (of the kind I7=U8=W8) distributed in this way:

- Each kind of the universal ones is present one time in the first group and one time in the second group.
- Each kind of the particular ones is present one time in the first group, one time in the second group and 2 times in the third group.

Note also that each particular derivation, like for Lines 3-4, the relative FI is connected with the conclusion and not with the second premise (as it happens for universal inferences). For instance, in Line 3 G8 is a generating statement of the conclusion **G** while is not connected with the premise A1.

It may be further noted that each FI of the second group and the corresponding of the first group are contradictory. For instance, the FI of Line 1 is L1=B2=C2 and that of Line 17 is T8=H7=K7 and L1-T8, B2-H7, C2-K7 are contradictory pairs.

A peculiarity of all *Bocardo*- and *Disamis*-like inferences (Lines 33, 34, 37, 3 41, 42, 45, and 46) is that they have a FI that shows a statement that pertains to the RGS of statement and a II that shows a statement that pertains to the LGS of the same statement (for instance, in Line 1 we have O8 and O2, respectively; in Line 2 we have I8 and I5 respectively). Moreover, all conclusions of the third group are particular.

The above identities suggest an interesting possibility: we can perform the same inference individuated by the same formal structures but following different paths in the three-dimensional logical space Moreover, knowing the GE, FI and II allows a different kind of mechanical computation of the inferences.

6.3 Sum Inferences

We can deal now with the sum derivations as shown in the following table (where there is no corresponding table number is when I did not perform an explicit inference):

Table 6.6 Summary of all sum inferences

Group		Table	Inference	GE	FI	II
	1	3.3	$(\mathbf{O} \vee \mathbf{K}) \leftarrow \mathbf{H}$	$(O8 + K7) = \mathbf{H}$	T8=H7=K7	O8=H8
	2	—	$(\mathbf{I} \vee \mathbf{K}) \leftarrow \mathbf{G}$	$(I8 + K8) = \mathbf{G}$	S8=G7=K8	I8=G8
	3	3.4	$(\mathbf{O} \vee \mathbf{F}) \leftarrow \mathbf{D}$	$(O8 + F4) = \mathbf{D}$	E1=D1=F1	D4=F4
	4	3.5	$(\mathbf{I} \vee \mathbf{F}) \leftarrow \mathbf{C}$	$(I8 + F7) = \mathbf{C}$	A1=C1=F2	C3=F7
	5	—	$(\mathbf{T} \vee \mathbf{J}) \leftarrow \mathbf{H}$	$(T8 + J7) = \mathbf{H}$	O8=H8=J7	T8=H7
	6	—	$(\mathbf{S} \vee \mathbf{J}) \leftarrow \mathbf{G}$	$(S8 + J8) = \mathbf{G}$	I8=G8=J8	S8=G7
	7	3.20	$(\mathbf{T} \vee \mathbf{B}) \leftarrow \mathbf{D}$	$(T8 + B3) = \mathbf{D}$	M1=B1=D	B3=D7
	8	3.21	$(\mathbf{S} \vee \mathbf{B}) \leftarrow \mathbf{C}$	$(S8 + B5) = \mathbf{C}$	L1=B2=C2	B5=C5
1st group	9	—	$(\mathbf{O} \vee \mathbf{\Omega}) \leftarrow \mathbf{V}$	$(O7 + \Omega 7) = \mathbf{V}$	$T7=V7=\Omega 7$	O7=V8
	10	—	$(\mathbf{I} \vee \mathbf{\Omega}) \leftarrow \mathbf{U}$	$(I7 + \Omega 8) = \mathbf{U}$	$S7=U7=\Omega 8$	I7=U8
	11	3.28	$(\mathbf{O} \vee \mathbf{R}) \leftarrow \mathbf{P}$	$(O7 + R6) = \mathbf{P}$	E2=P1=R1	P6=R6
	12	3.29	$(\mathbf{I} \vee \mathbf{R}) \leftarrow \mathbf{N}$	$(I7 + R8) = \mathbf{N}$	A2=N1=R2	N4=R8
	13	—	$(\mathbf{T} \vee \mathbf{W}) \leftarrow \mathbf{V}$	$(T7 + W7) = \mathbf{V}$	O7=V8=W	T7=V7
	14	—	$(\mathbf{S} \vee \mathbf{W}) \leftarrow \mathbf{U}$	$(S7 + W8) = \mathbf{U}$	I7=U8=W8	S7=U7
	15	3.12	$(\mathbf{T} \vee \mathbf{Q}) \leftarrow \mathbf{P}$	$(T7 + Q4) = \mathbf{P}$	M2=P2=Q	P8=Q4
	16	3.13	$(\mathbf{S} \vee \mathbf{Q}) \leftarrow \mathbf{N}$	$(S7 + Q7) = \mathbf{N}$	L2=N2=Q2	N7=Q7
	17	3.6	$(\mathbf{O} \vee \mathbf{C}) \leftarrow \mathbf{B}$	$(O8 + C5) = \mathbf{B}$	L1=B2=C2	B5=C5
	18	3.7	$(\mathbf{I} \vee \mathbf{D}) \leftarrow \mathbf{B}$	$(I8 + D7) = \mathbf{B}$	M1=B1=D2	B3=D7
	19	—	$(\mathbf{O} \vee \mathbf{G}) \leftarrow \mathbf{J}$	$(O8 + G8) = \mathbf{J}$	I8=G8=J8	O8=J7
	20	—	$(\mathbf{I} \vee \mathbf{H}) \leftarrow \mathbf{J}$	$(I8 + H8) = \mathbf{J}$	O8=H8=J7	I8=J8
	21	3.22	$(\mathbf{T} \vee \mathbf{C}) \leftarrow \mathbf{F}$	$(T8 + C3) = \mathbf{F}$	A1=C1=F2	C3=F7
	22	3.23	$(\mathbf{S} \vee \mathbf{D}) \leftarrow \mathbf{F}$	$(S8 + D4) = \mathbf{F}$	E1=D1=F1	D4=F4
	23	—	$(\mathbf{T} \vee \mathbf{G}) \leftarrow \mathbf{K}$	$(T8 + G7) = \mathbf{K}$	S8=G7=K8	T8=K7
	24	—	$(\mathbf{S} \vee \mathbf{H}) \leftarrow \mathbf{K}$	$(S8 + H7) = \mathbf{K}$	T8=H7=K7	S8=K8
2nd group	25	3.30	$(\mathbf{O} \vee \mathbf{N}) \leftarrow \mathbf{Q}$	$(O7 + N7) = \mathbf{Q}$	L2–N2=Q2	N7=Q7
	26	3.31	$(\mathbf{I} \vee \mathbf{P}) \leftarrow \mathbf{Q}$	$(I7 + P8) = \mathbf{Q}$	M2=P2=Q1	P8=Q4
	27	—	$(\mathbf{O} \vee \mathbf{U}) \leftarrow \mathbf{W}$	$(O7 + U8) = \mathbf{W}$	I7=U8=W8	O7=W7
	28	—	$(\mathbf{I} \vee \mathbf{V}) \leftarrow \mathbf{W}$	$(I7 + V8) = \mathbf{W}$	O7=V8=W7	I7=W8
	29	3.14	$(\mathbf{T} \vee \mathbf{N}) \leftarrow \mathbf{R}$	$(T7 + N4) = \mathbf{R}$	A2=N1=R2	N4=R8
	30	3.15	$(\mathbf{S} \vee \mathbf{P}) \leftarrow \mathbf{R}$	$(S7 + P6) = \mathbf{R}$	E2=P1=R1	P6=R6
	31	—	$(\mathbf{T} \vee \mathbf{U}) \leftarrow \mathbf{\Omega}$	$(T7 + U7) = \mathbf{\Omega}$	$S7=U7=\Omega 8$	$T7= \Omega 7$
	32	—	$(\mathbf{S} \vee \mathbf{V}) \leftarrow \mathbf{\Omega}$	$(S7 + V7) = \mathbf{\Omega}$	$T7=V7=\Omega 7$	$S7= \Omega 8$

(*Continued*)

Table 6.6 (*Continued*)

Group		Table	Inference	GE	FI	II
3rd group	33	3.8	(**C** ∨ **K**) ← **A**	(C6 + K7) = **A**	A1=C1=F2	A7=C6
	34	3.9	(**D** ∨ **K**) ← **E**	(D3 + K8) = **E**	E1=D1=F1	E4=D3
	35	3.10	(**G** ∨ **F**) ← **A**	(G8 + F5) = **A**	A1=C1=F2	A3=F5
	36	3.11	(**H** ∨ **F**) ← **E**	(H8 + F3) = **E**	E1=D1=F1	E3=F3
	37	3.24	(**C** ∨ **J**) ← **L**	(C8 + J7) = **L**	L1=B2=C2	L5=C8
	38	3.25	(**D** ∨ **J**) ← **M**	(D6 + J8) = **M**	M1=B1=D2	M3=D6
	39	3.26	(**G** ∨ **B**) ← **L**	(G7 + B8) = **L**	L1=B2=C2	L6=B8
	40	3.27	(**H** ∨ **B**) ← **M**	(H7 + B7) = **M**	M1=B1=D2	M7=B7
	41	3.32	(**N** ∨ **Ω**) ← **A**	(N5 + Ω7) = **A**	A2=N1=R2	A8=N5
	42	3.33	(**P** ∨ **Ω**) ← **E**	(P3 + Ω8) = **A**	E2=P1=R1	E6=P3
	43	3.34	(**U** ∨ **R**) ← **A**	(U8 + R4) = **A**	A2=N1=R2	A4=R4
	44	3.35	(**V** ∨ **R**) ← **E**	(V8 + R3) = **E**	E2=P1=R1	E5=R3
	45	3.16	(**N** ∨ **W**) ← **L**	(N8 + W7) = **L**	L2=N2=Q2	L7=N8
	46	3.17	(**P** ∨ **W**) ← **M**	(P5 + W8) = **M**	M2=P2=Q1	M4=P5
	47	3.18	(**U** ∨ **Q**) ← **L**	(U7 + Q8) = **L**	L2=N2=Q2	L8=Q8
	48	3.19	(**V** ∨ **Q**) ← **M**	(V7 + Q6) = **M**	M2=P2=Q1	M8=Q6

As for Table 6.5, inferences that are identical show the same FI, but obviously both the FI and the II of a derivation of Table 6.5 are different from the FI and II of the correspondent statement of Table 6.6: both the respective FI and the respective II are indeed contradictory pairs. For instance, let us take Lines 1 in the two derivations: The FI of the product inference is L1=B2=C2 while its II is A1=C1, whereas the FI of the sum inference is T8=H7=K7 (L1-T8, B2-K7, C2-H7 are contradictory pairs) and its II is O8=H8 (A1-O8 and C1-H8 are again contradictory pairs). Therefore, we have here 32 universal FI and 16 particular ones, distributed in this way:

- Each kind of the universal FI is present one time in the first group, one time in the second group and two times in the third group.
- Each kind of the particular FI is present one time in the first group and one time in the second group.

Moreover, each particular FI of the sum derivations occurs precisely at the place in which a universal FI occurs in the product derivations and vice versa, and this according to the countervalence relations established above. For instance, A1=C1=F2 and O8=H8=J7 are contradictory FI (that is, A1 and O8 are contradictory pairs and so C1 and H8 as well as F2 and J7).

This also implies that the premises of the counterpart of an universal derivation in the product table, for instance Line 1, contains here two particular statements (like O8 and K7) and the FI is related to the latter whilst the II to the former. On the contrary, inferences showing a mix of universal and particular statements (like Lines 3-4), are characterized by a II connected with one of the two premises and a FI connected with the conclusion (in Line 3 is related to the conclusion **D**).

I also recall the peculiarity showed by all the *Bocardo*- and *Disamis*-like inferences (Lines 33, 34, 37, 38, 41, 42, 45, and 46): here, we have the FI that shows a statement that pertains to the LGS of a statement and a II that shows a statement that pertains to the RGS of the same bold statement (for instance, in Line 33 we have A1 and A7, respectively; in Line 34 E1 and E4, respectively). Note also that all conclusions of the third group are universal.

6.4 Subtraction Inferences

I also summarize here the 48 derivations by means of subtractions:

Table 6.7 Summary of all subtraction inferences

Group		Table	Inference	GE	FI	II
1st group	1	4.2	(**A** ∧ ¬**K**) → **C**	(A1 - K7) = **C**	T8=H7=K7	A1=C1
	2	—	(**E** ∧ ¬**K**) → **D**	(E1 - K8) = **D**	S8=G7=K8	E1=D1
	3	4.3	(**A** ∧ ¬**F**) → **G**	(A1 - F4) = **G**	I8=G8=J8	F4=¬G5
	4	4.4	(**E** ∧ ¬**F**) → **H**	(E1 - F7) = **H**	O8=H8=J7	F7=¬H6
	5	—	(**L** ∧ ¬**J**) → **C**	(L1 - J7) = **C**	O8=H8=J7	L1=C2
	6	—	(**M** ∧ ¬**J**) → **D**	(M1 - J8) = **D**	I8=G8=J8	M1=D2
	7	4.19	(**L** ∧ ¬**B**) → **G**	(L1 - B3) = **G**	S8=G7=K8	B3=¬G2
	8	4.20	(**M** ∧ ¬**B**) → **H**	(M1 - B5) = **H**	T8=H7=K7	B5=¬H4
	9	—	(**A** ∧ ¬**Ω**) → **N**	(A2 - Ω7) = **N**	T7=V7=Ω7	A2=N1
	10	—	(**E** ∧ ¬**Ω**) → **P**	(E2 - Ω8) = **P**	S7=U7=Ω8	E2=P1
	11	4.27	(**A** ∧ ¬**R**) → **U**	(A2 - R6) = **U**	I7=U8=W8	R6=¬U3
	12	4.28	(**E** ∧ ¬**R**) → **V**	(E2 - R8) = **V**	O7=V8=W7	R8=¬V5

(*Continued*)

Table 6.7 (*Continued*)

Group		Table	Inference	GE	FI	II
	13	—	**(L ∧ ¬W) → N**	(L2 - W7) = **N**	O7=V8=W7	L2=N2
	14	—	**(M ∧ ¬W) → P**	(M2 - W8) = **P**	I7=U8=W8	M2=P2
	15	4.11	**(L ∧ ¬Q) → U**	(L2 - Q4) = **U**	S7=U7=Ω8	Q4=¬U1
	16	4.12	**(M ∧ ¬Q) → V**	(M2 - Q7) = **V**	T7=V7=Ω7	Q7=¬V2
	17	4.5	**(A ∧ ¬C) → K**	(A1 - C5) = **K**	T8=H7=K7	C5=¬K4
	18	4.6	**(E ∧ ¬D) → K**	(E1 - D7) = **K**	S8=G7=K8	D7=¬K6
	19	—	**(A ∧ ¬G) → F**	(A1 - G8) = **F**	I8=G8=J8	A1=F2
	20	—	**(E ∧ ¬H) → F**	(E1 - H8) = **F**	O8=H8=J7	E1=F1
	21	4.21	**(L ∧ ¬C) → J**	(L1 - C3) = **J**	O8=H8=J7	C3=¬J2
	22	4.22	**(M ∧ ¬D) → J**	(M1 - D4) = **J**	I8=G8=J8	D4=¬J5
	23	—	**(L ∧ ¬G) → B**	(L1 - G7) = **B**	S8=G7=K8	L1=B2
2nd group	24	—	**(M ∧ ¬H) → B**	(M1 - H7) = **B**	T8=H7=K7	M1=B1
	25	4.29	**(A ∧ ¬N) → Ω**	(A2 - N7) = **Ω**	T7=V7=Ω7	N7=¬V2=¬Ω2
	26	4.30	**(E ∧ ¬P) → Ω**	(E2 - P8) = **Ω**	S7=U7=Ω8	P8=¬U1=¬Ω5
	27	—	**(A ∧ ¬U) → R**	(A2 - U8) = **R**	I7=U8=W8	A2=R2
	28	—	**(E ∧ ¬V) → R**	(E2 - V8) = **R**	O7=V8=W7	E2=R1
	29	4.13	**(L ∧ ¬N) → W**	(L2 - N4) = **W**	O7=V8=W7	N4=¬W1
	30	4.14	**(M ∧ ¬P) → W**	(M2 - P6) = **W**	I7=U8=W8	P6=¬W3
	31	—	**(L ∧ ¬U) → Q**	(L2 - U7) = **Q**	S7=U7=Ω8	L2=Q2
	32	—	**(M ∧ ¬V) → Q**	(M2 - V7) = **Q**	T7=V7=Ω7	M2=Q1
	33	4.7	**(H ∧ ¬K) → O**	(H3 - K7) = **O**	O8=H8=J7	O2=H3
	34	4.8	**(G ∧ ¬K) → I**	(G6 - K8) = **I**	I8=G8=J8	I5=G6
	35	4.9	**(D ∧ ¬F) → O**	(D1 - F5) = **O**	O8=H8=J7	F5=¬J4
	36	4.10	**(C ∧ ¬F) → I**	(C1 - F3) = **I**	I8=G8=J8	F3=¬J6
	37	4.23	**(H ∧ ¬J) → T**	(H1 - J7) = **T**	T8=H7=K7	T4=H1
	38	4.24	**(G ∧ ¬J) → S**	(G3 - J8) = **S**	S8=G7=K8	S6=G3
	39	4.25	**(D ∧ ¬B) → T**	(D2 - B8) = **T**	T8=H7=K7	B8=¬T3=¬K1
	40	4.26	**(C ∧ ¬B) → S**	(C2 - B7) = **S**	S8=G7=K8	B7=¬S2=¬K2
3rd group	41	4.31	**(V ∧ ¬Ω) → O**	(V4 - Ω7) = **O**	T7=V7=Ω7	O1=V4
	42	4.32	**(U ∧ ¬Ω) → I**	(U6 - Ω8) = **I**	S7=U7=Ω8	I3=U6
	43	4.33	**(P ∧ ¬R) → O**	(P1 - R4) = **O**	O7=V8=W7	R4=¬W5=¬O5
	44	4.34	**(N ∧ ¬R) → I**	(N1 - R3) = **I**	I7=U8=W8	R3=¬W6=¬I4
	45	4.15	**(V ∧ ¬W) → T**	(V1 - W7) = **T**	O7=V8=W7	T2=V1
	46	4.16	**(U ∧ ¬W) → S**	(U4 - W8) = **S**	I7=U8=W8	S5=U4
	47	4.17	**(P ∧ ¬Q) → T**	(P2 - Q8) = **T**	T7=V7=Ω7	Q8=¬T1
	48	4.18	**(N ∧ ¬Q) → S**	(N2 - Q6) = **S**	S7=U7=Ω8	Q6=¬S1

It may be noted that (apart from 8 cases) either the FIs of the subtraction and product inferences are opposite or their II it is. In particular,

- In the case of what in the product table are considered universal inferences (for instance Lines 1 and 2) the II is the same in the two derivations but the FIs represent contradictory pairs.
- In the case of what in the product table are considered particular ones (like Lines 3-4), the two corresponding II represent contradictory pairs (consider for instance the third line in the two lists) but their FI is the same. In particular, the II is such that a premise is equal to the negation of the II of the relative product inference. It may be also noted that these inferences show here both premises as universal.

It may also be noted that all of the FI are particular. Moreover, there are 8 inferences in the third group (Lines 33, 34, 37, 38, 41, 42, 45, and 46) whose premises are both particular and whose FIs are connected with the conclusions. These are the *Bocardo*- and *Disamis*-like inferences showing a further peculiar features: Only in this case both the FI and the II are equal to the corresponding ones of the product inferences. Note also that are conclusions of the third group are particular as it happens for product inferences. Moreover, the conclusions of all inferences of the third group are particular statements of the ground type (**I, O, S, T**). At the opposite, the grounding universals (**A, E, L, M**) are absent as conclusions.

It is finally interesting to remark that all inference of the second group have the same three propositions as the inferences of the first group but with some permutation (and also interchange of signs). Compare for instance Lines 1 and 17.

6.5 Division Inferences

All division inferences may be summarized as in the following table:

Table 6.8 Summary of all division inferences

Group		Table	Inference	GE	FI	II
	1	5.3	(**O** ∨ ¬**B**) ← **H**	(O8 : B2)= **H**	L1=B2=C2	O8=H8
	2	—	(**I** ∨ ¬**B**) ← **G**	(I8 : B1)= **G**	M1=B1=D2	I8=G8
	3	5.4	(**O** ∨ ¬**J**) ← **D**	(O8 : J5)= **D**	E1=D1=F1	G5=J5=¬D4=¬F4
	4	5.5	(**I** ∨ ¬**J**) ← **C**	(I8 : J2)= **C**	A1=C1=F2	H6=J2=¬C3=¬F7
	5	—	(**T** ∨ ¬**F**) ← **H**	(T8 : F2)= **H**	A1=C1=F2	T8=H7
	6	—	(**S** ∨ ¬**F**) ← **G**	(S8 : F1)= **G**	E1=D1=F1	S8=G7
1st group	7	5.20	(**T** ∨ ¬**K**) ← **D**	(T8 : K6)= **D**	M1=B1=D2	G2=K6=¬D7=¬B3
	8	5.21	(**S** ∨ ¬**K**) ← **C**	(S8 : K4)= **C**	L1=B2=C2	H4=K4=¬B5=¬C5
	9	—	(**O** ∨ ¬**Q**) ← **V**	(O7 : Q2)= **V**	L2=N2=Q2	O7=V8
	10	—	(**I** ∨ ¬**Q**) ← **U**	(I7 : Q1)= **U**	M2=P2=Q1	I7=U8
	11	5.28	(**O** ∨ ¬**W**) ← **P**	(O7 : W3)= **P**	E2=P1=R1	U3=W3=¬P6=¬R6
	12	5.29	(**I** ∨ ¬**W**) ← **N**	(I7 : W1)= **N**	A2=N1=R2	V5=W1=¬N4=¬R8
	13	—	(**T** ∨ ¬**R**) ← **V**	(T7 : R2)= **V**	A2=N1=R2	T7=V7
	14	—	(**S** ∨ ¬**R**) ← **U**	(S7 : R1)= **U**	E2=P1=R1	S7=U7
	15	5.12	(**T** ∨ ¬**Ω**) ← **P**	(T7 : Ω5) = **P**	M2=P2=Q1	U1=Ω5=¬P8=¬Q4
	16	5.13	(**S** ∨ ¬**Ω**) ← **N**	(S7 : Ω2) = **N**	L2=N2=Q2	V2=Ω2=¬P7=¬Q7
	17	5.6	(**O** ∨ ¬**H**) ← **B**	(O8 : H4)= **B**	L1=B2=C2	H4=K4=¬C5=¬D5
	18	5.7	(**I** ∨ ¬**G**) ← **B**	(I8 : G2)= **B**	M1=B1=D2	G2=K6=¬B3=¬D7
	19	—	(**O** ∨ ¬**D**) ← **J**	(O8 : D2)= **J**	E1=D1=F1	O8=J7
	20	—	(**I** ∨ ¬**C**) ← **J**	(I8 : C1)= **J**	A1=C1=F2	I8=J8
	21	5.22	(**T** ∨ ¬**H**) ← **F**	(T8 : H6)= **F**	A1=C1=F2	H6=J2=¬C3=¬F7
	22	5.23	(**S** ∨ ¬**G**) ← **F**	(S8 : G5)= **F**	E1=D1=F1	G5=J5=¬D4=¬F4
2nd group	23	—	(**T** ∨ ¬**D**) ← **K**	(T8 : D2)= **K**	M1=B1=D2	T8=K7
	24	—	(**S** ∨ ¬**C**) ← **K**	(S8 : C2)= **K**	L1=B2=C2	S8=K8
	25	5.30	(**O** ∨ ¬**V**) ← **Q**	(O7 : V2)= **Q**	L2=N2=Q2	V2=Ω2=¬N7=¬Q7
	26	5.31	(**I** ∨ ¬**U**) ← **Q**	(I7 : U1)= **Q**	M2=P2=Q1	U1=Ω5=¬P8=¬Q4
	27	—	(**O** ∨ ¬**P**) ← **W**	(O7 : P1)= **W**	E2=P1=R1	O7=W7
	28	—	(**I** ∨ ¬**N**) ← **W**	(I7 : N1)= **W**	A2=N1=R2	I7=W8
	29	5.14	(**T** ∨ ¬**V**) ← **R**	(T7 : V5) = **R**	A2=N1=R2	V5=W1
	30	5.15	(**S** ∨ ¬**U**) ← **R**	(S7 : U3) = **R**	E2=P1=R1	U3=W3
	31	—	(**T** ∨ ¬**P**) ← **Ω**	(T7 : P2) = **Ω**	L2=N2=Q2	T7=Ω7
	32	—	(**S** ∨ ¬**N**) ← **Ω**	(S7 : N2) = **Ω**	M2=P2=Q1	S7=Ω8

Group		Table	Inference	GE	FI	II
3rd group	33	5.8	**(C** ∨ ¬**B)** ← **A**	(C6 : B2)= **A**	A1=C1=F2	A7=C6
	34	5.9	**(D** ∨ ¬**B)** ← **E**	(D3 : B1)= **E**	E1=D1=F1	E4=D3
	35	5.10	**(G** ∨ ¬**J)** ← **A**	(G8 : J4)= **A**	A1=C1=F2	O6=J4=¬A3=¬F5
	36	5.11	**(H** ∨ ¬**J)** ← **E**	(H8 : J6)= **E**	E1=D1=F1	I6=J6=¬E3=¬F3
	37	5.24	**(C** ∨ ¬**F)** ← **L**	(C8 : F2)= **L**	L1=B2=C2	L5=C3
	38	5.25	**(D** ∨ ¬**F)** ← **M**	(D6 : F1)= **M**	M1=B1=D2	M3=D6
	39	5.26	**(G** ∨ ¬**K)** ← **L**	(G7 : K1) = **L**	L1=B2=C2	T3=K1=¬L6=¬B8
	40	5.27	**(H** ∨ ¬**K)** ← **M**	(H7 : K2) = **M**	M1=B1=D2	S2=K2=¬M7=¬B7
	41	5.32	**(N** ∨ ¬**Q)** ← **A**	(N5 : Q2)= **A**	A2=N1=R2	A8=N5
	42	5.33	**(P** ∨ ¬**Q)** ← **E**	(P3 : Q1)= **E**	E2=P1=R1	E6=P3
	43	5.34	**(U** ∨ ¬**W)** ← **A**	(U8 : W5)= **A**	A2=N1=R2	O5=W5=¬A4=¬R4
	44	5.35	**(V** ∨ ¬**W)** ← **E**	(V8 : W6)= **E**	E2=P1=R1	I4=W6=¬E5=¬R3
	45	5.16	**(N** ∨ ¬**R)** ← **L**	(N8 : R2)= **L**	L2=N2=Q2	L7=N8
	46	5.17	**(P** ∨ ¬**R)** ← **M**	(P5 : R1)= **M**	M2=P2=Q1	M4=P5
	47	5.18	**(U** ∨ ¬**Ω)** ← **L**	(U7 : Ω1) = **L**	L2=N2=Q2	T1=Ω1=¬L8=¬Q8
	48	5.19	**(V** ∨ ¬**Ω)** ← **M**	(V7 : Ω3) = **M**	M2=P2=Q1	S1=Ω3=¬M8=¬Q6

It may be noted that all FIs are universal and together with their counterpart in the subtraction table constitute contradictory pairs. Also the II of subtraction inferences and division inferences constitute contradictory pairs. Note also that all conclusions of the third group are universal, as it happens for the sum inferences. Note also all the conclusions of the third group are universal statements of the grounding type **(A, E, L, M)**.

In conclusion, we may also remark some interesting relations. The product inferences have the following connection with division inferences: all the product inferences of the first group have the same three letters but in inverted order relative to all the division inferences of the third group, and vice versa all the product inferences of the third group have the same three letters but in inverted order relative to all the division inferences of the first group. This is not by chance since e.g. **(C** ∨ ¬**B)** ← **A** is logically equivalent to **(A** ∧ **B)** → **C**. On the other hand sum and subtraction inferences have the same relation among them as product and division inferences have, and again this is due to the equivalence of the logical relations. For example, **(C**∨**K)** ← **A** is logically equivalent to **(A** ∧ ¬**K)** → **C**. It is further worth noticing that the conclusions of all product inferences are precisely the same as the corresponding subtraction ones as well as the conclusions of all sum inferences are the same as those of the corresponding division inferences.

This examination so far allows for very interesting forms of cross–inferences. Let us consider the classical *Barbara* (Line 1 of Table 6.5):

$$(\mathbf{A} \wedge \mathbf{B}) \rightarrow \mathbf{C}. \tag{6.1}$$

Let us now consider two sum inferences whose conclusions are represented by the premises of *Barbara*, for instance, Lines 18 and 35 of Table 6.6. We can build a sum inference having the disjunction of these two inferences as premise and **C** as conclusion, i.e.

$$\{[(\mathbf{G} \vee \mathbf{F}) \leftarrow \mathbf{A}] \vee [(\mathbf{I} \vee \mathbf{D}) \leftarrow \mathbf{B}]\} \leftarrow \mathbf{C}, \tag{6.2}$$

We could also consider the inverse case, for instance a sum inference like Line 1 of Table 6.6, that is,

$$(\mathbf{O} \vee \mathbf{K}) \leftarrow \mathbf{H}. \tag{6.3}$$

Let us now consider two product inferences whose conclusion is represented by the negation of one of the two premises of *Barbara* (i.e. by **O** or **K**), for instance Lines 35 and 18 of Table 6.6. Then, we can consider the sum inference constituted by the disjunction between these two inferences with the conclusion **H**, i.e.

$$\{[(\mathbf{D} \wedge \mathbf{J}) \rightarrow \mathbf{O}] \vee [(\mathbf{E} \wedge \mathbf{G}) \rightarrow \mathbf{K}]\} \leftarrow \mathbf{H}, \tag{6.4}$$

which is again a tautology. We can also build a complex necessary–condition inference having the conjunction of latter two inference as premises and the conclusion of *Barbara* ((i.e. **C**) as a conclusion, i.e.:

$$\{[(\mathbf{D} \wedge \mathbf{J}) \rightarrow \mathbf{O}] \wedge [(\mathbf{E} \wedge \mathbf{G}) \rightarrow \mathbf{K}]\} \leftarrow \mathbf{C}, \tag{6.5}$$

which is again a tautology. Along these outlines many combinations can be built.

6.6 Simplified Summary of the Previous Inferences

In conclusion, I stress again that all the four kinds of inferences are fully equivalent so that inferences on the same row in the previous 4 tables could be considered as variation of the same inferential form.

This is even more true for product and subtraction on the one hand and for sum and division inferences on the other, which pairwise display precisely the same formal structure apart from a negative formulation of one of the premises. However, in order to make use of the computational potentialities of all the different operations (product, sum, subtraction and division) as well as to grasp the many formal consequences that can be drawn (some of them will be also clear in the following) this treatment is fundamental. It can be therefore helpful to summarize here the previous results. First, a helpful table can be a short summary of the 24 grounding statements:

Table 6.9 Summary of all grounding statements

A	$X \rightarrow Y$	Every X is Y
B	$Z \rightarrow X$	Every Z is X
C	$Z \rightarrow Y$	Every Z is Y
D	$Z \rightarrow \neg Y$	No Z is Y
E	$X \rightarrow \neg Y$	No X is Y
F	$Z \rightarrow \neg X$	No Z is X
G	$Z \wedge Y$	Some Z is Y
H	$Z \wedge \neg Y$	Some Z is not Y
I	$X \wedge Y$	Some X is Y
J	$Z \wedge X$	Some Z is X
K	$Z \wedge \neg X$	Some Z is not X
L	$\neg X \rightarrow Y$	Every non-X is Y
M	$\neg X \rightarrow \neg Y$	No non-X is Y
N	$\neg Z \rightarrow Y$	Every non-Z is Y
O	$X \wedge \neg Y$	Some X is not Y
P	$\neg Z \rightarrow \neg Y$	No non-Z is Y
Q	$\neg Z \rightarrow X$	Every non-Z is X
R	$\neg Z \rightarrow \neg X$	No non-Z is X
S	$\neg X \wedge Y$	Some non-X is Y
T	$\neg X \wedge \neg Y$	Some non-X is not Y
U	$\neg Z \wedge Y$	Some non-Z is Y
V	$\neg Z \wedge \neg Y$	Some non-Z is not Y
W	$\neg Z \wedge X$	Some non-Z is X
Ω	$\neg Z \wedge \neg X$	Some non-Z is not X

A helpful table is that of the product inferences and their translation in current language:

Table 6.10 Summary of the meaning of product inferences

$(A \wedge B) \rightarrow C$	If Every X is Y and Every Z is X, then Every Z is Y
$(E \wedge B) \rightarrow D$	If No X is Y and Every Z is X, then No Z is Y
$(A \wedge J) \rightarrow G$	If Every X is Y and Some Z is X, then Some Z is Y
$(E \wedge J) \rightarrow H$	If No X is Y and Some Z is X, then Some Z is not Y
$(L \wedge F) \rightarrow C$	If Every non-X is Y and No Z is X, then Every Z is Y
$(M \wedge F) \rightarrow D$	If No non-X is Y and No Z is X, then No Z is Y
$(L \wedge K) \rightarrow G$	If Every non-X is Y and Some Z is not X, then Some Z is Y
$(M \wedge K) \rightarrow H$	If No non-X is Y and Some Z is not X, then Some Z is not Y
$(A \wedge Q) \rightarrow N$	If Every X is Y and Every non-Z is X, then Every non-Z is Y
$(E \wedge Q) \rightarrow P$	If No X is Y and Every non-Z is X, then No non-Z is Y
$(A \wedge W) \rightarrow U$	If Every X is Y and Some non-Z is X, then Some non-Z is Y
$(E \wedge W) \rightarrow V$	If No X is Y and Some non-Z is X, then Some non-Z is not Y
$(L \wedge R) \rightarrow N$	If Every non-X is Y and No non-Z is X, then Every non-Z is Y
$(M \wedge R) \rightarrow P$	If No non-X is Y and No non-Z is X, then No non-Z is Y
$(L \wedge \Omega) \rightarrow U$	If Every non-X is Y and Some non-Z is not X, then Some non-Z is Y
$(M \wedge \Omega) \rightarrow V$	If No non-X is Y and Some non-Z is not X, then Some non-Z is not Y
$(A \wedge H) \rightarrow K$	If Every X is Y and Some Z is not Y, then Some Z is not X
$(E \wedge G) \rightarrow K$	If No X is Y and Some Z is Y, then Some Z is not X
$(A \wedge D) \rightarrow F$	If Every X is Y and No Z is Y, then No Z is X
$(E \wedge C) \rightarrow F$	If No X is Y and Every Z is Y, then No Z is X
$(L \wedge H) \rightarrow J$	If Every non-X is Y and Some Z is not Y, then Some Z is X
$(M \wedge G) \rightarrow J$	If No non-X is Y and Some Z is Y, then Some Z is X
$(L \wedge D) \rightarrow B$	If Every non-X is Y and No Z is Y, then Every Z is X
$(M \wedge C) \rightarrow B$	If No non-X is Y and Every Z is Y, then Every Z is X
$(A \wedge V) \rightarrow \Omega$	If Every X is Y and Some non-Z is not Y, then Some non-Z is not X
$(E \wedge U) \rightarrow \Omega$	If No X is Y and Some non-Z is Y, then Some non-Z is not X
$(A \wedge P) \rightarrow R$	If Every X is Y and No non-Z is Y, then No non-Z is X
$(E \wedge N) \rightarrow R$	If No X is Y and Every non-Z is Y, then No non-Z is X
$(L \wedge V) \rightarrow W$	If Every non-X is Y and Some non-Z is not Y, then Some non-Z is X
$(M \wedge U) \rightarrow W$	If No non-X is Y and Some non-Z is Y, then Some non-Z is X
$(L \wedge P) \rightarrow Q$	If Every non-X is Y and No non-Z is Y, then Every non-Z is X
$(M \wedge N) \rightarrow Q$	If No non-X is Y and Every non-Z is Y, then Every non-Z is X
$(H \wedge B) \rightarrow O$	If Some Z is not Y and Every Z is X, then Some X is not Y
$(G \wedge B) \rightarrow I$	If Some Z is Y and Every Z is X, then Some X is Y
$(D \wedge J) \rightarrow O$	If No Z is Y and Some Z is X, then Some X is not Y
$(C \wedge J) \rightarrow I$	If Every Z is Y and Some Z is X, then Some X is Y
$(H \wedge F) \rightarrow T$	If Some Z is not Y and No Z is X, then Some non-X is not Y
$(G \wedge F) \rightarrow S$	If Some Z is Y and No Z is X, then Some non-X is Y

$(\mathbf{D} \wedge \mathbf{K}) \rightarrow \mathbf{T}$	If No Z is Y and Some Z is not X, then Some non-X is not Y
$(\mathbf{C} \wedge \mathbf{K}) \rightarrow \mathbf{S}$	If Every Z is Y and Some Z is not X, then Some non-X is Y
$(\mathbf{V} \wedge \mathbf{Q}) \rightarrow \mathbf{O}$	If Some non-Z is not Y and Every non-Z is X, then Some X is not Y
$(U \wedge \mathbf{Q}) \rightarrow \mathbf{I}$	If Some non-Z is Y and Every non-Z is X, then Some X is Y
$(\mathbf{P} \wedge \mathbf{W}) \rightarrow \mathbf{O}$	If No non-Z is Y and Some non-Z is X, then Some X is not Y
$(\mathbf{N} \wedge \mathbf{W}) \rightarrow \mathbf{I}$	If Every non-Z is Y and Some non-Z is X, then Some X is Y
$(\mathbf{V} \wedge \mathbf{R}) \rightarrow \mathbf{T}$	If Some non-Z is not Y and No non-Z is X, then Some non-X is not Y
$(\mathbf{U} \wedge \mathbf{R}) \rightarrow \mathbf{S}$	If Some non-Z is Y and No non-Z is X, then Some non-X is Y
$(\mathbf{P} \wedge \mathbf{\Omega}) \rightarrow \mathbf{T}$	If No non-Z is Y and Some non-Z is not X, then Some non-X is not Y
$(\mathbf{N} \wedge \mathbf{\Omega}) \rightarrow \mathbf{S}$	If Every non-Z is Y and Some non-Z is not X, then Some non-X is Y

It is also helpful to give a summary of the sum inferences as in Table 6.11. Subtractions and divisions can easily be led back to the previous and this case, respectively, by assuming that the second premise of the sum inference and the second premise of the product inferences, respectively, is not true.

Table 6.11 Summary of the meaning of sum inferences

$(\mathbf{O} \vee \mathbf{K}) \leftarrow \mathbf{H}$	When Some X is not Y or Some Z is not X, we assume that Some Z is not Y
$(\mathbf{I} \vee \mathbf{K}) \leftarrow \mathbf{G}$	When Some X is Y or Some Z is not X, we assume that Some Z is Y
$(\mathbf{O} \vee \mathbf{F}) \leftarrow \mathbf{D}$	When Some X is not Y or No Z is X, we assume that No Z is Y
$(\mathbf{I} \vee \mathbf{F}) \leftarrow \mathbf{C}$	When Some X is Y or No Z is X, we assume that Every Z is Y
$(\mathbf{T} \vee \mathbf{J}) \leftarrow \mathbf{H}$	When Some non-X is not Y or some Z is X, we assume that Some Z is not Y
$(\mathbf{S} \vee \mathbf{J}) \leftarrow \mathbf{G}$	When Some non-X is Y or Some Z is X, we assume that Some Z is Y
$(\mathbf{T} \vee \mathbf{B}) \leftarrow \mathbf{D}$	When Some non-X is not Y or Every Z is X, we assume that No Z is Y
$(\mathbf{S} \vee \mathbf{B}) \leftarrow \mathbf{C}$	When Some non-X is Y or Every Z is X, we assume that Every Z is Y
$(\mathbf{O} \vee \mathbf{\Omega}) \leftarrow \mathbf{V}$	When Some X is not Y or Some non-Z is not X, we assume that Some non-Z is not Y
$(\mathbf{I} \vee \mathbf{\Omega}) \leftarrow \mathbf{U}$	When Some X is Y or Some non-Z is not X, we assume that Some non-Z is Y
$(\mathbf{O} \vee \mathbf{R}) \leftarrow \mathbf{P}$	When Some X is not Y or No non-Z is X, we assume that No non-Z is Y
$(\mathbf{I} \vee \mathbf{R}) \leftarrow \mathbf{N}$	When Some X is Y or No non-Z is X, we assume that Every non-Z is Y
$(\mathbf{T} \vee \mathbf{W}) \leftarrow \mathbf{V}$	When Some non-X is not Y or Some non-Z is X, we assume that Some non-Z is not Y
$(\mathbf{S} \vee \mathbf{W}) \leftarrow \mathbf{U}$	When Some non-X is Y or Some non-Z is X, we assume that Some non-Z is Y
$(\mathbf{T} \vee \mathbf{Q}) \leftarrow \mathbf{P}$	When Some non-X is not Y or Every non-Z is X, we assume that No non-Z is Y
$(\mathbf{S} \vee \mathbf{Q}) \leftarrow \mathbf{N}$	When Some non-X is Y or Every non-Z is X, we assume that Every non-Z is Y
$(\mathbf{O} \vee \mathbf{C}) \leftarrow \mathbf{B}$	When Some X is not Y or Every Z is Y, we assume that Every Z is X
$(\mathbf{I} \vee \mathbf{D}) \leftarrow \mathbf{B}$	When Some X is Y or No Z is Y, we assume that Every Z is X

(*Continued*)

Table 6.11 (*Continued*)

$(\mathbf{O} \vee \mathbf{G}) \leftarrow \mathbf{J}$	When Some *X* is not *Y* or Some *Z* is *Y*, we assume that Some *Z* is *X*
$(\mathbf{I} \vee \mathbf{H}) \leftarrow \mathbf{J}$	When Some *X* is *Y* or Some *Z* is not *Y*, we assume that Some *Z* is *X*
$(\mathbf{T} \vee \mathbf{C}) \leftarrow \mathbf{F}$	When Some non-*X* is not *Y* or Every *Z* is *Y*, we assume that No *Z* is *X*
$(\mathbf{S} \vee \mathbf{D}) \leftarrow \mathbf{F}$	When Some non-*X* is *Y* or No *Z* is *Y*, we assume that No *Z* is *X*
$(\mathbf{T} \vee \mathbf{G}) \leftarrow \mathbf{K}$	When Some non-*X* is not *Y* or Some *Z* is *Y*, we assume that Some *Z* is not *X*
$(\mathbf{S} \vee \mathbf{H}) \leftarrow \mathbf{K}$	When Some non-*X* is *Y* or Some *Z* is not *Y*, we assume that Some *Z* is not *X*
$(\mathbf{O} \vee \mathbf{N}) \leftarrow \mathbf{Q}$	When Some *X* is not *Y* or Every non-*Z* is *Y*, we assume that Every non-*Z* is *X*
$(\mathbf{I} \vee \mathbf{P}) \leftarrow \mathbf{Q}$	When Some *X* is *Y* or No non-*Z* is *Y*, we assume that Every non-*Z* is *X*
$(\mathbf{O} \vee \mathbf{U}) \leftarrow \mathbf{W}$	When Some *X* is not *Y* or Some non-*Z* is *Y*, we assume that Some non-*Z* is *X*
$(\mathbf{I} \vee \mathbf{V}) \leftarrow \mathbf{W}$	When Some *X* is *Y* or Some non-*Z* is not *Y*, we assume that Some non-*Z* is *X*
$(\mathbf{T} \vee \mathbf{N}) \leftarrow \mathbf{R}$	When Some non-*X* is not *Y* or Every non-*Z* is *Y*, we assume that No non-*Z* is *X*
$(\mathbf{S} \vee \mathbf{P}) \leftarrow \mathbf{R}$	When Some non-*X* is *Y* or No non-*Z* is *Y*, we assume that No non-*Z* is *X*
$(\mathbf{T} \vee \mathbf{U}) \leftarrow \mathbf{\Omega}$	When Some non-*X* is not *Y* or Some non-*Z* is *Y*, we assume that Some non-*Z* is not *X*
$(\mathbf{S} \vee \mathbf{V}) \leftarrow \mathbf{\Omega}$	When Some non-*X* is *Y* or Some non-*Z* is not *Y*, we assume that Some non-*Z* is not *X*
$(\mathbf{C} \vee \mathbf{K}) \leftarrow \mathbf{A}$	When Every *Z* is *Y* or Some *Z* is not *X*, we assume that Every *X* is *Y*
$(\mathbf{D} \vee \mathbf{K}) \leftarrow \mathbf{E}$	When No *Z* is *Y* or Some *Z* is not *X*, we assume that No *X* is *Y*
$(\mathbf{G} \vee \mathbf{F}) \leftarrow \mathbf{A}$	When Some *Z* is *Y* or No *Z* is *X*, we assume that Every *X* is *Y*
$(\mathbf{H} \vee \mathbf{F}) \leftarrow \mathbf{E}$	When Some *Z* is not *Y* or No *Z* is *X*, we assume that No *X* is *Y*
$(\mathbf{C} \vee \mathbf{J}) \leftarrow \mathbf{L}$	When Every *Z* is *Y* or Some *Z* is *X*, we assume that Every non-*X* is *Y*
$(\mathbf{D} \vee \mathbf{J}) \leftarrow \mathbf{M}$	When No *Z* is *Y* or Some *Z* is *X*, we assume that No non-*X* is *Y*
$(\mathbf{G} \vee \mathbf{B}) \leftarrow \mathbf{L}$	When Some *Z* is *Y* or Every *Z* is *X*, we assume that Every non-*X* is *Y*
$(\mathbf{H} \vee \mathbf{B}) \leftarrow \mathbf{M}$	When Some *Z* is not *Y* or Every *Z* is *X*, we assume that No non-*X* is *Y*
$(\mathbf{N} \vee \mathbf{\Omega}) \leftarrow \mathbf{A}$	When Every Non-*Z* is *Y* or Some non-*Z* is not *X*, we assume that Every *X* is *Y*
$(\mathbf{P} \vee \mathbf{\Omega}) \leftarrow \mathbf{E}$	When No non-*Z* is *Y* or Some non-*Z* is not *X*, we assume that No *X* is *Y*
$(\mathbf{U} \vee \mathbf{R}) \leftarrow \mathbf{A}$	When Some non-*Z* is *Y* or No non-*Z* is *X*, we assume that Every *X* is *Y*
$(\mathbf{V} \vee \mathbf{R}) \leftarrow \mathbf{E}$	When Some non-*Z* is not *Y* or No non-*Z* is *X*, we assume that No *X* is *Y*
$(\mathbf{N} \vee \mathbf{W}) \leftarrow \mathbf{L}$	When Every non-*Z* is *Y* or Some non-*Z* is *X*, we assume that Every non-*X* is *Y*
$(\mathbf{P} \vee \mathbf{W}) \leftarrow \mathbf{M}$	When No non-*Z* is *Y* or Some non-*Z* is *X*, we assume that No non-*X* is *Y*
$(\mathbf{U} \vee \mathbf{Q}) \leftarrow \mathbf{L}$	When Some non-*Z* is *Y* or Every non-*Z* is *X*, we assume that Every non-*X* is *Y*
$(\mathbf{V} \vee \mathbf{Q}) \leftarrow \mathbf{M}$	When Some non-*Z* is not *Y* or Every non-*Z* is *X*, we assume that No non-*X* is *Y*

Chapter 7

Generalized Representation and Structural Relations

Subtraction and division give us the possibility to consider the statements of this logical system as well as their mutual relations in a new way, expanding the customary understanding of the basic Lindenbaum–Tarski algebra. In other words, we pass here from a sort of product-sum "representation" to a sort of "division-subtraction representation". In the following I shall present some of the most insightful relations.

7.1 Subtractions

First of all, let us consider the possibility that all of the statements of Level 6-2 can be derived by subtracting statements of Level 1-7 (the second number in each pair in the second column of the following table) to statements of Level 7-1 (the first number in each pair in the second column). In other words, $1 - 7$ means that we are subtracting Statement 7 of Level 1-7 to Statement 1 of Level 7-1.

Note that every bold statement is generated by one of the two statements pertaining to its LGS (for **E** either E1 or E2). The other two terms contradict these two statements pertaining to its LGS.

Mechanical Logic in Three-Dimensional Space
Gennaro Auletta

ISBN 978-981-4411-50-9 (Hardcover), 978-981-4411-51-6 (eBook)
www.panstanford.com

Table 7.1 Derivation of statements of Level 6-2 through subtractions

Statements of Level 6-2	Calculation	Calculation in symbols	Result in symbols
1	1 – 7, 2 – 8	E1–I7=E2–I8	**E**
2	1 – 6, 3 – 8	F1–J7=F2–J8	**F**
3	2 – 6, 3 – 7	E2–O8=A1–I7	E2xA1
4	1 – 5, 4 – 8	D1–G7=D2–G8	**D**
5	2 – 5, 4 – 7	E2–S8=M1–I7	E2xM
6	3 – 5, 4 – 6	A1–S8=M1–O8	A1xM
7	1 – 4, 5 – 8	E1–O7=A2–I8	E1xA2
8	2 – 4, 5 – 7	R1–W7=R2–W8	**R**
9	3 – 4, 5 – 6	A1–O7=A2–O8	**A**
10	4 – 4, 5 – 5	M1–O7=A2–S8	M1xA
11	1 – 3, 6 – 8	E1–S7=M2–I7	E1xM
12	2 – 3, 6 – 7	P1–U7=P2–U8	**P**
13	3 – 3, 6 – 6	A1–S7=M2–O8	A1xM
14	4 – 3, 6 – 5	M1–S7=M2–S8	**M**
15	5 – 3, 6 – 4	A2–S7=M2–O7	A2xM
16	1 – 2, 7 – 8	E1–T8=L1–I8	E1xL1
17	2 – 2, 7 – 7	E2–T8=L1–I7	E2xL1
18	3 – 2, 7 – 6	C1–H7=C2–H8	**C**
19	4 – 2, 7 – 5	B1–K7=B2–K8	**B**
20	5 – 2, 7 – 4	A2–T8=L1–O7	A2xL1
21	6 – 2, 7 – 3	M2–T8=L1–S7	M2xL
22	1 – 1, 8 – 8	E1–T8=L2–I8	E1xL2
23	2 – 1, 8 – 7	E2–T7=L2–I7	E2xL2
24	3 – 1, 8 – 6	A1–T7=L2–O8	A1xL2
25	4 – 1, 8 – 5	M1–T7=L2–S8	M1xL
26	5 – 1, 8 – 4	N1–V7=N2–V8	**N**
27	6 – 1, 8 – 3	Q1–Ω7=Q2–Ω8	**Q**
28	7 – 1, 8 – 2	L1–T7=L2–T8	**L**

However, they are crossed (I7 is the negation of E2 but is paired here with E1 whilst I8 is the negation of E1 but is paired here with E2). Moreover, when the results are represented by cross-terms (like E2xA1, Line 3) the two possible derivations display precisely those statements (i.e. E2 and A1). Finally, I have not considered those subtractions that give again a statement of Level 7-1 as a result (for instance, E1-I8=E1).

Still more interesting is the derivation of statements of Level 5-3. Indeed, by subtracting to one of the grounding statements of Level 6-2 (like **A, B, C**, and so on) all of the statements (but the countervalent ones) of Level 1-7 we get all the elements of the RGSs of those statements. I shall focus therefore on these derivations and shall not care about other statements (also the order of the statements follows generating sets):

Table 7.2 Derivation of statements of Level 5-3 through subtractions

Level 5-3	Calculation	Calculation in symbols	Result	Level 5-3	Calculation	Calculation in symbols	Result
36	1 – 1	**E**–T7	E8	49	14 – 1	**M**–T7	M8
21	1 – 2	**E**–T8	E7	34	14 – 2	**M**–T8	M7
11	1 – 3	**E**–S7	E6	20	14 – 4	**M**–07	M6
5	1 – 4	**E**–07	E5	16	14 – 6	**M**–08	M5
2	1 – 5	**E**–S8	E4	15	14 – 7	**M**–I7	M4
1	1 – 6	**E**–08	E3	14	14 – 8	**M**–I8	M3
37	2 – 1	**F**–T7	F8	53	18 – 1	**C**–T7	C8
22	2 – 2	**F**–T8	F7	33	18 – 3	**C**–S7	C7
12	2 – 3	**F**–S7	F6	29	18 – 4	**C**–07	C6
6	2 – 4	**F**–07	F5	26	18 – 5	**C**–S8	C5
3	2 – 5	**F**–S8	F4	23	18 – 7	**C**–I7	C4
1	2 – 7	**F**–I7	F3	22	18 – 8	**C**–I8	C3
39	4 – 1	**D**–T7	D8	54	19 – 1	**B**–T7	B8
24	4 – 2	**D**–T8	D7	34	19 – 3	**B**–S7	B7
14	4 – 3	**D**–S7	D6	30	19 – 4	**B**–07	B6
8	4 – 4	**D**–07	D5	26	19 – 6	**B**–08	B5
3	4 – 6	**D**–08	D4	25	19 – 7	**B**–I7	B4
2	4 – 7	**D**–I7	D3	24	19 – 8	**B**–I8	B3
43	8 – 1	**R**–T7	R8	55	26 – 2	**N**–T8	N8
28	8 – 2	**R**–T8	R7	50	26 – 3	**N**–S7	N7
18	8 – 3	**R**–S7	R6	45	26 – 5	**N**–S8	N6
9	8 – 5	**R**–S8	R5	44	26 – 6	**N**–08	N5
7	8 – 6	**R**–08	R4	43	26 – 7	**N**–I7	N4
5	8 – 8	**R**–I8	R3	42	26 – 8	**N**–I8	N3
44	9 – 1	**A**–T7	A8	56	27 – 2	**Q**–8	Q8
29	9 – 2	**A**–T8	A7	50	27 – 4	**Q**–07	Q7
19	9 – 3	**A**–S7	A6	49	27 – 5	**Q**–S8	Q6
10	9 – 5	**A**–S8	A5	48	27 – 6	**Q**–08	Q5
7	9 – 7	**A**–I7	A4	47	27 – 7	**Q**–I7	Q4
6	9 – 8	**A**–I8	A3	46	27 – 8	**Q**–I8	Q3
47	12 – 1	**P**–T7	P8	56	28 – 3	**L**–S7	L8
32	12 – 2	**P**–T8	P7	55	28 – 4	**L**–07	L7
18	12 – 4	**P**–07	P6	54	28 – 5	**L**–S8	L6
15	12 – 5	**P**–S8	P5	53	28 – 6	**L**–08	L5
13	12 – 6	**P**–08	P4	52	28 – 7	**L**–I7	L4
11	12 – 8	**P**–I8	P3	51	28 – 8	**L**–I8	L3

It may be noted that the two particular statements of Level 1-7 that each time are excluded from calculation are those representing contradictory statements relative to the statements pertaining to the LGS of the involved universal statements (for instance, for the generation of the RGS of **E** propositions I7 and I8 are not included, which are indeed the negation of E1 and E2, respectively).

I do not consider here statements of Level 4-4, due to their weird nature (they represent a sort of noise, being situated at the middle level between universal and particular statements). Indeed, their form in the division-subtraction representation is weird. Later on, we shall consider some of these possibilities.

7.2 Divisions

Going to the levels under Level 4-4, it is convenient to make use of division in a way that is analogous to the interchange between product and sum that we have performed in Section 1.3. Moreover, I prefer to proceed always from the upper levels to the lower levels, although also the inverse route is quite possible. In doing so, I take for a moment for granted that we can derive through division lower levels.

In order to derive the statements of Level 3-5, we need to divide statements of Level 1-7 by statements of Level 6-2. Here, I shall only consider statements pertaining to the LGSs of particular statements:

Table 7.3 Derivation of statements of Level 3-5 through divisions

Level 5-3	Calculation	Calculation in symbols	Result	Level 5-3	Calculation	Calculation in symbols	Result
21	1 : 1	T7:**E**	I1	8	1 : 14	T7:**M**	S1
36	2 : 1	T8:**E**	I2	23	2 : 14	T8:**M**	S2
46	3 : 1	S7:**E**	I3	37	4 : 14	O7:**M**	S3
52	4 : 1	O7:**E**	I4	41	6 : 14	O8:**M**	S4
55	5 : 1	S8:**E**	I5	42	7 : 14	I7:**M**	S5
56	6 : 1	O8:**E**	I6	43	8 : 14	I8:**M**	S6
20	1 : 2	T7:**F**	J1	4	1 : 18	T7:**C**	H1
35	2 : 2	T8:**F**	J2	24	3 : 18	S7:**C**	H2

Level 5-3	Calculation	Calculation in symbols	Result	Level 5-3	Calculation	Calculation in symbols	Result
45	3 : 2	S7:**F**	J3	28	4 : 18	O7:**C**	H3
51	4 : 2	O7:**F**	J4	31	5 : 18	S8:**C**	H4
54	5 : 2	S8:**F**	J5	34	7 : 18	I7:**C**	H5
56	7 : 2	I7:**F**	J6	35	8 : 18	I8:**C**	H6
18	1 : 4	T7:**D**	G1	3	1 : 19	T7:**B**	K1
33	2 : 4	T8:**D**	G2	23	3 : 19	S7:**B**	K2
43	3 : 4	S7:**D**	G3	27	4 : 19	O7:**B**	K3
49	4 : 4	O7:**D**	G4	31	6 : 19	O8:**B**	K4
54	6 : 4	O8:**D**	G5	32	7 : 19	I7:**B**	K5
55	7 : 4	I7:**D**	G6	33	8 : 19	I8:**B**	K6
14	1 : 8	T7:**R**	W1	2	2 : 26	T8:**N**	V1
29	2 : 8	T8:**R**	W2	7	3 : 26	S7:**N**	V2
39	3 : 8	S7:**R**	W3	12	5 : 26	S8:**N**	V3
48	5 : 8	S8:**R**	W4	13	6 : 26	O8:**N**	V4
50	6 : 8	O8:**R**	W5	14	7 : 26	I7:**N**	V5
52	8 : 8	I8:**R**	W6	15	8 : 26	I8:**N**	V6
13	1 : 9	T7:**A**	O1	1	2 : 27	T8:**Q**	Ω1
28	2 : 9	T8:**A**	O2	7	4 : 27	O7:**Q**	Ω2
38	3 : 9	S7:**A**	O3	8	5 : 27	S8:**Q**	Ω3
47	5 : 9	S8:**A**	O4	9	6 : 27	O8:**Q**	Ω4
50	7 : 9	I7:**A**	O5	10	7 : 27	I7:**Q**	Ω5
51	8 : 9	I8:**A**	O6	11	8 : 27	I8:**Q**	Ω6
10	1 : 12	T7:**P**	U1	1	3 : 28	S7:**L**	T1
25	2 : 12	T8:**P**	U2	2	4.38	O7:**L**	T2
39	4 : 12	O7:**P**	U3	3	5 : 28	S8:**L**	T3
42	5 : 12	S8:**P**	U4	4	6 : 28	O8:**L**	T4
44	6 : 12	O8:**P**	U5	5	7 : 28	I7:**L**	T5
46	8 : 12	I8:**P**	U6	6	8 : 28	I8:**L**	T6

It may be noted that each statement in bold generates the whole LGS of its contradictory (for instance, **E** generates all I1-I6). However, again each statement in bold is not combined with any of the contradictory statements of its LGS statements (for instance, **E** does not combine with I7 and I8 that contradict E2 and E1, respectively). Finally, the statements of Level 2-6 can be derived by dividing statements of Level 1-7 by statements of Level 7-1:

Table 7.4 Derivation of statements of Level 2-6 through divisions

Statements of Level 6-2	Calculation	Calculation in symbols	Result in symbols
1	1 : 7, 2 : 8	T7:L1=T8:L2	**T**
2	1 : 6, 3 : 8	Ω7:Q1= Ω8:Q2	**Ω**
3	1 : 5, 4 : 8	V7:N1=V8:N2	**V**
4	1 : 4, 5 : 8	T7:M1=S8:L2	T7+S8
5	1 : 3, 6 : 8	T7:A1=O8:L2	T7+O8
6	1 : 2, 7 : 8	T7:E2=I7:L2	T7+I7
7	1 : 1, 8 : 8	T7:E1=I8:L2	T7+I8
8	2 : 6, 3 : 7	T8:M2=S7:L1	T8+S7
9	2 : 5, 4 : 7	T8:A2=O7:L1	T8+O7
10	2 : 4, 5 : 7	K7:B1=K8:B2	**K**
11	2 : 3, 6 : 7	H7:C1=H8:C2	**H**
12	2 : 2, 7 : 7	T8:E2=I7:L1	T8+I7
13	2 : 1, 8 : 7	T8:E1=I8:L1	T8+I8
14	3 : 5, 4 : 6	S7:A2=O7:M2	S7+O7
15	3 : 4, 5 : 6	S7:M1=S8:M2	**S**
16	3 : 3, 6 : 6	S7:A1=O8:M2	S7+O8
17	3 : 2, 7 : 6	U7:P1=U8:P2	**U**
18	3 : 1, 8 : 6	S7:E1=I8:M2	S7+I8
19	4 : 4, 5 : 5	O7:M1=S8:A2	O7+S8
20	4 : 3, 6 : 5	O7:A1=O8:A2	**O**
21	4 : 2, 7 : 5	W7:R1=W8:R2	**W**
22	4 : 1, 8 : 5	O7:E1=I8:A2	O7+I8
23	5 : 3, 6 : 4	S8:A1=O8:M1	S8+O8
24	5 : 2, 7 : 4	S8:E2=I7:M1	S8+I7
25	5 : 1, 8 : 4	G7:D1=G8:D2	**G**
26	6 : 2, 7 : 3	O8:E2=I7:A1	O8+I7
27	6 : 1, 8 : 3	J7:F1=J8:F2	**J**
28	7 : 1, 8 : 2	I7:E1=I8:E2	**I**

The previous derivations will be very helpful when dealing with generalized inferences. There are also further possibilities that I omit for the sake of simplicity.

7.3 Final Considerations

A word of warning is important here. Although, subtraction-division can be considered to a certain extent as an alternative representation of the logical space relatively to sum-product, there are also important differences that forbid to take the former representation as a complete structural description of the kind shown in Chapter 1. The reason is the (in Section 4.1) mentioned fact that subtraction and division are operations that are not symmetric under interchange of lines, what makes these two operations not very suitable for representing order (or structural) relations. In fact, if we try to combine statements of Level 3-5 in order to get uniform statements of whatever level, we shall immediately incur in difficulties. Indeed, a possible combination of statements of Level 3-5 is:

T1	11100000	–
T2	11010000	=
	00100000	

and here the result is a statement of level 1-7. However, a different combination is:

T1	11100000	–
O1	10010100	=
	01100000	

whose result is a statement of Level 6-2. Finally, we have:

T1	11100000	–
G4	00011001	=
	11100000	

whose result is a statement of Level 3-5 (T1 itself). Similar results are to a certain extent true for division. Indeed, a possible combination of statements of Level 6-2 is:

E	11111100	:
D	11110101	=
	11110101	

whose result is a statement of Level 6-2. However, we also have

E	11111100	:
M1xA2	11100111	=
	11100100	

whose result is a statement of Level 4-4. There are also additional problems: statements of a certain level can be derived by combining statements of different levels. For instance, to subtract statements of Level 1-7 to statements of Level 8-0 generates statements of level 7-1, to statements of Level 7-1 generates statements of Level 6-2, to statements of Level 6-2 generates statements of Level 5-3, and so on. However, statements of Level 5-3 are also generated by subtracting statements of Level 2-6 to statements of Level 7-1, statements of Level 4-4 when to those of Level 6-2 those of Level 2-6 are subtracted, and so on. Similarly in the case of division. For instance, statements of Level 5-3 can be generated in any one of these ways: by dividing either (i) statements of Level 1-7 by those of Level 2-6 or (ii) statements of Level 1-7 by those of Level 4-4.

Therefore, we discover that both we can generate statements of different levels by combining statements pertaining to the same level or even to two fixed levels and we can generates statements of the same level by combining statements of very different levels. We can then clearly see that the removal of an assumption through its explicit negation (although in different ways) in both subtraction and division is not precisely the same than the assumption of its countervalent statement but affects the way in which the logical operation, and therefore also the relative inference kind, is built.

Chapter 8

Generalized Inferences

For the intelligence of what follows it is important to distinguish between structural relations of the kind shown in Chapters 1 and 7 and inferential relations, as already mentioned in Section 7.3. Although structural relations can be to a certain extent considered kinds of inferences (since they show how a statement can be derived from other statements), the converse is not necessarily true, that is, inferential relations do not always possess an insightful structural meaning.

8.1 The Basic Forms of the Previous and of New Inferences

We can summarize all $48 \times 4 = 192$ closed inferences as in Fig. 8.1. It may be noted that each statement occurs 24 times in different combinations (i.e. different triplets) with other statements but never with those statements situated on the same octagon (with reference to Fig. 2.1). A useful table is the following (where I have taken as reference inference the classical *Barbara*):

Mechanical Logic in Three-Dimensional Space
Gennaro Auletta

ISBN 978-981-4411-50-9 (Hardcover), 978-981-4411-51-6 (eBook)
www.panstanford.com

	1st group				2nd group				3rd group			
	classical	1st mixed	2nd mixed	extension	classical	1st mixed	2nd mixed	extension	classical	1st mixed	2nd mixed	extension
PRODUCTS	ABC	LFC	AQN	LRN	AHK	LHJ	AVΩ	LVW	HBO	HFT	VQO	VRT
	EBD	MFD	EQP	MRP	EGK	MGJ	EUΩ	MUW	GBI	GFS	UQI	URS
	AJG	LKG	AWU	LΩU	ADF	LDB	APR	LPQ	DJO	DKT	PWO	PΩT
	EJH	MKH	EWV	MΩV	ECF	MCB	ENR	MNQ	CJI	CKS	NWI	NΩS
SUMS	OKH	TJH	OΩV	TWV	OCB	TCF	ONQ	TNR	CKA	CJL	NΩA	NWL
	IKG	SJG	IΩU	SWU	IDB	SDF	IPQ	SPR	DKE	DJM	PΩE	PWM
	OFD	TBD	ORP	TQP	OGJ	TGK	OUW	TUΩ	GFA	GBL	URA	UQL
	IFC	SBC	IRN	SQN	IHJ	SHK	IVW	SVΩ	HFE	HBM	VRE	VQM
SUBTRACTIONS	AKC	LJC	AΩN	LWN	ACK	LCJ	ANΩ	LNW	HKO	HJT	VΩO	VWT
	EKD	MJD	EΩP	MWP	EDK	MDJ	EPΩ	MPW	GKI	GJS	UΩI	UWS
	AFG	LBG	ARU	LQU	AGF	LGB	AUR	LUQ	DFO	DBT	PRO	PQT
	EFH	MBH	ERV	MQV	EHF	MHB	EVR	MVQ	CFI	CBS	NRI	NQS
DIVISIONS	OBH	TFH	OQV	TRV	OHB	THF	OVQ	TVR	CBA	CFL	NQA	NRL
	IBG	SFG	IQU	SRU	IGB	SGF	IUQ	SUR	DBE	DFM	PQE	PRM
	OJD	TKD	OWP	TΩP	ODJ	TDK	OPW	TPΩ	GJA	GKL	UWA	UΩL
	IJC	SKC	IWN	SΩN	ICJ	SCK	INW	SNΩ	HJE	HKM	VWE	VΩM

Figure 8.1 Summary of all inferences.

Table 8.1 Kinds of closed inferences

$\{\mathbf{A}\}_{\text{LGS}} \times \{\mathbf{B}\}_{\text{LGS}} = \{\ldots, \mathbf{C}, \ldots\}$	$\{\mathbf{O}\}_{\text{RGS}} + \{\mathbf{K}\}_{\text{RGS}} = \{\ldots, \mathbf{H}, \ldots\}$
$\{\mathbf{A}\}_{\text{LGS}} - \{\mathbf{K}\}_{\text{RGS}} = \{\ldots, \mathbf{C}, \ldots\}$	$\{\mathbf{O}\}_{\text{RGS}} : \{\mathbf{B}\}_{\text{LGS}} = \{\ldots, \mathbf{H}, \ldots\}$

It is understood here that the other statements that are omitted in the conclusion of the 4 kinds of inferences are those that each time follow when taking into account all derivations but that do not yield desired result. Obviously, the specific derivations that grounds the inferences, like e.g. the third one in Table 2.12 can be expressed in quite synthetic form as

$$\text{A1} \times \text{B2} = \mathbf{C}\,. \tag{8.1}$$

Similarly, we have for the other inference kinds (see Tables 3.3, 4.2, and 5.3)

$$\text{O8} + \text{K7} = \mathbf{H}\,,\ \text{A1} - \text{K7} = \mathbf{C}\,,\ \text{O8} : \text{B2} = \text{O8}/\text{B2} = \mathbf{H}. \tag{8.2}$$

This method allows for further possibilities of calculation. Indeed, from the equalities above, we also have (since A1=C1):

$$\text{C1} \times \text{B2} = \text{A1} - \text{K7}\,. \tag{8.3}$$

Far more interesting is:

$$\text{A1} = \mathbf{C}/\text{B2} \tag{8.4}$$

or

$$\text{O8} = \mathbf{H} \times \text{B2}\,. \tag{8.5}$$

These considerations are of absolute generality when considering the proper structural expressions and the proper derivations involving bold statements. For instance, the reader (taking advantage of Tables 2.1–2.3) can verify that we have the relations:

$$\mathbf{D} \times \mathbf{P} = \text{P3} \times \text{P5 and } \mathbf{P} \times \mathbf{N} = \text{P6} \times \text{P8}\,, \tag{8.6}$$

but also

$$\mathbf{P} = \text{P3} \times \text{P5} + \text{P6} \times \text{P8}\,, \tag{8.7}$$

so that we immediately have

$$\mathbf{P} = \mathbf{D} \times \mathbf{P} + \mathbf{P} \times \mathbf{N}\,. \tag{8.8}$$

Let us now exploring why we have these results. Let us consider *Barbara* and its equivalents in the 4 general forms of inferences (for the 1st and 2nd group). As it can be easily proved, we have

$$[(\mathbf{A} \times \mathbf{B}) : \mathbf{B}] = \mathbf{A}\,, \quad [(\mathbf{O} + \mathbf{K}) - \mathbf{K}] = \mathbf{O}\,, \tag{8.9a}$$

$$[(\mathbf{A} - \mathbf{K}) + \mathbf{K}] = \mathbf{A}\,, \quad [(\mathbf{O} : \mathbf{B}) \times \mathbf{B}] = \mathbf{O}\,. \tag{8.9b}$$

We have remarked that no logical operation is reversible in general. However, in the particular case of inferences this relations are reversible as displayed here. Why? What all product inferences involving only universal premises have in common is that the premises never share their zeros, while all sum inferences involving only particular premises never share their ones. Moreover, all subtraction inferences involving a first premise that is universal and a second premise that is a particular are such that the zeros of the universal coincide with some zeros of the particular, while all division inferences involving a first premise that is particular and a second premise that is an universal are such that the ones of the particular coincide with some ones of the universal. Since these relations hold for bold statements, they also hold for the generating sets displayed in Table 8.1 (since statements pertaining to a generating set of a bold statement displayed in Table 8.1 share with the latter the appropriate zeros or ones). Therefore, we do not only have that $\mathrm{A1} = \mathbf{C}/\mathrm{B2}$ but also $\mathrm{A2} = (\mathrm{A2} \times (\mathrm{B1} = \mathrm{M1}))/\mathrm{B1}$. Similar relations hold for all other derivations of the kind shown above.

In the case of derivations that are particular like *Darii* and its equivalents in the 4 forms of inference (for the 1st and 2nd group), the situation is less straight but similar relations can nevertheless be proved. Indeed, we need to consider one of the premises and the conclusion. Then, in the case of product and sum inferences we have:

$$[(\mathbf{A} - \mathbf{G}) + \mathbf{G}] = \mathbf{A}\,, \quad [(\mathbf{O} : \mathbf{D}) \times \mathbf{D}] = \mathbf{O}\,, \tag{8.10}$$

while the cases of *Darii* in the form of subtraction and division inference easily coincide with these two, and similarly for other particular inferences. Also in this case, we can use cross formulations, whose expression is however very cumbersome, and therefore I skip this part. Note that in the case of the third group, the form of the product inferences and of the sum inferences are interchanged.

Therefore, all kinds of combinations can be summarized as in the following table (where I recall that when we have two statements of different quantity, i.e. the one universal and the other particular, they represent one premise and the conclusion of the relative product or sum inference while the opposite holds true for subtraction and division inferences):

Table 8.2 Compatibility of the different statements in the four kinds of inferences

	In combination with universals	In combination with particulars
A	**B, D, P, Q**	**G, K, U, Ω**
E	**B, C, N, Q**	**H, K, V, Ω**
L	**D, F, P, R**	**G, J, U, W**
M	**C, F, N, R**	**H, J, V, W**
I	**B, C, N, Q**	**H, K, V, Ω**
O	**B, D, P, Q**	**G, K, U, Ω**
S	**C, F, N, R**	**H, J, V, W**
T	**D, F, P, R**	**G, J, U, W**

It may be remarked each universal statement builds a countervalent pair with one particular one on the same row, and that this combinations exhaust all allowed possibilities displayed in Fig. 2.1. For instance **A**, which is located in the inner octagon (as any other of the statements in the first column of Table 8.2), can be combined with the "segments" BQ, DP, GU, KΩ; **E** with BQ, CN, HV, KΩ; **L** with DP, FR, GU, JW; **M** with CN, FR, HV, JW; what cover both the middle and external octagons. It is easy to verify that the combinations allowed by **I**, **O**, **S**, and **T** also cover the latter two octagons. This examination shows two important features of the logical space: (i) valid inferences involve some kind of reversibility and (ii) the space is symmetric, since we could have taken as grounding statements for the inferences not the bold ones but the products and the sums shown in Table 6.2 apart from the 4 equivalences and negations of equivalences (Statements 10, 13, 17 and 22 of Level 6-2 and Statements 7, 12, 16 and 19 of Level 2-6), what covers all 28 statements of both Levels 6-2 and 2-6. In other words, an inference like (E2xA1 $\wedge$ E2xM1) $\rightarrow$ A1xM1 or

$$\{[X \rightarrow (Y \leftrightarrow Z)] \wedge [Y \rightarrow (X \leftrightarrow Z)]\} \rightarrow [Z \rightarrow (X \leftrightarrow Y)] \quad (8.11)$$

is perfectly valid (it is indeed in accordance with the above criteria and in fact is a tautology). We can also write down a particular inference like (E2xA1 $\wedge$ S8 + I7) $\rightarrow$ S8 + O8 or

$$\{[X \rightarrow (Y \leftrightarrow Z)] \wedge [Y \wedge \neg(Y \leftrightarrow Z)]\} \rightarrow [Z \wedge \neg(X \leftrightarrow Y)]\,, \quad (8.12)$$

which is again logically valid. Similarly, the inference (O8 + I7 $\vee$ S8 + I7) $\leftarrow$ S8 + O8 or

$$\{[X \wedge \neg(Y \leftrightarrow Z)] \vee [Y \wedge \neg(X \leftrightarrow Z)]\} \leftarrow [Z \wedge \neg(X \leftrightarrow Y)] \quad (8.13)$$

is also perfectly valid. Obviously, since these inferences always involve three variables in each premise and in the conclusion, we cannot make use here of the rules so far established for the derivations (see Sections 2.2 and 3.2). However, with this kind of inferences we can "read" the result directly in the premises. Therefore, we can build an alternative system by taking as primary e.g. the four statements, whose generating sets are

$$\{\text{E2xA1}\} = \{\{\text{E2, A1}\}_{\text{LGS}}, \{\text{E3, 4, A4, P4, C4, 38}\}_{\text{RGS}}\}$$
$$\{\text{E1xA2}\} = \{\{\text{E1, A2}\}_{\text{LGS}}, \{\text{E5, A3, D5, 17, 27, N3}\}_{\text{RGS}}\}$$
$$\{\text{M2xL1}\} = \{\{\text{M2, L1}\}_{\text{LGS}}, \{\text{31, P7, C7, M7, 35, L8}\}_{\text{RGS}}\}$$
$$\{\text{M1xL2}\} = \{\{\text{M1, L2}\}_{\text{LGS}}, \{\text{D8, 40, 41, N6, M8, L6}\}_{\text{RGS}}\}$$

and express all other ones in terms of these. Similarly, we can take four particular statements as primary wise generation sets are:

$$\{\text{O8+I7}\} = \{\{\text{19, H5, U5, O5, 53, I6}\}_{\text{LGS}}, \{\text{O8, I7}\}_{\text{RGS}}\}$$
$$\{\text{O7+I8}\} = \{\{\text{V6, 30, 40, G4, O6, I4}\}_{\text{LGS}}, \{\text{O7, I8}\}_{\text{RGS}}\}$$
$$\{\text{T8+S7}\} = \{\{\text{T1, 22, S2, H2, U2, 26}\}_{\text{LGS}}, \{\text{T8, S7}\}_{\text{RGS}}\}$$
$$\{\text{T7+S8}\} = \{\{\text{T3, S1, V3, 16, 17, G1}\}_{\text{LGS}}, \{\text{T7, S8}\}_{\text{RGS}}\}$$

Note that a pair of statements for each generating sets has not been named, since they are not part of any generating set of bold statements (see Tables 2.1 and 2.3). Then, we obtain Table 8.3.

Thanks to this table we can easily recover all variants of the above inferences. We may note that in both Table 8.2 and Table 8.3 each statement is combined with either universal or particular statements on the same line (and therefore the particular statements being countervalent relative to the universal ones and vice versa). However, there is also an important difference between

Table 8.3 Compatibility of the different cross statements of similar quantity

	In combination with universals	In combination with particulars
E2xA1	E2xM1, A1xL2, A1xM1, E2xL2	S8+I7, T7+O8, S8+O8, T7+I7
E1xA2	E1xM2, A2xL1, E1xL1, A2xM2	S7+I8, T8+O7, T8+I8, S7+O7
M2xL1	E1xM2, A2xL1, E1xL1, A2xM2	S7+I8, T8+O7, T8+I8, S7+O8
M1xL2	E2xM1, A1xL2, A1xM1, E2xL2	S8+I7, T7+O8, S8+O8, T7+I7
O8+I7	E2xM1, A1xL2, A1xM1, E2xL2	S8+I7, T7+O8, S8+O8, T7+I7
O7+I8	E1xM2, A2xL1, E1xL1, A2xM2	S7+I8, T8+O7, T8+I8, S7+O7
T8+S7	E1xM2, A2xL1, E1xL1, A2xM2	S7+I8, T8+O7, T8+I8, S7+O7
T7+S8	E2xM1, A1xL2, A1xM1, E2xL2	S8+I7, T7+O8, S8+O8, T7+I7

the two tables: in Table 8.3 we have essentially two lines that are repeated (this is due to the fact that we have here only statements with three variables) while in Table 8.2 only 2 over 4 statements are repeated from one line to the next one. The consequence is that we have 24 "universal" and 24 "particular" inferences, displaying all permutations of the three involved statements. For instance, we do not only have (E2xA1 $\wedge$ E2xM1) $\rightarrow$ A1xM1 but also (E2xA1 $\wedge$ A1xM1) $\rightarrow$ E2xM1 and (E2xM1$\wedge$A1xM1) $\rightarrow$ E2xA1. I finally remark that we can have mixed cases in which we use one statement from Table 8.2 and another from Table 8.3, like

$$\{(X \rightarrow Y) \wedge [Y \rightarrow (X \leftrightarrow Z)]\} \rightarrow [X \rightarrow (X \leftrightarrow Z)]\,, \qquad (8.14)$$

where the conclusion can be reduced to $X \rightarrow Z$. However, this adds nothing that would be conceptually new. In the following I shall no longer consider these inferences and limit my further examination to those presented in the previous chapters.

The above results establish criteria for verifying whether an inference is valid or not. There are also other procedures. Indeed, I remark that a product inference like $(\mathbf{A} \wedge \mathbf{B}) \rightarrow \mathbf{C}$ can be written as $\neg(\mathbf{A} \wedge \mathbf{B}) \vee \mathbf{C}$. Since the product between **A** and **B** is 10100011, its negation is 01011100, so that we can write

$$\begin{array}{ccc} \neg(\mathbf{A} \wedge \mathbf{B}) & 01011100 & + \\ \mathbf{C} & 10111011 & = \\ & 11111111 & \end{array}$$

which is a tautology, proving that the derivation is logically correct. Similar procedures apply to any kind of derivation. Note also that the negation of the conjunction of the premises and the conclusion share the truth of areas d and e. In the case of a product inference like *Darii*, there is only one area shared.

8.2 The Most General Forms of Closed Inference

All the derivations developed in the previous chapters have two fundamental kinds of parent derivations: *Barbara* and its analogues and *Darii* and its analogues. Indeed, all universal syllogisms in the product form like *Celarent*, *Cesare*, *Camestres*, and their analogues can be derived from *Barbara* thanks to some variable substitution. On the other hand, all syllogisms involving a particular premise, like *Ferio*, *Ferison*, *Datisi*, and so on, can be derived from *Darii* thanks to some substitution. The reason is the following: At least one universal premise is necessary for any inference (from particular premises only nothing follows). When we deal with three-statement inferences the other premise can either be itself universal or be a particular expressing some kind of experience that has denied a corresponding universal. Let us briefly show this. The ground of any inference (which may be called the logical principle: (Peirce, 1868a, p. 24; Peirce, 1881, p. 246) is that if one sign denotes generally everything denoted by a second, and this second one denotes generally everything denoted by a third, then the first one denotes generally everything denoted by the third (Peirce, 1869, pp. 243–47; *Aristotle Cat.*, 1b1012). This is also true of the symbols that we use in formal languages (Peirce, 1880, p. 166): any symbol of a symbol of a symbol is itself a symbol of the latter. Let us start from the basic expression

$$\text{If Every } X \text{ is } Y \text{ and Every } Z \text{ is } X\text{, then Every } Z \text{ is } Y,$$

which is an instantiation of the logical principle and has the form of *Barbara*. Since any of these statements has also

(1) Its negative counterpart (by substitution e.g. of the predicate Y by $\neg Y$) or
(2) Can be negated as a whole,

the different forms of reasoning can be derived. Both cases are indeed examples of substitution and the latter is a particular case of the logical principle. In Case 1, when putting **E** at the place of the first **A**, we obtain the basic form of *Celarent* (whose conclusion must also be an **E**):

If Every X is $\neg Y$ and Every Z is X, then Every Z is $\neg Y$.

In other words, keeping the previous second premise, if the class denoted by X is other than the class denoted by Y, which implies mutual exclusion (also Y is other than X, which is the rule of conversion), the conclusion will be that also Z is other than Y. This form is the only one allowed if we substitute **E** to **A** in one of the premises with the previous form. In fact, if we replace the second **A** by **E**, that is, if we start from the premises

Every X is Y and No Z is $\neg X$,

no conclusion can follow that can be derived according to logical laws. In Case 2, since **I** is the negation of **E**, we obtain the general form of which *Darii* (whose conclusion is the negation of the conclusion of *Celarent*) is an instantiation:

If Every X is Y and It is not true that Every Z is $\neg X$, then It is not true that Every Z is $\neg Y$.

It is evident that only this form will work while a substitution of the negation of **E** to the first premise will not. Instead, by substituting **E** to **A** in the general form of *Darii* we obtain the general form of *Ferio* (whose conclusion must be the negation of the conclusion of *Barbara*):

If Every X is $\neg Y$ and It is not true that Every Z is $\neg X$, then It is not true that Every Z is Y.

Also here a different substitution would hold no result. We can expand and formalize the above considerations in the following way by considering two further applications of the leading principle. Let us consider the three general forms (which are also the bases of the canonical forms of the three figures, i.e. *Barbara, Baroco* and *Bocardo*)

Table 8.4 The general forms of the three groups

First group	Second group	Third group
$X \to Y$	$X \to Y$	$Z \wedge \neg Y$ or $\neg(Z \to Y)$
$Z \to X$	$Z \wedge \neg Y$ or $\neg(Z \to Y)$	$Z \to X$
$Z \to Y$	$Z \wedge \neg X$ or $\neg(Z \to X)$	$X \wedge \neg Y$ or $\neg(X \to Y)$

We can summarize this table by saying (Peirce, 1865, pp. 186 and 283–84):

- The first group takes as predicate of the conclusion the predicate (Y) of a predicate (X) of another term (Z). All the different forms explored above are variants of this first form.
- The second group abolishes as predicate that of which the predicate of a subject cannot be a predicate. The symbol (X) of which something (i.e. Y) is a symbol which is not a symbol of Z, is not itself a symbol of Z.
- The third group denies of X a universal predicate Y which is not a predicate of what (i.e. Z) X is a predicate. In this case, Y is not a symbol of a symbol X which is a symbol of the object Z of which Y is not a symbol. In all three cases the logical principle is satisfied.

From these general forms all possible combinations (and therefore valid kinds of inference) can be derived in the same way I have shown here for the first group (it suffices to make use of the rule of substitution or deny one of the premises). Since I have previously shown that all kinds of inference (product, sum, subtraction and division) are equivalent, this examination is sufficient. In conclusion, we shall focus now on *Barbara* and *Darii* (as paradigms of universal and particular inferences, respectively) in order to extract some general considerations.

As we have seen above (Line 1 of Table 6.5), the inference *Barbara* in the product inference can be expressed as

$$(\mathbf{A} \wedge \mathbf{B}) \to \mathbf{C}\,, \tag{8.15}$$

where **A, B, C,** as we know, stand for some kinds of statements. Obviously, this implication needs to be a tautology (i.e. cannot be false) if it is a logical inference, and the same is true of similar forms. Let us now consider the involved statements not as molecular ones (involving two or more variables) but rather as atomic statements or hidden molecular ones, and therefore let us express the above derivation in the most general terms as a relation between variables (which could be or could be not statements themselves):

$$(X \wedge Y) \rightarrow Z \; . \tag{8.16}$$

A short logical calculation can show that this statement has the form of E2. Therefore, we can say that *Barbara* in the product form has the abstract logical form of E2. Now, what can be the abstract form of *Darii* as a product inference? According to the third line of Table 6.5 we can write *Darii* as:

$$(\mathbf{A} \wedge \mathbf{J}) \rightarrow \mathbf{G} \; . \tag{8.17}$$

However, if we like to follow our line of reasoning up to now, we need to maintain a certain structural similarity with the case of *Barbara* and therefore we need to make use of universal statements only. Indeed, if we would use both universal and particular statements, this would determine a certain ambiguity in the meaning of the variables X, Y, Z when using them at the place of bold statements. Now, both **J** and **G** represent negations of universal statements, in particular of **F** and **D**, respectively (what shows that also *Darii* can be derived from *Celarent* and therefore ultimately from *Barbara*, a consideration, however, that does not deprive these remarks of their validity, since I remark again that to deny an assumption has not always the same consequences as to assume its countervalent statement). Then, we can write *Darii* as

$$(\mathbf{A} \wedge \neg\mathbf{F}) \rightarrow \neg\mathbf{D} \; . \tag{8.18}$$

I have kept the premise **A** as fixed since this is what *Barbara* and *Darii* share. Then (since X, Y, Z are here kinds of "dummy" variables), the most abstract logical form of *Darii* is:

$$(X \wedge \neg Y) \rightarrow \neg Z \; , \tag{8.19}$$

which a logical calculation can show to be the statement A1. Considering now the sum inferences, *Barbara* (first line of Table 6.6) has the general form (which is indeed logically equivalent to the product form of *Barbara*):

$$(\mathbf{O} \vee \mathbf{K}) \leftarrow \mathbf{H} \; . \tag{8.20}$$

Let us again do not consider the specific statements here involved and let us rewrite this expression by making use of general variables:

$$(X \vee Y) \leftarrow Z\,, \tag{8.21}$$

where we do not need to worry with the problem that here particular statements are involved since we are interested in grasping the general form of this kind of inference (possible ambiguities can only arise in the context of the same form of inference). This expression is logically equivalent to

$$Z \rightarrow (X \vee Y)\,, \tag{8.22}$$

which can be reduced to the logical form of the statement L1. Let us now consider *Darii* in the sum form (Line 3 of Table 6.6):

$$(\mathbf{O} \vee \mathbf{F}) \leftarrow \mathbf{D}\,. \tag{8.23}$$

In order to avoid the use of universal propositions (since in the sum form of *Barbara* only particular ones occur), I recall that **F** and **D** constitute contradictory pairs with **J** and **G**, respectively. Then, we can reformulate the previous inference as (obviously, I keep the premise **O** as fixed since this is what *Barbara* and *Darii* share in the sum formulation):

$$(\mathbf{O} \vee \neg\mathbf{J}) \leftarrow \neg\mathbf{G}\,. \tag{8.24}$$

This can be reformulated in terms of generic variables as

$$(X \vee \neg Y) \leftarrow \neg Z\,, \tag{8.25}$$

which can be further transformed into

$$\neg Z \rightarrow (X \vee \neg Y)\,. \tag{8.26}$$

A short logical computation will show that this is a variant of statement M2. Let us now consider *Barbara* in the subtraction form (Line 1 of Table 6.7):

$$(\mathbf{A} \wedge \neg\mathbf{K}) \rightarrow \mathbf{C}\,. \tag{8.27}$$

We do not need to worry at this stage about the fact that both universal and particular statements occur here, since we are interested only in the general form of the inference and thus we only need to pay attention at the connection of further kinds of subtraction inferences with this form. Then, I express the above inference in terms of generic variables:

$$(X \wedge \neg Y) \rightarrow Z\ , \tag{8.28}$$

which can be reduced to the statement A2. *Darii* in the subtraction form (Line 3 of Table 6.7) is:

$$(\mathbf{A} \wedge \neg \mathbf{F}) \rightarrow \mathbf{G} \tag{8.29}$$

We already know that this can be expressed as (I again keep **A** since it is shared by *Barbara* and *Darii* in subtraction inferences):

$$(\mathbf{A} \wedge \mathbf{J}) \rightarrow \neg \mathbf{D}\ , \tag{8.30}$$

which presents universal and particular statements in the same places as *Barbara* in subtraction form. Then, the latter inference can be reformulated as

$$(X \wedge Y) \rightarrow \neg Z\ . \tag{8.31}$$

This expression can be easily reduced to the statement E1. Finally, the division form of *Barbara* (Line 1 of Table 6.8) is:

$$(\mathbf{O} \vee \neg \mathbf{B}) \leftarrow \mathbf{H}\ , \tag{8.32}$$

which can be reformulated in general logical terms as

$$(X \vee \neg Y) \leftarrow Z\ , \tag{8.33}$$

or

$$Z \rightarrow (X \vee \neg Y)\ . \tag{8.34}$$

It can be easily shown that this has the form of the statement M1. The division form of *Darii* is (Line 3 of Table 6.8):

$$(\mathbf{O} \vee \neg \mathbf{J}) \leftarrow \mathbf{D}\ , \tag{8.35}$$

which, according to the previously employed procedure can be reformulated as

$$(\mathbf{O} \vee \mathbf{F}) \leftarrow \neg \mathbf{G} \tag{8.36}$$

which can be expressed in general logical terms as:

$$(X \vee Y) \leftarrow \neg Z \tag{8.37}$$

or

$$\neg Z \rightarrow (X \vee Y)\ , \tag{8.38}$$

which has the logical form of the statement L2. In other words, we have expressed all derivations in terms of the eight statements of Level 7-1, which means that this level, which is the most universal one before the tautology is the level of the universal forms of derivations.

It is very interesting now to observe that all statements of Level 1-7, each of which constitutes a contradictory pair with one of the statements of Level 7-1, represent therefore negations of universal inferences. In other words, the lowest level of the logical space before reaching the contradiction represents a factual rejection of some inference, for instance by performing a kind of test. We can summarize the previous results in the following table (where on the same line are put contradictory pairs):

Table 8.5 Summary of all inference kinds

		Level 7-1			Level 1-7
Product	*Barbara*	$(X \wedge Y) \rightarrow Z$	E2	I7	$X \wedge Y \wedge \neg Z$
inferences	*Darii*	$(X \wedge \neg Y) \rightarrow \neg Z$	A1	O8	$X \wedge \neg Y \wedge Z$
Sum	*Barbara*	$(X \vee Y) \leftarrow Z$	L1	T8	$\neg X \wedge \neg Y \wedge Z$
inferences	*Darii*	$(X \vee \neg Y) \leftarrow \neg Z$	M2	S7	$\neg X \wedge Y \wedge \neg Z$
Subtraction	*Barbara*	$(X \wedge \neg Y) \rightarrow Z$	A2	O7	$X \wedge \neg Y \wedge \neg Z$
inferences	*Darii*	$(X \wedge Y) \rightarrow \neg Z$	E1	I8	$X \wedge Y \wedge Z$
Division	*Barbara*	$(X \vee \neg Y) \leftarrow Z$	M1	S8	$\neg X \wedge Y \wedge Z$
inferences	*Darii*	$(X \vee Y) \leftarrow \neg Z$	L2	T7	$\neg X \wedge \neg Y \wedge \neg Z$

It is helpful to remark that the many implications occurring at Level 5-3, like $X \rightarrow (\neg Y \wedge Z)$, which apparently are similar to those listed in the third column above, are in fact completely different. A comparison of many of these forms may be highly instructive. The previous considerations allow us to build connectives for three variables. We can distinguish single-variable or one-dimensional connectives (negation) and two-variable or two-dimensional connectives (disjunction, conjunction, implication, equivalence, countervalence). Actually, conjunction and disjunction (with negation) are sufficient. A more economic way is to use a single connective that makes also the introduction of negation superfluous: we can choose either the Peirce's functor (1000, i.e. $\neg X \wedge \neg Y$) or Sheffer's functor (1110, i.e. $\neg X \vee \neg Y$): see Fig. 1.3. For instance, using Peirce's formulation we can express basic logical relations in this way:

$$\neg X = \neg X \wedge \neg X,\ X \wedge Y = \neg(\neg X \wedge \neg X) \wedge \neg(\neg Y \wedge \neg Y),\ X \vee Y = \neg(\neg X \wedge \neg Y) \wedge \neg(\neg X \wedge \neg Y).$$

In analogy with these cases, we can chose either a family of three-dimensional connectives (for instance a choice could be Statement 2 of Level 7-1, Statement 6 of Level 6-2, Statement 26 of Level 5-3, Statement 31 of Level 4-4 and their negations) or choose a single connective, either 11111110 ($\neg X \vee \neg Y \vee \neg Z$, i.e. E1) or 10000000 ($\neg X \wedge \neg Y \wedge \neg Z$, i.e. T7). In other words, the general and abstract forms of inference turn out to be also the abstract forms of connectives.

8.3 The Results of the All Derivations

Now, it is interesting to consider from which derivations certain statements are derived. In other words, I shall consider the results of all derivations and not only of those that are the target of the relative inference. We know, that in the case of universal inferences these derivations are 4 while they are 12 in the case of particular inferences. In particular, it is interesting to consider how the statements of Level 6-2 and 2-6 are produced. I obviously skip all the cases in which the result banally repeats both premises or at least one of them. In the case of sum and product inferences we have the results shown in Tables 8.6–8.7 (data are extracted from Tables 6.5 and 6.6).

Table 8.6 Statements of Level 6-2 produced through sum derivation

1	11111100	**E**	O8=H8=J7+E3=F3, S8=G7=K8+E4=D3, S7=U7=Ω8+E6=P3, O7=V8=W7+E5=R3
2	11111010	**F**	O7=V8=W7+A3=F5, I7=U8=W8+E3=F3, T8=H7=K7+C3=F7, S8=G7=K8+D4=F4
3	11111001	E2xA1	I8=G8=J8+E3=F3, T8=H7=K7+C4, S7=U7=Ω8+P4, O7=V8=W7+A4=R4
4	11110110	**D**	O8=H8=J7+D4=F4, I7=U8=W8+E4=D3, T8=H7=K7+B3=D7, S7=U7=Ω8+M3=D6
5	11110101	E2xM1	I8=G8=J8+E4=D3, S7=U7=Ω8+M4=P5, T8=H7=K7+B4, O7=V8=W7+R5
6	11110011	A1xM1	T8=H7=K7+B5=C5, I8=G8=J8+D4=F4
7	11101110	E1xA2	O8=H8=J7+A3=F5, S8=G7=K8+D5, T7=V7=Ω7+N3, I7=U8=W8+E5=R3
8	11101101	**R**	T7=V7=Ω7+N4=R8, S7=U7=Ω8+P6=R6, O8=H8=J7+A4=R4, I8=G8=J8+E5=R3
9	11101011	**A**	I8=G8=J8+A3=F5, T8=H7=K7+A7=C6, T7=V7=Ω7+A8=N5, I7=U8=W8+A4=R4
10	11100111	M1xA2	I8=G8=J8+D5, T7=V7=Ω7+N6, T8=H7=K7+B6, I7=U8=W8+R5
11	11011110	E1xM2	O8=H8=J7+F6, T7=V7=Ω7+Q3, I7=U8=W8+E6=P3, S8=G7=K8+M3=D6
12	11011101	**P**	T7=V7=Ω7+P8=Q4, S8=G7=K8+M4=P5, O7=V8=W7+P6=R6, I8=G8=J8+E6=P3
13	11011011	A1xM2	I8=G8=J8+F6, T8=H7=K7+C7, T7=V7=Ω7+Q5, I7=U8=W8+P4
14	11010111	**M**	I8=G8=J8+M3=D6, T7=V7=Ω7+M8=Q6, I7=U8=W8+M4=P5, T8=H7=K7+M7=B7
15	11001111	A2xM2	T7=V7=Ω7+N7=Q7, I7=U8=W8+P6=R6
16	10111110	E1xL1	O8=H8=J7+C3=F7, S8=G7=K8+B3=D7
17	10111101	E2xL1	O8=H8=J7+C4, S7=U7=Ω8+P7, S8=G7=K8+B4, O7=V8=W7+R7, S7=U7=Ω8xP7
18	10111011	**C**	I8=G8=J8+C3=F7, O7=V8=W7+A7=C6, S8=G7=K8+B5=C5, T7=V7=Ω7+L5=C8
19	10110111	**B**	O8=H8=J7+B5=C5, I8=G8=J8+B3=D7, T7=V7=Ω7+L6=B8, S7=U7=Ω8+M7=B7
20	10101111	A2xL1	O8=H8=J7+A7=C6, T7=V7=Ω7+L7=N8, S8=G7=K8+B6, I7=U8=W8+R7
21	10011111	M2xL1	O8=H8=J7+C7, T7=V7=Ω7+L8=Q8, I7=U8=W8+P7, S8=G7=K8+M7=B7
22	01111110	E1xL2	O8=H8=J7+F8, S8=G7=K8+D8, S7=U7=Ω8+Q3, O7=V8=W7+N3
23	01111101	E2xL2	S7=U7=Ω8+P8=Q4, O7=V8=W7+N4=R8
24	01111011	A1xL2	I8=G8=J8+F8, T8=H7=K7+L5=C8, S7=U7=Ω8+Q5, O7=V8=W7+A8=N5
25	01110111	M1xL2	I8=G8=J8+D8, S7=U7=Ω8+M8=Q6, O7=V8=W7+N6, T8=H7=K7+L6=B8
26	01101111	**N**	S7=U7=Ω8+N7=Q7, T8=H7=K7+L7=N8, I7=U8=W8+N4=R8, O8=H8=J7+A8=N5
27	01011111	**Q**	T8=H7=K7+L8=Q8, S8=G7=K8+M8=Q6, O7=V8=W7+N7=Q7, I7=U8=W8+P8=Q4
28	00111111	**L**	O8=H8=J7+L5=C8, S7=U7=Ω8+L8=Q8, O7=V8=W7+L7=N8, S8=G7=K8+L6=B8

Table 8.7 Statements of Level 2-6 produced through product derivation

1	11000000	**T**	A1=C1=F2xT4=H1, M2=P2=Q1xT1=Ω1, A2=N1=R2xT2=V1, M1=B1=D2xT3=K1
2	10100000	**Ω**	L1=B2=C2xT1=Ω1, M1=B1=D2xS1=Ω3, A2=N1=R2xV2=Ω2, E2=P1=R1xU1=Ω5
3	10010000	**V**	M2=P2=Q1xV2=Ω2, L1=B2=C2xT2=V1, E2=P1=R1xV5=W1, A1=C1=F2xO1=V4
4	10001000	T7+S8	E1=D1=F1xG1, M2=P2=Q1xS1=Ω3, A2=N1=R2xV3, L1=B2=C2xT3=K1
5	10000100	T7+O8	E1=D1=F1xJ1, L1=B2=C2xT4=H1, M2=P2=Q1xΩ4, A2=N1=R2xO1=V4
6	10000010	T7+I7	M2=P2=Q1xU1=Ω5, A2=N1=R2xV5=W1
7	10000001	T7+I8	A1=C1=F2xJ1, M1=B1=D2xG1, A2=N1=R2xV6, M2=P2=Q1xΩ6
8	01100000	T8+S7	A1=C1=F2xH2, L2=N2=Q2xT1=Ω1, E2=P1=R1xU2, M1=B1=D2xS2=K2
9	01010000	T8+O7	A1=C1=F2xO2=H3, L2=N2=Q2xT2=V1, M1=B1=D2xK3, E2=P1=R1xW2
10	01001000	**K**	A1=C1=F2xH4=K4, E1=D1=F1xG2=K6, L2=N2=Q2xT3=K1, M2=P2=Q1xS2=K2
11	01000100	**H**	E1=D1=F1xH6=J2, A2=N1=R2xO2=H3, M1=B1=D2xH4=H4, L2=N2=Q2xT4=H1
12	01000010	T8+I7	A1=C1=F2xH5, M1=B1=D2xK5, A2=N1=R2xW2, M2=P2=Q1xU2
13	01000001	T8+I8	A1=C1=F2xH6=J2, M1=B1=D2xG2=K6
14	00110000	S7+O7	L2=N2=Q2xV2=Ω2, E2=P1=R1xU3=W3
15	00101000	**S**	E1=D1=F1xS6=G3, L2=N2=Q2xS1=Ω3, E2=P1=R1xS5=U4, L1=B2=C2xS2=K2
16	00100100	S7+O8	E1=D1=F1xJ3, L1=B2=C2xH2, L2=N2=Q2xΩ4, E2=P1=R1xU5
17	00100010	**U**	L2=N2=Q2xU1=Ω5, M1=B1=D2xS5=U4, A2=N1=R2xU3=W3, E1=D1=F1xI3=U6
18	00100001	S7+I8	A1=C1=F2xJ3, M1=B1=D2xS6=G3, L2=N2=Q2xΩ6, E2=P1=R1xI3=U6
19	00011000	O7+S8	E1=D1=F1xG4, L2=N2=Q2xV3, L1=B2=C2xK3, E2=P1=R1xW4
20	00010100	**O**	E1=D1=F1xO6=J4, L1=B2=C2xO2=H3, L2=N2=Q2xO1=V4, E2=P1=R1xO5=W5
21	00010010	**W**	L2=N2=Q2xV5=W1, M2=P2=Q1xU3=W3, A1=C1=F2xO5=W5, E1=D1=F1xI4=W6
22	00010001	O7+I8	A1=C1=F2xO6=J4, M1=B1=D2xG4, L2=N2=Q2xV6, E2=P1=R1xI4=W6
23	00001100	S8+O8	E1=D1=F1xG5=J5, L1=B2=C2xH4=K4
24	00001010	S8+I7	E1=D1=F1xI5=G6, M2=P2=Q1xS5=U4, L1=B2=C2xK5, A2=N1=R2xW4
25	00001001	**G**	A1=C1=F2xG5=J5, E2=P1=R1xI5=G6, L1=B2=C2xG2=K6, M2=P2=Q1xS6=G3
26	00000110	O8+I7	E1=D1=F1xI6=J6, L1=B2=C2xH5, M2=P2=Q1xU5, A2=N1=R2xO5=W5
27	00000101	**J**	A2=N1=R2xO6=J4, E2=P1=R1xI6=J6, L1=B2=C2xH6=J2, M1=B1=D2xG5=J5
28	00000011	**I**	A1=C1=F2xI6=J6, M1=B1=D2xI5=G6, M2=P2=Q1xI3=U6, A2=N1=R2xI4=W6

Table 8.8 Statements of Level 2-6 produced through subtraction derivations

1	11000000	**T**	A1=C1=F2–L5=C8, M2=P2=Q1–L8=Q8, M1=B1=D2–L6=B8, A2=N1=R2–L7=N8
2	10100000	**Ω**	L1=B2=C2–L8=Q8, M1=B1=D2–M8=Q6, E2=P1=R1–P8=Q4, A2=N1=R2–N7=Q7
3	10010000	**V**	M2=P2=Q1–N7=Q7, L1=B2=C2–L7=N8, E2=P1=R1–N4=R8, A1=C1=F2–A8=N5
4	10001000	T7+S8	E1=D1=F1–D8, M2=P2=Q1–M8=Q6, L1=B2=C2–L6=B8, A2=N1=R2–N6
5	10000100	T7+O8	E1=D1=F1–F8, M2=P2=Q1–Q5, L1=B2=C2–L5=C8, A2=N1=R2–A8=N5
6	10000010	T7+I7	M2=P2=Q1–P8=Q4, A2=N1=R2–N4=R8
7	10000001	T7+I8	A1=C1=F2–F8, M2=P2=Q1–Q3, M1=B1=D2–D8, A2=N1=R2–N3
8	01100000	T8+S7	A1=C1=F2–C7, L2=N2=Q2–L8=Q8, M1=B1=D2–M7=B7, E2=P1=R1–P7
9	01010000	T8+O7	A1=C1=F2–A7=C6, L2=N2=Q2–L7=N8, M1=B1=D2–B6, E2=P1=R1–R7
10	01001000	**K**	A1=C1=F2–B5=C5, E1=D1=F1–B3=D7, L2=N2=Q2–L6=B8, M2=P2=Q1–M7=B7
11	01000100	**H**	E1=D1=F1–C3=F7, A2=N1=R2–A7=C6, M1=B1=D2–B5=C5, L2=N2=Q2–L5=C8
12	01000010	T8+I7	A1=C1=F2–C4, M2=P2=Q1–P7, M1=B1=D2–B4, A2=N1=R2–R7
13	01000001	T8+I8	A1=C1=F2–C3=F7, M1=B1=D2–B3=D7
14	00110000	S7+O7	L2=N2=Q2–N7=Q7, E2=P1=R1–P6=R6
15	00101000	**S**	E1=D1=F1–M3=D6, L2=N2=Q2–M8=Q6, L1=B2=C2–M7=B7, E2=P1=R1–M4=P5
16	00100100	S7+O8	E1=D1=F1–F6, L2=N2=Q2–Q5, L1=B2=C2–C7, E2=P1=R1–P4
17	00100010	**U**	L2=N2=Q2–P8=Q4, M1=B1=D2–M4=P5, A2=N1=R2–P6=R6, E1=D1=F1–E6=P3
18	00100001	S7+I8	A1=C1=F2–F6, L2=N2=Q2–Q3, M1=B1=D2–M3=D6, E2=P1=R1–E6=P3
19	00011000	O7+S8	E1=D1=F1–D5, L2=N2=Q2–N6, L1=B2=C2–B6, E2=P1=R1–R5
20	00010100	**O**	E1=D1=F1–A3=F5, L2=N2=Q2–A8=N5, L1=B2=C2–A7=C6, E2=P1=R1–A4=R4
21	00010010	**W**	L2=N2=Q2–N4=R8, M2=P2=Q1–P6=R6, A1=C1=F2–A4=R4, E1=D1=F1–E5=R3
22	00010001	O7+I8	A1=C1=F2–A3=F5, L2=N2=Q2–N3, M1=B1=D2–D5, E2=P1=R1–E5=R3
23	00001100	S8+O8	E1=D1=F1–D4=F4, L1=B2=C2–B5=C5
24	00001010	S8+I7	E1=D1=F1–E4=D3, M2=P2=Q1–M4=P5, L1=B2=C2–B4, A2=N1=R2–R5
25	00001001	**G**	E2=P1=R1–E4=D3, A1=C1=F2–D4=F4, L1=B2=C2–B3=D7, M2=P2=Q1–M3=D6
26	00000110	O8+I7	E1=D1=F1–E3=F3, M2=P2=Q1–P4, L1=B2=C2–C4, A2=N1=R2–A4=R4
27	00000101	**J**	A2=N1=R2–A3=F5, E2=P1=R1–E3=F3, M1=B1=D2–D4=F4, L1=B2=C2–C3=F7
28	00000011	**I**	A1=C1=F2–E3=F3, M2=P2=Q1–E6=P3 , M1=B1=D2–E4=D3, A2=N1=R2–E5=R3

Table 8.9 Statements of Level 6-2 produced through division derivations

1	11111100	**E**	O8=H8=J7:I6=J6, S7=U7=Ω8:I3=U6, S8=G7=K8:I5=G6, O7=V8=W7:I4=W6
2	11111010	**F**	O7=V8=W7:O6=J4, I7=U8=W8:I6=J6, S8=G7=K8:G5=J5, T8=H7=K7:H6=J2
3	11111001	E2xA1	I8=G8=J8:I6=J6, S7=U7=Ω8:U5, T8=H7=K7:H5, O7=V8=W7:O5=W5
4	11110110	**D**	O8=H8=J7:G5=J5, I7=U8=W8:I5=G6, T8=H7=K7:G2=K6, S7=U7=Ω8:S6=G3
5	11110101	E2xM1	I8=G8=J8:I5=G6, S7=U7=Ω8:S5=U4, T8=H7=K7:K5, O7=V8=W7:W4
6	11110011	A1xM1	I8=G8=J8:G5=J5, T8=H7=K7:H4=K4
7	11101110	E1xA2	O8=H8=J7:O6=J4, T7=V7=Ω7:V6, S8=G7=K8:G4, I7=U8=W8:I4=W6
8	11101101	**R**	T7=V7=Ω7:V5=W1, S7=U7=Ω8:U3=W3, O8=H8=J7:O5=W5, I8=G8=J8:I4=W6
9	11101011	**A**	I8=G8=J8:O6=J4, T7=V7=Ω7:O1=V4, T8=H7=K7:O2=H3, I7=U8=W8:O5=W5
10	11100111	M1xA2	I8=G8=J8:G4, T7=V7=Ω7:V3, T8=H7=K7:K3, I7=U8=W8:W4
11	11011110	E1xM2	O8=H8=J7:J3, T7=V7=Ω7:Ω6, S8=G7=K8:S6=G3, I7=U8=W8:I3=U6
12	11011101	**P**	T7=V7=Ω7:U1=Ω5, S8=G7=K8:S5=U4, O7=V8=W7:U3=W3, I8=G8=J8:I3=U6
13	11011011	A1xM2	I8=G8=J8:J3, T7=V7=Ω7:Ω4, T8=H7=K7:H2, I7=U8=W8:U5
14	11010111	**M**	I8=G8=J8:S6=G3, T7=V7=Ω7:S1=Ω3, T8=H7=K7:S2=K2, I7=U8=W8:S5=U4
15	11001111	A2xM2	T7=V7=Ω7:V2=Ω2, I7=U8=W8:U3=W3
16	10111110	E1xL1	O8=H8=J7:H6=J2, S8=G7=K8:G2=K6
17	10111101	E2xL1	O8=H8=J7:H5, S7=U7=Ω8:U2, S8=G7=K8:K5, O7=V8=W7:W2
18	10111011	**C**	I8=G8=J8:H6=J2, O7=V8=W7:O2=H3, S8=G7=K8:H4=K4, T7=V7=Ω7:T4=H1
19	10110111	**B**	O8=H8=J7:H4=K4, I8=G8=J8:G2=K6, T7=V7=Ω7:T3=K1, S7=U7=Ω8:S2=K2
20	10101111	A2xL1	O8=H8=J7:O2=H3, T7=V7=Ω7:T2=V1, S8=G7=K8:K3, I7=U8=W8:W2
21	10011111	M2xL1	O8=H8=J7:H2, T7=V7=Ω7:T1=Ω1, S8=G7=K8:S2=K2, I7=U8=W8:U2
22	01111110	E1xL2	O8=H8=J7:J1, S7=U7=Ω8:Ω6, S8=G7=K8:G1, O7=V8=W7:V6
23	01111101	E2xL2	S7=U7=Ω8:U1=Ω5, O7=V8=W7:V5=W1
24	01111011	A1xL2	I8=G8=J8:J1, S7=U7=Ω8:Ω4, T8=H7=K7:T4=H1, O7=V8=W7:O1=V4
25	01110111	M1xL2	I8=G8=J8:G1, S7=U7=Ω8:S1=Ω3, T8=H7=K7:T3=K1, O7=V8=W7:V2=Ω2
26	01101111	**N**	S7=U7=Ω8:V2=Ω2, T8=H7=K7:T2=V1, I7=U8=W8:V5=W1, O8=H8=J7:O1=V4
27	01011111	**Q**	T8=H7=K7:T1=Ω1, S8=G7=K8: S1=Ω3, I7=U8=W8:U1=Ω5, O7=V8=W7:V2=Ω2
28	00111111	**L**	O8=H8=J7:T4=H1, S7=U7=Ω8:T1=Ω1, S8=G7=K8:T3=K1, O7=V8=W7:T2=V1

In the case of subtraction and division inferences, we have the results shown in Tables 8.8–8.9 (data are extracted from Tables 6.7 and 6.8).

Note that in each of the above 4 tables there are $24 \times 4 = 96$ plus $4 \times 2 = 8$ connections, what makes 104 relations as a whole for each table. It may be further noted that e.g. the statement **A** is sum-

Table 8.10 Statements of Level 6-2 produced through sum and division derivations

1	**E**	O8+E3=O8:I6, S8+E4=S8:I5, O7+E5=O7:I4, S7+E6=S7:I3
2	**F**	I7+F3=I7:J6, S8+F4=S8:J5, O7+F5=O7:J4, T8+F7=T8:J2
3	E2xA1	I8+E3=I8:I6, U7+P4=U7:U5, H7+C4=H7:H5, O7+A4=O7:O5
4	**D**	U8+D3=U8:G6, H8+D4=H8:G5, U7+D6=U7:G3, H7+D7=H7:G2
5	E2xM1	I8+E4=I8:I5, S7+M4=S7:S5, K7+B4=K7:K5, W7+R5=W7:W4
6	A1xM1	(K7=H7+B5=C5)=(K7=H7:K4=H4), (G8=J8+D4=F4)=(G8=J8:G5=J5)
7	E1xA2	O8+A3=O8:O6, G7+D5=G7:G4, V7+N3=V7:V6, W8+R3=W8:W6
8	**R**	J8+R3=J8:W6, J7+R4=J7:W5, Ω8+R6=Ω8:W3, Ω7+R8=Ω7:W1
9	**A**	I8+A3=I8:O6, I7+A4=I7:O5, T8+A7=T8:O2, T7+A8=T7:O1
10	M1xA2	G8+D5=G8:G4, V7+N6=V7:V3, K7+B6=K7:K3, W8+R5=W8:W4
11	E1xM2	J7+F6=J7:J3, Ω7+Q3=Ω7:Ω6, G7+D6=G7:G3, U8+P3=U8:U6
12	**P**	G8+P3=G8:U6, G7+P5=G7:U4, V8+P6=V8:U3, V7+P8=V7:U1
13	A1xM2	J8+F6=J8:J3, H7+C7=H7:H2, Ω7+Q5=Ω7:Ω4, U8+P4=U8:U5
14	**M**	I8+M3=I8:S6, I7+M4=I7:S5, T8+M7=T8:S2, T7+M8=T7:S1
15	A2xM2	(V7=Ω7+N7=Q7)=(V7=Ω7:V2=Ω2), (U8=W8+P6=R6)=(U8=W8:U3=W3)
16	E1xL1	(H8=J7+C3=F7)=(H8=J7:H6=J2), (G7=K8+D7=B3)=(G7=K8:G2=K6)
17	E2xL1	H8+C4=H8:H5, U7+P7=U7:U2, K8+B4=K8:K5, W7+R7=W7:W2
18	**C**	G8+C3=G8:H6, G7+C5=G7:H4, V6+C6=V8:H3, V7+C8=V7:H1
19	**B**	J8+B3=J8:K6, J7+B5=J7:K4, Ω8+B7=Ω8:K2, Ω7+B8=Ω7:K1
20	A2xL1	O8+A7=O8:O2, T7+L7=T7:T2, K8+B6=K8:K3, W8+R7=W8:W2
21	M2xL1	H8+C7=H8:H2, T7+L8=T7:T1, U8+P7=U8:U2, K8+B7=K8:K2
22	E1xL2	J7+F8=J7:J1, G7+D8=G7:G1, Ω8+Q3=Ω8:Ω6, V8+N3=V8:V6
23	E2xL2	(U7=Ω8+P8=Q4)=(U7=Ω8:U1=Ω5), (V8=W7+N4=R8)=(V8=W7:V5=W1)
24	A1xL2	J8+F8=J8:J1, T8+L5=T8:T4, Ω8+Q5=Ω8:Ω4, V8+N5=V8:V4
25	M1xL2	G8+D8=G8:G1, S7+M8=S7:S1, V8+N6=V8:V2, K7+B8=K7:K1
26	**N**	U8+N4=U8:V5, H8+N5=H8:V4, U7+N7=U7:V2, H7+N8=H7:V1
27	**Q**	W8+Q4=W8:Ω5, K8+Q6=K8:Ω3, W7+Q7=W7:Ω2, K7+Q8=K7:Ω1
28	**L**	O8+L5=O8:T4, S8+L6=S8:T3, O7+L7=O7:T2, S7+L8=S7:T1

Table 8.11 Statements of Level 2-6 produced through product and subtraction derivations

1	**T**	A1xT4=A1−L5, M1xT3=M1−L6, A2xT2=A2−L7, M2xT1=M2−L8
2	**Ω**	R1xΩ5=R1−Q4, B1xΩ3=B1−Q6, R2xΩ2=R2−Q7, B2xΩ1=B2−Q8
3	**V**	P1xV5=P1−N4, C2xV4=C2−N5, P2xV2=P2−N7, C2xV1=C2−N8
4	T7+S8	D1xG1=D1−D8, M2xS1=M2−M8, N1xV3=N1−N6, B2xK1=B2−B8
5	T7+O8	F1xJ1=F1−F8, L1xT4=l1−L5, Q1xΩ4=Q1−Q5, A2xO1=A2−A8
6	T7+I7	(P2=Q1xU1=Ω5)=(P2=Q1−P8=Q4), (N1=R2xV5=W1)=(N1=R2−N4=R8)
7	T7+I8	F2xJ1=F2−F8, D2xG1=D2−D8, N1xV6=N1−N3, Q1xΩ6=Q1-Q3
8	T8+S7	C1xH2=C1−C7, L2xT1=L2−L8, P1xU2=P1−P7, M1xS2=M1−M7
9	T8+O7	A1xO2=A1−A7, L2xT2=L2−L7, B1xK3=B1−B6, R1xW2=R1−R7
10	**K**	F1xK6=F1−B3, F2xK4=F2−B5, Q1xK2=Q1−B7, Q2xK1=Q2−B8
11	**H**	D1xH6=D1−C3, D2xH4=D2−C5, N1xH3=N1−C6, N2xH1=N2−C8
12	T8+I7	C1xH5=C1−C4, P2xU2=P2−P7, B1xK5=B1−B4, R2xW2=R2−R7
13	T8+I8	(C1=F2xH6=J2)=(C1=F2−C3=F7), (B1=D2xK6=G2)=(B1=D2−B3=D7)
14	S7+O7	(N2=Q2xV2=Ω2)=(N2=Q2−N7=Q7), (P1=R1xU3=W3)=(P1=R1−P6=R6)
15	**S**	E1xS6=E1−M3, E2xS5=E2−M4, L1xS2=L1−M7, L2xS1=L2−M8
16	S7+O8	F1xJ3=F1−F6, Q2xΩ3=Q2−Q5, C2xH2=C2−C7, P1xU5=P1−P4
17	**U**	D1xU6=D1−P3, D2xU4=D2−P5, N1xU3=N1−P6, N2xU1=N2−P8
18	S7+I8	F2xJ3=F2−F6, M1xS6=M1−M3, Q2xΩ6=Q2−Q3, E2xI3=E2−E6
19	O7+S8	D1xG4=D1−D5, N2xU3=N2−N6, B2xK3=B2−B6, R1xW4=R1−R5
20	**O**	E1xO6=E1−A3, E2xO5=E2−E4, L1xO2=L1−A7, L2xO1=L2−A8,
21	**W**	F1xW6=F1−R3, F2xW5=F2−R4, Q1xW3=Q1−R6, Q2xW1=Q2−R8
22	O7+I8	A1xO6=A1−A3, D2xG4=D2−D5, N2xV6=N2−N3, E2xI4=E2−E5
23	S8+O8	(D1=F1xG5=J5)=(D1=F1−D4=F4), (B2=C2xK4=H4)=(B2=C2−B5=C5)
24	S8+I7	E1xI5=E1−E4, M2xS5=M2−M4, B2xK5=B2−B4, R2xW4=R2−R5
25	**G**	P1xG6=P1−D3, C1xG5=C2−D4, P2xG3=P2−D6, C2xG2=C2−D7
26	O8+I7	E1xI6=E1−E3, C2xH5=C2−C4, P2xU5=P2−P4, A2xO5=A2−A4
27	**J**	R1xJ6=R1−F3, B1xJ5=B1−F4, R2xJ4=R2−F5, B2xJ2=B2−F7
28	**I**	A1xI6=A1−A3−E3, M1xI5=M1−E4, A2xI4=A2−E5, M2xI3=M2−E6

derived through any element of its RGS apart from A5 and A6, which are the only elements of the set that are not shared with the RGSs of other statements. The same is true for particular statements. For instance, the statement **V** is product-derived through any element of its LGS apart from V3 and V6 which are the only elements of the set to do not be shared. It may also be noted that each element of whatever set contributes to the derivation of one of the main statements either of Level 6-2 or 2-6 listed above only in a single product or sum combination with another kind of statement.

Tables 8.6–8.9 also allow to establish connections between all these different derivations, as displayed in Tables 8.10 and 8.11.

The previous two tables bring together all previous results. I remind that in Table 8.10 in the case of bold (grounding) statements we have in the sums all four elements pertaining to the RGS of the bold statement that are shared also with other RGSs. For instance, in the case of **F** these statements are: F3, F4, F5 and F7. Instead, the division expression contains a series of all four elements of the LGS of the contradictory statement that are again shared. Always in the case of **F**, we have: J6, J5, J4 and J2. Moreover, all 8 elements listed above (and similarly for all other bold statements) are ordered in contradictory pairs: F3 with J6, F4 with J5, F5 with J4 and F7 with J2. In the case of crossed expressions (like A2xL1 or S7+O7), note that there are two couples with connected elements (in some cases they coincide). The universal and particular statements occurring in these sums always represent statements pertaining to some set of two bold statements that are contradictory pairs although not being themselves such a pair. For instance, the expression E2xM1 shows the sum I8+E4. The divisions show pairs that pertain to the same set. For instance, the previous sum is connected with the division I8:I5. Similarly for Table 8.11.

The above relations allow a further and more interesting consideration: Tables 8.10 and 8.11 express kind of invariant inferences' results of the system. For instance, we can obtain **E** starting with O8 and by either summing to it E3 or dividing it by the countervalent statement, i.e. I6; we obtain **T** starting with A1 and by either multiplying it with T4 or subtracting to it the countervalent statement L5.

8.4 Cycles of Inferences

There are some remarkable connections among the different main statements that allow to build true reversible cycles across the logical space. Consider in particular the 6 propositions of Level 4-4 in the product-sum representation as displayed in Table 8.12 (have also a look at Tables 1.6 and 2.2).

Table 8.12 The six core statements of Level 4-4

#	ID	Symbol	Logical form
2	11101000	E3xE5=A3xA4=F3xF5=R3xR4=T1+T3=S1+S2=K1+K2=Ω1+Ω3	$\neg X$
7	11010100	E4xE6=M3xM4=D3xD6=P3xP5=T2+T4=O1+O2=H1+H3=V1+V4	$\neg Y$
18	10110010	B3xB5=C3xC5=D4xD7=F4xF7=U1+U3=V2+V5=W1+W3=Ω2+Ω5	$\neg Z$
53	01001101	P6xP8=N4xN7=Q4xQ7=R6xR8=G2+G5=H4+H6=J2+J5=K4+K6	Z
64	00101011	A7xA8=L5xL7=C6xC8=N5xN8=S5+S6=I3+I5=G3+G6=U4+U6	Y
69	00010111	M7xM8=L6xL8=B7xB8=Q6xQ8=O5+O6=I4+I6=J4+J6=W5+W6	X

Consider that all of the above statements already occur in the two-dimensional space. Now, these statements allow a circle between Levels 6-2 and 2-6. First of all, let us consider the fact that the above equalities involve LGSs and RGSs of all of the 12 universal statements as well as all of the 12 particular statements. According to Fig. 2.1. these 24 statements can be factorized in three different sets displayed by the three octagons with two subsets each (one for the universal, the other for the particular statements).

Table 8.13 The three sets of bold statements

Sets	Universal statements	Particular statements	Variables
First octagon	**A, E, L, M**	**I, O, S, T**	X, Y
Second octagon	**B, F, Q, R**	**J, K, W, Ω**	X, Z
Third octagon	**C, D, N, P**	**G, H, U, V**	Y, Z

Now, we shall consider the circle for the first set (Table 8.14).

Note that the statements of Level 4-4 are (from the left to the right): $\neg X$, $\neg Y$, X, and Y. What is relevant here is that from Level 4-4 we have two ascending and two descending transformations: We descend from Level 6-2 to Level 4-4 by multiplication of two of the universal statements and go back to Level 6-2 by summing the expression for two of the statements listed in Table 8.12. For instance, we get the statement **A** by summing the expression for

Table 8.14 First cycle in the product–sum representation

Level 6-2	**A**		**E**		**M**		**L**	
Flux	↓	↑	↓	↑	↓	↑	↓	↑
Operation	**A × E**	(A3xA4) + (A7xA8)	**E × M**	(E3xE5) + (E4xE6)	**M × L**	(M7xM8) + (M3xM4)	**L × A**	(L5xL7) + (L6xL8)
Flux	↓	↑	↓	↑	↓	↑	↓	↑
Level 4-4	A3xA4=E3xE5 T1+T3=S1+S2		E4xE6=M3xM4 O1+O2=T2+T4		M7xM8=L6xL8 I4+I6=O5+O6		L5xL7=A7xA8 S5+S6=I3+I5	
Flux	↓	↑	↓	↑	↓	↑	↓	↑
Operation	(T1+T3) × (T2+T4)	**T + S**	(O5+O6) × (O1+O2)	**O + T**	(I3+I5) × (I4+I6)	**I + O**	(S1+S2) × (S5+S6)	**S + I**
Flux	↓	↑	↓	↑	↓	↑	↓	↑
Level 2-6	**T**		**O**		**I**		**S**	

$\neg X$ and Y. The descending operation is simply the ordinary way to obtain in the logical space propositions of lower level from propositions of higher level (in other words we are embedding a two-dimensional space inside a three-dimensional space: see Table 1.3). Anyway, it corresponds to the classical operation of conjunction of statements that are held true or are assumed. The ascending operation is far more interesting since it corresponds to the classical operation called *Addition*. Now, thanks to the identities shown in Table 8.12, it is clear that to sum e.g. the statements $\neg X$ and Y to obtain **A** amounts to sum A3xA4 (since this is one of the ways in which we express $\neg X$) and A7xA8 (which is one of the ways in which we express Y). However, each pair of the involved statements is also able to generate **A** alone (the two couples A3-A4 and A7-A8 pertain indeed to **A**'s RGS), so that we can substitute to the expression (A3xA4)+(A7xA8) the sum A3+A8 or also A3+A5 or any other binary combination out of these four statements. In other words, we have performed a transformation of the kind:

A3xA4 maps to A3+A4 (i.e. **A**).

Therefore, we have as a general rule for the universal statements that if two statements pertaining to the RGS of such a statement are multiplied, we are always allowed to transform this in their sum (which generates the statement itself). Finally, note that the sum of statements pertaining to the RGS of universal statements (like A1, A2) does not work since it leads to the tautology 11111111.

For the particular statements, consider that the ascending movement is generated through sum of two statements. The net effect of each of these sums is to reduce a conjunction (like $X \wedge Y$) to one of its terms (either X or Y). Therefore, it corresponds to the classical rule of *Simplification*. We can establish as a general rule here for particular statements that when we have an expression like T1+T3, we are also allowed to derive T1+T3, which is equal to the statement **T** (indeed T1 and T3 pertain to the LGS of the particular statement **T**). In other words,

T1+T3 maps to T1$\times$T3 (i.e. **T**).

The product of two statements pertaining to the RGS of particular statements (like T7 and T8) does not work since it leads to the contradiction 00000000. The descending movement is generated here by multiplying two of the propositions of Level 4-4. This corresponds to the descending order of our logical space. For instance, multiplying $\neg X$ and $\neg Y$ gives rise to **T**. Note that these two expressions can be again formulated in terms of the couples T1-T3 and T2-T4. However, remark that this is not a logical rule but only an ordering relation. In other words, in the general case we cannot pass from a single term like X to its conjunction with Y. We discover again that structural relations and inferential relations are two distinct issues.

What we have considered here is that the two principles assumed at the start in Section 1.1, which tell us that we are allowed either to deny the component of some statement or to add some statement (but not to derive a conjunctions between two statements having a single one as a start) are to a certain extent justified here if not properly proved (there is indeed a certain circularity in the argument itself, which, however, I do not consider a defect but a factor contributing to the symmetric beauty of the system). Similar considerations apply to the second cycle (Table 8.15).

Note that the statements of Level 4-4 are (from the left to the right): $\neg Z$, $\neg X$, Z, X. Finally, the third cycle is shown in Table 8.16.

Note that here the statements of Level 4-4 are (from the left to the right): $\neg Z$, $\neg Y$, Z, Y. The previous considerations justify us in having assumed at the end of Section 1.2 that from the product between an implication and its antecedent we can derive its consequent (according to the logical rule of *modus ponens* or, alternatively, to the rule of *Disjunctive syllogism*). For instance,

$$\mathbf{A} \times (\text{M7} \times \text{M8}) = \mathbf{I}$$

Now, thanks to the above relations, from $\mathbf{I} = \text{I3} \times \text{I5}$ we can derive I3+I5. After this examination, it is also quite obvious that we are allowed to derive any particular statement like **I** from one of the statements of its RGS, like I7 or I8. It is only a very particular case of simplification. These considerations allow us also to understand all the sufficient-condition derivations (product and subtraction) listed in Table 8.5 as particular cases of *modus ponens* (having the inference plus the premises allows to derive the conclusion). The opposite is for necessary-conditions inferences (sum and division).

Table 8.15 Second cycle in the product–sum representation

Level 6-2	**B**		**F**		**R**		**Q**	
Flux	↓	↑	↓	↑	↓	↑	↓	↑
Operation	**B × F**	(B3xB5) + (B7xB8)	**F × R**	(F3xF5) + (F4xF7)	**R × Q**	(R3xR4) + (R6xR8)	**Q × B**	(Q4xQ7) + (Q6xQ8)
Flux	↓	↑	↓	↑	↓	↑	↓	↑
Level 4-4	B3xB5=F4xF7 Ω2+Ω5=W1+W3		F3xF5=R3xR4 K1+K2=Ω1+Ω3		R6xR8=Q4xQ7 J2+J5=K4+K6		Q6xQ8=B7xB8 W5+W6=J4+J6	
Flux	↓	↑	↓	↑	↓	↑	↓	↑
Operation	(Ω2+Ω5) × (Ω1+Ω3)	**Ω + W**	(K1+K2) × (K4+K6)	**K+ Ω**	(J2+J5) × (J4+J6)	**J + K**	(W1+W3) × (W5+W6)	**W + J**
Flux	↓	↑	↓	↑	↓	↑	↓	↑
Level 2-6	**Ω**		**K**		**J**		**W**	

Table 8.16 Third cycle in the product–sum representation

Level 6-2	**C**		**D**		**P**		**N**	
Flux	↓	↑	↓	↑	↓	↑	↓	↑
Operation	**C × D**	(C3xC5) + (C6xC8)	**D × P**	(D3xD6) + (D4xD7)	**P × N**	(P3xP5) + (P6xP8)	**N × C**	(N4xN7) + (N5xN8)
Flux	↓	↑	↓	↑	↓	↑	↓	↑
Level 4-4	C3xC5=D4xD7 V2+V5=U1+U3		D3xD6=P3xP5 H1+H3=V1+V4		P6xP8=N4xN7 G2+G5=H4+H6		N5xN8=C6xC8 U4+U6=G3+G6	
Flux	↓	↑	↓	↑	↓	↑	↓	↑
Operation	(V2+V5) × (V1+V4)	**V + U**	(H1+H3) × (H4+H6)	**H + V**	(G2+G5) × (G3+G6)	**G + H**	(U1+U3) × (U4+U6)	**U + G**
Flux	↓	↑	↓	↑	↓	↑	↓	↑
Level 2-6	**V**		**H**		**G**		**U**	

These results fully justify the notion of core elements of the space. In fact, we have embedded 3 two-dimensional logical spaces in a single three-dimensional space, as shown in Fig. 8.2 (see also Figs. 1.7–1.9). In a similar way, there are 2 one-dimensional spaces embedded in a two-dimensional one and 4 three-dimensional spaces embedded in a four-dimensional logical space. This allows us to build a combinatory of statements for any n-dimensional space as a combinatory of $(n - 1)$-dimensional spaces whose number is n.

There is here a consequence that is very relevant for the whole understanding of the logical space. In Section 1.2 I have stressed the circumstance that we cannot perform product inferences when particular statements only are involved since they could refer to different sets. This is the reason why I have suggested to write e.g. $X_1 \wedge Y_1$ and $X_2 \wedge Y_2$, where X_1 and X_2 can mean different subsets of a set X as well as Y_1 and Y_2 can mean different subsets of a set Y. The embedment of of one-dimensional logical spaces into two-dimensional spaces and of the latter into a three-dimensional one clarifies this point. For instance, the conjunctions $X \wedge Y$ and $Z \wedge X$ pertain to two different two-dimensional spaces (displayed by the first and the third couples of sets), so that X denotes here two different sets (sets 1 and 2). It is only when we pair a particular statement with an universal one that we can reach the level (i.e. Level 6-2) in which inferences involving these three variables become possible. This is why product inferences involve lowering generating sets and at least one of the two premises is an universal (pertains to Level 6-2) as it is displayed in Table 6.5. Obviously, we have several time remarked that not any combination of this kind of statements can give rise to a true inference. It is indeed, only at the level of highest generality (i.e. Level 7-1) that the three variables are connected in a way that is always logically sound, and this explains why the statements of this level represent the most general forms of admissible inferences, as shown in Section 8.2 (and, reciprocally, why their negations of Level 1-7 represent the most general forms to deny the validity of an inference). All the subspaces are given by Table 8.17.

This allows us to introduce now sum and product over logical spaces, that can be again considered in terms of sets and their parts.

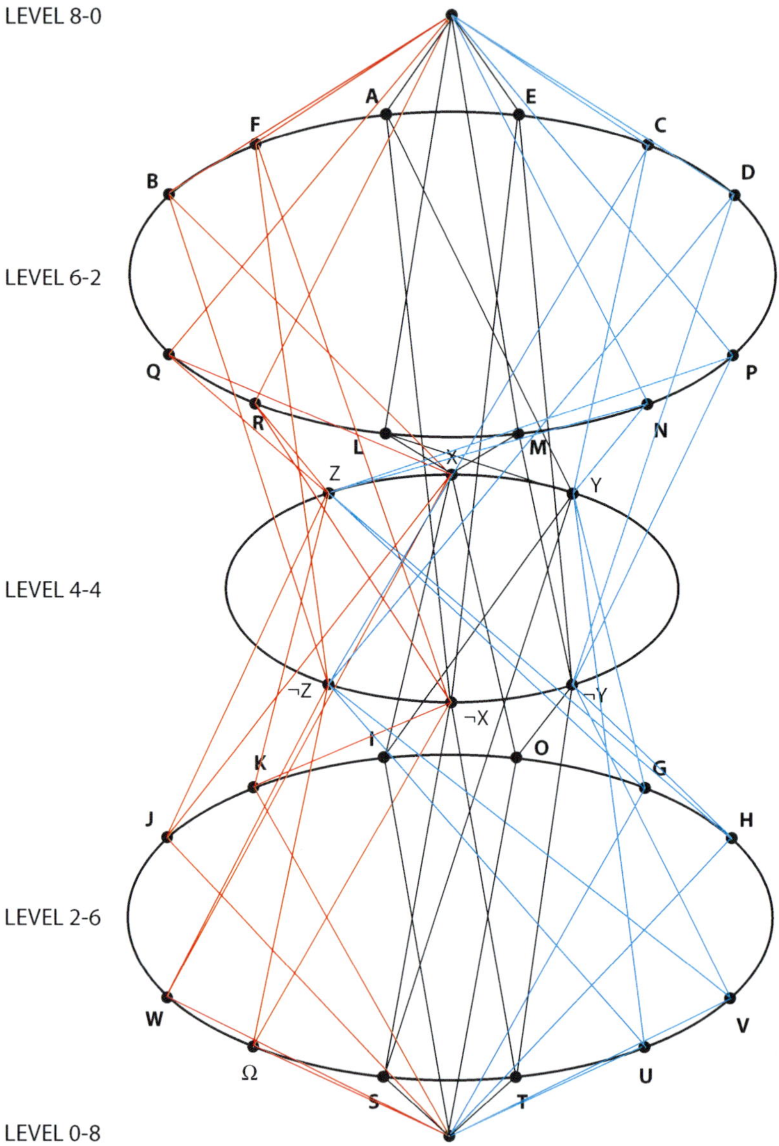

Figure 8.2 The 3 two-dimensional logical spaces are indicated in three different colors. For practical purposes I have not included statements $X \leftrightarrow Y$ and its negation (Statements 15 and 56 of Level 4-4, respectively). Obviously, the three main levels here are embedded in the space between the 8 highest (Level 7-1) and the 8 lowest (Level 1-7) statements.

Table 8.17 One-dimensional subspaces in the tridimensional space

	a	b	c	d	e	f	g	h
X	1 7	1 7	1 8	2 7	1 8	2 7	2 8	2 8
Y	3 9	3 9	4 10	3 9	4 10	3 9	4 10	4 10
Z	5 11	6 12	5 11	5 11	6 12	6 12	5 11	6 12

Let us consider that the whole union of the classes X and Y. This can be written as $2 \oplus 4$ or $1 \oplus 4$. This is typical for statements of Level 6-2 like $X \rightarrow Y$ or $\neg X \rightarrow Y$. However, when we consider statements of Level 2-6, they only deal with small intersections of classes. We could write in such a case $2 \otimes 3$ or also $1 \otimes 3$. Statements of the first kind are $X \wedge \neg Y$ or $\neg X \wedge \neg Y$. For the full intelligence of what follows, it is suitable to consider only statements that show a certain regularity, i.e. the statements in the first rows of Tables 1.25, 1.28, 1.35, 1.38 (see also Fig. 1.13). I have not indicated a privileged set of statements for Level 4-4, since the forms at this level are very mixed.

In analogy with traditional logical expressions, the following formula can be helpful (in analogy with well know logical expressions):

- $[(1 = 7 \otimes 3 = 9 \otimes 5 = 11) \oplus (1 = 7 \otimes 3 = 9 \otimes 6 = 12)] = 1 = 7 \otimes 3 = 9 = 1 \otimes 3$.
- $[(1 = 7 \otimes 3 = 9 \otimes 5 = 11) \oplus (1 = 7 \otimes 3 = 9 \otimes 6 = 12)] = 1 = 7 \otimes [(3 = 9 \otimes 5 = 11) \oplus (4 = 10 \otimes 6 = 12)] = 1 = 7 \otimes [(9 \otimes 11) \oplus (10 \otimes 12)]$.
- $[(1 = 7 \otimes 3 = 9) \oplus (1 = 7 \otimes 5 = 11)] = 1 = 7 \otimes (9 \oplus 11)$.
- $[(1 = 7 \oplus 3 = 9) \otimes (1 = 7 \oplus 5 = 11)] = 1 = 7 \oplus (9 \otimes 11)$.
- $[(1 = 7 \oplus 3 = 9 \oplus 5 = 11) \otimes (1 = 7 \oplus 3 = 9 \oplus 6 = 12)] = 1 = 7 \oplus [(9 \oplus 11) \otimes (10 \oplus 12)]$.
- $[(1 = 7 \oplus 3 = 9 \oplus 5 = 11) \otimes (1 = 7 \oplus 3 = 9 \oplus 6 = 12)] = 1 \oplus 3$.

These operations display the fact that the concept of subspace has certain analogies with that of logical expression. However, it is also slightly different. It is clear that at the lowest level before contradiction (Level 1-7) we have no longer sets but only parts or areas, like a, b, c, ... (see Fig. 1.6 and Table 1.12). This shows that this level is rather the proper propositional level and fully justifies Peirce's functor and its three-dimensional analogue (Sheffer's functor although formally correct, being the negation of Peirce's functor, does not catch this specific meaning) (see previous section). This also explains why we have cross terms at level 2-6 (see Table 6.2): they are combinations of areas pertaining to different classes. For instance, Statement 24 (i.e. $Y \wedge \neg(X \leftrightarrow Z)$) combines (sums) areas e and g that obviously represent a partial intersection of Z and Y and a partial intersection of Z and X. Instead, Statement 27 (i.e. $Z \wedge X$) is a combination (a sum) of areas f and h that represent the full intersection of Z and X. Obviously, we could have defined new (but less natural) sets that represent new combinations of areas, like the set constituted by the sum of: d, e, f, h (which is Statement 67 of Level 4-4)—see also Table 8.3.

To a certain extent the opposite situation is true for sum inferences. Here, only raising generating sets are involved, and therefore at least one of the premises needs to be a particular, as it is evident from Table 6.6. The reason is that we have projections to common sets only at the lowest logical levels (from Level 2-6 downwards, and at Level 1-7 we only have the areas a, b, c, ...), so that we have a kind of inverted schema relatively to pervious examination (note indeed that Fig. 8.2 shows convergence both at the top and at the bottom, according to whether we are considering the raising or the lowering order). Again different are the cases of subtraction and division inferences. As it is evident from Table 8.1, subtraction inferences involve a LGS for the first premise and a RGS for the second premise. This means that it is necessary that the first premise is universal or the second premise is particular (in the counterparts of the so-called universal inferences we have both), as Table 6.7 shows. Since division inferences also involve a LGS and a RGS but in a reversed order relative to subtraction inferences, we have here the opposite situation, as displayed in Table 6.8. This examination also shows an interesting possibility: I guess that for subtraction and division inferences we could have taken as grounding statements neither the bold ones nor the crossed ones of Levels 6-2 and 2-6 (see Tables 8.2 and 8.3) but (i) 12 out of the 24 statements of Level 5-3, whose general form is shown in the first row of Table 1.28, and (ii) 12 out of the 24 statements of Level 3-5, whose general form is shown the first row of Table 1.35.

This allows us to express the relations among subspaces in the following terms:

Table 8.18 Two-dimensional subspaces in the tridimensional space

	a	b	c	d	e	f	g	h
X	1	1	1	2	1	2	2	2
Y	3	3	4	3	4	3	4	4
X	7	7	7	8	7	8	8	8
Z	5	6	5	5	6	6	5	6
Y	9	9	10	9	10	9	10	10
Z	11	12	11	11	12	12	11	12

Note that the class of the objects X (the subspace 2) and the class of objects that are not X (the subspace 1) are in themselves neither universal, nor particular. It is only the disjunction or conjunction with other classes of objects that determines these relations in one way or the other. For X and Y, the following couples are built: a-b, c-e, d-f, g-h; for X and Z we have: a-c, b-e, d-g, f-h; for Y and Z: a-d, b-f, c-g, e-h. Note that these couples represent either cases in which the two values are both zero (for Level 6-2) or are both one (for Level 2-6), as displayed by Tables 1.24 and 1.37, respectively (taking one element from each couple we get the bidimensional case). For instance, Statement 28 of Level 2-6 displays two ones in positions g and h: 00000011, while Statement 1 of Level 6-2 displays two zeros in the same position (11111100), i.e. both correspond to the latter two columns of the first block of Table 8.18. It is interesting to consider any inference as a combinations of subspaces. For instance, *Barbara* and *Ferio* can be expresses as:

$$[(1 = 7 \odot 4 = 10) \wedge (5{=}11 \odot 2{=}8)] \rightarrow (11 \odot 10) \text{ and}$$
$$[(1 = 7 \odot 3{=}9) \wedge (6{=}12 \odot 2{=}8)] \rightarrow (12 \odot 9)\,, \qquad (8.39)$$

respectively, where $\odot$ can mean either product or sum of subspaces. Note that the premises deal with the whole three-dimensional space, while the conclusion only with the proper bidimensional space. It may be further noted that, according to Table 8.2, e.g. Statement **A** ($X \rightarrow Y$) occurs always with one of the universal premises **B** or **Q** (for variables X, Z: Z or $\neg Z$ implies X) as well as **D** or **P** (for variables Y, Z: Y implies either Z or $\neg Z$). This means, that the couple d-f is compatible with a-c or b-e for X-Z and with c-g or e-h for Y-Z. Similarly, **A** (in product inferences) occurs with one of the particular premises **J** or **W** (for X-Z: X is conjoint either with Z or $\neg Z$) as well as **H** or **V** (for Y-Z: $\neg Y$ is conjoint either with Z or $\neg Z$). In other words, d-f is compatible with d-g or f-h for X-Z and with a-d or b-f for Y-Z. This means that **A** ($1 = 7 \oplus 4 = 10$) occurs either with one of these universal combinations: $1 = 7 \oplus 5 = 11$, $1 = 7 \oplus 6 = 12$, $4 = 10 \oplus 5 = 11$, $4 = 10 \oplus 6 = 12$, or with one of these particular combinations: $2 = 8 \otimes 5 = 11$, $2 = 8 \otimes 6 = 12$, $3 = 9 \otimes 5 = 11$, $3 = 9 \otimes 6 = 12$. It is then relatively easy to single out which kind of statement a combination of subspaces means when occurring in an inference.

8.5 Open Inferences with Two and More Variables

I am now interested in generalizing the previous theory of inferences to forms that do not require a predetermined number of assumptions or variables, neither a pre-specified general form of conclusion. Let us first consider some very elementary examples in the two-dimensional logical space. Supposes that we wish to prove that from the following expression

$$(X \vee \neg Y) \rightarrow [(X \wedge Y) \rightarrow Y]$$

the tautology $X \vee \neg X$ can be derived. In this case, we write

$$(X \vee \neg Y) \rightarrow [(X \wedge Y) \rightarrow Y] \quad / X \vee \neg X$$

For performing calculation with the numeric ID, it is suitable to try to reduce all expressions to a basic form in which only negations, conjunctions and disjunctions are present (this may be further reduced to the so-called conjunctive normal form or disjunctive normal form). In this case we obtain

$$(\neg X \wedge Y) \vee (\neg X \vee \neg Y \vee Y)\,.$$

It is easy to see that second term already contains a tautology. Nevertheless, an explicit calculations can show that

$$\begin{array}{ll} 0100 & + \\ 1111 & = \\ 1111 & \end{array}$$

In other words, a disjunction of an arbitrary proposition with a tautology reduces to a tautology. A more complex example is the following one:

$$(X \vee \neg Y) \rightarrow [(X \wedge \neg Y) \rightarrow Y] \qquad /\neg X \vee Y$$

We reduce again this single premise to

$$(\neg X \wedge Y) \vee (\neg X \vee Y \vee Y)\,.$$

Since in the second expression we have two times Y, this can be reduced to a single occurrence (rule of *Idempotency*: to perform a product of an expression with itself produces the same expression). Then we have:

$$\begin{array}{ll} 0100 & + \\ 1101 & = \\ 1101 & \end{array}$$

which proves the conclusion.

Let us now consider some examples in the three-dimensional logical space. Let us consider the following example in particular:

$$\begin{array}{ll} X \rightarrow (Y \rightarrow Z) & \\ \neg Z & \quad / \, X \rightarrow \neg Y \end{array}$$

The desired conclusion is obviously **E**, while the expression between parentheses in the first premise is **P**. Let us transform the first premise into $\neg X \vee (Y \rightarrow Z)$. Then, the term $\neg X$ is denoted by the product E3xE5 while the term $\neg Z$ is denoted by the product F4xF7. Therefore, we have:

E3xE5	11101000	+
P	11011101	=
E2	11111101	×
F4xF7	10110010	=
E	11111100	

This result can be also understood in this way: F4xF7 = **F** = (F1=E1)x(F3=E3). However, E1xE3xE2 = **E**. Another example is the following one:

$$[(X \vee Z) \rightarrow Y] \rightarrow Z$$
$$\neg Z \qquad\qquad / \neg Y$$

The first premise can be transformed into

$$[(X \vee Z) \wedge \neg Y] \vee Z \ .$$

Recalling that $\neg Y = \mathrm{O1} + \mathrm{O2}$, $Z = \mathrm{Q4xQ7}$, and $Z = \mathrm{D4xD7}$, this allows us to compute:

Q	01011111	×
O1+O2	11010100	=
O2	01010100	+
Q4xQ7	01001101	=
Q4=P8	01011101	×
D4xD7	10110010	=
O7	00010000	

Now, O7 is itself a double conjunction: $X \wedge \neg Y \wedge \neg Z$. We already have $\neg Z$ (the second premise) so that we can simplify this expression to

$$\mathbf{O} = \mathrm{O1xO2} \quad 00010100$$

which maps to

$$\mathrm{O1} + \mathrm{O2} \quad 11010100$$

It is very important to do not make confusion between components of premises and premise or results. In the previous derivation, $\neg Y$ was not derived until the last step.

More complex are those cases in which we have 4 and even more variables. One possibility is to develop a n-dimensional logical space for the case of n variables. This is something that should be done in the future. However, it is also not so easy to manage. In the case of only four variables, we have 16 truth-value assignments that produce $2^{16} = 65,536$ different propositions (IDs), 17 different levels and the considerable amount of 1,048,576 relations. It is therefore suitable to explore alternative options. In order to this, we should prove that any m-variable inference and any n-premise inference can be reduced to the case of 3-variable and 2-premise inferences treated up to now. This was already one of Peirce's conjectures (Peirce, 1903). He also presented a general proof (Peirce, 1880, pp. 168–69). This can also be accomplished in a relative straightforward (but slightly different) way on the basis that we have built up to now (I limit my considerations to sufficient-condition inferences but this can be easily extended to necessary-condition ones). Consider the n-premise inference:

$$(\mathrm{P1} \wedge \mathrm{P2} \wedge \ldots \wedge \mathrm{P}_n) \rightarrow \mathrm{C}\,, \tag{8.40}$$

where C is some conclusion. Let us first treat the 4-premise case. The reader can verify that the following expressions

$$(\mathrm{P1} \wedge \mathrm{P2} \wedge \mathrm{P3} \wedge \mathrm{P4}) \rightarrow \mathrm{C} \tag{8.41}$$

and

$$\{[(\mathrm{P1} \wedge \mathrm{P2}) \rightarrow X] \wedge [(X \wedge \mathrm{P3}) \rightarrow Y] \wedge [(Y \wedge \mathrm{P4}) \rightarrow Z]\} \rightarrow \mathrm{C} \tag{8.42}$$

are equivalent. I am assuming that X is a true logical consequence of the premises P1 and P2, Y is a true logical consequence of premises X and P3, Z is a true logical consequence of premises Y and P4. In other words, the two inferences (8.41) and (8.42) are falsified only in one case, that is, when the conjunction of the premises is true but the conclusion is false. In the case of inference (8.41) this implies that each of the premises P1, P2, P3, P4 holds true but the conclusion C be false. Now, if P1 and P2 are true, given the previous assumption, we also have that X must be true, which together with P3 implies that also Y is true, which finally implies with P4 that also Z is true. Then, the truth of P1, P2, P3, and P4 implies the truth of all premises in inference (8.42). Therefore, if the conclusion C is false, both inferences are false. Given this result, we can proceed step-wise and first deduce X from premises P1 and P2, then deduce Y from the premises X and P3, and so on. Each time we have again 3 variables. Obviously, we can face the situation in which at least one of the premises P1, P2, P3, ... is itself a complex expression involving more variables. In such a case, we need to reformulate such an expression in a way that is analogue to an implication of the previous kinds and to proceed again recursively. In this way we can also solve the case of inferences involving *m* variables.

This procedure can be easily generalized by mathematical induction (if it is true for 4 premises it is true for any $4 + n$ cases). Indeed, let us consider here the case of a a 7-premise inference. Let us denote with P' a logical consequence of the whole antecedent of implication (8.42). Then, we proceed in such a way:

$$\{[(\mathrm{P}' \wedge \mathrm{P5}) \rightarrow X'] \wedge [(X' \wedge \mathrm{P6}) \rightarrow Y'] \wedge [(Y' \wedge \mathrm{P7}) \rightarrow Z']\} \rightarrow \mathrm{C}\,, \tag{8.43}$$

where X', Y', Z' are again three propositional variables that logically follow from the relative premises as before. Now the reader may worry that we have used 6 variables instead of 3. However, we are proceeding in a recursive and hierarchical way, and in each step of the inferential reasoning we use no more than 3 variables. Since each subsegment of an inference of this kind can be led to one of the canonical forms explored before (have especially a look at Table 8.5), it is clear that any open inference can be treated in a mechanical way. It is obvious that here and in the following I shall treat the variables always as meaning the same sets: this is a consequence of the fact that I shall deal only with propositions of the propositional calculus. In the case in which, we have first-order logic we need in some cases the sophistications mentioned in the previous section. Since, however, each first-order sentence can be appropriately transformed in the language used in this book (an issue about which I shall have also something to add), this is a problem that can be bypassed in such a context.

According to the previous examination, I proceed now to factorize the premises in groups such that each group does not exceed three variables (better if only two) and to proceed with step-wise variable substitutions. Obviously, the way in which we sort the premises is arbitrary. Let me give here a concrete example drawn from the ordinary propositional calculus:

$$
\begin{array}{ll}
p \vee \neg r & \\
r \vee s & \\
\neg s & \\
p \rightarrow t & /t
\end{array}
$$

and let us proceed in this way:

Step 1 We sort the premises $\neg s$ and $r \vee s$ and rewrite them as $\neg Y$ and $X \vee Y$ and proceed as usual:

$$
\begin{array}{ll}
1010 & \times \\
0111 & = \\
0010 &
\end{array}
$$

which allows to derive $X (= r)$ by simplification.

Step 2 We maintain now our previous result $r = X$ and select premise $p \vee \neg r$, which we reformulate as $Y \vee \neg X$. Then, we have:

$$
\begin{array}{ll}
0011 & \times \\
1101 & = \\
0001 &
\end{array}
$$

from which we can derive $Y (= p)$ by simplification.

Step 3 We keep $p = Y$ and select the last premise $p \rightarrow t$, which we rewrite as $Y \vee \neg X$. I stress again that the substitutions hold only stepwise apart from the single statement that we pass from one step to another. Therefore, $Y \vee \neg X$ means something different relative to the same expression occurring in Step 2. Then, we have

$$
\begin{array}{ll}
0101 & \times \\
1011 & = \\
0001 &
\end{array}
$$

from which t=X can be derived by simplification.

We can reformulate the previous example in slightly more complex terms by making use of five variables:

$$\begin{array}{c} p \vee \neg r \\ r \vee s \\ \neg s \\ p \rightarrow (\neg t \vee u) \\ \mathrm{t} \qquad\qquad / u \end{array}$$

Although the first steps are the same, we reformulate now the whole in three-dimensional terms.

Step 1 We sort the premises $\neg s$ and $r \vee s$ and rewrite them as $\neg Y$ and $X \vee Y$:

E4xE6	11010100	×
L	00111111	=
O=O5xO6	00010100	

The latter step maps to O5+O6, and in this way we have derived $X(= r)$.

Step 2 We maintain now our previous result $r = X$ and select premise $p \vee \neg r$, which we reformulate as $Y \vee \neg X$. Then, we have:

L6xL8	00010111	×
A	11101011	=
I=I3xI5	00000011	

which maps to I3+I5, that is, to $Y(= p)$.

Step 3 This is the more complex step. Since we need to consider the premise $p \rightarrow (\neg t \vee u)$ which contains three variables it is suitable to involve both p from the last step and the last premise, i.e. t. We substitute X to t and transform the implication into $Y \rightarrow (\neg X \vee Z)$. It is not difficult to see that this is the statement E2, which is the general inference shown in the first line of Table 8.5:

$$(X \wedge Y) \rightarrow Z \ .$$

Therefore, having X as one premise and Y as a result of previous computational steps, we can derive $Z(= u)$, which is the desired conclusion. In other words, we have reduced this complex inference to the canonical closed forms.

A slightly more complex example is the following:

$$(p \vee q) \rightarrow [r \vee (s \wedge t)]$$
$$\neg s$$
$$q$$
$$\neg r \vee u$$
$$w \rightarrow \neg u \qquad\qquad / \neg w$$

Again we proceed stepwise. Since the involved variables are seven, it is suitable to substitute entire portions of propositions. The most suitable example is the first premise and it is in particular convenient to substitute a single variable to the whole consequent.

Step 1 If we reformulate the antecedent of the implication of the first premise as $X \vee Y$, we can substitute Z to the consequent $[r \vee (s \wedge t)]$, so that we can reformulate the first premise as $(X \vee Y) \rightarrow Z$. Let us also consider the third premise (which is now Y). We know that we are always allowed to proceed as follows:

L5xL7 maps to L5+L7=**L**.

We know that L5xL7 is Y, and the above operation is the Addition that gives **L** as a result. Now, the latter is the antecedent of the implication of the first premise so that we are allowed to make use of the *modus ponens* to derive Z (the consequent).

Step 2 Now, we keep the previous result (Z) that we back translate into its original form: $r \vee (s \wedge t)$, and consider the second premise: $\neg s$. We substitute X to r, Y to s and Z to t and, by reformulating the consequent of the implication represented by the first premise as $X \vee (Y \wedge Z)$, proceed in this way:

L6xL8	00010111	+
G	00001001	=
L8	00011111	×
O1+O2	11010100	=
O=O5xO6	00010100	

I recall that $\neg Y$ is indeed O1+O2. Now, O5xO6 maps to O5+O6 (a Simplification), which is $X (= r)$.

Step 3 We keep $r = X$ and consider the fourth premise which we reformulate as $\neg X \vee Y$. Then,

O5+O6	00010111	$\times$
A	11101011	=
I=I3xI5	00000011	

The latter statement maps to I3+I5 (again we are applying a simplification), which is $Y(= u)$.

Step 4 We keep the previous result $u = Y$ and reformulate the latter premise as: $\neg Z \vee \neg Y$:

I3+I5	00101011	$\times$
D	11110110	=
U	00100010	

It is easy to show that we can then obtain by simplification $\neg Z(= \neg w)$. Indeed, **U**=U1xU3 that maps to U1+U3, which is $\neg Z$.

8.6 Mereological Inferences and Related Ones

One of the problems of classical calculations is the following. It is quite logical to say:

All lions are vertebrates
All vertebrates have a head
Then, all lion's heads are vertebrates' heads.

However, the difficulty is how to derive through mechanical means such a conclusion that can be said intuitively to be from a part to a whole (this is the significance of the term "mereological"). As a matter of fact the previous inference was not considered a syllogism (*De Morgan*, 1847, p. 114). Obviously, the first problem is how to correctly put in good symbolic form the previous inference. A reasonable translation in traditional sufficient-condition inference is:

$$[(\forall x)(Lx \rightarrow Vx) \wedge (\forall x)(Vx \rightarrow Hx)] \rightarrow \neg(\exists x)[(Lx \wedge Hx) \wedge \neg(Vx \wedge Hx)]\,. \tag{8.44}$$

Once that I have set the issue in formal terms by using quantifiers, I like to make use again of a formalism that allows as to avoid their use (see also Section 1.2). In this case, we have:

$$[(X \rightarrow Y) \wedge (Z \rightarrow X)] \rightarrow [(Z \wedge Y) \rightarrow (X \wedge Y)]\,, \qquad (8.45)$$

where I have chosen X for "being vertebrate", Y for "having a head", and Z for "to be a lion". The advantage of this expression is that it makes immediately clear that the problem is not with mereological inferences in particular but with any kind of inference having this general form (i.e. a conjunction of implications implies an implication of conjunctions), whatever the meaning of X, Y and Z can be. What I like to show in the following is that our logical language is sufficiently powerful for dealing with this problem and even to deal with a whole class of similar problems. First, let us remark that the two premises are those of *Barbara* (Table 2.12), i.e. **A** and **B**. Instead, the conclusion is nothing else than Statement M1 (see Table 8.5). Indeed, it can be easily shown that

$$(Z \wedge Y) \rightarrow (X \wedge Y) \qquad (8.46)$$

is logically equivalent to

$$X \vee \neg Y \vee \neg Z\,. \qquad (8.47)$$

Now, I recall that one of the two statement of the LGS of **B** (i.e. $X \vee \neg Z$) is precisely B1=M1, so that it is fully understandable that by applying the rule of addition to **B** we obtain M1. A more interesting way to proceed is to find a straight and mechanical calculation that allows us to find the whole class of similar inferences. Then, our problem is how to derive M1 from a product between **A** and **B**. I remark that we have:

$$\{(\mathbf{A} \times \mathbf{B} \times \mathbf{C}) : [(\mathbf{A} \times \mathbf{C})=\text{A7}]\}=\text{M1}.$$

The reader could take this inference as being the result of a chance. To show that it represents a new form of drawing inferences that are all logically correct, we need first to deal with a set of statements of Level 4-4. I first draw a summary table in which each statement is considered as the result of a product of three statements from above or as a sum of three statements from below (and is therefore also presented in two distinct but equivalent formulations):

Table 8.19 Statements of Level 4-4 involved in sums and products

#	ID	Statements	Logical form	Derivation in which the statements occur
4	11100010	D4=F4xA3=F5xD5xA5 **A × D × F**	$(X \to \neg Z) \wedge (X \to Y) \wedge (Y \to \neg Z)$	Table 6.5, Inf. 19
		T+U+ Ω T1=Ω1+T5+U1=Ω5+U2	$(\neg X \wedge \neg Y) \vee (Y \wedge \neg Z)$	Table 6.6, Inf. 31
8	11010010	D4=F4xF6xM3=D6xM5 **M × F × D**	$(Z \to \neg X) \wedge (Z \to \neg Y) \wedge (\neg X \to \neg Y)$	Table 211, Inf. 6
		T+W+V T2=V1+T5+V5=W1+W2	$(X \wedge \neg Z) \vee (\neg X \wedge \neg Y)$	Table 6.6, Inf. 13
12	11001001	A4=R4xP4xP6=R6xA6 **A × P × R**	$(\neg X \to \neg Y) \wedge (\neg Z \to \neg X) \wedge (\neg Z \to \neg Y)$	Table 211, Inf. 27
		T+G+K T3=K1+T6+G1+G2=K6	$(\neg X \wedge \neg Y) \vee (Y \wedge Z)$	Table 6.6, Inf. 23
14	11000101	R5xM4xP6=R6xM6 **M × R × P**	$(Y \to Z) \wedge (Y \to X) \wedge (X \to Z)$	Table 6.5, Inf. 14
		T+J+H T4=H1+T6+J1+H6=J2	$(\neg X \wedge \neg Y) \vee (X \wedge Z)$	Table 6.6, Inf. 5
16	10111000	E3=F3xE7xC3=F7xC4 **E × C × F**	$(Z \to \neg X) \wedge (Z \to Y) \wedge (Y \to \neg X)$	Table 6.5, Inf. 20
		S+V+ Ω V2=Ω2+S1=Ω3+V3+S3	$(\neg X \wedge Y) \vee (\neg Y \wedge \neg Z)$	Table 6.6, Inf. 32

17	10110100	E4=D3xE7xB3=D7xC4 **E × B × D**	$(Z \to \neg Y) \wedge (Z \to X) \wedge (X \to \neg Y)$	Table 6.5, Inf. 2
		O+ Ω +V V2=Ω2+Ω4+O1=V4+O3	$(X \wedge \neg Y) \vee (\neg X \wedge \neg Z)$	Table 6.6, Inf. 9
25	10100011	A5xB5=C5xA7=C6xB6 **A × B × C**	$(Z \to Y) \wedge (Z \to X) \wedge (X \to Y)$	Table 6.5, Inf. 1
		I+ Ω +U U1=Ω5+Ω6+I1+I3=U6	$(X \wedge Y) \vee (\neg X \wedge \neg Z)$	Table 6.6, Inf. 10
31	10010011	M5xB5=C5xC7xM7=B7 **M × C × B**	$(Z \to X) \wedge (Z \to Y) \wedge (Y \to X)$	Table 6.5, Inf. 24
		I+V+W V5=W1+V6+I1+I4=W6	$(X \wedge Y) \vee (\neg Y \wedge \neg Z)$	Table 6.6, Inf. 28
40	01101100	E5=R3xE8xN3xN4=R8 **E × N × R**	$(X \to Z) \wedge (X \to \neg Y) \wedge (\neg Y \to Z)$	Table 6.5, Inf. 28
		S+H+K S2=K2+H2+H4=K4+S4	$(\neg X \wedge Y) \vee (\neg Y \wedge Z)$	Table 6.6, Inf. 24
46	01011100	E6=P3xE8xQ3xP8=Q4 **E × Q × P**	$(\neg Z \to \neg Y) \wedge (\neg Z \to X) \wedge (X \to \neg Y)$	Table 6.5, Inf. 10
		O+K+H K3+O2=H3+H4=K4+O4	$(X \wedge \neg Y) \vee (\neg X \wedge Z)$	Table 6.6, Inf. 1
54	01001011	A6xA8=N5xQ5xN7=Q7 **A × Q × N**	$(\neg Z \to Y) \wedge (\neg Z \to X) \wedge (X \to Y)$	Table 6.5, Inf. 9
		I+K+G K2+G2=K6+I2+I5=G6	$(X \wedge Y) \vee (\neg X \wedge Z)$	Table 6.6, Inf. 2

(Continued)

Table 8.19 (*Continued*)

#	ID	Statements	Logical form	Derivation in which the statements occur
55	01000111	M6xN6xM8=Q6xN7=Q7 **M × N × Q**	$(\neg X \rightarrow Z) \wedge (\neg X \rightarrow \neg Y) \wedge (\neg Y \rightarrow Z)$	Table 6.5, Inf. 32
		I+H+J H5+H6=J2+I2+I6=J6	$(X \wedge Y) \vee (\neg Y \wedge Z)$	Table 6.6, Inf. 20
57	00111010	C3=F7xF8xL3xL5=C8 **L × F × C**	$(\neg Y \rightarrow \neg Z) \wedge (\neg Y \rightarrow X) \wedge (X \rightarrow \neg Z)$	Table 6.5, Inf. 5
		S+W+U S3+U3=W3+S5=U4+W4	$(\neg X \wedge Y) \vee (X \wedge \neg Z)$	Table 6.6, Inf. 14
59	00110110	B3=D7xD8xL3xL6=B8 **L × D × B**	$(\neg X \rightarrow \neg Z) \wedge (\neg X \rightarrow Y) \wedge (Y \rightarrow \neg Z)$	Table 6.5, Inf. 23
		O+U+W O3+U3=W3+U5+O5=W5	$(X \wedge \neg Y) \vee (Y \wedge \neg Z)$	Table 6.6, Inf. 27
63	00101101	R7xN4=R8xL4xL7 **L × R × N**	$(\neg Z \rightarrow Y) \wedge (\neg Z \rightarrow \neg X) \wedge (\neg X \rightarrow Y)$	Table 6.5, Inf. 13
		S+J+G S4+S6=G3+J3+G5=J5	$(\neg X \wedge Y) \vee (X \wedge Z)$	Table 6.6, Inf. 6
67	00011101	P7xP8=Q4xL4xL8=Q8 **L × P × Q**	$(\neg X \rightarrow Z) \wedge (\neg X \rightarrow Y) \wedge (Y \rightarrow Z)$	Table 6.5, Inf. 31
		O+G+J O4+G4+O6=J4+G5=J5	$(X \wedge \neg Y) \vee (Y \wedge Z)$	Table 6.6, Inf. 19

The implications that occur in each of the above statements are present in a product derivation whilst the two conjunctions occur as premises of the corespondent sum derivation, as I have indicated in the last column. It is interesting to remark that only universal inferences of the kind *Barbara* and *Celarent* (and their variants) for the first group as well as of the kind *Camestres* and *Cesare* (and their variants) for the second group occur (but some inferences of the third group of subtraction and division inferences are represented although with a permutation of the involved statements). Note that Statements 3,6, 22, 30, 41, 49, 65, 68 of Level 4-4, although having the same form of the statements listed here, do not posses the logical requirements for being part of this group. For instance, Statement 3 can be certainly considered as the following product:

$$\mathbf{E} \times \mathbf{R} \times \mathbf{D}.$$

Indeed, Statement 3 (according to Table 1.33) can be understood as a product of the following statements of Level 5-3: 2, 5, 8, 9, so that (according to Table 1.27) we are allowed to write:

$$\mathrm{E4{=}D3 \times E5{=}R3 \times D5 \times R5},$$

which fully justifies the previous product. Now, although this is logically right, these three statements (**E, B, R**) cannot be put in any relation that allows an inference and indeed they never occur together in the previous considered cases.

I remark that both the above product and sum forms show in their decomposition two premises that are not shared and these are associated with the premise of the relative inference. For instance, the product **A** × **D** × **F** displays the RGS elements A5 and D5 that are not shared with elements of other RGSs of universal statements. Now, the antecedents of Inference 19, Table 6.5, are precisely **A** and **D**. The same is true for the relative sum inference (Inf. 31, Table 6.6), whose premises are **T** and **U**: indeed, the two not-shared terms are here T5 and U2. We shall deal below with the summation inferences (whose summary table is the 6.6). Consider also that the only involved inferences are the universal ones in both the product and division forms: 16 from I and II group (8 each) of the product forms and 8 from the third group of the division forms. Now, what we proceed as follows:

Table 8.20 "Mereological" product inferences

1	From **A** and **B**, M1 follows	Level 4-4, # 25	$\mathbf{A} \times \mathbf{B} \times \mathbf{C}$	:	
		Level 5-3, # 29	A7=$\mathbf{A} \times \mathbf{C}$	=	M1
	$[(X \to Y) \wedge (Z \to X)] \to [(Z \wedge Y) \to (X \wedge Y)]$				
2	From **E** and **B**, L1 follows	Level 4-4, # 17	$\mathbf{E} \times \mathbf{B} \times \mathbf{D}$	:	
		Level 5-3, # 2	E4=$\mathbf{E} \times \mathbf{D}$	=	L1
	$[(X \to \neg Y) \wedge (Z \to X)] \to [(Z \wedge \neg Y) \to (X \wedge \neg Y)]$				
3	From **L** and **F**, E1 follows	Level 4-4, # 57	$\mathbf{L} \times \mathbf{F} \times \mathbf{C}$	:	
		Level 5-3, # 53	L5=$\mathbf{L} \times \mathbf{C}$	=	E1
	$[(\neg X \to Y) \wedge (Z \to \neg X)] \to [(Z \wedge Y) \to (\neg X \wedge Y)]$				
4	From **M** and **F**, A1 follows	Level 4-4, # 8	$\mathbf{M} \times \mathbf{F} \times \mathbf{D}$	:	
		Level 5-3, # 14	M3=$\mathbf{M} \times \mathbf{D}$	=	A1
	$[(\neg X \to \neg Y) \wedge (Z \to \neg X)] \to [(Z \wedge \neg Y) \to (\neg X \wedge \neg Y)]$				
5	From **A** and **Q**, M2 follows	Level 4-4, # 54	$\mathbf{A} \times \mathbf{Q} \times \mathbf{N}$	:	
		Level 5-3, # 44	A8=$\mathbf{M} \times \mathbf{D}$	=	M2
	$[(X \to Y) \wedge (\neg Z \to X)] \to [(\neg Z \wedge Y) \to (X \wedge Y)]$				
6	From **E** and **Q**, L2 follows	Level 4-4, # 46	$\mathbf{E} \times \mathbf{Q} \times \mathbf{P}$	:	
		Level 5-3, # 11	E6=$\mathbf{E} \times \mathbf{P}$	=	L2
	$[(X \to \neg Y) \wedge (\neg Z \to X)] \to [(\neg Z \wedge \neg Y) \to (X \wedge \neg Y)]$				
7	From **L** and **R**, E2 follows	Level 4-4, # 15	$\mathbf{L} \times \mathbf{R} \times \mathbf{N}$	:	
		Level 5-3, # 55	L7=$\mathbf{L} \times \mathbf{N}$	=	E2
	$[(\neg X \to Y) \wedge (\neg Z \to \neg X)] \to [(\neg Z \wedge Y) \to (\neg X \wedge Y)]$				
8	From **M** and **R**, A2 follows	Level 4-4, # 14	$\mathbf{M} \times \mathbf{R} \times \mathbf{P}$	:	
		Level 5-3, # 55	M4=$\mathbf{M} \times \mathbf{P}$	=	A2
	$[(\neg X \to \neg Y) \wedge (\neg Z \to \neg X)] \to [(\neg Z \wedge \neg Y) \to (\neg X \wedge \neg Y)]$				
9	From **A** and **D**, M1 follows	Level 4-4, # 4	$\mathbf{A} \times \mathbf{D} \times \mathbf{F}$	:	
		Level 5-3, # 6	A3=$\mathbf{A} \times \mathbf{F}$	=	M1
	$[(X \to Y) \wedge (Z \to \neg Y)] \to [(Z \wedge \neg X) \to (\neg Y \wedge \neg X)]$				
10	From **E** and **C**, L1 follows	Level 4-4, # 16	$\mathbf{E} \times \mathbf{C} \times \mathbf{F}$	:	
		Level 5-3, # 1	E3=$\mathbf{E} \times \mathbf{F}$	=	L1
	$[(X \to \neg Y) \wedge (Z \to Y)] \to [(Z \wedge \neg X) \to (Y \wedge \neg X)]$				
11	From **L** and **D**, E1 follows	Level 4-4, # 59	$\mathbf{L} \times \mathbf{D} \times \mathbf{B}$	:	
		Level 5-3, # 54	L6=$\mathbf{L} \times \mathbf{D}$	=	E1
	$[(\neg X \to Y) \wedge (Z \to \neg Y)] \to [(Z \wedge X) \to (\neg Y \wedge X)]$				
12	From **M** and **C**, A1 follows	Level 4-4, # 31	$\mathbf{M} \times \mathbf{C} \times \mathbf{B}$	:	
		Level 5-3, # 34	M7=$\mathbf{M} \times \mathbf{B}$	=	A1
	$[(\neg X \to \neg Y) \wedge (Z \to Y)] \to [(Z \wedge X) \to (Y \wedge X)]$				
13	From **A** and **P**, M2 follows	Level 4-4, # 12	$\mathbf{A} \times \mathbf{P} \times \mathbf{R}$	:	
		Level 5-3, # 7	A4=$\mathbf{A} \times \mathbf{R}$	=	M2
	$[(X \to Y) \wedge (\neg Z \to \neg Y)] \to [(\neg Z \wedge \neg X) \to (\neg Y \wedge \neg X)]$				

14	From **E** and **N**, L2 follows	Level 4-4, # 40	$\mathbf{E} \times \mathbf{N} \times \mathbf{R}$	:	
		Level 5-3, # 5	E5=$\mathbf{E} \times \mathbf{N}$	=	L2
	$[(X \to \neg Y) \wedge (\neg Z \to Y)] \to [(\neg Z \wedge \neg X) \to (Y \wedge \neg X)]$				
15	From **L** and **P**, E2 follows	Level 4-4, # 67	$\mathbf{L} \times \mathbf{P} \times \mathbf{Q}$	:	
		Level 5-3, # 56	L8=$\mathbf{L} \times \mathbf{Q}$	=	E2
	$[(\neg X \to Y) \wedge (\neg Z \to \neg Y)] \to [(\neg Z \wedge X) \to (\neg Y \wedge X)]$				
16	From **M** and **N**, A2 follows	Level 4-4, # 55	$\mathbf{M} \times \mathbf{N} \times \mathbf{Q}$	:	
		Level 5-3, # 49	M8=$\mathbf{M} \times \mathbf{Q}$	=	A2
	$[(\neg X \to \neg Y) \wedge (\neg Z \to Y)] \to [(\neg Z \wedge X) \to (Y \wedge X)]$				

The term "product inferences" may give rise to some confusion since the inferences here are in fact divisions. The problem is that division is used also in other kinds of inference to be treated below. Moreover, any kind of logical inference treated in this section is identified by the logical operation that, in the first line of each derivation, connects the two statements that are also present in the second line (For instance, **A** and **B** in the first inference here) and the third one (i.e. **C**). In this case it is indeed a product.

It may interesting to consider an example of the previous derivations, for instance of the second inference above:

No mammals is cold-blooded
All lions are mammals
Then, every lion that is also not cold-blooded
Is a mammal that is not cold-blooded.

It is also interesting to consider explicitly the way in which we compute any of the above inferences. As an example, let us choose the 13th one:

A	A1xA2	11101011	×
R	E2xA2	11101101	=
$\mathbf{A} \times \mathbf{R}$	A4	11101001	×
P	E2xM2	11011101	=
A4=R4xP4xP6=R6xA6	$\mathbf{A}\times\mathbf{P}\times\mathbf{R}$	11001001	:
	A4	11101001	=
	M2	11011111	

What is relevant is that each line is connected with the previous one by sharing some part of the previous statement. The interest of all previous inferences lies in the fact that any of them implies both its conclusion and the conclusion of the relative original inference. For instance, let us consider again the first derivation above that is connected with $(\mathbf{A} \wedge \mathbf{B}) \rightarrow \mathbf{C}$. This inference can be written as:

$$[(X \rightarrow Y) \wedge (Z \rightarrow X)] \rightarrow (Z \rightarrow Y) \,. \tag{8.48}$$

Now, what I am claiming is that we have

$$[(X \rightarrow Y) \wedge (Z \rightarrow X)] \rightarrow \{[(Z \wedge Y) \rightarrow (X \wedge Y)] \wedge (Z \rightarrow Y)\} \tag{8.49}$$

or, more synthetically,

$$(\mathbf{A} \wedge \mathbf{B}) \rightarrow (\mathrm{M1} \wedge \mathbf{C}). \tag{8.50}$$

We may consider this as an expanded syllogism that although new has the same general form of the closed inferences examined before and can be therefore very helpful for dealing with open inferences (see also the end of Section 6.5).

A considerable step further is made when taking also sum derivations into account. Let us consider as an example the first derivation of Table 6.6, i.e.

$$(\mathbf{O} \vee \mathbf{K}) \leftarrow \mathbf{H} \,. \tag{8.51}$$

A possible interpretation is:

When there are viruses without DNA or
There are organisms that are not viruses,
We can assume that there organisms that do not have DNA.

In order to compute the enlarged form of inference, we can proceed in analogy with the previous cases (but by interchanging sums with products and subtractions with divisions) and write:

O	O7+O8	00010100	+
H	O8+T8	01000100	=
O+H	O2	01010100	+
K	T8+S8	01001000	=
K3+O2=H3+H4=K4+O4	**O+K+H**	01011100	–
	O2	01010100	=
	S8	00001000	

Note that the statement **O+K+H** is the sum-form of number 46 of Level 4-4 in the 6th derivation of Table 8.19. Note also that Statement S8 contradicts Statement M1 (see Table 8.5), as expected since the first derivation of Table 6.6 is the counterpart of the first derivation of Table 6.5. The logical form of the above inference is therefore:

$$[(X \wedge \neg Y) \vee (Z \wedge \neg X)] \leftarrow \neg[(Z \wedge Y) \rightarrow (X \wedge Y)] \qquad (8.52)$$

or

$$[(X \wedge \neg Y) \vee (Z \wedge \neg X)] \leftarrow [(Z \wedge Y) \wedge \neg(X \wedge Y)] \qquad (8.53)$$

A possible interpretation of the above inference could be the following:

When there are viruses without DNA or
There are organisms that are not viruses,
We can assume that both there are organisms with DNA
But there are no viruses with DNA.

Such an inference makes even more explicit the inductive character of sum-inferences (see Section 3.1). Indeed the conclusion represents a kind of generalization (the first premise is a particular statement while the conclusion tells us that there are no viruses with DNA). As before, I like to stress that the previous inference allows us to write

$$\{[(X \wedge \neg Y) \vee (Z \wedge \neg X)] \leftarrow [(Z \wedge Y) \wedge \neg(X \wedge Y)]\} \leftarrow (X \wedge \neg Y) \quad (8.54)$$

or, more synthetically,

$$[(\mathbf{O} \vee \mathbf{K}) \leftarrow \mathrm{S8}] \leftarrow \mathbf{H}\,. \qquad (8.55)$$

I also remark here that we can invert the previous reasoning and start from the conclusion assuming the form of a sufficient condition of what was previously a premise. In such a case, we start from S8 as a test that has disproved some previous assumption:

$$\mathrm{S8} \rightarrow (\mathbf{O} \vee \mathbf{K})\,, \qquad (8.56)$$

which could be interpreted as:

The fact that there are organisms with DNA but no viruses with DNA implies that
Viruses do not have DNA or organisms are not viruses.

This looks much more familiar. Obviously, what is interesting is that we can consider Statement **H** to be a necessary condition of the previous inference in the following crossed mode:

$$[(\mathbf{O} \vee \mathbf{K}) \rightarrow \mathrm{S8}] \leftarrow \mathbf{H}\,. \tag{8.57}$$

On such bases we can draw following table:

Table 8.21 "Mereological" sum inferences

1	When **O** or **K**, we assume S8	Level 4-4, # 46	**O** + **K** + **H**	–	
		Level 3-5, # 28	O2=**O** + **K**	=	S8
	$[(X \wedge \neg Y) \vee (Z \wedge \neg X)] \leftarrow [(Z \wedge Y) \wedge \neg(X \wedge Y)]$				
2	When **I** or **K**, we assume T8	Level 4-4, # 54	**I** + **K** + **G**	–	
		Level 3-5, # 55	I5=**I** + **G**	=	T8
	$[(X \wedge Y) \vee (Z \wedge \neg X)] \leftarrow [(Z \wedge \neg Y) \wedge \neg(X \wedge \neg Y)]$				
3	When **T** or **J**, we assume I8	Level 4-4, # 14	**T** + **J** + **H**	–	
		Level 3-5, # 4	T4=**T** + **H**	=	I8
	$[(\neg X \wedge \neg Y) \vee (Z \wedge X)] \leftarrow [(Z \wedge Y) \wedge \neg(\neg X \wedge Y)]$				
4	When **S** or **J**, we assume O8	Level 4-4, # 63	**S** + **J** + **G**	–	
		Level 3-5, # 43	S6=**S** + **G**	=	O8
	$[(\neg X \wedge Y) \vee (Z \wedge X)] \leftarrow [(Z \wedge \neg Y) \wedge \neg(\neg X \wedge \neg Y)]$				
5	When **O** or **Ω**, we assume S7	Level 4-4, # 17	**O** + **Ω** + **V**	–	
		Level 3-5, # 13	O1=**O** + **V**	=	S7
	$[(X \wedge \neg Y) \vee (\neg Z \wedge \neg X)] \leftarrow [(\neg Z \wedge Y) \wedge \neg(X \wedge Y)]$				
6	When **I** or **Ω**, we assume T7	Level 4-4, # 25	**I** + **Ω** + **U**	–	
		Level 3-5, # 46	I3=**I** + **U**	=	T7
	$[(X \wedge Y) \vee (\neg Z \wedge \neg X)] \leftarrow [(\neg Z \wedge \neg Y) \wedge \neg(X \wedge \neg Y)]$				
7	When **T** or **W**, we assume I7	Level 4-4, # 8	**T** + **W** + **V**	–	
		Level 3-5, # 2	T2=**T** + **V**	=	I7
	$[(\neg X \wedge \neg Y) \vee (\neg Z \wedge X)] \leftarrow [(\neg Z \wedge Y) \wedge \neg(\neg X \wedge Y)]$				
8	When **S** or **W**, we assume O7	Level 4-4, # 57	**S** + **W** + **U**	–	
		Level 3-5, # 42	S5=**S** + **U**	=	O7
	$[(\neg X \wedge Y) \vee (\neg Z \wedge X)] \leftarrow [(\neg Z \wedge \neg Y) \wedge \neg(\neg X \wedge \neg Y)]$				
9	When **O** or **G**, we assume S8	Level 4-4, # 67	**O** + **G** + **J**	–	
		Level 3-5, # 51	O6=**O** + **J**	=	S8
	$[(X \wedge \neg Y) \vee (Z \wedge Y)] \leftarrow [(Z \wedge \neg X) \wedge \neg(\neg Y \wedge \neg X)]$				
10	When **I** or **H**, we assume T8	Level 4-4, # 55	**I** + **H** + **J**	–	
		Level 3-5, # 56	I6=**I** + **J**	=	T8
	$[(X \wedge Y) \vee (Z \wedge \neg Y)] \leftarrow [(Z \wedge \neg X) \wedge \neg(Y \wedge \neg X)]$				

11	When **T** or **G**, we assume I8	Level 4-4, # 12	$\mathbf{T}+\mathbf{G}+\mathbf{K}$	–	
		Level 3-5, # 3	T3=$\mathbf{T}+\mathbf{K}$	=	I8
	$[(\neg X \wedge \neg Y) \vee (Z \wedge Y)] \leftarrow [(Z \wedge X) \wedge \neg(\neg Y \wedge X)]$				
12	When **S** or **H**, we assume O8	Level 4-4, # 40	$\mathbf{S}+\mathbf{H}+\mathbf{K}$	–	
		Level 3-5, # 23	S2=$\mathbf{S}+\mathbf{K}$	=	O8
	$[(\neg X \wedge Y) \vee (Z \wedge \neg Y)] \leftarrow [(Z \wedge X) \wedge \neg(Y \wedge X)]$				
13	When **O** or **U**, we assume S7	Level 4-4, # 59	$\mathbf{O}+\mathbf{U}+\mathbf{W}$	–	
		Level 3-5, # 50	O5=$\mathbf{O}+\mathbf{W}$	=	S7
	$[(X \wedge \neg Y) \vee (\neg Z \wedge Y)] \leftarrow [(\neg Z \wedge \neg X) \wedge \neg(\neg Y \wedge \neg X)]$				
14	When **I** or **V**, we assume T7	Level 4-4, # 31	$\mathbf{I}+\mathbf{V}+\mathbf{W}$	–	
		Level 3-5, # 52	I4=$\mathbf{I}+\mathbf{W}$	=	T7
	$[(X \wedge Y) \vee (\neg Z \wedge \neg Y)] \leftarrow [(\neg Z \wedge \neg X) \wedge \neg(Y \wedge \neg X)]$				
15	When **T** or **U**, we assume I7	Level 4-4, # 4	$\mathbf{T}+\mathbf{U}+\boldsymbol{\Omega}$	–	
		Level 3-5, # 1	T1=$\mathbf{T}+\boldsymbol{\Omega}$	=	I7
	$[(\neg X \wedge \neg Y) \vee (\neg Z \wedge Y)] \leftarrow [(\neg Z \wedge X) \wedge \neg(\neg Y \wedge X)]$				
16	When **S** or **V**, we assume O7	Level 4-4, # 16	$\mathbf{S}+\mathbf{V}+\boldsymbol{\Omega}$	–	
		Level 3-5, # 8	S1=$\mathbf{S}+\boldsymbol{\Omega}$	=	O7
	$[(\neg X \wedge Y) \vee (\neg Z \wedge \neg Y)] \leftarrow [(\neg Z \wedge X) \wedge \neg(Y \wedge X)]$				

Again, I recall that the term mereological "sum" inferences is helpful only for denoting this particular kind of inferences although the main operation is here a subtraction.

Similar relations can be built in the division-subtraction representation, whose logical meaning is less usual. First, allow me to reformulate Table 8.19 in terms of this representation by making use of hybrid forms (that preserve two statements over three of that table and simultaneously display many potentialities of this kind of computation), as displayed in Table 8.22.

I have not considered the generating statements of the previous forms due to their hybrid nature.

As said, subtraction is not so easy to understand. Let us consider the statements involved in the first derivation of Table 6.7. In such a case it can be proved that we have also the inference:

$$[(X \rightarrow Y) \wedge \neg(Z \wedge \neg X)] \rightarrow [(Z \wedge Y) \rightarrow (X \wedge Y)]\,. \qquad (8.58)$$

Table 8.22 Statements of Level 4-4 involved in subtractions and divisions

#	ID	Statements	Logical form	Derivation in which the statements occur
4	11100010	**(A × F) − G**	$(X \to Y) \wedge (Z \to \neg X) \wedge \neg(Z \wedge Y)$	Table 6.7, Inf. 19
		D − (O + J)	$(Z \to \neg Y) \wedge \neg[(X \wedge \neg Y) \vee (Z \wedge X)]$	Table 6.8, Inf. 31
8	11010010	**(M × D) − J**	$(\neg X \to \neg Y) \wedge (Z \to \neg Y) \wedge \neg(Z \wedge X)$	Table 6.7, Inf. 6
		F − (S + G)	$(Z \to \neg X) \wedge \neg[(\neg X \wedge Y) \vee (Z \wedge Y)]$	Table 6.8, Inf. 13
12	11001001	**(A × R) − U**	$(X \to Y) \wedge (\neg Z \to \neg X) \wedge \neg(\neg Z \wedge Y)$	Table 6.7, Inf. 27
		P − (O + W)	$(\neg Z \to \neg Y) \wedge \neg[(X \wedge \neg Y) \vee (\neg Z \wedge X)]$	Table 6.8, Inf. 23
14	11000101	**(M × P) − W**	$(\neg X \to \neg Y) \wedge (\neg Z \to \neg Y) \wedge \neg(\neg Z \wedge X)$	Table 6.7, Inf. 14
		R − (S + U)	$(\neg Z \to \neg X) \wedge \neg[(\neg X \wedge Y) \vee (\neg Z \wedge Y)]$	Table 6.8, Inf. 5
16	10111000	**(E × F) − H**	$(X \to \neg Y) \wedge (Z \to \neg X) \wedge \neg(Z \wedge \neg Y)$	Table 6.7, Inf. 20
		C − (I + J)	$(Z \to Y) \wedge \neg[(X \wedge Y) \vee (Z \wedge X)]$	Table 6.8, Inf. 32
17	10110100	**(E × D) − K**	$(X \to \neg Y) \wedge (Z \to \neg Y) \wedge \neg(Z \wedge \neg X)$	Table 6.7, Inf. 2
		B − (I + G)	$(Z \to X) \wedge \neg[(X \wedge Y) \vee (Z \wedge Y)]$	Table 6.8, Inf. 2
25	10100011	**(A × C) − K**	$(X \to Y) \wedge (Z \to Y) \wedge \neg(Z \wedge \neg X)$	Table 6.7, Inf. 1
		B − (O + H)	$(Z \to X) \wedge \neg[(X \wedge \neg Y) \vee (Z \wedge \neg Y)]$	Table 6.8, Inf. 1
31	10010011	**(M × B) − H**	$(\neg X \to \neg Y) \wedge (Z \to X) \wedge \neg(Z \wedge \neg Y)$	Table 6.7, Inf. 24
		C − (S + K)	$(Z \to Y) \wedge \neg[(\neg X \wedge Y) \vee (Z \wedge \neg X)]$	Table 6.8, Inf. 28
40	01101100	**(E × R) − V**	$(X \to \neg Y) \wedge (\neg Z \to \neg X) \wedge \neg(\neg Z \wedge \neg Y)$	Table 6.7, Inf. 28
		N − (I + W)	$(\neg Z \to Y) \wedge \neg[(X \wedge Y) \vee (\neg Z \wedge X)]$	Table 6.8, Inf. 24
46	01011100	**(E × P) − Ω**	$(X \to \neg Y) \wedge (\neg Z \to \neg Y) \wedge \neg(\neg Z \wedge \neg X)$	Table 6.7, Inf. 10
		Q − (I + U)	$(\neg Z \to X) \wedge \neg[(X \wedge Y) \vee (\neg Z \wedge Y)]$	Table 6.8, Inf.1
54	01001011	**(A × N) − Ω**	$(X \to Y) \wedge (\neg Z \to Y) \wedge \neg(\neg Z \wedge \neg X)$	Table 6.7, Inf. 9
		Q − (O + V)	$(\neg Z \to X) \wedge \neg[(X \wedge \neg Y) \vee (\neg Z \wedge \neg Y)]$	Table 6.8, Inf. 2
55	01000111	**(M × Q) − V**	$(\neg X \to \neg Y) \wedge (\neg Z \to X) \wedge \neg(\neg Z \wedge \neg Y)$	Table 6.7, Inf. 32
		N − (S + Ω)	$(\neg Z \to Y) \wedge \neg[(\neg X \wedge Y) \vee (\neg Z \wedge \neg X)]$	Table 6.8, Inf. 20
57	00111010	**(L × C) − J**	$(\neg X \to Y) \wedge (Z \to Y) \wedge \neg(Z \wedge X)$	Table 6.7, Inf. 5
		F − (T + H)	$(Z \to \neg X) \wedge \neg[(\neg X \wedge \neg Y) \vee (Z \wedge \neg Y)]$	Table 6.8, Inf. 5
59	00110110	**(L × B) − G**	$(\neg X \to Y) \wedge (Z \to X) \wedge \neg(Z \wedge Y)$	Table 6.7, Inf. 23
		D − (T + K)	$(Z \to \neg Y) \wedge \neg[(\neg X \wedge \neg Y) \vee (Z \wedge \neg X)]$	Table 6.8, Inf. 27
63	00101101	**(L × N) − W**	$(\neg X \to Y) \wedge (\neg Z \to Y) \wedge \neg(\neg Z \wedge X)$	Table 6.7, Inf. 13
		R − (T + V)	$(\neg Z \to \neg X) \wedge \neg[(\neg X \wedge \neg Y) \vee (\neg Z \wedge \neg Y)]$	Table 6.8, Inf. 6
67	00011101	**(L × Q) − U**	$(\neg X \to Y) \wedge (\neg Z \to X) \wedge \neg(\neg Z \wedge Y)$	Table 6.7, Inf. 31
		P − (T + Ω)	$(\neg Z \to \neg Y) \wedge \neg[(\neg X \wedge \neg Y) \vee (\neg Z \wedge \neg X)]$	Table 6.8, Inf. 19

A possible translation in ordinary language could be the following:

If all Italian citizens are European citizens and
There are no people voting in Italy who are not Italian citizens,
It follows that to be both a voter in Italy and a European citizen
implies
To be both an Italian and a European citizen.

We can say that the previous inference allows:

$$\{[(X \to Y) \wedge \neg(Z \wedge \neg X)] \to [(Z \wedge Y) \to (X \wedge Y)]\} \to (Z \to Y) \tag{8.59}$$

or

$$(\mathbf{A} \wedge \neg\mathbf{K}) \to (\mathrm{M1} \wedge \mathbf{C}). \tag{8.60}$$

In the case of subtraction the main procedure is e.g.

$$\{[(\mathbf{A} \times \mathbf{C}) - \mathbf{K}] : (\mathbf{A} \times \mathbf{C})\} = \mathrm{M1}\,, \tag{8.61}$$

so that we have the following table:

Table 8.23 "Mereological" subtraction inferences

1	From **A** and not-**K**, M1 follows	Level 4-4, # 25	$(\mathbf{A} \times \mathbf{C}) - \mathbf{K}$	:	M1
		Level 5-3, # 29	A7=$\mathbf{A} \times \mathbf{C}$	=	
		$[(X \to Y) \wedge \neg(Z \wedge \neg X)] \to [(Z \wedge Y) \to (X \wedge Y)]$			
2	From **E** and not-**K**, L1 follows	Level 4-4, # 17	$(\mathbf{E} \times \mathbf{D}) - \mathbf{K}$	:	L1
		Level 5-3, # 2	E4=$\mathbf{E} \times \mathbf{D}$	=	
		$[(X \to \neg Y) \wedge \neg(Z \wedge \neg X)] \to [(Z \wedge \neg Y) \to (X \wedge \neg Y)]$			
3	From **L** and not-**J**, E1 follows	Level 4-4, # 57	$(\mathbf{L} \times \mathbf{C}) - \mathbf{J}$	:	E1
		Level 5-3, # 53	L5=$\mathbf{L} \times \mathrm{C}$	=	
		$[(\neg X \to Y) \wedge \neg(Z \wedge X)] \to [(Z \wedge Y) \to (\neg X \wedge Y)]$			
4	From **M** and not-**J**, A1 follows	Level 4-4, # 8	$(\mathbf{M} \times \mathbf{D}) - \mathbf{J}$	:	A1
		Level 5-3, # 14	M3=$\mathbf{M} \times \mathbf{D}$	=	
		$[(\neg X \to \neg Y) \wedge \neg(Z \wedge X)] \to [(Z \wedge \neg Y) \to (\neg X \wedge \neg Y)]$			
5	From **A** and not-**Ω**, M2 follows	Level 4-4, # 54	$(\mathbf{A} \times \mathbf{N}) - \mathbf{\Omega}$	:	M2
		Level 5-3, # 44	A8=$\mathbf{A} \times \mathbf{N}$	=	
		$[(X \to Y) \wedge \neg(\neg Z \wedge \neg X)] \to [(\neg Z \wedge Y) \to (X \wedge Y)]$			
6	From **E** and not-**Ω**, L2 follows	Level 4-4, # 46	$(\mathbf{E} \times \mathbf{P}) - \mathbf{\Omega}$	:	L2
		Level 5-3, # 11	E6=$\mathbf{E} \times \mathbf{P}$	=	
		$[(X \to \neg Y) \wedge \neg(\neg Z \wedge \neg X)] \to [(\neg Z \wedge \neg Y) \to (X \wedge \neg Y)]$			

(Continued)

Table 8.23 (*Continued*)

7	From **L** and not-**W**, E2 follows	Level 4-4, # 63	(**L** × **N**) − **W**	:	
		Level 5-3, # 55	L7=**L** × **N**	=	E2
	$[(\neg X \to Y) \wedge \neg(\neg Z \wedge X)] \to [(\neg Z \wedge Y) \to (\neg X \wedge Y)]$				
8	From **M** and not-**W**, A2 follows	Level 4-4, # 14	(**M** × **P**) − **W**	:	
		Level 5-3, # 15	M4=**M** × **P**	=	A2
	$[(\neg X \to \neg Y) \wedge \neg(\neg Z \wedge X)] \to [(\neg Z \wedge \neg Y) \to (\neg X \wedge \neg Y)]$				
9	From **A** and not-**G**, M1 follows	Level 4-4, # 4	(**A** × **F**) − **G**	:	
		Level 5-3, # 6	A3=**A** × **F**	=	M1
	$[(X \to Y) \wedge \neg(Z \wedge Y)] \to [(Z \wedge \neg X) \to (\neg Y \wedge \neg X)]$				
10	From **E** and not-**H**, L1 follows	Level 4-4, # 16	(**E** × **F**) − **H**	:	
		Level 5-3, # 1	E3=**E** × **F**	=	L1
	$[(X \to \neg Y) \wedge \neg(Z \wedge \neg Y)] \to [(Z \wedge \neg X) \to (Y \wedge \neg X)]$				
11	From **L** and not-**G**, E1 follows	Level 4-4, # 59	(**L** × **B**) − **G**	:	
		Level 5-3, # 54	L6=**L** × **B**	=	E1
	$[(\neg X \to Y) \wedge \neg(Z \wedge Y)] \to [(Z \wedge X) \to (\neg Y \wedge X)]$				
12	From **M** and not-**H**, A1 follows	Level 4-4, # 31	(**M** × **B**) − **H**	:	
		Level 5-3, # 34	M7=**M** × **B**	=	A1
	$[(\neg X \to \neg Y) \wedge \neg(Z \wedge \neg Y)] \to [(Z \wedge X) \to (Y \wedge X)]$				
13	From **A** and not-**U**, M2 follows	Level 4-4, # 12	(**A** × **R**) − **U**	:	
		Level 5-3, # 7	A4=**A** × **R**	=	M2
	$[(X \to Y) \wedge \neg(\neg Z \wedge Y)] \to [(\neg Z \wedge \neg X) \to (\neg Y \wedge \neg X)]$				
14	From **E** and not-**V**, L2 follows	Level 4-4, # 40	(**E** × **R**) − **V**	:	
		Level 5-3, # 5	E5=**E** × **R**	=	L2
	$[(X \to \neg Y) \wedge \neg(\neg Z \wedge \neg Y)] \to [(\neg Z \wedge \neg X) \to (Y \wedge \neg X)]$				
15	From **L** and not-**U**, E2 follows	Level 4-4, # 67	(**L** × **Q**) − **U**	:	
		Level 5-3, # 56	L8=**L** × **Q**	=	E2
	$[(\neg X \to Y) \wedge \neg(\neg Z \wedge Y)] \to [(\neg Z \wedge X) \to (\neg Y \wedge X)]$				
16	From **M** and not-**V**, A2 follows	Level 4-4, # 55	(×) −	:	
		Level 5-3, # 49	M8=**M** × **Q**	=	A2
	$[(\neg X \to \neg Y) \wedge \neg(\neg Z \wedge \neg Y)] \to [(\neg Z \wedge X) \to (Y \wedge X)]$				

Note that the statements of Level 5-3 are the same as those for the product inferences (Table 8.20). Even more difficult are division inferences. Let us consider the first inference of Table 6.8, i.e.

$$(\mathbf{O} \vee \neg \mathbf{B}) \leftarrow \mathbf{H}\,. \tag{8.62}$$

For instance, this inference could be interpreted as:

When there are complex systems without metabolism
Or it is not true that all physical systems are complex,
We can assume that there are physical systems without metabolism.

Now, it can be proved that the kind of inference that corresponds to the previous ones for division inferences is:

When there are complex systems without metabolism
Or it is not true that all physical systems are complex,
We can assume both that there are physical systems with metabolism
But complex system with metabolism do not exist.

In formal terms, we have:

$$[(X \wedge \neg Y) \vee \neg(Z \rightarrow X)] \leftarrow [(Z \wedge Y) \wedge \neg(X \wedge Y)]\,, \tag{8.63}$$

where X is to be complex, Y to have a metabolism, and Z to be a physical system. Note that this kind of inference highlights the abductive nature of division inferences (see Section 5.1). Indeed, we are postulating complex systems without metabolism, that is, systems that are not characterized by this property. In other words, we are postulating the existence of a new equivalence class of objects (objects sharing many different properties (Auletta, 2009)), which is a sub-class of physical systems, whose further properties, however, we do not know but that is by now individuated by absence of metabolism. Note that the first premise does not necessarily say that to do not have a metabolism individuates an equivalence sub-class of complex systems. For instance, to say that there are Germans who do not drink beer is a pure statistical fact and does not select any specific class that is individuated by sharing many different properties. As before we can write:

$$\{[(X \wedge \neg Y) \vee \neg(Z \rightarrow X)] \leftarrow [(Z \wedge Y) \wedge \neg(X \wedge Y)]\} \leftarrow (Z \wedge \neg Y) \tag{8.64}$$

or

$$[(\mathbf{O} \vee \neg\mathbf{B}) \leftarrow \mathrm{S8}] \leftarrow \mathbf{H}\,. \tag{8.65}$$

As before, we could invert the order of the antecedent of the previous inference and consider S8 as test that has disproved some previous assumptions:

$$\mathrm{S8} \rightarrow (\mathbf{O} \vee \neg\mathbf{B})\,. \tag{8.66}$$

A possible interpretation could be:

The fact that both there are physical systems with metabolism
But complex systems with metabolism cannot be found implies that
There are complex systems without metabolism
Or it is not true that all physical systems are complex.

Again, crossed forms of inference are interesting, like:

$$[\text{S8} \rightarrow (\mathbf{O} \vee \neg\mathbf{B})] \leftarrow \mathbf{H}\,. \tag{8.67}$$

In order to perform division inferences of this kind we need to follow this simple rule that I apply to the first inference as example (I again introduce hybrid forms):

$$\{[(\mathbf{O}+\mathbf{H}) : \mathbf{B}] - (\mathbf{O}+\mathbf{H})\} = \text{S8},$$

so that we can draw following table:

Table 8.24 "Mereological" division inferences.

1	When **O** or not–**B**, we assume S8	Level 4-4, # 25	$(\mathbf{O}+\mathbf{H}) : \mathbf{B}$	–	
		Level 3-5, # 28	$\text{O2} = \mathbf{O}+\mathbf{H}$	=	S8
	$[(X \wedge \neg Y) \vee \neg(Z \rightarrow X)] \leftarrow [(Z \wedge Y) \wedge \neg(X \wedge Y)]$				
2	When **I** or not–**B**, we assume T8	Level 4-4, # 17	$(\mathbf{I}+\mathbf{G}) : \mathbf{B}$	–	
		Level 3-5, # 55	$\text{I5} = \mathbf{I}+\mathbf{G}$	=	T8
	$[(X \wedge Y) \vee \neg(Z \rightarrow X)] \leftarrow [(Z \wedge \neg Y) \wedge \neg(X \wedge \neg Y)]$				
3	When **T** or not–**F**, we assume I8	Level 4-4, # 57	$(\mathbf{T}+\mathbf{H}) : \mathbf{F}$	–	
		Level 3-5, # 4	$\text{T4} = \mathbf{T}+\mathbf{H}$	=	I8
	$[(\neg X \wedge \neg Y) \vee \neg(Z \rightarrow \neg X)] \leftarrow [(Z \wedge Y) \wedge \neg(\neg X \wedge Y)]$				
4	When **S** or not–**F**, we assume O8	Level 4-4, # 8	$(\mathbf{S}+\mathbf{G}) : \mathbf{F}$	–	
		Level 3-5, # 43	$\text{S6} = \mathbf{S}+\mathbf{G}$	=	O8
	$[(\neg X \wedge Y) \vee \neg(Z \rightarrow \neg X)] \leftarrow [(Z \wedge \neg Y) \wedge \neg(\neg X \wedge \neg Y)]$				
5	When **O** or not–**Q**, we assume S7	Level 4-4, # 54	$(\mathbf{O}+\mathbf{V}) : \mathbf{Q}$	–	
		Level 3-5, # 13	$\text{O1} = \mathbf{O}+\mathbf{V}$	=	S7
	$[(X \wedge \neg Y) \vee \neg(\neg Z \rightarrow X)] \leftarrow [(\neg Z \wedge Y) \wedge \neg(X \wedge Y)]$				
6	When **I** or not–**Q**, we assume T7	Level 4-4, # 46	$(\mathbf{I}+\mathbf{U}) : \mathbf{Q}$	–	
		Level 3-5, # 46	$\text{I3} = \mathbf{I}+\mathbf{U}$	=	T7
	$[(X \wedge Y) \vee \neg(\neg Z \rightarrow X)] \leftarrow [(\neg Z \wedge \neg Y) \wedge \neg(X \wedge \neg Y)]$				
7	When **T** or not–**R**, we assume I7	Level 4-4, # 63	$(\mathbf{T}+\mathbf{V}) : \mathbf{R}$	–	
		Level 3-5, # 2	$\text{T2} = \mathbf{T}+\mathbf{V}$	=	I7
	$[(\neg X \wedge \neg Y) \vee \neg(\neg Z \rightarrow \neg X)] \leftarrow [(\neg Z \wedge Y) \wedge \neg(\neg X \wedge Y)]$				

8	When **S** or not–**R**, we assume O7	Level 4-4, # 14	$(\mathbf{S}+\mathbf{U}):\mathbf{R}$	–	
		Level 3-5, # 42	S5=$\mathbf{S}+\mathbf{U}$	=	O7
	$[(\neg X \wedge Y) \vee \neg(\neg Z \rightarrow \neg X)] \leftarrow [(\neg Z \wedge \neg Y) \wedge \neg(\neg X \wedge \neg Y)]$				
9	When **O** or not–**D**, we assume S8	Level 4-4, # 4	$(\mathbf{O}+\mathbf{J}):\mathbf{D}$	–	
		Level 3-5, # 51	O6=$\mathbf{O}+\mathbf{J}$	=	S8
	$[(X \wedge \neg Y) \vee \neg(Z \rightarrow \neg Y)] \leftarrow [(Z \wedge \neg X) \wedge \neg(\neg Y \wedge \neg X)]$				
10	When **I** or not–**C**, we assume T8	Level 4-4, # 16	$(\mathbf{I}+\mathbf{J}):\mathbf{C}$	–	
		Level 3-5, # 56	I6=$\mathbf{I}+\mathbf{J}$	=	T8
	$[(X \wedge Y) \vee \neg(Z \rightarrow Y)] \leftarrow [(Z \wedge \neg X) \wedge \neg(Y \wedge \neg X)]$				
11	When **T** or not–**D**, we assume I8	Level 4-4, # 59	$(\mathbf{T}+\mathbf{K}):\mathbf{D}$	–	
		Level 3-5, # 3	T3=$\mathbf{T}+\mathbf{K}$	=	I8
	$[(\neg X \wedge \neg Y) \vee \neg(Z \rightarrow \neg Y)] \leftarrow [(Z \wedge X) \wedge \neg(\neg Y \wedge X)]$				
12	When **S** or not–**C**, we assume O8	Level 4-4, # 31	$(\mathbf{S}+\mathbf{K}):\mathbf{C}$	–	
		Level 3-5, # 23	S2=$\mathbf{S}+\mathbf{K}$	=	O8
	$[(\neg X \wedge Y) \vee \neg(Z \rightarrow Y)] \leftarrow [(Z \wedge X) \wedge \neg(Y \wedge X)]$				
13	When **O** or not–**P**, we assume S7	Level 4-4, # 12	$(\mathbf{O}+\mathbf{W}):\mathbf{P}$	–	
		Level 3-5, # 50	O5=$\mathbf{O}+\mathbf{W}$	=	S7
	$[(X \wedge \neg Y) \vee \neg(\neg Z \rightarrow \neg Y)] \leftarrow [(\neg Z \wedge \neg X) \wedge \neg(\neg Y \wedge \neg X)]$				
14	When **I** or not–**N**, we assume T7	Level 4-4, # 40	$(\mathbf{I}+\mathbf{W}):\mathbf{N}$	–	
		Level 3-5, # 52	I4=$\mathbf{I}+\mathbf{W}$	=	T7
	$[(X \wedge Y) \vee \neg(\neg Z \rightarrow Y)] \leftarrow [(\neg Z \wedge \neg X) \wedge \neg(Y \wedge \neg X)]$				
15	When **T** or not–**P**, we assume I7	Level 4-4, # 67	$(\mathbf{T}+\mathbf{\Omega}):\mathbf{P}$	–	
		Level 3-5, # 1	T1=$\mathbf{T}+\mathbf{\Omega}$	=	I7
	$[(\neg X \wedge \neg Y) \vee \neg(\neg Z \rightarrow \neg Y)] \leftarrow [(\neg Z \wedge X) \wedge \neg(\neg Y \wedge X)]$				
16	When **S** or not–**N**, we assume O7	Level 4-4, # 55	$(\mathbf{S}+\mathbf{\Omega}):\mathbf{N}$	–	
		Level 3-5, # 8	S1=$\mathbf{S}+\mathbf{\Omega}$	=	O7
	$[(\neg X \wedge Y) \vee \neg(\neg Z \rightarrow Y)] \leftarrow [(\neg Z \wedge X) \wedge \neg(Y \wedge X)]$				

In conclusion, it may be interesting to remark that all the above inferences have either division or subtraction as the main operations (connecting the two line of each inference): product and division inferences are characterized by division (Tables 8.20 and 8.24) while sum and subtraction are characterized by subtraction (Tables 8.21 and 8.23). It is also possible to show that the inferences of the kind of *Darii* as well as relative variants and counterparts can be put in some of the previous forms (consider the results of Section 8.1). For instance, we have for the inferences corresponding to the classical

Darii and *Ferio*:

$$\{(\mathbf{A}-\mathbf{J}-\mathbf{G}):[(\mathbf{A}-\mathbf{J})=\mathrm{A3}]\}=\mathrm{M1}\ \text{and}$$
$$\{(\mathbf{E}-\mathbf{J}-\mathbf{H}):[(\mathbf{E}-\mathbf{J})=\mathrm{E3}]\}=\mathrm{L1}\ , \tag{8.68}$$

respectively. However, this kind of results do not add something conceptually new to the previous ones (they are even quite difficult to interpret) although being very lengthy, and therefore I skip this part.

8.7 Open Inferences in General

As anticipated, the analysis developed in the previous section allows us to go back to the examination of the open inferences with many variables. In Section 8.5, we have met this inference:

$$\begin{array}{cc} p \vee \neg r & \\ r \vee s & \\ \neg s & \\ p \rightarrow (\neg t \vee u) & \\ t & /\, u \end{array}$$

Thanks to the reduction to variables $r = X$, $p = Y$ and $u = Z$ in subsequent steps and the already performed three-dimensional analysis, we can reduce this inference to the following terms:

From X, Y and $(X \wedge Y) \rightarrow Z$, the conclusion Z follows,

which has the expanded syllogistic form (on these expanded forms see also the conclusive remarks in Section 6.5):

$$(\mathbf{I} \wedge \mathrm{E2}) \rightarrow (\mathrm{N4xN7})\ . \tag{8.69}$$

Indeed, $\mathbf{I} \times \mathrm{E2} = \mathrm{I8}$. However, $\mathrm{I8} = X \wedge Y \wedge Z$ and since we already have $X \wedge Y$, we can derive Z (i.e. N4xN7) by simplification. Another example is:

$$\begin{array}{cc} (p \vee q) \rightarrow [r \vee (s \wedge t)] & \\ \neg s & \\ q & \\ \neg r \vee u & \\ w \rightarrow \neg u & /\neg w \end{array}$$

This example is more complex. The four steps followed in Section 8.5 can be resumed as follows (here, $r = X$, $u = Y$ and $w = Z$):

(1) $\{[(\mathrm{L6} \times \mathrm{L8})+\mathbf{G}]=\mathrm{L8}\} \times (\mathrm{O1}+\mathrm{O2}) = \mathbf{O}=\mathrm{O5} \times \mathrm{O6}$,
(2) $(\mathrm{O5}+\mathrm{O6}) \times \mathbf{A} = \mathbf{I}=\mathrm{I3} \times \mathrm{I5}$,
(3) $(\mathrm{I3}+\mathrm{I5}) \times \mathbf{D} = \mathbf{U}=\mathrm{U1} \times \mathrm{U3}$,
(4) $\mathrm{U1}+\mathrm{U3}$

This summary does not strictly follow the four steps of Section 8.5 (in fact, I have introduced the further simplification to shrink the first three steps there to the first two here and have factorized the last step there in Step 3 and 4 here). Such a summary can be reduced to:

$$\{[(\mathrm{L8} \wedge \mathrm{O}) \rightarrow (\mathrm{O5} + \mathrm{O6})] \wedge \{[(\mathrm{O5} + \mathrm{O6}) \wedge \mathbf{A}] \rightarrow (\mathrm{I3} + \mathrm{I5})\} \wedge \{[(\mathrm{I3} + \mathrm{I5}) \wedge \mathbf{D}] \rightarrow (\mathrm{U1} \times \mathrm{U3})\}\} \rightarrow (\mathrm{U1} + \mathrm{U3}), \tag{8.70}$$

which can reformulated as

$$\{\{\{[(\mathbf{O} + \mathbf{I}) = (\mathrm{O5} + \mathrm{O6})] \times \mathbf{A}\} = \mathbf{I}\} \times \mathbf{D}\}=\mathrm{U8} =\mathrm{I7},$$

or

$$(\mathbf{I} \wedge \mathbf{D}) \rightarrow \mathrm{I7}.$$

Therefore, I have shown how to reduce open inference to enlarged closed ones, which makes much easier mechanical calculations.

8.8 Why Three?

I have therefore justified (at least to a certain extent) the generality of a three-dimensional space and therefore also of the three-statement kind of inference, we need to explore now the reasons for this. There is indeed a fundamental conceptual reason why the three-statement inferences are sufficiently general to cover any kind of inference, at least when not particular kinds of relations are involved. Any inference is the drawing of some conclusion out of some premises. Then, it must necessarily deal with a rule (Peirce, 1865, pp. 259–61)(Peirce, 1866, pp. 365–69) (Auletta, 2009). Indeed, the absence of a rule would compromise the very essence of inference, since it would transform it in a free mental association (like those occurring when we recall things of the past), and no conclusion whatsoever could follow in these conditions. However, an inference cannot consist in a rule alone, since this would be at most the definition of a rule but not a derivation of some conclusion from some premises. Therefore, we must also be able to consider some application domain of the rule, that is, some case to which, even hypothetically, such a rule could be applied. However, if a set of statements were constituted of these two judgments only, it would not be an inference but at most the enunciation of a law or rule and the exhibition of a possible example or application domain. As said, from an inference we expect also some consequence out of the first two statements, i.e. a conclusion.

Then, an inference is necessarily made of these three propositions, which may also be called thesis, condition and conclusion or even rule, subsumption under the rule, result (Peirce, 1865, p. 259) (Peirce, 1866, pp. 362–63). I have shown that we cannot have a smaller number of propositions, but reduction to n-variables inferences to 3-variable inferences also shows that it is immaterial to have more since this would only be a multiplication of one of the above statements (e.g. by conjoining predicates or disjoining subjects).

Another relevant consideration is the following. The three terms occurring in every basic inference have a very different meaning (Peirce, 1868b, pp. 72–74 and 83). Indeed, the most general term that appears as predicate in the conclusion is the connotative term $\mathcal{C}$ denoting the class of all things that share that common property (that connotation). On the contrary, the particular term that appears as subject in the conclusion is the denotative term $\mathcal{D}$, denoting a class of objects that share many properties, namely an equivalence class of objects. It is likely that the above distinction between extension and intension does not coincide with the current one in logical studies. Indeed, Peirce distinguishes among connotation, denotation and information (or distinctness) and notes that connotation is the inverse of denotation only if the information remains the same (Peirce, 1865, pp. 187–89 and 284–86). This implies that we can increase the informational content (or determination) of a concept by refining its connotation without appreciably change its denotation; or vice versa, we can increase its information by discovering new domains of applicability, i.e. by increasing its variety and therefore denotation, without appreciably change its connotation.

Now, the term that appears in the premises but not in the conclusion of the general forms of inference is the informative term $\mathcal{I}$ and is what establishes a connection between connotation and denotation in the conclusion and owes its name to the circumstance that it has a connection with some kind of (not necessarily empirical) experience, representing the amount of knowledge that we believe to have among certain facts or relations: it can express explanations of causal type but also mathematical knowledge. For this reason, the conclusion can also be called an *informative* statement, the premise in which the connotative term appears a *connotative* statement and the premise in which the denotative term is present a *denotative* statement. In general, the denotative statement expresses the application domain of a law while the connotative statement the rule or law.

In general, the subject of a proposition is the term determined in connotation and determining denotation, while its predicate is the term determined in denotation and determining in connotation (Peirce, 1865, pp. 273, 277–79 and 288):

- A subject can be either denotative or informative, whilst
- The predicate can be either connotative or informative.

Then, we have three basic kinds of judgment (the connection between subject and predicate is left indeterminate since it can be universal or particular, affirmative or negative):

$\mathcal{IC}$: Subject informative and predicate connotative (connotative judgments),

$\mathcal{DI}$: Subject denotative and predicate informative (denotative judgments)

$\mathcal{DC}$: Subject denotative and predicate connotative (informative judgments)

In the first case the connotation of the symbol forming the subject is explicated by the predicate. Such propositions are analytical (a particular case of analytical propositions is represented by identities of the kind All X are X). In the second case, the thing that the subject denotes is offered as an example of the application of the symbol which forms the predicate. In the third case the thing that is denoted by the subject is said to embody the form connoted by the predicate.

Since connotative terms in general denote nothing in particular, they are called by Peirce pure forms or icons (Peirce, 1866, pp. 467–68). Denotative terms in general connote nothing in particular since they refer to (or even are) tokens, and therefore can be taken to be indexes referring to objects (Peirce, 1885a, pp. 162–64). Then, the subject of any judgment must contain some index while the predicate is an icon (Peirce, 1895, pp. 20–21). In sum, indexes are marks for individuating objects (establishing denotation) while predicates or universal concepts are icons expressing the resemblance (or some connotative relation) with something else. Simple terms allow to connote what one denotes, in which case, as said, they constitute informative propositions. They can be considered symbols in a strict sense (or also in a perfect sense). A symbol in this strict sense declares that the set of objects, which is denoted by whatever set of indices may be in certain ways attached to it, is represented by an icon associated with it (Peirce, 1865, p. 287)(Peirce, 1895, p. 17).

When symbols are combined together in extension (when we combine different classes of objects), their sum possesses denotation but no connotation (they are enumerative or *disjunctive terms*). When symbols are combined together in intension or comprehension (when we put together predicates), they possess connotation but no denotation (can be called *conjunctive* terms). Enumerative terms cannot form the subject of a proposition (logically this would represent a set of different propositions) but only the predicate, in which case the proposition is connotative. Conjunctive terms cannot form the predicate of a proposition (since it is like to predicate two unconnected terms) but only the subject, in which case the proposition is denotative.

Due to the intrinsic limitations of human knowledge, these two types of propositions are kinds of degenerate forms relative to the standard ones. In fact, conjunctive terms appear to denote a class of objects but actually they only represent intersections of properties (like hippogriffs). On the other hand, disjunctive terms are larger set of disparate objects (as it is necessarily the case for a connotation) but these objects do not share a common property (like in clusters of disparate objects). At the opposite, symbols possessing either proper connotation or proper denotation are called simple terms and they can constitute informative propositions. In other words:

- We have informative statements when both terms are simple, i.e. a proper connotative term as subject and a proper denotative term as predicate.
- Connotative terms can be either conjunctive terms or proper connotative terms (in this case they are simple terms). Thus, conjunctive terms give rise to denotative statements whilst proper connotative terms to connotative or informative statements.
- Similarly, denotative terms can be either enumerative terms or proper denotative terms (in this case they are simple term). This means that enumerative terms give rise to connotative statements whilst proper denotative terms to denotative or informative statements.

Summarizing, in the three different groups in any kind of inference the three terms always show the following general structure (see also Table 8.4), where I again stress that the connection can be universal or particular as well as positive or negative):

Table 8.25 The terms and the three groups

1st group	2nd group	3rd group
$\mathcal{IC}$	$\mathcal{IC}$	$\mathcal{DC}$
$\mathcal{DI}$	$\mathcal{DC}$	$\mathcal{DI}$
$\mathcal{DC}$	$\mathcal{DI}$	$\mathcal{IC}$

Consider in particular that the archetype of the first group as shown here is that of the product inferences, the archetype of the second group is that of the division inferences and the archetype of the third group that of sum inferences. As remarked, subtraction inferences are somehow a mixed form of product and sum inferences. As the reader may recall, while deductions have the general form of product inferences, inductions have the general form of sum inferences, and abductions the general form of division inferences (see Sections

3.1 and 5.1). Peirce calls deduction inference a priori, abduction inference a posteriori, and induction inference *a particularis* (Peirce, 1865, pp. 184, 267, 281). Deduction is reasoning from determinant to determinate and is related to symbols, since the definition of a symbol alone determines its character (as it is clear by the informative conclusion). A simple term occurring as the subject of a description can be taken to be a general description or definition (Peirce, 1881, p. 249). Then, a deduction consists in constructing an icon $\mathcal{C}$ the relation of whose constituent parts shall present a complete analogy with those of the characters of the object of reasoning denoted by $\mathcal{D}$ (Peirce, 1885a, p. 164), as it is clear by the conclusion of the first group. On the contrary, reasoning a posteriori must proceed from determinate to determinant. When an inference is about pure forms or connotations (i.e. it is an abduction), it may also be called inference of a character ((Peirce, 1878, p. 330), since it is a certain information determining a denotation (like the conclusion $\mathcal{DI}$ of the second group). The principle of inductive inference must go from parts to whole and is therefore related to things (referents). Since abduction stresses the iconic aspect, it deals with analogy or resemblance among objects (metaphor). At the opposite, since induction stresses the indexical aspect, it deals with contiguity (metonymy) among objects and therefore all the disparate objects that share some information $\mathcal{I}$ come to share only the connotation $\mathcal{C}$, as in the conclusion of the third group (Peirce, 1865, pp. 257–78, 272, 278–79) (Peirce, 1866, p. 485) (Peirce, 1901, pp. 96–97 and 106) (*Auletta*, 2011, Sections 16.5, 18.4 and 20.4–20.6).

This allows me to try to lead all the three forms of inference under a common heuristic principle, which in opposition to the logical principle (Section 8.2) is not foundational but is derived from the above examination of the procedures adopted in the three forms of inference and their interconnections. We may even call this a restatement of Leibnizs famous principle of sufficient reason, which in the traditional formulation can be enunciated as: Everything has a reason to be as it is (Leibniz, 1710, Par. 44). The reformulation that I propose is the following one: Of everything a ground can be found that enables some results that we hold to be true. It is not difficult to see that this formulation covers both anterograde (deductions) and retrograde (abductions and inductions) inferences without any foundational aim, in accordance with the character of this logical system.

8.9 Predication and Relations

With this logical system we can easily treat problems of objects and their properties. However, a possible worry is the connection between predication and theory of relations. I do not wish to enter here a difficult domain with many philosophical implications but simply point out some basic issues. It is clear that if we take many statements in which relations occur they are irreducible to factorized predicates attributed to the objects involved in the same statements. For instance, the statement

Rome is situated approximately halfway between Florence and Naples

expresses a relation between three towns (Rome, Florence and Naples) that results irreducible to any attribution of factorized predicates to one or two of these objects. However, at a very basic level this problem arises when one considers predicates as expressing properties that somehow intrinsically pertain to a certain object. If, on the contrary, one assumes that properties consist in general in relations that objects have with other objects, then this objection loses most of its power. As a matter of fact, Peirce speaks of a term as expressing a relation of a quality, character, fact, or predicate to its subject (Peirce, 1885a, p. 185). For instance, color is a property of many objects but it depends on the environment in which light is propagated and reflected (*Auletta*, 2011, Chap. 4). Another example is constituted by weight: it depends on the location that a certain object has relative to some gravitational field. It is clear that the personal constitution of an individual seems something intrinsic. However, it can be shown that every organism is as it is thanks to a very complex interaction with the environment during development (*Auletta*, 2011, Chapter 11) that can even involve cultural aspects in the case of humans (*Auletta*, 2011, Part 3). I do not need to insist on the fact that every material object has its own "history" through which it is become as it currently is.

Therefore, I see no particular problem in considering the relation "To be situated approximately halfway between Florence and Naples" as a whole as a property of the town Rome. Obviously, also the objects Florence and Naples shall have certain properties expressing their location relative to Rome as well as between them. This does not interfere with the attribution of that property, not less than when attributing a color to an object. Indeed, also in such a case this color is related to the color of other objects nearby. If this is true for individual objects, it is much more true for classes of objects, the true scope of logic.

Although the problem is conceptually solved, it is not yet clear at a formal level whether predicates can or should be reduced to relations or vice versa. In particular, some specific problems may occur when we have relations in the context of the predicate calculus. Let us explore this problem step by step by considering several examples and trying to introduce further sophistications when needed (a systematic exposition of the logical theory of relations can be found in (*Schröder*, 1890–95, iii v.)). Identity is an important relation that can be dealt with in the second-order logic. Let us therefore consider as an example of second-order expression the identity relation (the so-called Leibniz's identity of indiscernibles (Leibniz, 1689, pp. 519–20)):

$$(\forall X)(\forall Y)(\forall t)(\forall w)\{[(Xt \rightarrow Xw) \wedge (Yw \rightarrow Yt)] \rightarrow Itw\}, \quad (8.71)$$

where it is said that if the objects t's having the properties X's implies the objects w's having the same properties as well as if the objects w's having the properties Y's implies the objects t's having the same properties, then the t's and w's are identical. It is not possible to translate this expression into the language used in this book (without quantifiers) if not using some further sophistication. A possible translation of the previous expression would be the following:

$$[(X^2 \rightarrow X^2) \wedge (Y^2 \rightarrow Y^2)] \rightarrow Z, \quad (8.72)$$

where Z is understood as a non-intrinsic predicate of t (relative to w) or of w (relative to t) according to the previous examination of relations (we could also use I in oder to let clearly understand that is an identity by convention). In other words, we express here the predicate variables in terms of squares while constant predicates in terms of ordinary variables with power 1. We can remark that we have here the same variable (X or Y) occurring in the same implication. It is understood that the object variables (t and w) must be different. Indeed, an expression like the following:

$$(\forall X)(\forall Y)(\forall t)(\forall w)\{[(Xt \rightarrow Xt) \wedge (Yw \rightarrow Yw)] \rightarrow Z\} \qquad (8.73)$$

seems to be not semantically interesting and it is likely to be also syntactically not well formed, whatever the form of Z is. Said this, I further remark that it is irrelevant in which order we put the object variables (in the same way in which it is irrelevant if we interpret Z as a relation of t with w or vice versa). A further complication may arise when considering that quantifications over predicates can be either universal or existential. Let us consider a quite common expression like the following:

$$(\forall X)(\forall t)(\exists w)[Xt \rightarrow (Itw \vee Rtw)]. \qquad (8.74)$$

This could be interpreted as expressing that the object t having any property X implies that there is at least one object w that is either identical with t (Itw) or has with this some other kind of relation R. This expression could be translated as

$$X^2 \rightarrow (I \vee R). \qquad (8.75)$$

Now, the question is: how can we understand that there is existential quantification of one of the two first-order variables? The expression could mean that if an object has the predicate X it either is identical with any other object or has with it any relation R. This interpretation is however excluded by the fact that R (like I) is a constant here (although by convention connecting two objects) and therefore it means a particular relation. Then, the second object cannot be universally quantified, since there is no particular non-empty relation R such that every object is either identical with itself or has the relation R with any other object (see Axiom 2 in Tarski, 1941). However, this also shows that the universal quantification of the predicate X is not correct, since there is no object t such that for every possible predicate X and without any further specification, it is identical with some object w or has a certain relation R with the latter. Then, the statement (8.74) needs to be rewritten as a first-order one, like $X \rightarrow (I \vee R)$, showing that this language helps for disambiguation.

Obviously there can be cases that are more difficult to solve, especially when more complex relations are involved. My guess is that we could solve some of these problems by using a higher-level logical space. In such a case, at least when there is universal quantification, we consider X^2 (as well as Y^2) as a predicate acting as a set of the kind considered in this book in a four-dimensional space, in which several different three-dimensional logical spaces are embedded, while Z is confined to a three-dimensional space only. We could also introduce the further sophistication to have different powers ($X^2, X^3, \ldots$) according to the number of first-order variables of which they are predicates. In this way, X and Y are treated as ordinary predicates in each subspace. In other words, we can deal with this problem in analogy with two-dimensional statements in a three-dimensional space, as shown in Section 8.4. What happens now when predicates or relations are quantified existentially? For instance, let us consider the expression

$$(\exists R)\,\{(\forall t)(\forall u)(\forall w)[(Rtu \wedge Ruw) \rightarrow Rtw] \wedge (\forall t)(\neg Rtt \wedge (\exists u)Rtu)\}\,, \tag{8.76}$$

which means that there is a non-reflexive transitive relation R, such that it is always possible to find an object u such that any object t has such a relation with but the latter cannot have this relation with itself. This could be translated as:

$$\left[(R_1^3 \wedge R_1^3) \rightarrow R_1^3\right] \wedge (\neg R_1 \wedge R_1^2), \tag{8.77}$$

where R_1 is a non-empty subset of R according to the convention of Section 1.2 (and therefore denotes existential quantification: note that the main logical sign is a conjunction) while the parentheses allow us also to modularize different segments of quantification (note that also that the last two expression are connected by a conjunction). By convention, any relation R occurring without exponent (like in the last parenthesis) is self-relative (Peirce, 1883a, p. 457). There could nevertheless be still an ambiguity, since the above expression could be interpreted as:

$$(\exists R)\,\{(\forall t)(\forall u)(\forall w)[(Rtu \wedge Rvw) \rightarrow Rtw] \wedge (\forall t)(\neg Rtt \wedge (\exists u)Rtu)\}\,. \tag{8.78}$$

When considering the left part, such a relation could only mean either equivalence or identity between at least two different objects. However, the right part tells us that R cannot be interpreted in this sense. Then, the ambiguity is removed. However, how can we disambiguate the case of a relation R^2 from a predicate X^2 referring to two distinct classes of objects? We can again do this by making us of subspaces and relative sets. In such a case the lower indexes could label specific subspaces so that even double-indexed expressions like R_{14} would be allowed, so that we can rewrite the expression (8.76) as

$$\left[(R^3_{x4} \wedge R^3_{46}) \rightarrow R^3_{x6}\right] \wedge (\neg R_x \wedge R^2_{x4}), \tag{8.79}$$

where the lower index x means that this kinds of object are taken in the whole logical space.

In solving this kind of problems, it is obvious that the recursive methodology shown in Section 8.5 needs to be extensively applied. However, the question is whether it is sufficiently general.

In fact, this examination allows us a deeper understanding of the relations among spaces of different dimensions. Consider again Fig. 8.2. If we consider X, Y and Z either as standing for propositions or as first-order (constant) predicates (recall what has been said in Section 1.2), it is quite natural the conclusion that a bidimensional space has at most the power to cover (a part of) the propositional calculus. Here, these symbols mean propositions. Instead, when no relations are involved, a three-dimensional space is sufficiently powerful to cover a first-order logic (and here symbols are interpreted as constant predicates) and at least a class of problems of a second-order predicate calculus. Indeed, it is customary to interpret a syllogism in terms of the first-order calculus as follows:

$$(\forall t)\{[(Xt \rightarrow Yt) \wedge (Zt \rightarrow Xt)] \rightarrow (Zt \rightarrow Yt)\}, \tag{8.80}$$

where X, Y, Z are taken as constant predicates and the main implication needs to be a tautology. However, it seems to be much more in accordance with Aristotle's point of view to consider also the predicates as variables (*Aristotle An. Pr., An. Post.*). In this case we should rather write:

$$(\forall X)(\forall Y)(\forall Z)(\forall t)(\forall w)\{[(Xt \rightarrow Yt) \wedge (Zw \rightarrow Xw)] \rightarrow (Zw \rightarrow Yw)\}. \tag{8.81}$$

We can translate this according to the previous formalism as:

$$[(X^2 \to Y^2) \wedge (Z^2 \to X^2)] \to (Z^2 \to Y^2). \tag{8.82}$$

However, since no relations are involved here, this expressions can also be simplified to:

$$[(X \to Y) \wedge (Z \to X)] \to (Z \to Y), \tag{8.83}$$

without any loss of meaning. We could even say that any inference represents a kind of relation instantiation, as the comments to Table 8.4 show, and this brings again predication and relation very close. Indeed, we can express an inference like *Barbara* as

$$(R_{xy} \wedge R_{zx}) \to R_{zy}, \tag{8.84}$$

where the relation R can mean "To be a symbol of", "To stand for", "To be a member of the class", and so on (Peirce, 1865, pp. 186 and 283–84). An inference like *Celarent* can be written as

$$(\neg R_{xy} \wedge R_{zx}) \to \neg R_{zy}, \tag{8.85}$$

where $\neg R_{xy}$ means that no X has the relation R with any Y, while an inference like *Darii* takes the form

$$(R_{xy} \wedge R_{62}) \to R_{64}. \tag{8.86}$$

Similarly, we have for *Ferio*

$$(\neg R_{xy} \wedge R_{62}) \to \neg R_{54}. \tag{8.87}$$

Far more interesting are the inferences considered in the previous section. For instance, let us focus on the first inference of Table 8.20, which can be expressed as:

$$(R_{xy} \wedge R_{zx}) \to (R_{64} \to R_{24}) . \tag{8.88}$$

This examination allows us also to express the transitive relation displayed in Eq. (8.79) as

$$\neg R_{xx} \to \left[(R_{xy} \wedge R_{yz}) \to R_{xz}\right], \tag{8.89}$$

which looks quite similar to the previous expressions. I have made use of the implication as main connective otherwise the expression would have been always false (another case of disambiguation). Another possibility is to make use of Table 8.18 and express explicitly the subspaces involved (dropping any variable altogether).

For instance (in accordance with the expressions (8.39)), *Barbara* becomes: $(R_{14} \wedge R_{52}) \rightarrow R_{54}$, or *Ferio* becomes: $(R_{13} \wedge R_{62}) \rightarrow R_{63}$, where the particular kinds of relations (whether negative or affirmative, universal or particular) are already coded in the relations displayed by Table 8.18 as explained in the subsequent comment. We can proceed similarly for mereological inferences. Each of the inferences shown here (in whatever form) could be taken as a definition of a particular ternary relation since it establishes an universal or particular connection among the variables X, Y, and Z. For instance, (8.84) could be a definition of a universal relation R_{xyz}, whilst (8.86) of a particular relation R_{321}.

It is quite easy to associate each (binary or ternary) relation with some numeric ID of the tridimensional space, and many alternative solutions are possible. However, a tridimensional space only allows for dealing with a generic binary or ternary relation R. If we like to distinguish particular kinds of relations, we need however higher orders of logical spaces. In fact, a four-dimensional space allows for four three-dimensional logical subspaces and $4 \times 3 = 12$ bidimensional (and 24 one-dimensional) subspaces. Of these 12 bidimensional subspaces, there are two sets that involve the same variables, what allows to discriminate among two kinds of particular binary relations, for instance A_{xy}, B_{xy} (obviously, these different forms are distinguished thanks to the different subspaces involved, i.e. the numeric values taken by the variables x and y). However, we need a pentadimensional logical space for dealing with different ternary relations. In this case, we have five four-dimensional logical subspaces and $5 \times 4 = 20$ tridimensional subspaces, allowing to discriminate among two different relations among the same variables (e.g. A_{xyz}, B_{xyz}, where again are the subspaces to make the difference here) and the considerable number of 60 different binary relations as a whole (and 6 different binary relation for each couple of variables). This should largely suffice to deal with practically any kind of inference.

Then, I have shown that binary relations and predication can be brought under a common denominator. In particular, it is quite reasonable to finally express any logical connection (included predication) in terms of relations and not vice versa (predication is implemented by binary relations of the same kind R). Although less convenient that the reverse reduction, the advantage here is twofold: we can make use of the same formalism everywhere and it is likely that a three-dimensional space would suffice, since a three-dimensional relation can be expressed as R_{xyz} (and Peirce's guess is that higher-order relations can be reduced to ternary ones). Summarizing so far, a two-dimensional space is sufficient for dealing with a propositional calculus when only rules and no inferences are involved. A three-dimensional space is necessary for dealing with inferences (with three or more variables), where it does not matter whether they are interpreted in terms of propositional calculus, first-order or second-order logic, although the straight interpretation is in terms of first-order logic. It is likely that this is also sufficient for some subclasses of basic ternary relations of generic kind. However, higher orders become necessary when we like to distinguish among particular kinds of relations.

Chapter 9

Applications

9.1 Artificial Intelligence

For what concerns possible applications to AI, I stress that this system allows to substitute a fixed matrix or a predetermined set of operations with a dynamical approach that is capable of processing information building each time its logical rule and the necessary inferences.

The main idea is that with this system we no longer need to build fixed logical circuits (instantiating logical rules) in order to let an intelligent system work. The two general principles for building the logical space combined with the two principles of closed derivations and the previous treatment of open inferences are sufficient to build any possible logical connection in due course according to the inputs received through the running program. In other words, this system allows to build a complete new generation of dynamical intelligent systems that perform themselves the needed inferences and make use of appropriate logical rules at the same time in which they perform the inference. Moreover, this plasticity allows for using specific logical rules according to the procedure needed. Therefore, several and alternative logics can be instantiated on the same

Mechanical Logic in Three-Dimensional Space
Gennaro Auletta

ISBN 978-981-4411-50-9 (Hardcover), 978-981-4411-51-6 (eBook)
www.panstanford.com

general logical scheme (the nodes in the logical space represented by their numerical ID).

9.2 Classical Computing

It is interesting to remark that although the ID here has a logical meaning and not an arithmetic one, an interesting connection can be established between this logical space and the ordinary binary-numbers codification. This will allow to couple in an easy way a processor built with this system with a classical processor or to let a software built with these principles run on a classical processor. First, let us consider following table:

Table 9.1 Connection with binary numbers

Level	#	ID	Number	Level	#	ID	Number
					0-8	00000000	0
1-7	8	00000001	1	1-7	7	00000010	2
2-6	28	00000011	3	1-7	6	00000100	4
2-6	27	00000101	5	2-6	26	00000110	6
3-5	56	00000111	7	1-7	5	00001000	8
2-6	25	00001001	9	2-6	24	00001010	10
3-5	55	00001011	11	2-6	23	00001100	12
3-5	54	00001101	13	3-5	53	00001110	14
4-4	70	00001111	15	1-7	4	00010000	16
2-6	22	00010001	17	2-6	21	00010010	18
3-5	52	00010011	19	2-6	20	00010100	20
3-5	51	00010101	21	3-5	50	00010110	22
4-4	69	00010111	23	2-6	19	00011000	24
3-5	49	00011001	25	3-5	48	00011010	26
4-4	68	00011011	27	3-5	47	00011100	28
4-4	67	00011101	29	4-4	66	00011110	30
5-3	56	00011111	31	1-7	3	00100000	32
2-6	18	00100001	33	2-6	17	00100010	34
3-5	46	00100011	35	2-6	16	00100100	36
3-5	45	00100101	37	3-5	44	00100110	38
4-4	65	00100111	39	2-6	15	00101000	40
3-5	43	00101001	41	3-5	42	00101010	42
4-4	64	00101011	43	3-5	41	00101100	44
4-4	63	00101101	45	4-4	62	00101110	46
5-3	55	00101111	47	2-6	14	00110000	48

Level	#	ID	Number	Level	#	ID	Number
3-5	40	00110001	49	3-5	39	00110010	50
4-4	61	00110011	51	3-5	38	00110100	52
4-4	60	00110101	53	4-4	59	00110110	54
5-3	54	00110111	55	3-5	37	00111000	56
4-4	58	00111001	57	4-4	57	00111010	58
5-3	53	00111011	59	4-4	56	00111100	60
5-3	52	00111101	61	5-3	51	00111110	62
6-2	28	00111111	63	1-7	2	01000000	64
2-6	13	01000001	65	2-6	12	01000010	66
3-5	36	01000011	67	2-6	11	01000100	68
3-5	35	01000101	69	3-5	34	01000110	70
4-4	55	01000111	71	2-6	10	01001000	72
3-5	33	01001001	73	3-5	32	01001010	74
4-4	54	01001011	75	3-5	31	01001100	76
4-4	53	01001101	77	4-4	52	01001110	78
5-3	50	01001111	79	2-6	9	01010000	80
3-5	30	01010001	81	3-5	29	01010010	82
4-4	51	01010011	83	3-5	28	01010100	84
4-4	50	01010101	85	4-4	49	01010110	86
5-3	49	01010111	87	3-5	27	01011000	88
4-4	48	01011001	89	4-4	47	01011010	90
5-3	48	01011011	91	4-4	46	01011100	92
5-3	47	01011101	93	5-3	46	01011110	94
6-2	27	01011111	95	2-6	8	01100000	96
3-5	26	01100001	97	3-5	25	01100010	98
4-4	45	01100011	99	3-5	24	01100100	100
4-4	44	01100101	101	4-4	43	01100110	102
5-3	45	01100111	103	3-5	23	01101000	104
4-4	42	01101001	105	4-4	41	01101010	106
5-3	44	01101011	107	4-4	40	01101100	108
5-3	43	01101101	109	5-3	42	01101110	110
6-2	26	01101111	111	3-5	22	01110000	112
4-4	39	01110001	113	4-4	38	01110010	114
5-3	41	01110011	115	4-4	37	01110100	116
5-3	40	01110101	117	5-3	39	01110110	118
6-2	25	01110111	119	4-4	36	01111000	120
5-3	38	01111001	121	5-3	37	01111010	122
6-2	24	01111011	123	5-3	36	01111100	124
6-2	23	01111101	125	6-2	22	01111110	126
7-1	8	01111111	127	1-7	1	10000000	128

(*Continued*)

Table 9.1 (*Continued*)

Level	#	ID	Number	Level	#	ID	Number
2-6	7	10000001	129	2-6	6	10000010	130
3-5	21	10000011	131	2-6	5	10000100	132
3-5	20	10000101	133	3-5	19	10000110	134
4-4	35	10000111	135	2-6	4	10001000	136
3-5	18	10001001	137	3-5	17	10001010	138
4-4	34	10001011	139	3-5	16	10001100	140
4-4	33	10001101	141	4-4	32	10001110	142
5-3	35	10001111	143	2-6	3	10010000	144
3-5	15	10010001	145	3-5	14	10010010	146
4-4	31	10010011	147	3-5	13	10010100	148
4-4	30	10010101	149	4-4	29	10010110	150
5-3	34	10010111	151	3-5	12	10011000	152
4-4	28	10011001	153	4-4	27	10011010	154
5-3	33	10011011	155	4-4	26	10011100	156
5-3	32	10011101	157	5-3	31	10011110	158
6-2	21	10011111	159	2-6	2	10100000	160
3-5	11	10100001	161	3-5	10	10100010	162
4-4	25	10100011	163	3-5	9	10100100	164
4-4	24	10100101	165	4-4	23	10100110	166
5-3	30	10100111	167	3-5	8	10101000	168
4-4	22	10101001	169	4-4	21	10101010	170
5-3	29	10101011	171	4-4	20	10101100	172
5-3	28	10101101	173	5-3	27	10101110	174
6-2	20	10101111	175	3-5	7	10110000	176
4-4	19	10110001	177	4-4	18	10110010	178
5-3	26	10110011	179	4-4	17	10110100	180
5-3	25	10110101	181	5-3	24	10110110	182
6-2	19	10110111	183	4-4	16	10111000	184
5-3	23	10111001	185	5-3	22	10111010	186
6-2	18	10111011	187	5-3	21	10111100	188
6-2	17	10111101	189	6-2	16	10111110	190
7-1	7	10111111	191	2-6	1	11000000	192
3-5	6	11000001	193	3-5	5	11000010	194
4-4	15	11000011	195	3-5	4	11000100	196
4-4	14	11000101	197	4-4	13	11000110	198
5-3	20	11000111	199	3-5	3	11001000	200
4-4	12	11001001	201	4-4	11	11001010	202
5-3	19	11001011	203	4-4	10	11001100	204
5-3	18	11001101	205	5-3	17	11001110	206
6-2	15	11001111	207	3-5	2	11010000	208

Level	#	ID	Number	Level	#	ID	Number
4-4	9	11010001	209	4-4	8	11010010	210
5-3	16	11010011	211	4-4	7	11010100	212
5-3	15	11010101	213	5-3	14	11010110	214
6-2	14	11010111	215	4-4	6	11011000	216
5-3	13	11011001	217	5-3	12	11011010	218
6-2	13	11011011	219	5-3	11	11011100	220
6-2	12	11011101	221	6-2	11	11011110	222
7-1	6	11011111	223	3-5	1	11100000	224
4-4	5	11100001	225	4-4	4	11100010	226
5-3	10	11100011	227	4-4	3	11100100	228
5-3	9	11100101	229	5-3	8	11100110	230
6-2	10	11100111	231	4-4	2	11101000	232
5-3	7	11101001	233	5-3	6	11101010	234
6-2	9	11101011	235	5-3	5	11101100	236
6-2	8	11101101	237	6-2	7	11101110	238
7-1	5	11101111	239	4-4	1	11110000	240
5-3	4	11110001	241	5-3	3	11110010	242
6-2	6	11110011	243	5-3	2	11110100	244
6-2	5	11110101	245	6-2	4	11110110	246
7-1	4	11110111	247	5-3	1	11111000	248
6-2	3	11111001	249	6-2	2	11111010	250
7-1	3	11111011	251	6-2	1	11111100	252
7-1	2	11111101	253	7-1	1	11111110	254
8-0		11111111	255				

All IDs are generated either by shift of 1 (as for Level 1-7) or through addition of some previous IDs. It may be clearly seen that there is a specific algorithm that rules the generation of statements. Indeed, it may be noted that there are quadruplets of statements' ID such that we have one statement of a certain level that is followed by two statements of the subsequent level and in turn the latter is followed by a statement of the subsequent level (always in progression from below). Moreover, the quadruplets follow a specific rule:

Table 9.2 Quadruplets of numbers

Level 0-8	Level 1-7	Level 2-6	Level 3-5	Level 4-4	Level 5-3	Level 6-2	Level 7-1	Level 8-0	Numbers
1	2	1							0-3
	1	2	1						4-7
	1	2	1						8-11
		1	2	1					12-15
	1	2	1						16-19
		1	2	1					20-23
		1	2	1					24-27
			1	2	1				28-31
	1	2	1						32-35
		1	2	1					36-39
		1	2	1					40-43
			1	2	1				44-47
		1	2	1					48-51
			1	2	1				52-55
			1	2	1				56-59
				1	2	1			60-63
	1	2	1						64-67
		1	2	1					68-71
		1	2	1					72-75
			1	2	1				76-79
		1	2	1					80-83
			1	2	1				84-87
			1	2	1				88-91
				1	2	1			92-95
		1	2	1					96-99
			1	2	1				100-103
			1	2	1				104-107
				1	2	1			108-111
			1	2	1				112-115
				1	2	1			116-119
				1	2	1			120-123
					1	2	1		124-127
	1	2	1						128-131
		1	2	1					132-135
		1	2	1					136-139
			1	2	1				140-143
		1	2	1					144-147
			1	2	1				148-151
			1	2	1				152-155

Level 0-8	Level 1-7	Level 2-6	Level 3-5	Level 4-4	Level 5-3	Level 6-2	Level 7-1	Level 8-0	Numbers
				1	2	1			156-159
		1	2	1					160-163
			1	2	1				164-167
			1	2	1				168-171
				1	2	1			172-175
			1	2	1				176-179
				1	2	1			180-183
				1	2	1			184-187
					1	2	1		188-191
		1	2	1					192-195
			1	2	1				196-199
			1	2	1				200-203
				1	2	1			204-207
			1	2	1				208-211
				1	2	1			212-215
				1	2	1			216-219
					1	2	1		220-223
			1	2	1				224-227
				1	2	1			228-231
				1	2	1			232-235
					1	2	1		236-239
				1	2	1			240-243
					1	2	1		244-247
					1	2	1		248-251
						1	2	1	252-255

Note that the quadruplet sequence has not only the same length but also the same structure, so that each quadruplet can be taken as a unit. Then, what does really matter (i.e. makes the difference) is only the level of the logical space at which the sequence begins (or, equivalently, ends). If we consider only the first number of the two characteristic numbers of any logical-space level (e.g. 1 for Level 1-7), we obtain the following series of 64 numbers (I have put doublets in bold):

0**11**21**22**31**22**32**33**41**22**32**33**42**33**43**44**51**22**32**33**42**33**43**44**52**33**43**44**53**44**54**55**6

We may see that there is a characteristic fractal-like algorithm, with doubled numbers interspaced with 2 single units with recurrent form. As it is evident, the spaces between a development of the series and its going back to the initial case become longer and longer (according to the series 0-2-4-8-16). Note indeed that the following structure recurs several times and with spatial shifts:

Table 9.3 Recurrent structure

1	2	1			
	1	2	1		
	1	2	1		
		1	2	1	
	1	2	1		
		1	2	1	
		1	2	1	
			1	2	1

Therefore, the whole has also the structure of a spiral, as shown in Fig. 9.1. We can introduce a further simplification and consider only the doublets, so that we obtain the following series of 16 numbers:

1**22**32**33**42**33**43**44**5.

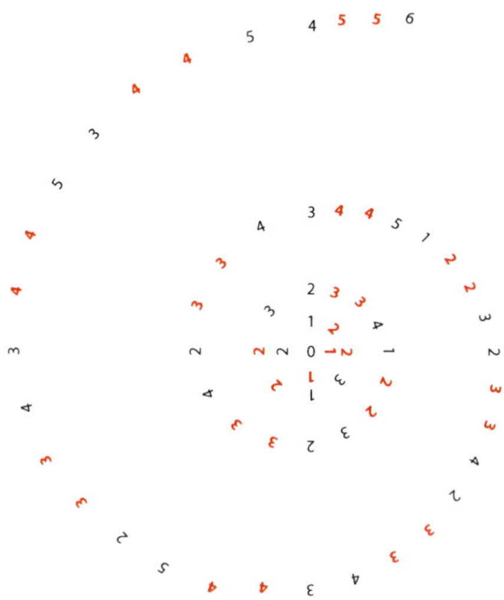

Figure 9.1 The spiral form of the series of statements. Doublets are in bold gray.

Taking into account Table 9.3, this structure can be reduced to its first half (23=8 elements), so that we obtain:

1**22**32**33**4,

which can be considered as the generating algorithm of three-dimensional sequence of statements. Let us consider 4 subsequent numbers: j, k, l, m. Then, we have:

$$j, k, k, l, k, l, l, m \, .$$

Therefore, any substitution of the first number will involve a relative shift of all other numbers. For instance, let us consider the first 8 numbers of the initial sequence:

0**11**21**22**3.

It is easy to see that 0 takes the place of 1 as first number and all others are shifted accordingly. The subsequent 8 numbers constitute the generating sequence itself, and the same is true of the subsequent 8 numbers. The fourth part of the initial sequence is

2**33**43**44**5,

which is the generating sequence shifted of a number. The fifth part is:

1**22**32**33**4,

i.e. again the generating sequence. The next 8 numbers are

2**33**43**44**5,

which is again the generating sequence shifted of one number. This is repeated by the next 8 numbers. Finally the 8th and last part is

3**44**54**55**6,

which is the generated sequence shifted of two numbers. Therefore, the whole series of 64 numbers is itself an instance of its generating sequence, confirming the fractal nature of the algorithm (each of the terms above must be expanded to a whole sequence whose starting elements is given by the term that is expanded, and so on). In other words, the whole series can be represented as:

$$(j, k, k, l, k, l, l, m), (k, l, l, m, l, m, m, n),$$
$$(k, l, l, m, l, m, m, n), (l, m, m, n, m, n, n, o) \, , \qquad (9.1)$$

and so on, where *n* and *o* are successors of *m*. Consider the first term in each parenthesis, and you will have the same algorithm: *j*, *k*, *k*, *l*, ... An interesting issue is both the reduction and the expansion of this fractal structure to any *n*-dimensional logical space. For instance, let us consider the two-dimensional logical space. It is easy to see that we get here recurrent doublets instead of quadruplets:

Table 9.4 Recurrent structure for the two-dimensional logic

Level 0-4	Level 1-3	Level 2-2	Level 3-1	Level 4-0	Numbers
1	1				0-1
	1	1			2-3
	1	1			4-5
		1	1		6-7
	1	1			8-9
		1	1		10-11
		1	1		12-13
			1	1	14-15

Then, an easy calculation shows that we obtain following generating sequence (I again use the numbers of the following table as referring to the first number of the couple denoting a level of the logical space and this time I start with 0):

0**11**2,

which is the first part of our generating series (4=22). Obviously, this generating sequence corresponds also to the quadruplets of Table 9.2. In other words, the basic form assumed by our algorithm is here:

$$j, k, k, l\ , \tag{9.2}$$

where each expansion must have the same form. This, thanks to the one-dimensional logical space, can be further reduced to the binary couple 01 (j, k). We can therefore assume that a four-dimensional logic space will show a generating series of sixteen (24) elements that are the first sixteen of the sequence from which we started (see Fig. 9.2):

0**11**21**22**31**22**32**33**4.

This examination is interesting since it represents a practical and relatively easy way to generate high-level logical spaces when needed (see Section 8.9). Moreover, as mentioned, this correspondence between logical space and binary numbers allows very easily the coordination of a classical processor and of a new-generation one or even the implementation on a classical core of a logical-space image.

9.3 Quantum Computing: Raising and Lowering Operators

The previous formalism can be easily used for implementing quantum computation too (*Auletta et al.*, 2009, Chapter 17). In particular, we can represent the three sets X, Y and Z as three harmonic oscillators that can be in ground (0) or excited (1) state each. What we need then is a family of raising and lowering operators allowing us to climb or descend the ladder of the possible states. We focus on the tridimensional logic (the results can be easily extracted for the one- and bidimensional case). Obviously, the rules for product and sum established previously need always to be taken into account. We can build a family of raising operators whose combination can give rise to any of the passages from one

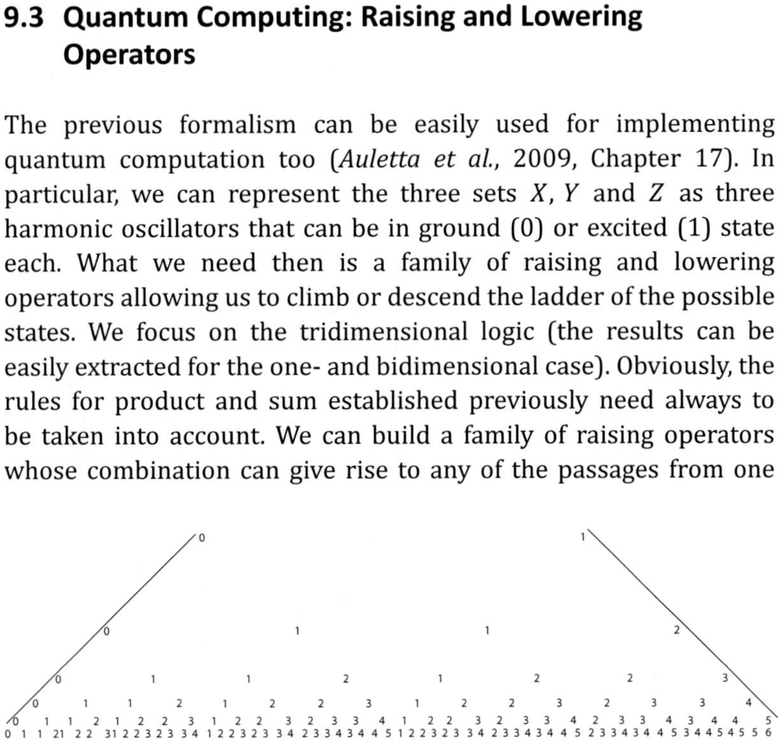

Figure 9.2 The fractal generation of sequences from above to below. The last line is the 64-elements series from which we started. Obviously, there is no limit on the generation of the series. It is possible to put a 0 on the top, what would make the above fractal fully symmetric: note that each left "diagonal" repeats always the same number (starting from 0) whilst all the right "diagonals" follow the numerical series (the last one on the right is the series form 1 or 0 until 6). However, this would be the result of a retroactive inference starting from some generating series.

level to the next higher one (see Fig. 9.3). Any of these operators can be assumed to act on columnar vectors represented by the ID sequence of each statement of the starting level and produces other columnar vectors of the upper level as output (repetitions are not considered as well as results that are identical to the input).

Mathematically speaking, we cannot act with a raising operator on a vector composed only of zeros (Level 0-8). However, this can be easily done by adding a qubit representing the environment and keeping it constant (=1) so that it is irrelevant for the operations inside the logical space. Having said this, in the following I shall no longer deal with this problem.

A family of raising operators from Level 1-7 to Level 2-6 is shown in Fig. 9.4. They are the result of the combination of the previously shown operators. Starting from the top line from the left to right (see Tables 1.19 and 1.20):

- The first operator allows the generation of Level 2-6 Statements 1, 8, 14, 19, 23, 26, and 28. Consider that the number of the propositions perfectly correspond to the number of the operator in the series displayed in Fig. 9.3 and the same is true for each of the subsequent transformations. This is due to the fact that each of these statements is an eigenvector of this operator and the same is true for the following transformations.
- The second one the generation of Statements 2, 9, 15, 20, 24, and 27.
- The third the generation of Statements 3, 10, 16, 21, 25.
- The fourth (the first on left in the bottom line) the generation of Statements 4, 11, 17, and 22.
- The fifth the generation of Statements 5, 12, 18.
- The sixth the generation of Statements 6 and 13.
- Finally, the last operator generates Statement 7.

$$\begin{bmatrix}1&0&0&0&0&0&0&0\\1&1&0&0&0&0&0&0\\0&0&1&0&0&0&0&0\\0&0&0&1&0&0&0&0\\0&0&0&0&1&0&0&0\\0&0&0&0&0&1&0&0\\0&0&0&0&0&0&1&0\\0&0&0&0&0&0&0&1\end{bmatrix}\begin{bmatrix}1&0&0&0&0&0&0&0\\0&1&0&0&0&0&0&0\\1&0&1&0&0&0&0&0\\0&0&0&1&0&0&0&0\\0&0&0&0&1&0&0&0\\0&0&0&0&0&1&0&0\\0&0&0&0&0&0&1&0\\0&0&0&0&0&0&0&1\end{bmatrix}\begin{bmatrix}1&0&0&0&0&0&0&0\\0&1&0&0&0&0&0&0\\0&0&1&0&0&0&0&0\\1&0&0&1&0&0&0&0\\0&0&0&0&1&0&0&0\\0&0&0&0&0&1&0&0\\0&0&0&0&0&0&1&0\\0&0&0&0&0&0&0&1\end{bmatrix}\begin{bmatrix}1&0&0&0&0&0&0&0\\0&1&0&0&0&0&0&0\\0&0&1&0&0&0&0&0\\0&0&0&1&0&0&0&0\\1&0&0&0&1&0&0&0\\0&0&0&0&0&1&0&0\\0&0&0&0&0&0&1&0\\0&0&0&0&0&0&0&1\end{bmatrix}$$

$$\begin{bmatrix}1&0&0&0&0&0&0&0\\0&1&0&0&0&0&0&0\\0&0&1&0&0&0&0&0\\0&0&0&1&0&0&0&0\\0&0&0&0&1&0&0&0\\1&0&0&0&0&1&0&0\\0&0&0&0&0&0&1&0\\0&0&0&0&0&0&0&1\end{bmatrix}\begin{bmatrix}1&0&0&0&0&0&0&0\\0&1&0&0&0&0&0&0\\0&0&1&0&0&0&0&0\\0&0&0&1&0&0&0&0\\0&0&0&0&1&0&0&0\\0&0&0&0&0&1&0&0\\1&0&0&0&0&0&1&0\\0&0&0&0&0&0&0&1\end{bmatrix}\begin{bmatrix}1&0&0&0&0&0&0&0\\0&1&0&0&0&0&0&0\\0&0&1&0&0&0&0&0\\0&0&0&1&0&0&0&0\\0&0&0&0&1&0&0&0\\0&0&0&0&0&1&0&0\\0&0&0&0&0&0&1&0\\1&0&0&0&0&0&0&1\end{bmatrix}\begin{bmatrix}1&0&0&0&0&0&0&0\\0&1&0&0&0&0&0&0\\0&1&1&0&0&0&0&0\\0&0&0&1&0&0&0&0\\0&0&0&0&1&0&0&0\\0&0&0&0&0&1&0&0\\0&0&0&0&0&0&1&0\\0&0&0&0&0&0&0&1\end{bmatrix}$$

$$\begin{bmatrix}1&0&0&0&0&0&0&0\\0&1&0&0&0&0&0&0\\0&0&1&0&0&0&0&0\\0&1&0&1&0&0&0&0\\0&0&0&0&1&0&0&0\\0&0&0&0&0&1&0&0\\0&0&0&0&0&0&1&0\\0&0&0&0&0&0&0&1\end{bmatrix}\begin{bmatrix}1&0&0&0&0&0&0&0\\0&1&0&0&0&0&0&0\\0&0&1&0&0&0&0&0\\0&0&0&1&0&0&0&0\\0&1&0&0&1&0&0&0\\0&0&0&0&0&1&0&0\\0&0&0&0&0&0&1&0\\0&0&0&0&0&0&0&1\end{bmatrix}\begin{bmatrix}1&0&0&0&0&0&0&0\\0&1&0&0&0&0&0&0\\0&0&1&0&0&0&0&0\\0&0&0&1&0&0&0&0\\0&0&0&0&1&0&0&0\\0&1&0&0&0&1&0&0\\0&0&0&0&0&0&1&0\\0&0&0&0&0&0&0&1\end{bmatrix}\begin{bmatrix}1&0&0&0&0&0&0&0\\0&1&0&0&0&0&0&0\\0&0&1&0&0&0&0&0\\0&0&0&1&0&0&0&0\\0&0&0&0&1&0&0&0\\0&0&0&0&0&1&0&0\\0&1&0&0&0&0&1&0\\0&0&0&0&0&0&0&1\end{bmatrix}$$

$$\begin{bmatrix}1&0&0&0&0&0&0&0\\0&1&0&0&0&0&0&0\\0&0&1&0&0&0&0&0\\0&0&0&1&0&0&0&0\\0&0&0&0&1&0&0&0\\0&0&0&0&0&1&0&0\\0&0&0&0&0&0&1&0\\0&1&0&0&0&0&0&1\end{bmatrix}\begin{bmatrix}1&0&0&0&0&0&0&0\\0&1&0&0&0&0&0&0\\0&0&1&0&0&0&0&0\\0&0&1&1&0&0&0&0\\0&0&0&0&1&0&0&0\\0&0&0&0&0&1&0&0\\0&0&0&0&0&0&1&0\\0&0&0&0&0&0&0&1\end{bmatrix}\begin{bmatrix}1&0&0&0&0&0&0&0\\0&1&0&0&0&0&0&0\\0&0&1&0&0&0&0&0\\0&0&0&1&0&0&0&0\\0&0&1&0&1&0&0&0\\0&0&0&0&0&1&0&0\\0&0&0&0&0&0&1&0\\0&0&0&0&0&0&0&1\end{bmatrix}\begin{bmatrix}1&0&0&0&0&0&0&0\\0&1&0&0&0&0&0&0\\0&0&1&0&0&0&0&0\\0&0&0&1&0&0&0&0\\0&0&0&0&1&0&0&0\\0&0&1&0&0&1&0&0\\0&0&0&0&0&0&1&0\\0&0&0&0&0&0&0&1\end{bmatrix}$$

$$\begin{bmatrix}1&0&0&0&0&0&0&0\\0&1&0&0&0&0&0&0\\0&0&1&0&0&0&0&0\\0&0&0&1&0&0&0&0\\0&0&0&0&1&0&0&0\\0&0&0&0&0&1&0&0\\0&0&1&0&0&0&1&0\\0&0&0&0&0&0&0&1\end{bmatrix}\begin{bmatrix}1&0&0&0&0&0&0&0\\0&1&0&0&0&0&0&0\\0&0&1&0&0&0&0&0\\0&0&0&1&0&0&0&0\\0&0&0&0&1&0&0&0\\0&0&0&0&0&1&0&0\\0&0&0&0&0&0&1&0\\0&0&1&0&0&0&0&1\end{bmatrix}\begin{bmatrix}1&0&0&0&0&0&0&0\\0&1&0&0&0&0&0&0\\0&0&1&0&0&0&0&0\\0&0&0&1&0&0&0&0\\0&0&0&1&1&0&0&0\\0&0&0&0&0&1&0&0\\0&0&0&0&0&0&1&0\\0&0&0&0&0&0&0&1\end{bmatrix}\begin{bmatrix}1&0&0&0&0&0&0&0\\0&1&0&0&0&0&0&0\\0&0&1&0&0&0&0&0\\0&0&0&1&0&0&0&0\\0&0&0&0&1&0&0&0\\0&0&0&1&0&1&0&0\\0&0&0&0&0&0&1&0\\0&0&0&0&0&0&0&1\end{bmatrix}$$

$$\begin{bmatrix}1&0&0&0&0&0&0&0\\0&1&0&0&0&0&0&0\\0&0&1&0&0&0&0&0\\0&0&0&1&0&0&0&0\\0&0&0&0&1&0&0&0\\0&0&0&0&0&1&0&0\\0&0&0&1&0&0&1&0\\0&0&0&0&0&0&0&1\end{bmatrix}\begin{bmatrix}1&0&0&0&0&0&0&0\\0&1&0&0&0&0&0&0\\0&0&1&0&0&0&0&0\\0&0&0&1&0&0&0&0\\0&0&0&0&1&0&0&0\\0&0&0&0&0&1&0&0\\0&0&0&0&0&0&1&0\\0&0&0&1&0&0&0&1\end{bmatrix}\begin{bmatrix}1&0&0&0&0&0&0&0\\0&1&0&0&0&0&0&0\\0&0&1&0&0&0&0&0\\0&0&0&1&0&0&0&0\\0&0&0&0&1&0&0&0\\0&0&0&0&1&1&0&0\\0&0&0&0&0&0&1&0\\0&0&0&0&0&0&0&1\end{bmatrix}\begin{bmatrix}1&0&0&0&0&0&0&0\\0&1&0&0&0&0&0&0\\0&0&1&0&0&0&0&0\\0&0&0&1&0&0&0&0\\0&0&0&0&1&0&0&0\\0&0&0&0&0&1&0&0\\0&0&0&0&1&0&1&0\\0&0&0&0&0&0&0&1\end{bmatrix}$$

$$\begin{bmatrix}1&0&0&0&0&0&0&0\\0&1&0&0&0&0&0&0\\0&0&1&0&0&0&0&0\\0&0&0&1&0&0&0&0\\0&0&0&0&1&0&0&0\\0&0&0&0&0&1&0&0\\0&0&0&0&0&0&1&0\\0&0&0&0&1&0&0&1\end{bmatrix}\begin{bmatrix}1&0&0&0&0&0&0&0\\0&1&0&0&0&0&0&0\\0&0&1&0&0&0&0&0\\0&0&0&1&0&0&0&0\\0&0&0&0&1&0&0&0\\0&0&0&0&0&1&0&0\\0&0&0&0&0&1&1&0\\0&0&0&0&0&0&0&1\end{bmatrix}\begin{bmatrix}1&0&0&0&0&0&0&0\\0&1&0&0&0&0&0&0\\0&0&1&0&0&0&0&0\\0&0&0&1&0&0&0&0\\0&0&0&0&1&0&0&0\\0&0&0&0&0&1&0&0\\0&0&0&0&0&0&1&0\\0&0&0&0&0&1&0&1\end{bmatrix}\begin{bmatrix}1&0&0&0&0&0&0&0\\0&1&0&0&0&0&0&0\\0&0&1&0&0&0&0&0\\0&0&0&1&0&0&0&0\\0&0&0&0&1&0&0&0\\0&0&0&0&0&1&0&0\\0&0&0&0&0&0&1&0\\0&0&0&0&0&0&1&1\end{bmatrix}$$

Figure 9.3 The 28 raising operators allowing the passage from any level to the next higher one.

$$\begin{bmatrix} 1&0&0&0&0&0&0&0 \\ 1&1&0&0&0&0&0&0 \\ 0&1&1&0&0&0&0&0 \\ 0&0&1&1&0&0&0&0 \\ 0&0&0&1&1&0&0&0 \\ 0&0&0&0&1&1&0&0 \\ 0&0&0&0&0&1&1&0 \\ 0&0&0&0&0&0&1&1 \end{bmatrix} \quad \begin{bmatrix} 1&0&0&0&0&0&0&0 \\ 0&1&0&0&0&0&0&0 \\ 1&0&1&0&0&0&0&0 \\ 0&1&0&1&0&0&0&0 \\ 0&0&1&0&1&0&0&0 \\ 0&0&0&1&0&1&0&0 \\ 0&0&0&0&1&0&1&0 \\ 0&0&0&0&0&1&0&1 \end{bmatrix} \quad \begin{bmatrix} 1&0&0&0&0&0&0&0 \\ 0&1&0&0&0&0&0&0 \\ 0&0&1&0&0&0&0&0 \\ 1&0&0&1&0&0&0&0 \\ 0&1&0&0&1&0&0&0 \\ 0&0&1&0&0&1&0&0 \\ 0&0&0&1&0&0&1&0 \\ 0&0&0&0&1&0&0&1 \end{bmatrix}$$

$$\begin{bmatrix} 1&0&0&0&0&0&0&0 \\ 0&1&0&0&0&0&0&0 \\ 0&0&1&0&0&0&0&0 \\ 0&0&0&1&0&0&0&0 \\ 1&0&0&0&1&0&0&0 \\ 0&1&0&0&0&1&0&0 \\ 0&0&1&0&0&0&1&0 \\ 0&0&0&1&0&0&0&1 \end{bmatrix} \quad \begin{bmatrix} 1&0&0&0&0&0&0&0 \\ 0&1&0&0&0&0&0&0 \\ 0&0&1&0&0&0&0&0 \\ 0&0&0&1&0&0&0&0 \\ 0&0&0&0&1&0&0&0 \\ 1&0&0&0&0&1&0&0 \\ 0&1&0&0&0&0&1&0 \\ 0&0&1&0&0&0&0&1 \end{bmatrix} \quad \begin{bmatrix} 1&0&0&0&0&0&0&0 \\ 0&1&0&0&0&0&0&0 \\ 0&0&1&0&0&0&0&0 \\ 0&0&0&1&0&0&0&0 \\ 0&0&0&0&1&0&0&0 \\ 0&0&0&0&0&1&0&0 \\ 1&0&0&0&0&0&1&0 \\ 0&1&0&0&0&0&0&1 \end{bmatrix} \quad \begin{bmatrix} 1&0&0&0&0&0&0&0 \\ 0&1&0&0&0&0&0&0 \\ 0&0&1&0&0&0&0&0 \\ 0&0&0&1&0&0&0&0 \\ 0&0&0&0&1&0&0&0 \\ 0&0&0&0&0&1&0&0 \\ 0&0&0&0&0&0&1&0 \\ 1&0&0&0&0&0&0&1 \end{bmatrix}$$

Figure 9.4 The seven raising operators allowing the passage from Level 1-7 to Level 2-6.

The raising operators allowing the ascension from Level 2-6 to Level 3-5 (see Tables 1.18 and 1.19) are represented in Fig. 9.3 (always starting from the top line from the left to right):

- Operator 1 generates Level 5-3 Statements 1, 2, 3, 4, 5, 6.
- Operator 2 generates Statements 7, 8, 9, 10, 11.
- Operator 3 generates Statements 12, 13, 14, 15.
- Operator 4 generates statements 16, 17, 18.
- Operator 5 generates Statements 19, 20.
- Operator 6 generates Statement 21. Operator 7 is redundant.
- Operator 8 generates Statements 22, 23, 24, 25, 26.
- Operator 9 generates Statements 27, 28, 29, 30.
- Operator 10 generates Statements 31, 32, 33.
- Operator 11 generates Statements 34 and 34.
- Operator 12 generates Statement 36. Operator 13 is redundant.
- Operator 14 generates Statements 37, 38, 39, 40.
- Operator 15 generates Statements 41, 42, 43.

- Operator 16 generates Statements 44 and 45.
- Operator 17 generates Statement 46. Operator 18 is redundant.
- Operator 19 generates Statements 47, 48, 49.
- Operator 20 generates Statements 50 and 51.
- Operator 21 generates Statement 52. Operator 22 is redundant.
- Operator 23 generates Statements 53 and 54.
- Operator 24 generates Statement 55. Operator 25 is redundant.
- Operator 26 generates Statement 56. Operators 27 and 28 are redundant.

Again I have not considered repetition, so that "later" operators are more diminished in their generating capacity than is actually the case. We can reiterate this procedure and generate any subsequent level.

The lowering operators that bring back statements from any given level to the next lower one can be built as in Fig. 9.5. It is clear

$$\begin{bmatrix}0&0&0&0&0&0&0&0\\0&1&0&0&0&0&0&0\\0&0&1&0&0&0&0&0\\0&0&0&1&0&0&0&0\\0&0&0&0&1&0&0&0\\0&0&0&0&0&1&0&0\\0&0&0&0&0&0&1&0\\0&0&0&0&0&0&0&1\end{bmatrix}\quad\begin{bmatrix}1&0&0&0&0&0&0&0\\0&0&0&0&0&0&0&0\\0&0&1&0&0&0&0&0\\0&0&0&1&0&0&0&0\\0&0&0&0&1&0&0&0\\0&0&0&0&0&1&0&0\\0&0&0&0&0&0&1&0\\0&0&0&0&0&0&0&1\end{bmatrix}\quad\begin{bmatrix}1&0&0&0&0&0&0&0\\0&1&0&0&0&0&0&0\\0&0&0&0&0&0&0&0\\0&0&0&1&0&0&0&0\\0&0&0&0&1&0&0&0\\0&0&0&0&0&1&0&0\\0&0&0&0&0&0&1&0\\0&0&0&0&0&0&0&1\end{bmatrix}\quad\begin{bmatrix}1&0&0&0&0&0&0&0\\0&1&0&0&0&0&0&0\\0&0&1&0&0&0&0&0\\0&0&0&0&0&0&0&0\\0&0&0&0&1&0&0&0\\0&0&0&0&0&1&0&0\\0&0&0&0&0&0&1&0\\0&0&0&0&0&0&0&1\end{bmatrix}$$

$$\begin{bmatrix}1&0&0&0&0&0&0&0\\0&1&0&0&0&0&0&0\\0&0&1&0&0&0&0&0\\0&0&0&1&0&0&0&0\\0&0&0&0&0&0&0&0\\0&0&0&0&0&1&0&0\\0&0&0&0&0&0&1&0\\0&0&0&0&0&0&0&1\end{bmatrix}\quad\begin{bmatrix}1&0&0&0&0&0&0&0\\0&1&0&0&0&0&0&0\\0&0&1&0&0&0&0&0\\0&0&0&1&0&0&0&0\\0&0&0&0&1&0&0&0\\0&0&0&0&0&0&0&0\\0&0&0&0&0&0&1&0\\0&0&0&0&0&0&0&1\end{bmatrix}\quad\begin{bmatrix}1&0&0&0&0&0&0&0\\0&1&0&0&0&0&0&0\\0&0&1&0&0&0&0&0\\0&0&0&1&0&0&0&0\\0&0&0&0&1&0&0&0\\0&0&0&0&0&1&0&0\\0&0&0&0&0&0&0&0\\0&0&0&0&0&0&0&1\end{bmatrix}\quad\begin{bmatrix}1&0&0&0&0&0&0&0\\0&1&0&0&0&0&0&0\\0&0&1&0&0&0&0&0\\0&0&0&1&0&0&0&0\\0&0&0&0&1&0&0&0\\0&0&0&0&0&1&0&0\\0&0&0&0&0&0&1&0\\0&0&0&0&0&0&0&0\end{bmatrix}$$

Figure 9.5 The eight lowering operators allowing the passage from any level to the next lower level.

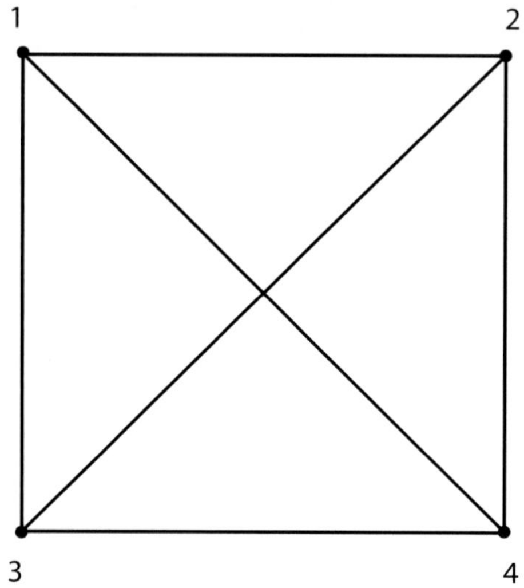

Figure 9.6 Two-dimensional space.

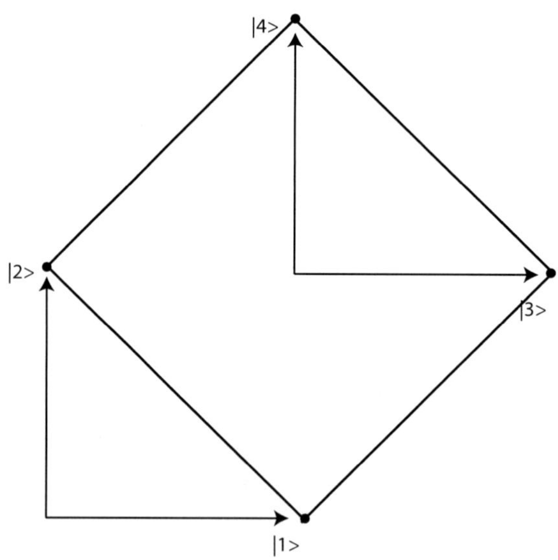

Figure 9.7 Relations between 2D vectorial spaces and 2D logical subspaces.

that the first one (always starting from the top line from the left to right) annihilates the term a in any statement, the second the term b, and so on. They are sort of negative projectors: instead of projecting e.g. on the component a they project on not-a.

9.4 A Vectorial Representation

An interesting possibility is to conceive all logical statements and also subspaces in vectorial terms (as for quantum systems). We can indeed represent the 8 value assignments in Table 1.12 in terms of the following orthogonal basis:

$$|a\rangle = \begin{pmatrix} 1 \\ 0 \\ 0 \\ 0 \\ 0 \\ 0 \\ 0 \\ 0 \end{pmatrix}, \quad |b\rangle = \begin{pmatrix} 0 \\ 1 \\ 0 \\ 0 \\ 0 \\ 0 \\ 0 \\ 0 \end{pmatrix}, \quad |c\rangle = \begin{pmatrix} 0 \\ 0 \\ 1 \\ 0 \\ 0 \\ 0 \\ 0 \\ 0 \end{pmatrix}, \quad |d\rangle = \begin{pmatrix} 0 \\ 0 \\ 0 \\ 1 \\ 0 \\ 0 \\ 0 \\ 0 \end{pmatrix},$$

$$|e\rangle = \begin{pmatrix} 0 \\ 0 \\ 0 \\ 0 \\ 1 \\ 0 \\ 0 \\ 0 \end{pmatrix}, \quad |f\rangle = \begin{pmatrix} 0 \\ 0 \\ 0 \\ 0 \\ 0 \\ 1 \\ 0 \\ 0 \end{pmatrix}, \quad |g\rangle = \begin{pmatrix} 0 \\ 0 \\ 0 \\ 0 \\ 0 \\ 0 \\ 1 \\ 0 \end{pmatrix}, \quad |h\rangle = \begin{pmatrix} 0 \\ 0 \\ 0 \\ 0 \\ 0 \\ 0 \\ 0 \\ 1 \end{pmatrix}. \quad (9.3)$$

We are now in position to write any statement in the 3D space as a combination of these basis vectors. However, also the 0D components of any subspace can be written in this terms, since we have (for the codes used for the subspaces see end of Section 8.4 and especially Tables 8.17–8.18):

$$|1\rangle = |7\rangle = \begin{pmatrix} 1 \\ 1 \\ 1 \\ 0 \\ 1 \\ 0 \\ 0 \\ 0 \end{pmatrix}, \quad |3\rangle = |9\rangle = \begin{pmatrix} 1 \\ 1 \\ 0 \\ 1 \\ 0 \\ 1 \\ 0 \\ 0 \end{pmatrix}, \quad |5\rangle = |11\rangle = \begin{pmatrix} 1 \\ 0 \\ 1 \\ 1 \\ 0 \\ 0 \\ 1 \\ 0 \end{pmatrix},$$

$$|2\rangle = |8\rangle = \begin{pmatrix} 0 \\ 0 \\ 0 \\ 1 \\ 0 \\ 1 \\ 1 \\ 1 \end{pmatrix}, \quad |4\rangle = |10\rangle = \begin{pmatrix} 0 \\ 0 \\ 1 \\ 0 \\ 1 \\ 0 \\ 1 \\ 1 \end{pmatrix}, \quad |6\rangle = |12\rangle = \begin{pmatrix} 0 \\ 1 \\ 0 \\ 0 \\ 1 \\ 1 \\ 0 \\ 1 \end{pmatrix}. \quad (9.4)$$

Note that $|1\rangle$ ($|7\rangle$) and $|2\rangle$ ($|8\rangle$) constitute an orthogonal basis for the one-dimensional logical space and similarly for the other variables. Instead, all vectors on the first row are parallel (as well as those in the second row), what can be seen by the fact that they pairwise share 4 values out of 8 (2 out of the first 4 numbers and the other 2 out of the last 4 numbers of each column vector). This means that the two-dimensional reference frame whose axes are $|3\rangle$ and $|4\rangle$ is displaced of some length relative to reference frame constituted by axes $|1\rangle$ and $|2\rangle$. This in turn means that the line connecting the points individuated by $|1\rangle$ and $|2\rangle$ and the line connecting the points individuated by $|3\rangle$ and $|4\rangle$ are parallel, what allows to recover the plane shown in Fig. 9.6, as displayed in Fig. 9.7. Similarly, also the reference frame whose axes are $|11\rangle$ and $|12\rangle$ is displaced of same length relative to the reference frame $|9\rangle - |10\rangle$ as well as the he reference frame whose axes are $|5\rangle$ and $|6\rangle$ is displaced of same length relative to the reference frame $|7\rangle - |8\rangle$. This allows to fully recover the 3D logical space of Fig. 9.8, as displayed in Fig. 9.9. Note that at least one of the reference frames (here $|1\rangle - |2\rangle$) needs to be displaced along two directions, one for each of the other two reference frames.

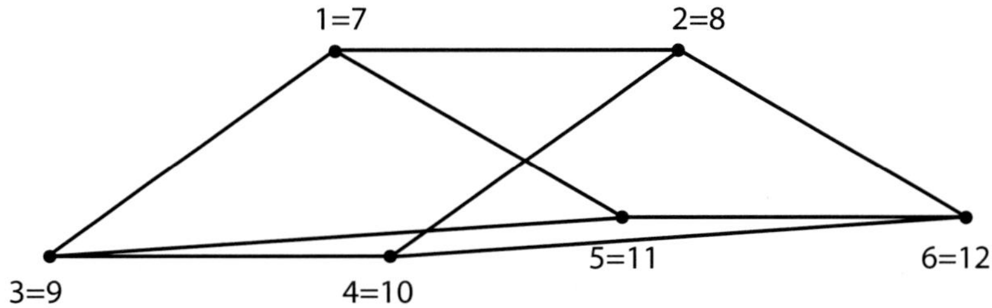

Figure 9.8 Three-dimensional space. Only the edges of the pentahedron are shown for the sake of representation.

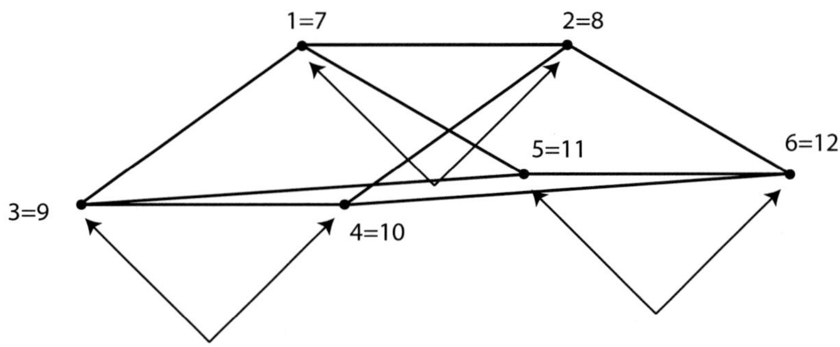

Figure 9.9 Relations between 3D vectorial spaces and 3D logical space.

This representation allows us to write a derivation like *Barbara* in this terms:

$$\mathrm{Tr}_X \left(| 1 = 7 \rangle \oplus | 4 = 10 \rangle \otimes | 2 = 8 \rangle \oplus | 5 = 11 \rangle \right) = | 11 \rangle \oplus | 10 \rangle \,, \tag{9.5}$$

where Tr_X symbolizes an analogous of the mathematical operation of tracing the "system" X out. However, considered that the nature of the connections (that have the value of operations on logical subspaces) here need to follow the previous logical rules and not the traditional mathematical form. Therefore, the connection of the two kets in the output state is dependent on that rules, that allow for following definitions:

$$|1 \oplus 4\rangle = \begin{pmatrix} 1 \\ 1 \\ 1 \\ 0 \\ 1 \\ 0 \\ 1 \\ 1 \end{pmatrix}, \ |8 \oplus 5\rangle = \begin{pmatrix} 1 \\ 0 \\ 1 \\ 1 \\ 0 \\ 1 \\ 1 \\ 1 \end{pmatrix}, \ |11 \oplus 10\rangle = \begin{pmatrix} 1 \\ 0 \\ 1 \\ 1 \\ 1 \\ 0 \\ 1 \\ 1 \end{pmatrix}. \quad (9.6)$$

Analogously, we can define similar vectors in the case of subspaces product, for instance:

$$|1 \otimes 4\rangle = \begin{pmatrix} 0 \\ 0 \\ 1 \\ 0 \\ 1 \\ 0 \\ 0 \\ 0 \end{pmatrix}, \ |8 \otimes 5\rangle = \begin{pmatrix} 0 \\ 0 \\ 0 \\ 1 \\ 0 \\ 0 \\ 1 \\ 0 \end{pmatrix}, \ |11 \otimes 10\rangle = \begin{pmatrix} 0 \\ 0 \\ 1 \\ 0 \\ 0 \\ 0 \\ 1 \\ 0 \end{pmatrix}. \quad (9.7)$$

Chapter 10

Conclusions

The logical system proposed here presents some interesting properties:

(1) The system displays a beautiful symmetry expressed in the position of the statements and their mutual relations.
(2) Any proposition is identified through a sequence of numbers that both tells the ways in which it can be generated as well as the number of further propositions to which it can give rise.
(3) Any n-dimensional logical space is reversible in itself. I have indeed shown how many cycles there are.
(4) The three-dimensional logical space allows logical calculations of propositional and predicative kind with any number of variables.
(5) Any relation between nodes in any n-dimensional logical space is given through arithmetic sums, products, subtractions or divisions among two (or more) numbers in the same column.
(6) Logical rules are not necessary but are a consequence of the "path" chosen in any n-dimensional logical space.
(7) The systems displays structural relations also when treated in the subtraction-division "representation".

Mechanical Logic in Three-Dimensional Space
Gennaro Auletta

ISBN 978-981-4411-50-9 (Hardcover), 978-981-4411-51-6 (eBook)
www.panstanford.com

(8) The logical space displays interesting invariants (for instance, statements that lead to the same result whether we perform a product or a division).

(9) The tautology and the contradiction can be understood as limiting cases when the series of 1s or 0s, respectively, tend to infinity. In this way there is a smooth and pure mathematical transition between the contingent propositions constituting any logical space and these two limiting cases.

(10) This establishes an interesting connection between logic and information (both classical and quantum). Indeed, the lowering order can be understood as displayed through selection acts, while the raising one through information sharing (see Auletta 2005).

The main accomplishments of these book can be therefore considered the following ones:

- The system provides to logic a kind of numerical substrate analogous to that of arithmetics relative to algebra. The great advantage of arithmetic over logic until now was precisely the constitution of a mathematical space (the line for natural numbers) that allows univocal position and relations of all numbers. Now, with this language this is the case also for logic.
- In this way, I have reduced logic to a branch of the combinatorial calculus.
- I have introduced kind of arithmetic operations on sequence of numbers instead of on the logical statement themselves, which is a far more general procedure.
- As mentioned, I have introduced not only product and sum that could be understood as analogues of conjunction (AND) and disjunction (OR), but also subtraction (AND NOT) and division (OR NOT). This expands considerably the space of the possible operations.
- This allows two kind of inferences (based on sum and division) that are retrograde since they represent a necessary-condition reasoning instead of a sufficient-condition one.
- Each statements has a Lowering Generating Set and a Raising Generating Set. Treating any kind of statement as

derived from other ones allows us to mechanically compute any kind of inference (the computation is performed on the generating statements).

- I have proved that any inference with n variables can be reduced to an inference with three ones.
- I have shown the existence of logical cycles across the logical space.
- The system can deal with mereological inferences.
- We can easily deal with first-order and second-order logic, when no relations are considered.
- Such a logical system can be implemented easily in classical computing thanks to the established connection with binary numbers.
- It can be applied to quantum computation thanks to the possibility to generate statements through quantum oscillators or interferometry and spin-flip devices.
- Artificial intelligence remains the field in which I expect in the future the most important applications and developments of this approach.

Bibliography

[*Aristotle An. Pr., An. Post.*]	Aristotle, *Analytica Priora et Posteriora* (Ed. W. D. Ross and L. Minio–Paluello), Oxford, 1964, 1968, 1978, 1982, 1986.
[*Aristotle Cat.*]	Aristotle, *Categoriæ*, in (*Aristotle Cat-Int.*).
[*Aristotle Cat-Int.*]	Aristotle, *Categoriæ et Liber de Interpretatione* (Ed. L. Minio–Paluello), Oxford, 1949, 1986.
[Auletta 2005]	Auletta, Gennaro, "Quantum Information as a General Paradigm," *Foundations of Physics* **35**: 787–815.
[Auletta 2009]	Auletta, Gennaro, "What About the Three Forms of Inference?," *Acta Philosophica* **1.18**: 59–74.
[Auletta *et al.*, 2009]	Auletta, G., Fortunato, M., and Parisi, G., *Quantum Mechanics*, Cambridge, University Press.
[Auletta 2011]	Auletta, Gennaro, *Cognitive Biology: Dealing with Information from Bacteria to Minds*, Oxford, University Press.
[Auletta 2013]	Auletta, Gennaro, "Aristotle's Syllogistic Revisited," *Epistemologia* **36**: in press.
[*Boole* 1854]	Boole, George, *An Investigation of the Laws of Thought, on which are Founded the Mathematical Theories of Logic and Probabilities*, London, Walton and Maberly, 1854; New York, Dover, 1958.
[*De Morgan* 1847]	De Morgan, Augustus, *Formal Logic or the Calculus of Inference, Necessary and Probable*, London, Taylor and Walton, 1847; Chicago, 1926.

[*Frege* 1884] Frege, Gottlob, *Die Grundlagen der Arithmetik: eine logisch-mathematische Untersuchung über den Begriff der Zahl*, Breslau, W. Koebner.

[Gödel 1931] Gödel, Kurt, "Über formal unentscheidbar Sätze der *Principia Mathematica* und verwandter Systeme. I," *Monatshefte für Mathematik und Physik* **38**: 173–98.

[Hobbes 1655] Hobbes, Thomas, *Elementorum philosophiae sectio prima de corpore*, London, 1655; rep. in (*Hobbes*, 1839–1845, Vol. I).

[Hobbes 1839–1845] Hobbes, Thomas, *Opera Philosophica quae latine scripsit Omnia*, London, J. Bohn, 1839–1845.

[*Leibniz* 1666] Leibniz, Gottfried W., *Dissertatio de Arte Combinatoria*; Leipzig, 1666; rep. in (*Leibniz PS*, IV, 27–104).

[Leibniz 1689] Leibniz, Gottfried W., "Primæ veritates," manuscript published in (*Leibniz C*, 518–23).

[Leibniz 1710] Leibniz, Gottfried W., *Essais de Théodicée*, 1710; in (*Leibniz PS*, VI, 21–436).

[*Leibniz C*] Leibniz, Gottfried W., *Opuscules et fragments inédits* (Ed. Couturat), Paris, Alcan, 1903; rep. Hildesheim, Olms, 1988.

[*Leibniz PS*] Leibniz, Gottfried W., *Philosophische Schriften* (Ed. Gerhardt), Halle, 1875; rep. Hildesheim, Olms, 1978.

[Llul 1305] Llul, Ramon, *Ars magna*, 1305.

[Peirce 1865] Peirce, Charles S., "On the Logic of Science: Harvard Lectures," in (*Peirce W*, I, 161–302). Manuscript of 1865.

[Peirce 1866] Peirce, Charles S., "The Logic of Science or Induction and Hypothesis: Lowell Lectures," in (*Peirce W*, I, 357–504). (Manuscript).

[Peirce 1868a] Peirce, Charles S., "On the Natural Classifications of Arguments," *Proceedings of the American Academy of Arts and Sciences* **7**: 261–87; (*Peirce W*, II, 23–48).

[Peirce 1868b] Peirce, Charles S., "Upon Logical Comprehension and Extension," *Proceedings of the American Academy of Arts and Sciences* **7**: 416–32; (*Peirce W*, II, 70–86).

[Peirce 1869] Peirce, Charles S., "Grounds of Validity of the Laws of Logic: Further Consequences of Four Incapacities," *Journal of Speculative Philosophy* **2**: 193–208; in (*Peirce W*, II, 242–72).

[Peirce 1878] Peirce, Charles S., "Deduction, Induction, and Hypothesis," *Popular Science Monthly* **13**: 470–82; in (*Peirce W*, III, 323–38).

[Peirce 1880] Peirce, Charles S., "On the Algebra of Logic," *American Journal of Mathematics* **3**: 15–57; (*Peirce W*, IV, 163–209).

[Peirce 1881] Peirce, Charles S., "Methods of Reasoning," in (*Peirce W*, IV, 245–56). (Manuscript).

[Peirce 1883a] Peirce, Charles S., "Note B: The Logic of Relatives," in (*Peirce*, 1883b, 187–203). Rep. in (*Peirce W*, IV, 453–66).

[*Peirce* 1883b] Peirce, Charles S. (Ed.), *Studies in Logic: By Members of the John Hopkins University*, Boston, Little, Brown, and Company.

[Peirce 1885a] Peirce, Charles S., "On the Algebra of Logic: A Contribution to the Philosophy of Notation," *American Journal of Mathematics* **7**: 180–202; in (*Peirce W*, V, 162–90).

[Peirce 1885b] Peirce, Charles S., "Studies in Logical Algebra," manuscript published in (*Peirce W*, V, 204–220).

[Peirce 1895] Peirce, Charles S., "Of Reasoning in General," in (*Peirce EP*, II, 11–26). (Manuscript).

[Peirce 1901] Peirce, Charles S., "On the Logic of Drawing History from Ancient Documents, Especially from Testimonies," in (*Peirce EP*, II, 75–114). Manuscript of 1901.

[Peirce 1903] Peirce, Charles S., "The Categories Defended," in (*Peirce EP*, II, 160–78). Manuscript of 1903.

[*Peirce EP*] Peirce, Charles S., *The Essential Peirce*, Bloomington, Indiana University Press, 1998, vols. I–II.

[*Peirce W*] Peirce, Charles S., *Writings*, Bloomington, Indiana University Press, 1982–.

[*Petrus Hispanus SL*] Petrus Hispanus, *SummulæLogicales*, Torino, Marietti, 1947.

[Schröder 1890–95] Schröder, Ernst, *Vorlesungen über die Algebra der Logik: Exakte Logik*, Leipzig, Teubner.

[Tarski 1935a] Tarski, Alfred Zur Grundlegung de Booleschen Algebra. IFundamenta Mathematicæ 24177–98; rep. in English as "On the Foundations of Boolean Algebra" in (*Tarski*, 1956, 320–41).

[Tarski 1935b] Tarski, Alfred, "Grundzüge des Systemenkalkül, Erster Teil," *Fundamenta Mathematicæ* **25**: 503–26; rep. in English as "Foundations of the Calculus of Systems" in (*Tarski*, 1956, 342–64).

[Tarski 1936] Tarski, Alfred, "Grundzüge des Systemenkalkül, Zweiter Teil," *Fundamenta Mathematicæ* **26**: 283–301; rep. in English as "Foundations of the Calculus of Systems" in (*Tarski*, 1956, 364–83).

[Tarski 1941] Tarski, Alfred, "On the Calculus of Relations," *Journal of Symbolic Logic* **6**: 73–89.

[*Tarski* 1956] Tarski, Alfred, *Logic, Semantics, Meta–Mathematics*, Oxford, University Press, 1956; Indianapolis, Hackett, 1983; 2nd ed. 1990.

[*William of Shyreswood IL*] William of Shyreswood, *Introductiones in logicam*, München, 1937.

Author Index

Subject Index